Lecture Notes in Business Information Processing 565

LNBIP reports state-of-the-art results in areas related to business information systems and industrial application software development – timely, at a high level, and in both printed and electronic form.

The type of material published includes

- Proceedings (published in time for the respective event)
- Postproceedings (consisting of thoroughly revised and/or extended final papers)
- Other edited monographs (such as, for example, project reports or invited volumes)
- Tutorials (coherently integrated collections of lectures given at advanced courses, seminars, schools, etc.)
- Award-winning or exceptional theses

LNBIP is abstracted/indexed in DBLP, EI and Scopus. LNBIP volumes are also submitted for the inclusion in ISI Proceedings.

Mahendrawathi ER · Avigdor Gal ·
Thomas Grisold · Flavia Santoro ·
Mathias Weske · Remco M. Dijkman ·
Dimka Karastoyanova · Banu Aysolmaz ·
Wasana Bandara · Kate Revoredo
Editors

Business Process Management

Responsible BPM Forum, Process Technology Forum, Educators Forum

BPM 2025 RBPM, PT, and Educators Forum
Seville, Spain, August 31 – September 5, 2025
Proceedings

Springer

Editors
Mahendrawathi ER
Institut Teknologi Sepuluh Nopember
Surabaya, Indonesia

Avigdor Gal
Technion - Israel Institute of Technology
Haifa, Israel

Thomas Grisold
WU Vienna
St. Gallen, Switzerland

Flavia Santoro
Universidade Estado do Rio de Janeiro
Rio de Janeiro, Brazil

Mathias Weske
Hasso Plattner Institute, University of Potsdam
Potsdam, Germany

Remco M. Dijkman
Faculty of Informatics, Atlas 5.402
Eindhoven University of Technology
Eindhoven, The Netherlands

Dimka Karastoyanova
University of Groningen
Groningen, The Netherlands

Banu Aysolmaz
Eindhoven University of Technology
Eindhoven, The Netherlands

Wasana Bandara
Queensland University of Technology
Brisbane, QLD, Australia

Kate Revoredo
Humboldt-Universität zu Berlin
Berlin, Germany

ISSN 1865-1348 ISSN 1865-1356 (electronic)
Lecture Notes in Business Information Processing
ISBN 978-3-032-02935-5 ISBN 978-3-032-02936-2 (eBook)
https://doi.org/10.1007/978-3-032-02936-2

This Springer imprint is published by the registered company Springer Nature Switzerland AG
The registered company address is: Gewerbestrasse 11, 6330 Cham, Switzerland

Preface

The Business Process Management Conference was founded 23 years ago to bring together pioneering researchers and industry leaders to explore innovations, share best practices, and advance the field of process management through collaboration and knowledge exchange. BPM 2025 also featured specialized forums to present advancements and discuss topics on specific topics related to business process management. All forums used a double-blind review process in which submissions received three reviews each, on average.

The Responsible Business Process Management (RBPM) Forum is driven by the need to extend the traditional efficiency-oriented focus of BPM to encompass broader societal and environmental responsibilities. Responsible BPM is the practice of applying BPM by accounting for the needs of stakeholders involved in business processes (e.g., workers and customers), the broader environment (e.g., regulators and natural resources), and society in more general terms (e.g., grand challenges). The RBPM Forum provides a platform for exploring how to integrate responsibility into BPM practices, addressing challenges from three perspectives: technological, managerial, and conceptual. This first edition of the Responsible BPM Forum received 21 submissions, of which 10 papers were selected for presentation at the venue and for inclusion in these proceedings. These papers offer diverse and original perspectives on the broader ethical, societal, and environmental implications of BPM, providing fertile ground for further research on this concept in the future.

The Process Technology Forum covers technical and engineering aspects related to the study, design, development, and maintenance of software systems that support the modeling, simulation, analysis, enactment, and monitoring of business processes, "process systems" in short. It combines advanced computer science methods and techniques with real-world applications that impact nearly every industry, helping enterprises to streamline their business operations, enhance process efficiency, and respond to the needs of ever-changing markets. The Process Technology Forum received 20 submissions, of which 10 were accepted for presentation. Furthermore, Stefanie Rinderle-Ma was invited as a keynote speaker.

The Educators Forum aims to be the meeting point for educators within the BPM community to share and improve BPM teaching in a dynamic and practice-oriented way, acknowledging that BPM has become part of the curricula across a number of primary, secondary, professional, online, and in-company education programs. Now in its third edition, the forum introduced a refined structure of six contribution types—BPM Education Research, BPM Teaching Cases, Educational Practice Insights, Tools/Resources Contributions, Assignment Contributions, and Lightning Contributions—capturing the breadth of innovation in BPM education. The forum received a total of 24 submissions, of which 12 were accepted. Contributions spanned five types, reflecting a dynamic and evolving field. Key emerging themes included the integration of project-based and experiential learning, the use of generative AI and low-code platforms in instructional design and assessment, curriculum innovation aligned with next-generation BPM drifts, and the bridging of academia and industry through collaborative assignments and platforms. The forum also featured a keynote by Deniz Iren, including a hands-on experience with an AI-based tutor curated by educators to enhance learning responsibly.

We hope that the reader finds the selection of papers and presentations on these specific topics inspiring.

September 2025

Mahendrawathi ER
Avigdor Gal
Thomas Grisold
Flavia Santoro
Mathias Weske
Remco M. Dijkman
Dimka Karastoyanova
Banu Aysolmaz
Wasana Bandara
Kate Revoredo

Organization

Responsible BPM Forum Chairs

Mahendrawathi ER	Institut Teknologi Sepuluh Nopember, Indonesia
Avigdor Gal	Technion – Israel Institute of Technology, Israel
Thomas Grisold	University of St. Gallen, Switzerland
Flavia Santoro	Instituto de Tecnologia e Liderança, Brazil

Process Technology Forum Chairs

Mathias Weske	HPI, University of Potsdam, Germany
Remco M. Dijkman	Eindhoven University of Technology, The Netherlands
Dimka Karastoyanova	University of Groningen, The Netherlands

Educators Forum Chairs

Banu Aysolmaz	Eindhoven University of Technology, The Netherlands
Wasana Bandara	Queensland University of Technology, Australia
Kate Revoredo	Humboldt-Universität zu Berlin, Germany

Responsible BPM Forum Program Committee

Banu Aysolmaz	Eindhoven University of Technology, The Netherlands
Marco Comuzzi	Ulsan National Institute of Science and Technology, South Korea
Laura Dobusch	Vienna University of Economics and Business, Austria
Djordje Djurica	Vienna University of Economics and Business, Austria
Stephan A. Fahrenkrog-Petersen	Weizenbaum-Institut, Germany
Jens Gulden	Utrecht University, The Netherlands
Jan Mendling	Humboldt-Universität zu Berlin, Germany

Michael Rosemann	Queensland University of Technology, Australia
Maximilian Röglinger	University of Bayreuth, Germany
Johannes Schneider	University of Liechtenstein, Liechtenstein
Stefan Schönig	University of Regensburg, Germany
Rehan Syed	Queensland University of Technology, Australia
Lucinéia Heloisa Thom	Federal University of Rio Grande do Sul, Brazil
Oktay Türetken	Eindhoven University of Technology, The Netherlands
Amy Van Looy	Ghent University, Belgium
Arif Wibisono	Institut Teknologi Sepuluh Nopember, Indonesia
Anna Maria Oberländer	University of Bayreuth, Germany

Process Technology Forum Program Committee

Andreas Metzger	Universität Duisburg-Essen, Germany
Andrés Jiménez Ramírez	Universidad de Sevilla, Spain
Barbara Weber	Universität St. Gallen, Switzerland
Barbara Re	University of Camerino, Italy
Bernhard Rumpe	RWTH Aachen, Germany
Carl Corea	Universität Koblenz, Germany
Chiara Di Francescomarino	University of Trento, Italy
Cristina Cabanillas	University of Seville, Spain
Dimitris Karagiannis	Universität Wien, Austria
Dominik Bork	TU Wien, Austria
Gregor Engels	Universität Paderborn, Germany
Ingo Weber	Technische Universität München, Germany
Irene Vanderfeesten	KU Leuven, Belgium
Jochen Küster	Hochschule Bielefeld, Germany
Julius Köpke	Universität Klagenfurt, Austria
Luise Pufahl	Technische Universität München, Germany
Manfred Reichert	Universität Ulm, Germany
Matthias Weidlich	HU Berlin, Germany
Mirela Riveni	University of Groningen, The Netherlands
Monica Vitali	Politecnico di Milano, Italy
Orlenys López-Pintado	University of Tartu, Estonia
Patrick Delfmann	Universität Koblenz, Germany
Peter Fettke	Universität des Saarlandes, DFKI, Germany
Peter Loos	Universität des Saarlandes, DFKI, Germany
Pierluigi Plebani	Politecnico di Milano, Italy
Raimundas Matulevicius	University of Tartu, Estonia

Rik Eshuis	Eindhoven University of Technology, The Netherlands
Robin Bergenthum	Fernuniversität Hagen, Germany
Ronny Seiger	Universität St. Gallen, Switzerland
Sander Leemans	RWTH Aachen, Germany
Shazia Sadiq	University of Queensland, Australia
Simon Hacks	Stockholm University, Sweden
Stefan Schönig	Universität Regensburg, Germany
Stefanie Rinderle-Ma	Technische Universität München, Germany
Chun Ouyang	Queensland University of Technology, Australia

Educators Forum Program Committee

Amy van Looy	Ghent University, Belgium
Andrea Magalhaes	Dheka, Brazil
Fernanda Baião	PUC-Rio, Brazil
Flavia Santoro	Inteli, Brazil
Ilia Maslov	KU Leuven, Belgium
Iris Beerepoot	Utrecht University, The Netherlands
Jorge Munoz-Gama	Pontificia Universidad Católica de Chile, Chile
Lucineia Heloisa Thom	Federal University of Rio Grande do Sul, Brazil
Mahendrawathi ER	Institut Teknologi Sepuluh Nopember, Indonesia
Manual Seneviratne	University of Moratuwa, Sri Lanka
Marleen Voss	University of Münster, Germany
Marzena Grzesiak	Gdańsk University of Technology, Poland
Matthijs Berkhout	Utrecht University of Applied Sciences, The Netherlands
Michael Leyer	University of Rostock, Germany
Niels Martin	Hasselt University, Belgium
Oktay Turetken	Eindhoven University of Technology, The Netherlands
Patrick Delfmann	Koblenz University, Germany
Prasanna Illankoon	University of Moratuwa, Sri Lanka
Saimir Bala	Humboldt-Universität zu Berlin, Germany
Sareh Sadeghianasl	Queensland University of Technology, Australia
Shahrzad Roohy Gohar	University of Queensland, Australia
Silvia Inês Dallavalle de Pádua	University of São Paulo, Brazil
Thomas Grisold	University of St. Gallen, Switzerland
Vinicius Stein Dani	Utrecht University, The Netherlands

Additional Reviewers

Andree, Kerstin
Burke, Adam
Bursch, Tom
Ehrendorfer, Matthias
Gavric, Aleksandar
Klessascheck, Finn
Lüpges, Alexander
Mangler, Juergen
Motnikar, Lenart
Poss, Leo

Contents

Responsible BPM Forum

Educators Forum

Responsible BPM Forum

Preface of the Responsible BPM Forum

Traditionally, BPM has emphasized efficiency and effectiveness as core motivation. In light of the many societal and environmental challenges we are facing today, this forum fosters academic exchange around Responsible Business Process Management (RBPM).

We define Responsible BPM as the practice of applying BPM by accounting for the needs of stakeholders involved in business processes (e.g., workers and customers), the broader environment (e.g., regulators and natural resources) and society in more general terms (e.g., grand challenges). RBPM can pertain to different activities, such as responsibly designing, implementing and monitoring business processes, as well as developing and using BPM systems in responsible ways. Along these lines, RBPM can be realized to different degrees. At least, RBPM accounts for relevant needs by not producing harm, but it can also imply intentional pursuit of positive effects, such as by proactively promoting the well-being of workers and customers, fostering sustainability or addressing societal concerns. RBPM faces challenges from three perspectives: technological, managerial, and conceptual. From a technological standpoint, issues arise around data security, privacy, and the ethical use of AI and automation. The integration of these technologies often involves handling sensitive information, raising concerns about transparency, accountability, and algorithmic bias. In terms of managerial issues, there is a challenge in aligning organizational goals with ethical and sustainable practices, as managers should balance efficiency and profitability with social responsibility. This often includes navigating regulatory requirements and maintaining stakeholder trust. From a conceptual angle, there is the difficulty of embedding ethical principles into BPM frameworks, as traditional models are often focused on efficiency and cost-effectiveness rather than ethical outcomes. Defining and measuring "responsibility" in processes, as well as understanding the long-term impact of decisions on society and the environment, further complicate the development of a truly responsible BPM approach.

The 1st RBPM Forum of the 23rd Business Process Management aims to provide a forum to tackle these challenges. As its first edition, the forum attracted 21 international submissions around a diverse range of topics. All submissions are reviewed by at least two Program Committee members, which led to the acceptance of 10 papers which we summarize below (in alphabetical order of the first authors).

The paper "**Designing Business Processes for Good: A Canvas-Based Framework for Routinized Benevolence**" by Chandrasiri et al. discusses the growing need to embed ethical intent directly into for-profit business operations. The paper introduces the Routinized Benevolence (RB) Canvas, a structured design artefact that supports organizations systematically integrating voluntary, customer-focused benevolent actions into routine processes. Using a Design Science Research approach grounded in real-world cases, the study defines RB as mutualistic, repeatable practices that align customer well-being with long-term organizational benefit. By capturing drivers, actions, impacts, and risks, the RB Canvas illustrates Responsible BPM by operationalizing stakeholder-centric values within everyday workflows. The work shows how organizations can translate benevolent intent into actionable routines, bridging the gap between aspirational ethics and practical process design, and thus advances the RBPM goal of integrating societal value creation with sustainable profitability.

The study "**Operationalizing Responsible BPM: A Method for Value-based Process Redesign**" by Djurica et al. departs from the question of how responsible BPM can be realized. Specifically, the authors investigate how the concept can be embedded in actual business processes. Drawing from Value-based Engineering—an approach from the systems and software engineering domain—the authors propose a four-step model for responsibility-aware business process redesign. Accordingly, redesign initiatives should (1) elucidate and cluster the values of relevant stakeholders, (2) prioritize and conceptualize values, (3) identify requirements for the consideration of values in business processes, and (4) assess and validate the impact of such initiatives. The method is demonstrated through a fictional redesign of a hospital discharge process, incorporating values like patient well-being and sustainability.

In their paper "**The Impact of Event Data Partitioning on Privacy-aware Process Discovery**", Lim et al. addresses the fundamental tension between organizational learning and individual privacy protection in process mining applications. The paper proposes an innovative pipeline that combines event data partitioning with anonymization techniques, using event abstraction to segment logs into multiple parts that can be anonymized separately, thereby preserving privacy while mitigating utility loss in process discovery. This research exemplifies responsible BPM by tackling the technological challenge of data security and privacy protection while maintaining analytical value for organizations. By demonstrating that event partitioning can improve process discovery utility for directly-follows-based anonymization techniques, the work shows how technical innovation can serve both organizational needs and individual privacy rights, embodying the RBPM principle of avoiding harm to stakeholders while enabling legitimate business process analysis.

The study "**Bridging BPM and Sustainability Journeys in Navigating Sustainability Paradoxical Tensions**" by Nurkasanah and van Looy explores how BPM and sustainability initiatives can be combined. Drawing from two case studies, the authors find that organizations encounter shared tensions—learning, organizing, performing, and belonging—as they align BPM with sustainability. Applying the lenses of paradox theory and BPM capability frameworks, the paper identifies how BPM capability areas (e.g., governance, strategy, culture, IT) can help to address these paradoxes. The authors propose a framework for navigating these tensions and advancing sustainable BPM.

The paper "**Measuring Moral Performance: A Framework for Ethical Business Process Management**" by Poss et al. addresses the lack of systematic methods for integrating and assessing ethics in BPM. The paper proposes a framework and prototype that introduce explicit ethical perspective questionnaires and value assessments throughout the BPM lifecycle. By enabling real-time annotation, user feedback, and visualization of ethical performance data, the framework makes moral performance measurable and manageable in daily operations. Evaluated with domain experts, it shows how BPM can move beyond efficiency to embrace transparency, fairness, and stakeholder trust. This research operationalizes ethics as quantifiable metrics, indicating how socio-technical systems can align automation with moral accountability, thus embodying the RBPM principle of enhancing value while preventing harm.

In their study "**Anchoring ESG Goals in Business Processes: Engagement Patterns in Polish Organizations**", Potoczek et al. stress the significance of responsible BPM in organizations. The authors examine how Polish enterprises integrate ESG (Environmental, Social, Governance) goals into their business processes in the context of evolving EU regulations on sustainability, especially the 2023 CSRD directive. Presenting the results of a survey of 547 organizations in Poland, the authors report strong links between ESG strategies and business process changes, especially in large and listed companies. However, public sector organizations show limited ESG awareness and support. The research emphasizes the need for broader institutional support and well-designed managerial and technological support to advance responsible business process management and achieve sustainable development objectives effectively across sectors.

The paper "**ReDPT-CT: Designing Responsible Digital Process Twins for Clinical Trials**" by Ribeiro et al. addresses the fundamental tension between operational efficiency and ethical responsibility in the digital transformation of clinical trials. The paper proposes a structured framework that integrates Responsible BPM principles with Digital Process Twin (DPT) design, mapping fifteen ethical responsibility aspects, such as transparency, informed consent, and scientific integrity, the onto core, support, and management processes in Clinical Research Centers. By aligning these aspects with concrete DPT features like real-time monitoring, traceable data flows, and AI governance, the framework translates abstract ethical principles into actionable digital practices. This research makes a novel contribution to Responsible BPM by undertaking the complex challenge of embedding participant protection, data privacy, and accountability within data-intensive clinical trial environments. By showing how DPTs can enhance process compliance while safeguarding ethical standards, the work demonstrates how digital innovation can balance efficiency with the RBPM goal of avoiding harm and fostering trust in sensitive, stakeholder-dependent processes.

In the paper "**AI-driven BPM: Dehumanization of the Firm?**", Simonis and Overes explore a critical but often overlooked consequence of AI adoption in business processes, namely its impact on employee-organization relationships and stakeholder well-being. Through an experimental study comparing AI-driven versus human-informed BPM approaches, the research reveals that employees of companies adopting AI-driven BPM are more likely to dehumanize the firm, resulting in decreased attribution of moral agency, reduced organizational trust, and increased turnover intentions. This work directly challenges traditional BPM's efficiency-focused paradigm by demonstrating how technological advancement can inadvertently harm stakeholder relationships, aligning perfectly with RBPM's emphasis on accounting for worker needs and avoiding harm. The findings underscore the necessity of incorporating stakeholder theory into BPM evolution, advocating that responsible autonomization must consider the well-being of all stakeholders rather than prioritizing process performance over human factors.

The study "**Motivational Process Mining: A Conceptualization By Adding Self-Determination Theory**" by van Looy and Vlerick aims to complement the primarily technical research on process mining with human-related implications. They introduce motivational process mining as a novel concept to underscore the human-centered and psychological aspects of process mining initiatives. Drawing from self-determination theory, the authors suggest how process mining can be used with an eye on employee well-being and motivation. At the core of their observations is that managers and analysts should consider employees' needs for autonomy, relatedness, competence and intrinsic motivation to ensure that employees accept and support process mining initiatives.

Agentic Business Process Management: Practitioner Perspectives on Agent Governance in Business Processes by Vu et al. presents a comprehensive examination of how AI agents can be responsibly integrated into business processes through the lens of practitioner insights. Based on interviews with BPM practitioners, the study reveals a dual landscape where agentic AI promises enhanced efficiency, improved data quality, and better compliance, while simultaneously raising concerns about bias, over-reliance, cybersecurity threats, and job displacement. The paper directly addresses core challenges of RBPM by proposing key recommendations for responsible adoption: defining clear business goals, establishing legal and ethical guardrails, enabling effective human-agent collaboration, customizing agent behavior, managing operational risks, and ensuring safe integration with fallback mechanisms. This work exemplifies the technological challenge of ethically integrating AI, the managerial challenge of balancing efficiency with stakeholder well-being, and the conceptual challenge of embedding responsibility into automated decision-making systems while preserving human oversight and organizational trust.

Taken together, the papers accepted for the first edition of the Responsible BPM Forum provide a variety of original perspectives to consider the broader ethical, societal and environmental implications of BPM. They provide a fertile ground to further investigate this concept in the future.

Organization

Program Chairs

Mahendrawathi ER	Institut Teknologi Sepuluh Nopember, Indonesia
Avigdor Gal	Technion – Israel Institute of Technology, Israel
Thomas Grisold	University of St. Gallen, Switzerland
Flavia Santoro	Inteli – Instituto de Tecnologia e Liderança, Brazil

Program Committee

Banu Aysolmaz	Eindhoven University of Technology, The Netherlands
Marco Comuzzi	Ulsan National Institute of Science and Technology, South Korea
Laura Dobusch	Vienna University of Economics and Business, Austria
Djordje Djurica	Vienna University of Economics and Business, Austria
Stephan A. Fahrenkrog-Petersen	Weizenbaum-Institut, Germany
Jens Gulden	Utrecht University, The Netherlands
Jan Mendling	Humboldt-Universität zu Berlin, Germany
Michael Rosemann	Queensland University of Technology, Australia
Maximilian Röglinger	University of Bayreuth, Germany
Johannes Schneider	University of Liechtenstein, Liechtenstein
Stefan Schönig	University of Regensburg, Germany
Rehan Syed	Queensland University of Technology, Australia
Lucinéia Heloisa Thom	Federal University of Rio Grande do Sul, Brasil
Oktay Türetken	Eindhoven University of Technology, Netherlands
Amy Van Looy	Ghent University, Belgium
Arif Wibisono	Institut Teknologi Sepuluh Nopember, Indonesia
Anna Maria Oberländer	University of Bayreuth, Germany

The Impact of Event Data Partitioning on Privacy-Aware Process Discovery

Jungeun Lim[1], Stephan A. Fahrenkrog-Petersen[2,3](✉), Xixi Lu[4], Jan Mendling[2,3], and Minseok Song[1]

[1] Pohang University of Science and Technology, Pohang, South Korea
[2] Weizenbaum Institute, Berlin, Germany
[3] Humboldt-Universität zu Berlin, Berlin, Germany
stephan.fahrenkrog-petersen@hu-berlin.de
[4] Utrecht University, Utrecht, The Netherlands

Abstract. Information systems support the execution of business processes. The event logs of these executions generally contain sensitive information about customers, patients, and employees. The corresponding privacy challenges can be addressed by anonymizing the event logs while still retaining *utility* for process discovery. However, trading off utility and privacy is difficult: the higher the complexity of event log, the higher the loss of utility by anonymization. In this work, we propose a pipeline that combines anonymization and *event data partitioning*, where event abstraction is utilized for partitioning. By leveraging event abstraction, event logs can be segmented into multiple parts, allowing each sublog to be anonymized separately. This pipeline preserves privacy while mitigating the loss of utility. To validate our approach, we study the impact of event partitioning on two anonymization techniques using three real-world event logs and two process discovery techniques. Our results demonstrate that event partitioning can bring improvements in process discovery utility for directly-follows-based anonymization techniques.

Keywords: Privacy-preserving Process Mining · Event Log Anonymization · Event Log Pre-processing

1 Introduction

Event logs recorded by information systems are the backbone of process mining. Typically, these event logs contain personal information [32] that need to be protected in terms of privacy [10]. Such protection can be achieved through the anonymization of event logs [12]. Anonymization aims to provide a formal privacy guarantee for the data and, at the same time, maximize the *utility* of the anonymized event log. One way utility can be defined is by evaluating the quality of process models discovered using the anonymized log [27].

Many anonymization techniques work through the insertion of noise [14] with the aim of providing *differential privacy* [6]. This noise is inserted into the result

M. ER et al. (Eds.): BPM 2025, LNBIP 565, pp. 7–22, 2026.
https://doi.org/10.1007/978-3-032-02936-2_1

of queries that determine how often a control-flow behavior appears in the original event log. The query can either return a count of the trace-variants or the directly-follows relations within an event log [24]. A common goal of anonymization is to minimize the inserted noise while maximizing the preserved utility. A high-level of utility is especially hard to reach for less-structured processes.

In this paper, we build on the hypothesis that event data partitioning could facilitate a better trade-off between privacy and utility by reducing the amount of required noise. As an example, consider a process where sub-process A consists of activities a, b, and c, and sub-process B consists of activities d, e, and f, as shown in Fig. 1 (a). If a privacy-preserving technique injects noise into the directly-follows relations without considering the process structure, the resulting noise can be extensive, as illustrated in Fig. 1 (b) in blue. However, when the structure is considered, as in Fig. 1 (c) in blue, the noise has only to be applied to each sub-process independently. The additional noise in Fig. 1 (b) could introduce unrealistic process behaviors, impacting process discovery utility.

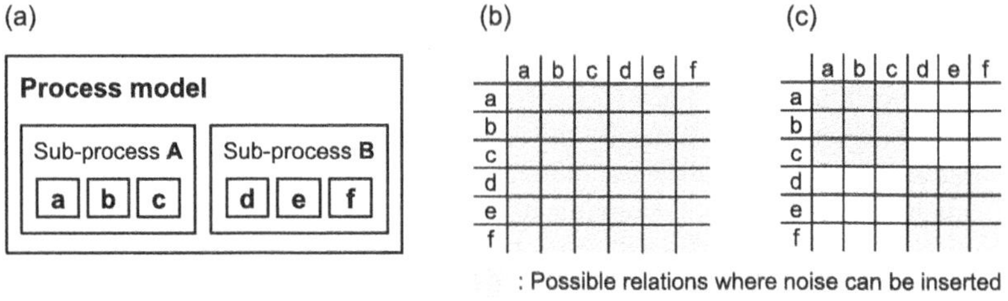

Fig. 1. Noise insertion differences between considering and not considering process structure. (a): Structure of process model, (b): Noise insertion areas when not considering process structure, (c): Noise insertion areas when considering process structure.

In line with this argument, we propose a pipeline that first partitions an event log and then applies anonymization. The anonymized logs are subsequently used for process discovery. The utility of the anonymized log is assessed based on the quality of the discovered models. First, we assess its effectiveness by measuring the utility of process models derived from sub-logs. We compare the utility of these models with those obtained from only anonymized event logs. Second, we evaluate whether event data partitioning always provides utility benefits. To do so, we compare the utility of process models obtained by applying partitioning before anonymization versus those obtained by applying partitioning after anonymization. We evaluated these questions for different types of anonymization, considering both directly-follows-based and trace-variant-based anonymization. Our results show that performing event partitioning before anonymization leads to better process discovery utility for directly-follows-based anonymization.

The remainder of this paper is structured as follows: Sect. 2 reviews existing research on privacy-preserving process mining and event log abstraction.

Section 3 introduces our proposed pipeline, and Sect. 4 presents the evaluation. Finally, Sect. 5 summarizes the research and concludes the paper.

2 Related Work

We outline the related work in this section. First, we give a short overview of privacy-preserving process mining in Sect. 2.1 and outline related work in the area of event log abstraction in Sect. 2.2.

2.1 Privacy-Aware Process Mining

Differential privacy [6] has been widely studied in the process mining literature [11,14,28]. One big advantage of differential privacy is that it is immune to post-processing. This means that process models generated from anonymized event logs are also secure against attacks like the one introduced by Kirchmann et al. [19]. The literature knows two types of mechanisms for control-flow anonymization: They anonymize either the *directly-follows distribution* of an event log or the *trace-variant distribution.* Most anonymization techniques are based on noise insertion. While many techniques exist for anonymization, the impact of pre-processing was not studied in the literature. The only exception is work by Elkoumy et al. [8] that implemented an approach of privacy amplification. The idea is that through sub-sampling of traces a higher privacy guarantee can be given with less noise than if a differential privacy mechanism would be applied directly. However, this approach does not use process specific knowledge. Through our work, we present the first approach that utilizes event data partitioning in the pre-processing to achieve better utility for process discovery.

A direction differing from the work on differential privacy is focused on group-based privacy guarantees. The main line of research here, focuses on generalization [17], substitution [26], or the merging of events [13]. While we study our approach in combination with one differential privacy mechanism, it could also be applied in combination with one of these group-based privacy approaches.

Furthermore, non-anonymization approaches for privacy have been explored. These are usually focused on deriving process models in settings where the event logs are distributed over multiple data owners. These approaches use techniques such as multi-party computation [9] and trusted execution environments [15]. In contrast to these approaches, our technique can be applied to event logs stored by a single data owner.

2.2 Event Log Abstraction

Event log abstraction involves identifying sub-processes and their instances from low-level events [22]. Various studies have proposed approaches for this, often addressing the identification of sub-processes and their instances separately. For identifying sub-processes, Baier et al. [2] assumed that low-level activities with similar names originate from the same sub-process. Günther et al. [16] assumed that low-level activities appearing close together in an event log likely originate from the same higher-level activity. Lu et al. [23] proposed techniques that use domain knowledge and clustering.

Identifying sub-process instances has also been studied. A straightforward approach is to classify consecutive events with the same sub-process as originating from the same activity instance [7,16]. Other approaches include assuming that a sub-process can only be executed once [23] or setting a time limit for the execution of a single instance to distinguish between instances [2].

In conclusion, the privacy-aware process mining literature did not focus on how pre-processing can be helpful for better utility. However, it was shown in the literature that pre-processing such as event log abstraction can lead to better process discovery utility. In this work, we address this gap in the literature by studying how event log abstraction impacts privacy-aware process discovery.

3 Anonymization Utilizing Event Data Partitioning

In Fig. 2 we give an overview of the anonymization pipeline that allows us to combine event data partitioning and anonymization. In the remainder of this section, we elaborate on the different components of the pipeline.

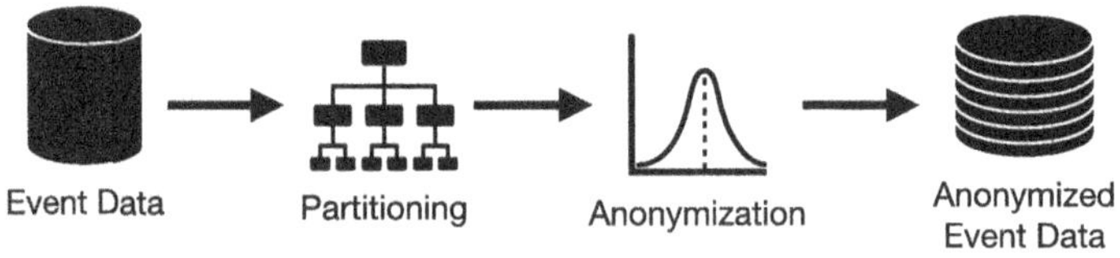

Fig. 2. Anonymization pipeline.

Event Data. Let us define an event as the execution of an activity $\alpha \in \mathcal{A}$, with $\mathcal{A}$ being the universe of all activities. Let a trace $\sigma \in \mathcal{A}^*$ be a sequence of events. Each trace represents the execution of a process. The sequence of a trace indicates in which order the events have been executed. We define an event log $L \in \mathcal{L}$ as a multi-set of traces.

Partitioning. We turn to event data partitioning. First, we need to define event abstraction as a function $\psi : \mathcal{A} \mapsto \mathcal{A}$. Let $\mathcal{A}$ always contain $\top$, which is the root activity, the highest-level activity within event abstraction. ψ allows us to abstract low-level activities, such as the reading of a sensor, into higher-level activities. Note that multiple low-level activities can be mapped onto the same higher-level activity. We assume that an activity a can not be abstracted to itself $\psi(a) \neq a$, with the exception of $\psi(\top) = \top$. The hierarchy ψ can be either user-defined or derived through automated approaches such as clustering. We assume that when multiple low-level activities are abstracted into the same higher-level activity α, the higher-level activity α represents a sub-process composed of these lower-level activities. A hierarchy ψ has to be defined such there are no cycles, e.g. $\psi(a) = \alpha$, $\psi(\alpha) = a$, and every activity can eventually be abstracted to the root activity $\top$, e.g. $\psi(a) = \alpha$, $\psi(\alpha) = \top$.

Our running example shown in Fig. 3, illustrates such a process with such an event abstraction hierarchy. Here, the activities $\{a, b, c\}$ can be abstracted to the activity A. Now we could apply the function ψ to every event within a trace σ to create an abstracted trace σ'. In such a scenario, all information that can be derived from the low-level activities would be lost. Therefore, it would be better to *decompose* a log L into multiple sub-logs: $\{L_\psi, L_{\alpha_1}, L_{\alpha_2}\}$ where L_ψ is the event log consisting of the higher-level activities while L_{α_1} contains all the activities belonging to sub-process α_1. These composed logs can be generated by the *event data partitioning* function $\psi_L : \mathcal{L} \mapsto 2^{\mathcal{L}}$.

In Algorithm 1, we outline how ψ_L works. First, we initialize the result set of sub-logs S and the abstracted event log L_ψ (see line 1-2). Next, we iterate over every trace σ in our event log L (see line 3). Next, we create the set C that is used to store traces of the sub-processes (see line 4. Now, we check for every activity of trace σ if it can be abstracted using ψ (see line 5-6). If an activity $\sigma(i)$ can be abstracted, we abstract it within the trace σ (see line 7 & line 10) and save the sub-process information belonging to the sub-process α' in a corresponding trace σ'_α (see line 8-9). Such a procedure, leads to many occurrences of the sub-process activity α' in the trace σ. We apply a *compression function* that removes every occurrence except the first and last of activity α' from the trace σ, signaling the start and end of the sub-process (see line 13), while the actual information is encoded in the sub-process trace $\sigma_{\alpha'}$. We add the trace ψ to the abstracted event log L (see line 14) and add the sub-process traces stored in the set C to the set of event logs S (see line 15).

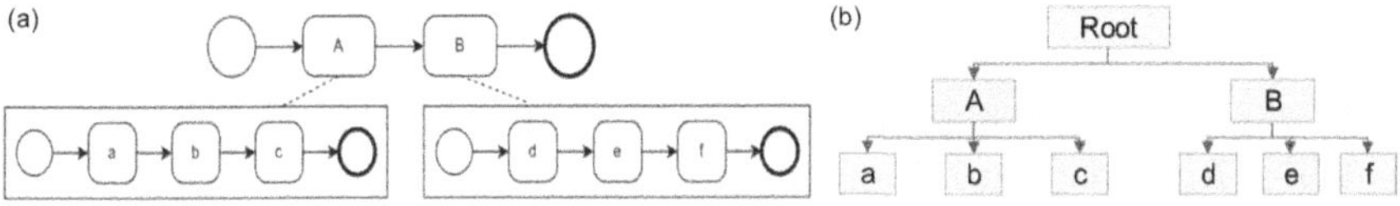

Fig. 3. Running example for process with sub-processes.

Anonymization. Next, we anonymize the partitioned event data. We apply the notion of differential privacy. We define $\gamma : \mathcal{L} \mapsto \mathcal{L}$ as an anonymization function that transforms an event log into an anonymized event log. Further, let us define $img(\gamma) \subseteq \mathcal{L}$ as the *image* of γ, i.e., the set of all event logs that may be returned by γ. Finally, we define two event logs $L_1, L_2 \in \mathcal{L}$ to be *neighboring*, if they differ by exactly the data of one individual. In our setting, this corresponds to one trace, i.e., $|L_1 \setminus L_2| + |L_2 \setminus L_1| = 1$. Based on [6], we use the following definition for differential privacy:

Definition 1 (Differential Privacy). *Given an anonymization function γ and privacy parameter $\epsilon \in \mathbb{R}$, function γ provides ϵ-differential privacy, if for all*

Algorithm 1: Event Log Partitioning function ψ_L

Input: An event log L
Output: Set of event logs $S \in 2^{\mathcal{L}}$

$S \leftarrow \emptyset$; `// Defines the empty set of new event logs`
$L_\psi \leftarrow \emptyset$;
for $\sigma \in L$ **do**
  $C \leftarrow \emptyset$;
  for $i \in [1, \ldots, |\sigma|]$ **do**
    if $\psi(\sigma(i)) \neq \emptyset$ **then**
      $\alpha' \leftarrow \psi(\sigma(i))$;
      $\sigma_{\alpha'} \leftarrow C(\alpha')$ if $\sigma_{\alpha'} \in C$; otherwise $\sigma_{\alpha'} \leftarrow \emptyset$;
      $C \leftarrow (C \setminus \sigma_\alpha) \cup \sigma'_\alpha \circ \sigma(i))$;
      $\sigma(i) \leftarrow \alpha'$;
    end
  end
  $compress(\sigma)$;
  $L_\psi \leftarrow L_\psi \cup \sigma$;
  $S.update(C)$;
end
$S \leftarrow S \cup L_\psi$;
return S

neighboring pairs of event logs $L_1, L_2 \in \mathcal{L}$ *and all subsets* $\rho_\gamma \subseteq img(\alpha)$, *it holds that:*

$$Pr[\gamma(L_1) \in \rho_\gamma] \leq e^\epsilon \times Pr[\gamma(L_2) \in \rho_\gamma]$$

where the probability is taken over the randomness introduced by the anonymization function γ.

We can apply any implementation of γ that gives the differential privacy guarantee. An important rule of differential privacy is parallel composition. This rule states that different datasets that have no intersection can be anonymized independently with ϵ-differential privacy their combination also is ϵ-differential private. Therefore, we can anonymize the different event logs created by the event log partitioning independently.

Anonymized Event Data. After anonymization, we end up with multiple anonymized event logs, i.e., L'_ψ, L'_A and L'_B. The anonymized log L'_ψ shows the process consisting of activities of higher level. The other logs contain the data of the anonymized sub-processes. For each of these logs, we apply a discovery algorithm and evaluate the quality of the discovered process model to measure the *utility*. Note that the other way around is also possible, namely, we first anonymize the event log and then partition the log into multiple sub logs. In the evaluation, we investigate the difference in the resulting utilities (RQ. 2).

4 Evaluation

In this section, we evaluate our proposed pipeline. Specifically, we examine the results for both directly-follows-based and trace-variant-based anonymization techniques. We focus specifically on the following two aspects:

RQ1 How does performing partitioning before anonymization affect the utility of anonymized logs for process discovery?

RQ2 How does the order of partitioning and anonymization impact process discovery utility?

In Sect. 4.1, we outline our experimental design and the datasets we employ. In Sect. 4.2, we present the results of our experiments. Finally, in Sect. 4.3, we provide a qualitative discussion on the trade-offs of combining partitioning with anonymization.

4.1 Experimental Setup

We provide an overview of our experimental setup in Fig. 4. The first question (RQ1) aims to investigate the effect of our proposed pipeline (i.e., first partitioning and then anonymization, highlighted in orange in Fig. 4) on utility (i.e., the quality of the process models discovered using anonymized logs). One may argue it is unfair to compare the utility of a set of logs to one log. Therefore, in the second question (RQ2), we examine whether any observed improvement is solely due to the introduction of partitioning rather than its specific placement in the pipeline. To investigate this, we compare the resulting utility of partitioning before (highlighted in orange) and after (highlighted in black) anonymization.

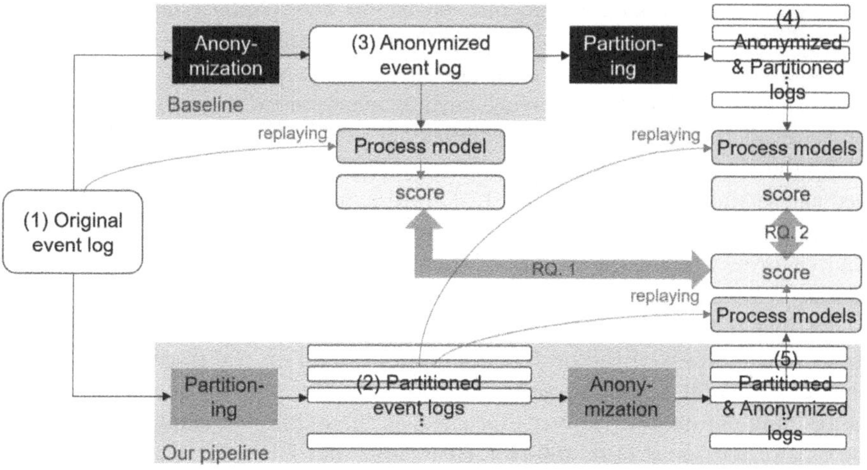

Fig. 4. Experimental setup overview.

Datasets. For our evaluation, we used the BPIC2012 [4], BPIC2015 [1][1], and BPIC2017 [5] event logs. All three event logs stem from real-world business processes. Both BPIC2012 and BPIC2017 describe a loan application process. BPIC2015 describes a permit application process used by Dutch municipalities. All of these logs are known to consist of events from multiple sub-processes and have therefore been widely used in various event abstraction studies [21,23,29]. In Table 1, we provide descriptive statistics of these datasets.

Table 1. Statistical description of the datasets.

Data	#case	#events	#acts	avg. e/c	#trace variants
BPIC2012	13,087	262,200	24	20	4,366
BPIC2015	902	21,656	70	24	295
BPIC2017	31,509	1,202,267	26	38	15,930

Applied Techniques for Event Data Partitioning. During event data partitioning, we utilized three techniques for the higher-level activity mapping function (ψ in Sect. 3). The first technique developed by Lu et al. [23] integrates domain knowledge into event abstraction (FHM, FlexHMiner). Second, we use a technique by Günther et al. [16] based on activity clustering (AC). Finally, we employ a random clustering that allows us to study if we also experience benefits with less sophisticated abstraction techniques. We assumed that a high-level activity could only be executed once and distinguished between the start and end of higher-level activities, following the approach in [23].

Applied Techniques for Anonymization. For anonymization, we used two anonymization techniques: one based on the Laplace mechanism (DF-Laplace) [24] (since it does not require an additional parameter, as does the technique based on the exponential mechanism such as SaPa [14]); the other one based on SaCoFa [14] for the trace-variant query, since it provides the highest utility of all techniques that add additional behavior to anonymized logs. We used public implementations of both techniques [18]. For differential privacy, we used the settings of $\varepsilon = \{0.01, 0.1, 1.0\}$ to test different orders of magnitudes for privacy guarantee. Lower values of ε mean stronger privacy protection. In our result figures, we plot the results of these settings within one bin, due to space limitations.

SaCoFa anonymizes trace-variants by constructing a prefix tree that accounts for the harmfulness of the prefix. The upper bound on the trace-variant length (l) was set to 50 across all settings. However, the privacy guarantee (ε) and the pruning parameter (p) were evaluated at three different combinations: (0.01, 300), (0.1, 200), and (1.0, 100) for BPIC2012 and BPIC2017, and (0.01, 100),

[1] We used filtered log described in [1]. Although five datasets are available, we only used the first one, BPIC_1f.

(0.1, 100), and (1.0, 50) for BPIC2015. To account for non-determinism of noise injection, we repeated the anonymization process 10 times for each setting.

Applied Techniques for Process Discovery. For process discovery, we used Inductive Miner [20] and Heuristic Miner [33], setting the noise threshold to 0.2 for both techniques. Both techniques are widely used process discovery algorithms and we used their implementation in PM4Py [3].

Evaluation Measures. We evaluated the utility of the process models using four metrics. For fitness, we utilized token-based replay fitness [31]. For precision, we employed ETC Precision [25]. We also calculated the F1-score, which is the harmonic mean of fitness and precision. Finally, for generalization, we adapted the technique from [1] with $k = 3$.

When evaluating the multi-level process model, we followed the approach in [23]. This involves calculating the performance for each sub-process in the multi-level process model and then averaging them.

4.2 Results

RQ1: Effect of Partitioning on DF-Laplace. Figure 5 compares the utility of performing only anonymization with that of performing anonymization after partitioning, categorized by dataset, applied technique in event partitioning, and process discovery technique. The fitness was higher when only anonymization was applied, for almost all datasets. On the other hand, precision exhibited significant benefits from event partitioning: it was below 0.2 when only anonymization was performed, but increased to as high as 0.6 after partitioning at most. Overall, the increase in precision contributed to higher F1-score values when anonymization was preceded by event partitioning. Generalization showed good utility in most cases, with values exceeding 0.8 regardless of whether partitioning was applied, and differences remained within 0.2. These results indicate that incorporating partitioning, particularly with the DF-Laplace anonymization technique, can enhance precision while maintaining good levels of fitness and generalization.

RQ1: Effect of Partitioning on SaCoFa. When SaCoFa was used for anonymization, the results differed from those observed with DF-Laplace. While partitioning tended to improve the utility of anonymized logs when DF-Laplace was used, it appeared to have little to no effect on utility when SaCoFa was applied. As shown in Fig. 6, there was generally little difference across all metrics between applying anonymization alone and applying partitioning before anonymization. Specifically, when DF-Laplace was used, precision showed a significant improvement after partitioning, but with SaCoFa, the difference in precision between logs with and without partitioning was minimal.

This outcome can be attributed to the fact that DF-Laplace significantly reduces precision when applied to the full log, whereas SaCoFa maintains relatively stable precision levels, and in some cases even improves them.

For the BPIC2017 dataset, however, precision showed an improvement compared to other datasets. This can be explained by the relatively low precision

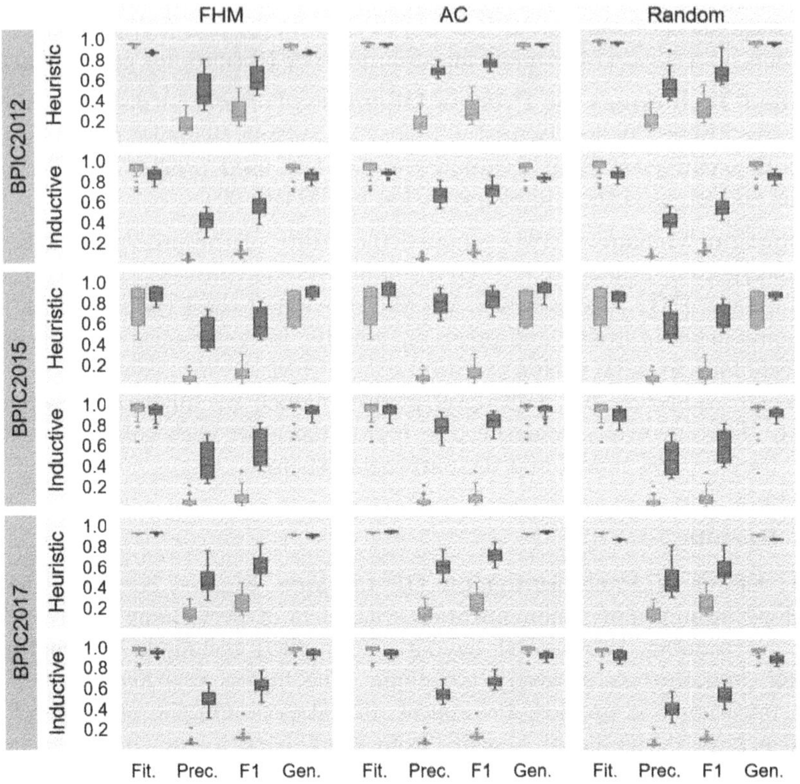

Fig. 5. Utility comparison with DF-Laplace: Anonymization only (gray) vs. Partitioning and Anonymization (blue). (Color figure online)

of the BPIC2017 log after anonymization, allowing partitioning to have a more noticeable effect.

For the BPIC2015 dataset, precision itself did not improve significantly, but partitioning effectively reduced its variation. This suggests that while anonymization alone results in good precision on average, there are cases where its quality is not consistently guaranteed. In such instances, partitioning can help stabilize the quality of anonymization and ensure more reliable results.

RQ2: Effect of Partitioning Order on DF-Laplace. Figure 7 compares the utility of performing partitioning after anonymization and the utility of performing partitioning before anonymization, categorized by dataset, applied abstraction technique in event partitioning, and process discovery technique. Overall, across all datasets, performing partitioning before anonymization yielded better performance in all metrics compared to partitioning after anonymization. However, fitness and generalization generally exceeded 0.8 in both cases, resulting in relatively minor differences, while precision showed a more pronounced disparity. For example, when partitioning was performed using random clus-

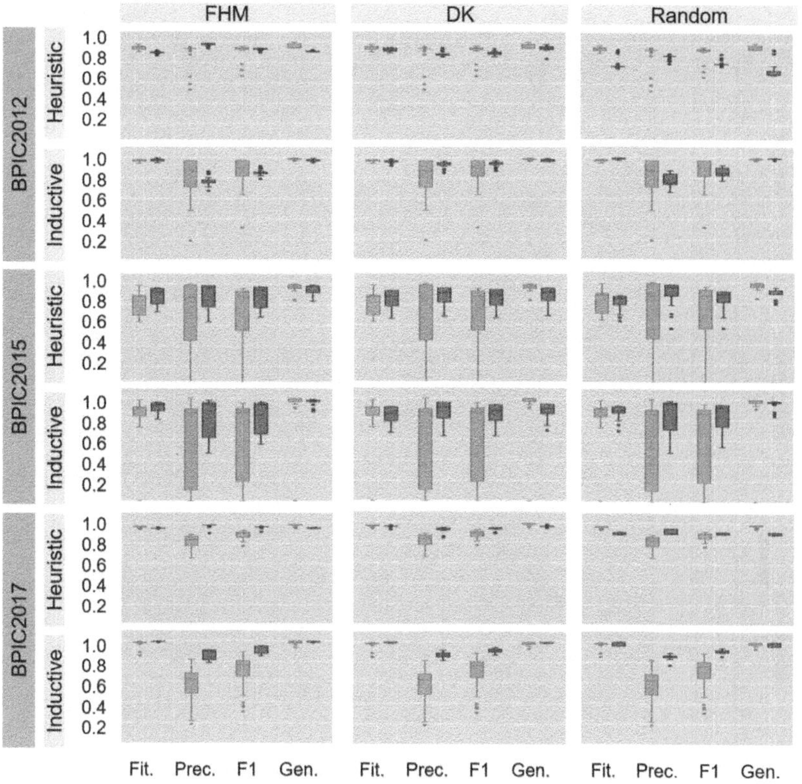

Fig. 6. Utility comparison with SaCoFa: Anonymization only (gray) vs. Partitioning and Anonymization (blue). (Color figure online)

tering on BPIC2015 and the process model was discovered using the heuristic algorithm, the precision improved significantly–from approximately 0.3 when partitioning was performed after anonymization to 0.8 when partitioning was performed beforehand. In conclusion, these results demonstrate that when using DF-Laplace for anonymization, partitioning before anonymization helps to preserve more information in the data, thereby enhancing process discovery performance.

When partitioning was performed using FHM on BPIC2015, the results for precision differed somewhat from other observations. While precision generally improved significantly when partitioning was performed first, in this case it decreased. Considering that the precision was high when FHM was performed on BPIC2015 without anonymization (0.98 with the heuristic algorithm and 0.89 with the inductive algorithm), this result can be interpreted as the impact of data distortion from anonymization being more pronounced in the sub-logs.

RQ2: Effect of Partitioning Order on SaCoFa. For BPIC2015 and BPIC2017, partitioning first showed slightly better performance, but overall the

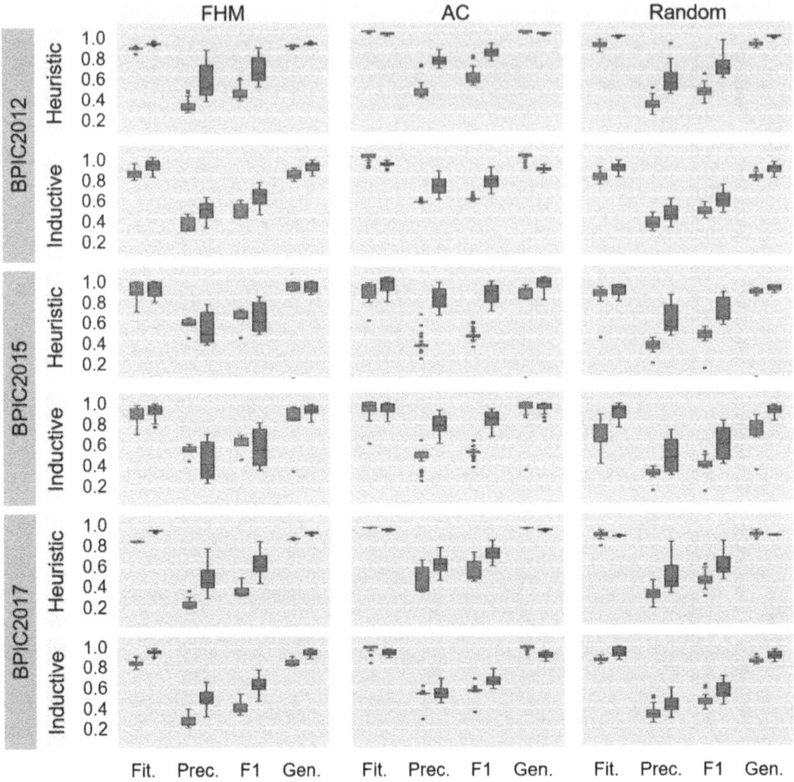

Fig. 7. Utility comparison with DF-Laplace: Anonymization and Partitioning (red) vs. Partitioning and Anonymization (blue). (Color figure online)

differences were minimal. In BPIC2012, partitioning later performed slightly better. Looking back at the earlier results with DF-Laplace, where applying partitioning last significantly reduced precision–leading to much better performance when partitioning was done beforehand–this outcome can be interpreted differently for SaCoFa. Since SaCoFa does not significantly degrade precision even when anonymization is applied without partitioning, this likely explains the observed results.

However, when applying FHM to BPIC2015, partitioning first resulted in significantly better utility compared to partitioning later. This is an unusual result, considering that for the same dataset, when using Günther's technique or Random clustering, there was little difference between partitioning first and partitioning later. Comparing this with the results of using SaCoFa for anonymization alone (Fig. 6), we see that when FHM was applied to anonymized data, precision dropped significantly. In contrast, when other techniques were applied, the precision remained relatively stable. This suggests that the effectiveness of an abstraction technique used for partitioning can vary.

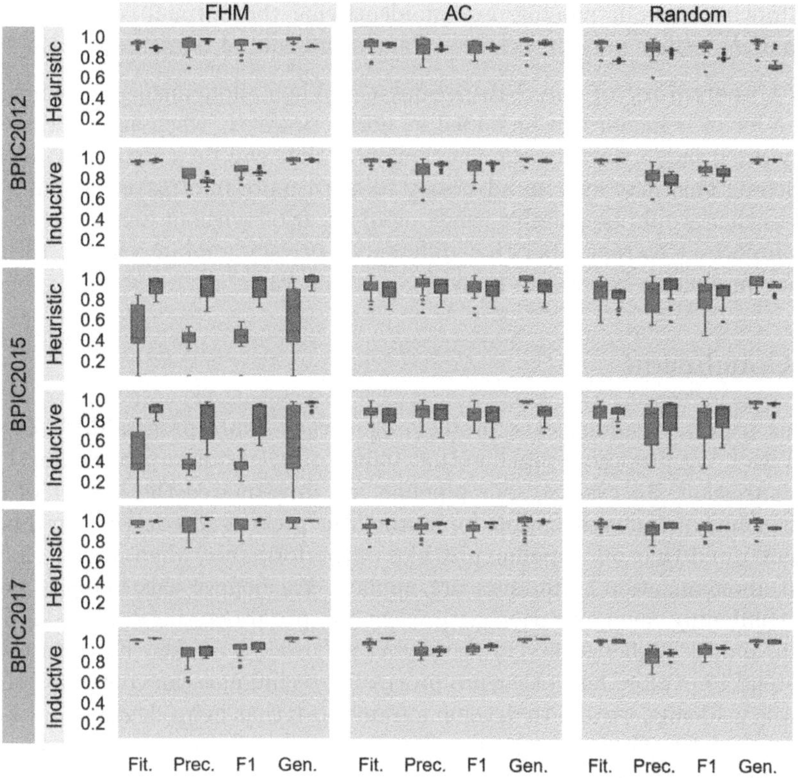

Fig. 8. Utility comparison with SaCoFa: Anonymization and Partitioning (red) vs. Partitioning and Anonymization (blue). (Color figure online)

4.3 Trade-off Discussion

In this section, we discuss trade-offs [30] that come from integrating event data partitioning into the anonymization process.

Loss of Information due to Partitioning. Event log partitioning can remove information that is useful for the analysis of the business process. Therefore, event partitioning can lead to an additional information loss that is difficult to quantify. This trade-off has to be considered when weighing the utility gains from lower noise addition.

Explosion of Pipeline Options. Choosing the right anonymization technique is challenging as there is no standardized guide. This creates a risk that users may apply a suboptimal anonymization technique for their analytical needs. In this paper, we introduce a pipeline that further increases the complexity of this decision. Instead of selecting only an anonymization technique, users must also choose an abstraction technique for event data partitioning. As a result, one drawback of our pipeline is that identifying the optimal configuration for maximizing utility in each scenario requires substantial domain expertise.

Less Uncertainty of the Sub-processes. When anonymizing the whole process, a lot of behavior can be added as noise. However, when sub-processes are anonymized independently, the new behavior that can be added is significantly restricted. This may lead an adversary to learn more information about a sub-process than would be possible using anonymization without event partitioning. It is important to remember that differential privacy guarantees that the same privacy protection is given to individuals either way.

5 Conclusion

In this paper, we address the problem of privacy-aware process discovery in a novel way. We propose a pipeline that builds on event data partitioning before anonymization. To evaluate our pipeline, we investigated the impact of event partitioning on anonymization for automatic process discovery. Our findings demonstrated that our pipeline provides outperformance when directly-follows-based anonymization techniques are applied. We believe this findings suggest that combining anonymization with event log pre-processing has the opportunity to unlock higher utility in privacy-aware process mining. For future work, we intend to explore how other pre-processing techniques can contribute to better utility. Ideally, we aim to develop a framework that helps determine the most effective combination of pre-processing and anonymization techniques based on event log characteristics.

References

1. Augusto, A., et al.: Automated discovery of process models from event logs: review and benchmark. IEEE Trans. Knowl. Data Eng. **31**(4), 686–705 (2019)
2. Baier, T., Mendling, J., Weske, M.: Bridging abstraction layers in process mining. Inf. Syst. **46**, 123–139 (2014)
3. Berti, A., Van Zelst, S.J., van der Aalst, W.: Process mining for python (pm4py): bridging the gap between process-and data science. arXiv preprint arXiv:1905.06169 (2019)
4. van Dongen, B.: Bpi challenge 2012 (2012). https://data.4tu.nl/articles/dataset/BPI_Challenge_2012/12689204/1
5. van Dongen, B.: Bpi challenge 2017 (2017). https://data.4tu.nl/articles/dataset/BPI_Challenge_2017/12696884/1
6. Dwork, C.: Differential privacy. In: Bugliesi, M., Preneel, B., Sassone, V., Wegener, I. (eds.) ICALP 2006. LNCS, vol. 4052, pp. 1–12. Springer, Heidelberg (2006). https://doi.org/10.1007/11787006_1
7. van Eck, M.L., Sidorova, N., van der Aalst, W.M.P.: Enabling process mining on sensor data from smart products. In: RCIS, pp. 1–12. IEEE (2016)
8. Elkoumy, G., Dumas, M.: Libra: high-utility anonymization of event logs for process mining via subsampling. In: Burattin, A., Polyvyanyy, A., Weber, B. (eds.) ICPM, pp. 144–151. IEEE (2022)

9. Elkoumy, G., Fahrenkrog-Petersen, S.A., Dumas, M., Laud, P., Pankova, A., Weidlich, M.: Secure multi-party computation for inter-organizational process mining. In: Nurcan, S., Reinhartz-Berger, I., Soffer, P., Zdravkovic, J. (eds.) BPMDS/EMMSAD -2020. LNBIP, vol. 387, pp. 166–181. Springer, Cham (2020). https://doi.org/10.1007/978-3-030-49418-6_11
10. Elkoumy, G., et al.: Privacy and confidentiality in process mining: threats and research challenges. ACM TMIS (2021)
11. Elkoumy, G., Pankova, A., Dumas, M.: Differentially private release of event logs for process mining. Inf. Syst. **115**, 102161 (2023)
12. Fahrenkrog-Petersen, S.A., van der Aa, H., Weidlich, M.: PRIPEL: privacy-preserving event log publishing including contextual information. In: Fahland, D., Ghidini, C., Becker, J., Dumas, M. (eds.) BPM 2020. LNCS, vol. 12168, pp. 111–128. Springer, Cham (2020). https://doi.org/10.1007/978-3-030-58666-9_7
13. Fahrenkrog-Petersen, S.A., van der Aa, H., Weidlich, M.: Optimal event log sanitization for privacy-preserving process mining. DKE **145**, 102175 (2023)
14. Fahrenkrog-Petersen, S.A., Kabierski, M., van der Aa, H., Weidlich, M.: Semantics-aware mechanisms for control-flow anonymization in process mining. Inf. Syst. **114**, 102169 (2023)
15. Goretti, V., Basile, D., Barbaro, L., Ciccio, C.D.: Trusted execution environment for decentralized process mining. In: Guizzardi, G., Santoro, F.M., Mouratidis, H., Soffer, P. (eds.) CAiSE (2024)
16. Günther, C.W., Rozinat, A., van der Aalst, W.M.P.: Activity mining by global trace segmentation. In: BPM Workshops (2009)
17. Hildebrant, R., Fahrenkrog-Petersen, S.A., Weidlich, M., Ren, S.: PMDG: privacy for multi-perspective process mining through data generalization. In: CAiSE. Springer (2023). https://doi.org/10.1007/978-3-031-34560-9_30
18. Kirchmann, H., Fahrenkrog-Petersen, S.A., Kabierski, M., van der Aa, H., Weidlich, M.: Privacy-preserving process mining with pm4py (extended abstract). In: ICPM Demo. CEUR-WS.org (2022)
19. Kirchmann, H., Fahrenkrog-Petersen, S.A., Mannhardt, F., Weidlich, M.: Control-flow reconstruction attacks on business process models. CoRR abs/arXiv: 2409.10986 (2024)
20. Leemans, S.J.J., Fahland, D., van der Aalst, W.M.P.: Discovering block-structured process models from event logs containing infrequent behaviour. In: Lohmann, N., Song, M., Wohed, P. (eds.) BPM 2013. LNBIP, vol. 171, pp. 66–78. Springer, Cham (2014). https://doi.org/10.1007/978-3-319-06257-0_6
21. Li, C.Y., van Zelst, S.J., van der Aalst, W.M.: Event abstraction for partial order patterns. In: BPM, pp. 38–54. Springer (2023). https://doi.org/10.1007/978-3-031-41620-0_3
22. Lim, J., Song, M.: Navigating event abstraction in process mining: A comprehensive analysis of sub-problems, data, and process characteristic considerations. In: BPM Workshops. pp. 174–185. Springer (2024). https://doi.org/10.1007/978-3-031-50974-2_14
23. Lu, X., Gal, A., Reijers, H.A.: Discovering hierarchical processes using flexible activity trees for event abstraction. In: ICPM, pp. 145–152. IEEE (2020)
24. Mannhardt, F., Koschmider, A., Baracaldo, N., Weidlich, M., Michael, J.: Privacy-preserving process mining: differential privacy for event logs. BISE **61**, 595–614 (2019)
25. Muñoz-Gama, J., Carmona, J.: A fresh look at precision in process conformance. In: Hull, R., Mendling, J., Tai, S. (eds.) BPM 2010. LNCS, vol. 6336, pp. 211–226. Springer, Heidelberg (2010). https://doi.org/10.1007/978-3-642-15618-2_16

26. Rafiei, M., van der Aalst, W.M.P.: Group-based privacy preservation techniques for process mining. DKE **134**, 101908 (2021)
27. Rafiei, M., van der Aalst, W.M.P.: Towards quantifying privacy in process mining. In: Leemans, S., Leopold, H. (eds.) ICPM 2020. LNBIP, vol. 406, pp. 385–397. Springer, Cham (2021). https://doi.org/10.1007/978-3-030-72693-5_29
28. Rafiei, M., Wangelik, F., Pourbafrani, M., van der Aalst, W.M.P.: Travag: differentially private trace variant generation using gans. In: RCIS. Lecture Notes in Business Information Processing, vol. 476, pp. 415–431. Springer (2023). https://doi.org/10.1007/978-3-031-33080-3_25
29. Rebmann, A., Weidlich, M., van der Aa, H.: GECCO: constraint-driven abstraction of low-level event logs. In: ICDE, pp. 150–163. IEEE (2022)
30. Robillard, M.P., et al.: Communicating study design trade-offs in software engineering. ACM TOSEM (2024)
31. Van Der Aalst, W., van der Aalst, W.: Data science in action. Springer (2016)
32. Maatouk, K., Mannhardt, F.: Quantifying the re-identification risk in published process models. In: Munoz-Gama, J., Lu, X. (eds.) ICPM 2021. LNBIP, vol. 433, pp. 382–394. Springer, Cham (2022). https://doi.org/10.1007/978-3-030-98581-3_28
33. Weijters, A., Ribeiro, J.: Flexible heuristics miner (fhm). In: CIDM, pp. 310–317 (2011)

AI-Driven BPM: Dehumanization of the Firm?

Paul Simonis(✉) and Ed Overes

Zuyd University of Applied Sciences, HDI-lab, Heerlen/Maastricht, The Netherlands
paul.simonis@zuyd.nl, ed.overes@zuyd.nl

Abstract. This research-in-progress paper addresses the question what consequences the use of Artificial Intelligence (AI) in Business Process Management (BPM) has on the relationships between companies and their employees. Our pilot study compared an AI-driven BPM approach with a human-informed approach with regards to dehumanization, moral agency, trust, and turnover intention. Preliminary results suggest that employees of companies that adopt an AI-driven approach are more likely to dehumanize the firm, resulting in a decrease in the attribution of moral agency towards the firm. Also, results suggest that those employees are less likely to trust the company, resulting in an increased probability of leaving the firm. Our findings emphasize the need for including stakeholder effects when introducing AI in developing responsible business processes.

Keywords: Responsible Business Process Management · Process Autonomization · Artificial Intelligence · Stakeholder Theory

1 Introduction

Artificial Intelligence (AI) is one of the fast emerging technologies that created new conditions for newly or reinterpreted Business Process Management (BPM) capabilities [18]. AI refers to machines performing cognitive functions that are typically linked to human intelligence [29] and can both substitute and complement human capabilities [31]. Within BPM, AI might introduce Generative Process Science [5] as the fourth type of Process Science. The distinction with the three earlier distinguished types (i.e., descriptive, explanatory, and prescriptive) [9]) is that the next stage of the deployment of technology to BPM enables delegation of the governance of a process to machines [32], giving rise to AI-driven BPM. AI-driven BPM makes it possible to continuously adapt and improve business processes [28] in which tasks, flow, interactions and governance is carried out by AI [32]. Allowing AI to govern processes relates to one of the three important drifts in contemporary and emerging BPM: process autonomization [32]. Although the next generation of BPM embraces the possibilities enabled by AI, concerns arise about the effects of process autonomization on stakeholder well-being [32]. In this pilot study, we research the attitudes of the employee towards

M. ER et al. (Eds.): BPM 2025, LNBIP 565, pp. 23–28, 2026.
https://doi.org/10.1007/978-3-032-02936-2_2

a firm that adopts AI-driven BPM. We research the impact on employees of companies on both ends of the spectrum task automation - process autonomization [32] and are particularly interested in the impact of process autonomization on dehumanization, moral agency, trust in the firm, and employee turnover intention.

2 Related Work

The position of the employee in BPM has gained interest of several researchers from the perspective of diversity and inclusion [25], well-being [8], and visual literacy [30]. The development of AI-augmented Business Process Management Systems changes the role of the employee in BPM [10] and adds the role of trust [1] to the variables when mining for well-being [8]. Investigation of these concerns through the lens of stakeholder theory [13] contributes to the development of BPM's underexposed dimension of process ethics [32]. Stakeholder theory advocates that firms should create value for all stakeholders [13], and that firms should deal with stakeholders based on a principle of fairness [13], moral concerns [20] and ethical principles [21]. In line with this, scholars have argued that the BPM discipline should reach beyond maximizing efficiency, rooted in the traditional core of the field. Currently, BPM is often labeled as prioritizing process performance over stakeholder well-being [24]. To move away from this stigma, we propose that scholars should incorporate principles of morality, fairness, and ethics in the development of next generation business process management [32]. In the next section, we develop two hypotheses that focus on the effect of AI-driven BPM on a firm's employees.

3 Hypotheses

AI-Driven BPM, Dehumanization and Moral Agency. Researchers are increasingly interested in psychological factors underlying attitudes towards AI, including dehumanization [14]. Relevant attributions for dehumanization are mind attributions [16], consisting of experience and agency [19]. AI lacks experience [17] as it does not signal capacity for feeling and emotions. Also, people attribute less agency and experience to decision making by advanced machines versus humans [6]. Consequently, we expect that employees of firms that adopt AI-driven BPM are more likely to dehumanize the firm, as they elicit less mind to the firm. Next, we expect that dehumanization will decrease the likelihood of attributing moral agency towards the firm. The attribution of agency is important in being able to make moral decisions [12] as morality is an important aspect of human beings [22]. The above reasoning culminates in our first hypothesis:

H1. There is a significant negative indirect effect of an AI-driven BPM approach (compared to a human-informed BPM approach) on the attribution of moral agency toward the firm, through increased dehumanization of the firm.

AI-Driven BPM, Trust and Employee Turnover-Intention. Another often studied psychological mechanism underlying peoples reaction towards AI is trust [15] [1], which derives from accepting vulnerability [33]. As AI involves a black-box problem in which observers can view input and output but not internal processes [11], employees are less likely to accept vulnerability because AI lacks transparency [15]. Furthermore, unfairness is also considered a great challenge for AI, hindering the development of trust [34] [37]. Additionally, we expect that employees who are less likely to trust the firm construct higher turnover intentions as negative work attitudes can lead to thinking about quitting a job [27]. Consequently, we hypothesize the following:

H2. There is a significant positive indirect effect of an AI-driven BPM approach (compared to a human-informed BPM approach) on employee turnover-intention, through the reduction of organizational trust.

4 Research Design and Results

Sample. Assuming a medium effect size and aiming for sufficient power $B = 0.80$ and $\alpha = .05$, we calculated with G*Power that we needed 204 participants. Our sample consisted of undergraduate students from a large University of Applied Sciences who participated in a classroom-based experiment. As the use of undergraduate students can harm data quality [26], we recruited a total of 269 participants. Of 269 participants, we excluded 70 participants who did not finish the experiment. We also excluded 10 participants because they spent an average time of less than 2 s or more than 20 s per item [7]. We retained a sample of 189 participants (Mage = 21.24; SD = 1.74, 31% female). ***Procedure.*** We used an experimental vignette study to test our hypotheses, which is an often used methodology in organizational research [2]. After administering written consent, we randomly allocated students towards vignettes in which the independent variable (i.e., a firm's BPM approach) was manipulated. Before reading these vignettes, participants were asked imagine being an employee of a company. In one vignette, we described a firm that employs an AI-driven tool to apply BPM. In the other vignette, we described a firm in which employees improve processes using BPM. Subsequently, we measured participants' trust, turnover intention, dehumanization, and moral agency. ***Measures.*** All measures were administered using a 7-point Likert scale. We randomly presented all items within each measure and all measured were counterbalanced to account for order effects To measure dehumanization of the firm ($\alpha = 0.81$; $M = 3.74$; $SD = 1.15$), we adapted items from a mind attribution subscale [36]. We measured moral agency ($\alpha = 0.91$; $M = 4.41$; $SD = 1.29$) with the use of an adapted morality scale [4]. Employee turnover intention ($\alpha = 0.83$; $M = 4.11$; $SD = 1.02$) was measured with the use of an adapted scale [35]. Trust in the firm ($\alpha = 0.69$; $M = 3.39$; $SD = .99$) was measured with the use of an adapted trust scale [3]. All supplementary materials are available at https://doi.org/10.6084/m9.figshare.29108873.v1. ***Manipulation Check and Perceived Realism.*** To ensure successful manipulation, we presented the participants a manipulation check. The results of an ANOVA suggest that our manipulation was reasonably successful but with room for improvement. Results suggest a significance difference between the mean scores of the AI-driven vignette ($M = 15.57$; $SD = 23.67$) and the human-informed vignette ($M = 4.03$; $SD = 24.18$); $F(1{,}187) = 10.98$, $p = .001$, $f = 0.24$. The participants of both vignette perceived the vignettes as equally realistic, $F(1{,}187) = 1.2$, $p = .28$, $f < .01$. ***Mediation models.*** We used model 4 of Hayes' PROCESS macro for SPSS with 5,000 bootstrap samples to run two mediation models and interpreted effects as statistically significant if the 95% confidence intervals did not include zero. The results suggest that AI-driven BPM negatively affects humanization of the firm ($b = -.63$, $SE = .17$, $p < .001$) and that this negatively affects the attribution of moral agency ($b = .25$, $SE = .083$, $p = .0029$). We also found results that suggest a negative indirect effect of AI-driven BPM on the attribution of moral agency through the increased likelihood of dehumanization ($b = -.16$, $SE = .072$, 95% CI [–.31; –.035]). Results of the second mediation analysis suggest that AI-driven BPM negatively effects trust in the firm ($b = -.31$, $SE = .14$, $p = .035$) and that this positively effects employee turnover intention ($b = -.27$, $SE = .07$, $p < .001$). Also, results suggest an indirect effect of AI-driven BPM on employee turnover intention, via a decreased likelihood of trust in the firm ($b = .082$, $SE = .044$, 95% CI [.0049; .18]).

5 Discussion

In the evolution of process autonomization, BPM scholars have called to explore responsible autonomization to advance capabilities for next-generation BPM [32]. In the current research-in-progress paper, we point towards stakeholder theory as a potential valuable source for the development of responsible autonomization. More specifically, in order to be responsible, we argue that autonomized BPM should consider the well-being of all a firm's stakeholders. Paradoxically, our preliminary results of a pilot study suggest that the use of AI in the autonomization of BPM may negatively impact employees. Companies that adopt AI-driven BPM are more likely to be dehumanized, less likely to be perceived as moral agents, less likely to be trusted, and more likely to face higher turnover intentions. This calls for further research. We plan to introduce AI complementation, next to AI substitution, and investigate behavioral differences of stakeholders towards these approaches. Also, to increase the external validity we will expand our research with more experienced employees opposed to students. Additionally, future research and multiple experiments are needed to gain a stronger indication that the expected relationships are viable, as this research-in-progress paper reports preliminary findings from a single classroom-based experiment and refrains from making strong causal claims. However, our findings suggest a clear direction: next generation BPM needs to take the impact of stakeholders into account in order to develop responsible and ethical processes.

References

1. Acitelli, G., Agostinelli, S., Casciani, A., Marrella, A.: The role of trust in AI-augmented business process management systems. In: International Conference on Business Process Management, pp. 5-17 (2024)
2. Aguinis, H., Bradley, K.J.: Best practice recommendations for designing and implementing experimental vignette methodology studies. Organ. Res. Methods **17**(4), 351–371 (2014)
3. Baer, M.D., Van Der Werff, L., Colquitt, J.A., Rodell, J.B., Zipay, K. P., Buckley, F.: Trusting the "Look and Feel: Situational Normality, Situational Aesthetics, and the Perceived Trustworthiness of Organizations (2018)
4. Banks, J.: A perceived moral agency scale: development and validation of a metric for humans and social machines. Comput. Hum. Behav. **90**, 363–371 (2019)
5. Beheshti, A., et al.: ProcessGPT: transforming business process management with generative artificial intelligence. In: Proceedings of the IEEE Conference on Web Services (IEEE ICWS 2023), Chicago USA, 2–8 July 2023, pp. 731–739 (2023)
6. Bigman, Y.E., Gray, K.: People are averse to machines making moral decisions. Cognition **181**, 21–34 (2018)
7. Bowling, N.A., Huang, J.L., Brower, C.K., Bragg, C.B.: The quick and the careless: the construct validity of page time as a measure of insufficient effort responding to surveys. Organ. Res. Methods **26**(2), 323–352 (2023)
8. Braakman, M., et al.: Mining for Well-Being: the potential of process mining for evaluating employee well-being. In: International In: Conference on Business Process Management, pp. 180-191 (2024)
9. vom Brocke, J, et al.: Process science: the interdisciplinary study of continuous change. SSRN Electron J, pp. 1–9 (2021)
10. Dumas, M., Fournier, F., Marrella, A., Montali, M., Rehse, J.: AI-augmented business process management systems: a research manifesto. ACM Trans. Manag. Inf. Syst. **14**(1), 1–19 (2023)
11. Von Eschenbach, W.J.: Transparency and the black box problem: why we do not trust AI. Philos. Technol. **34**(4), 1607–1622 (2021)
12. Floridi, L., Sanders, J.W.: On the morality of artificial agents. Mind. Mach. **14**(3), 349–379 (2004)
13. Freeman, R. E., Harrison, J. S., Wicks, A. C., Parmar, B. L., de Colle, S.: Stakeholder theory: the state of the art. Cambridge University Press (2010)
14. De Freitas, J., Agarwal, S., Schmitt, B.: Psychological factors underlying attitudes toward AI tools. Nat. Hum. Behav. **7**, 1845–1854 (2023)
15. Glikson, E., Woolley, A.W.: Human trust in artificial intelligence: review of empirical research. Acad. Manag. Ann. **14**(2), 627–660 (2020)
16. Gray, K., Young, L., Waytz, A.: Mind perception is the essence of morality. Psychol. Inquiry **23**, 101–124 (2012)
17. Gray, K., Wegner, D.M.: Feeling robots and human zombies: mind perception and the uncanny valley. Cognition **125**(1), 125–130 (2012)
18. Grisold, T., vom Brocke, J., Klatsch, W., Mendling, J., Vidgof, M.: Using large language models in business processes, Proceedings of the 21st International Conference of Business Process Management, Tutorial, Utrecht (2023)
19. Haslam, N.: Dehumanization: an integrative review. Person. Soc. Psychol. Rev. **10**, 252–264 (2006)
20. Harrison, J.S., Freeman, R.E., Abreu, M.C.S.D.: Stakeholder theory as an ethical approach to effective management: applying the theory to multiple contexts. Revista brasileira de gestão de negócios **17**(55), 858–869 (2015)

21. Jones, T.M.: Instrumental stakeholder theory: a synthesis of ethics and economics. Acad. Manag. Rev. **20**(2), 404–437 (1995)
22. Kteily, N.S., Landry, A.P.: Dehumanization: trends, insights, and challenges. Trends Cogn. Sci. **26**(3), 222–240 (2022)
23. Li, J., Bonn, M.A., Ye, B.H.: Hotel employee's artificial intelligence and robotics awareness and its impact on turnover intention: the moderating roles of perceived organizational support and competitive psychological climate. Tour. Manag. **73**, 172–181 (2019)
24. Van Looy, A., Shafagatova, A.: Business process performance measurement: a structured literature review of indicators, measures and metrics. Springerplus **5**(1), 1–24 (2016). https://doi.org/10.1186/s40064-016-3498-1
25. Mahendrawathi, E.: How Can We Develop Inclusive Business Process Management?. In: International Conference on Business Process Management, pp. 370-375 (2024)
26. Meade, A.W., Craig, S.B.: Identifying careless responses in survey data. Psychol. Methods **17**(3), 437–455 (2012)
27. Mobley, W.: Intermediate linkages in the relationship between job satisfaction and employee turnover. J. Appl. Psychol. **62**, 237–240 (1977)
28. Montali, M.: Constraints for process framing in AI-augmented BPM. In: Cabanillas, C., Garmann-Johnsen, N.F., Koschmider, A. (eds.) Business Process Management Workshops. BPM 2022. LNBIP, vol. 460. Springer, Cham (2023). https://doi.org/10.1007/978-3-031-25383-6_1
29. Nilsson, N.J.: Problem-solving methods in artificial intelligence. McGraw-Hill, New York, NY (1971)
30. Overes, E., Santoro, F.: The Effectiveness of Business Process Visualisations: a Systematic Literature Review. arXiv preprint arXiv:2504.10971 (2025)
31. Raisch, S., Krakowski, S.: Artificial intelligence and management: the automation-augmentationparadox. Acad. Manag. Rev. **46**(1), 192–210 (2021)
32. Rosemann, M., Brocke, J. V., Van Looy, A., Santoro, F.: Business process management in the age of AI–three essential drifts. Inform. Syst. e-Business Manag., 1-15 (2024)
33. Rousseau, D.M., Sitkin, S.B., Burt, R.S., Camerer, C.: Not so different after all: a cross-discipline view of trust. Acad. Manag. Rev. **23**(3), 393–404 (1998)
34. Sullivan, Y., Bourmont, M., Dunaway, M.: Appraisals of harms and injustice trigger an eerie feeling that decreases trust in artificial intelligence systems. Ann. Oper. Res. **308**(1), 525–548 (2022)
35. Walsh, J.P., Ashford, S.J., Hill, T.E.: Feedback obstruction: the influence of the information environment on employee turnover intentions. Human Relations **38**, 23–36 (1985)
36. Waytz, A., Cacioppo, J., Epley, N.: Who sees human? the stability and importance of individual differences in anthropomorphism. Perspect. Psychol. Sci. **5**, 219–232 (2010)
37. Zhou, J., Verma, S., Mittal, M., Chen, F.: Understanding relations between perception of fairness and trust in algorithmic decision making. In: 2021 8th International Conference on Behavioral and Social Computing (BESC), pp. 1–5 (2021)

Agentic Business Process Management: Practitioner Perspectives on Agent Governance in Business Processes

Hoang Vu[1](✉), Nataliia Klievtsova[2], Henrik Leopold[3], Stefanie Rinderle-Ma[2], and Timotheus Kampik[1,4]

[1] SAP, Berlin, Germany
{h.vu,timotheus.kampik}@sap.com
[2] TUM School of Computation, Information and Technology, Technical University of Munich, Garching, Germany
{nataliia.klievtsova,stefanie.rinderle-ma}@tum.de
[3] Kühne Logistics University, Hamburg, Germany
henrik.leopold@klu.org
[4] Umeå University, Umeå, Sweden

Abstract. With the rise of generative AI, industry interest in software *agents* is growing. Given the stochastic nature of generative AI-based agents, their effective and safe deployment in organizations requires robust governance, which can be facilitated by *agentic* business process management. However, given the nascence of this new-generation *agent* notion, it is not clear what BPM practitioners consider to be an agent, and what benefits, risks and governance challenges they associate with agent deployments. To investigate how organizations can effectively govern AI agents, we conducted a qualitative study involving semi-structured interviews with 22 BPM practitioners from diverse industries. They anticipate that agents will enhance efficiency, improve data quality, ensure better compliance, and boost scalability through automation, while also cautioning against risks such as bias, over-reliance, cybersecurity threats, job displacement, and ambiguous decision-making. To address these challenges, the study presents six key recommendations for the responsible adoption of AI agents: define clear business goals, set legal and ethical guardrails, establish human-agent collaboration, customize agent behavior, manage risks, and ensure safe integration with fallback options. Additionally, the paper outlines actions to align traditional BPM with agentic AI, including balancing human and agent roles, redefining human involvement, adapting process structures, and introducing performance metrics. These insights provide a practical foundation for integrating AI agents into business processes while preserving oversight, flexibility, and trust.

Keywords: Autonomous agents · Generative AI · Business Process Management

M. ER et al. (Eds.): BPM 2025, LNBIP 565, pp. 29–43, 2026.
https://doi.org/10.1007/978-3-032-02936-2_3

1 Introduction

For over three decades, *agents* have periodically surged in Business Process Management (BPM). The 1990s saw early excitement around goal-oriented software agents [14,15], followed by the rise of Robotic Process Automation (RPA) in the late 2010s [8,9,12], promising efficiency gains for knowledge work. However, high initial expectations and maintenance costs led to mixed experiences in RPA adoption [13].

Large Language Model (LLM)-based agents fuel another wave of optimism [3,6]. However, the stochastic nature of LLMs and related generative AI (genAI) technologies raises concerns, particularly regarding the need to govern the artifacts, decisions, and behaviors they produce. This prompts a critical question: how can BPM support the responsible integration of AI agents into organizations, ensuring both effective business outcomes and social accountability?

To explore this challenge from a practical perspective, we conducted interviews with BPM practitioners, capturing their insights on the potential implications of agents on organizations. The results reveal a dual landscape of opportunities and risks. On the one hand, practitioners anticipate benefits such as improved efficiency, predictive insights, and proactive decision-making. On the other hand, they express concerns about bias, over-reliance on automation, lack of transparency, and potential job displacement. This paper aims to address these concerns by providing recommendations practitioners should consider when introducing autonomous agents in business processes.

The rest of the paper is organized as follows. Section 2 reviews the history and evolution of agents in BPM, RPA, generative AI in BPM and defines the concept of *agentic BPM*. Section 3 describes the qualitative research design. Section 4 presents findings from the study with BPM practitioners on agents, while Sect. 5 highlights the key recommendations proposed for AI agent adoption. Section 6 and 7 discusses the limitation of the study and concludes by highlighting the need for further research into governance frameworks for agents within business processes.

2 Background

In this section, we provide a brief overview of the history and evolution of agents in BPM. We highlight both the continuous research progress in areas applying agent technologies in business processes (Sect. 2.1), as well as RPA (Sect. 2.2). Finally, we discuss genAI in BPM (Sect. 2.3) and define agentic BPM as the governed introduction of AI agents into business processes (Sect. 2.4).

2.1 Agents in Business Process Management

The integration of agent technologies into business process execution and coordination began in the early 1990s. An agent is typically defined as a computer program that "*operates autonomously, perceives its environment, endures over time, adapts to change, and pursues goals*" [26, pp.21–22][1]. In the context of Multi-Agent Systems (MAS)[2], a business process can be viewed as a set of interacting agents coordinated by common organizational goals [15,16]; accoridngly Agent-Oriented Programming (AOP) focuses, in its application proposals to BPM, primarily on business process execution [14]). Over the decades, researchers have applied MAS to BPM to create more dynamic, cross-organizational processes while maintaining overall control. However, early implementations often faced challenges, including weak system control and issues with trust and delegation [11]. Nonetheless, such processes, as well as the MAS executing them, need to be managed to align with organizational and broader societal requirements.

In process modeling, *process choreographies* describe how independent agents coordinate–following the BPMN standard [24]–by exchanging messages in a defined sequence and under specific conditions to accomplish a shared goal [5]. Similarly, subject-oriented BPM shifts the focus from traditional process flows to subjects and their interactions, where subjects (e.g., humans, software, or agents) autonomously manage their tasks and decisions [10]. With the rise of blockchain technologies, research further explored how networks of autonomous agents could underpin data sharing and decision logic through smart contracts [21,23]. Although blockchain-based BPM gained traction as a potential future framework, it did not move autonomous agents into mainstream BPM practice.

In process analysis, agent-based methods have been developed for tasks such as process mining and simulation [29,30], including data-driven simulations that utilize event logs [18]. *Agent System Mining* (ASM) combines process mining with agent-based modeling to derive MAS models from real-world data [28, 31,31,32]. *Reinforcement learning* approaches have also been applied, enabling agents to test and refine different process variants through trial and error [19,27].

2.2 Robotic Process Automation

While agents and MAS address complex decision-making and coordination challenges, RPA aims to efficiently automate simple, rule-based tasks on a large scale. Such tasks represent a considerable amount of routine business activities [9].

Since 2015, research on RPA has highlighted numerous examples of how business process automation can significantly enhance performance [12]. When combined with BPM, RPA offers several advantages, including scalability, improved process accuracy, greater transparency and traceability, cost savings, and increased job satisfaction [8,9]. However, successfully applying RPA remains challenging, with 30-50% of initial projects failing during implementation [13].

---

[1] Definition of *agents* vary; discussing these definitions is beyond the scope of this paper.

[2] We use the terms *agents* and *MAS* interchangeably, considering a collection of agents as a MAS.

While the primary causes of RPA failures result from technical limitations, there is also the problem of resistance against automation. Automated decision-making must comply with regulations, as the absence of human oversight can lead to legal and ethical risks. Automation may also create a black box effect, making it difficult for employees to understand the logic behind automated decisions [12]. Therefore, it is crucial to consider the effects of human-automation interaction alongside the technical aspects of introducing new technology [34].

Even if successful, RPA is not suitable for full-scale business process automation as it focuses on automating repetitive tasks rather than orchestrating end-to-end processes [20]. Since workflow automation and Integration platform as a service (iPaaS) are better suited for human-in-the-loop and API integration scenarios respectively, the automation use cases for RPA are limited [35].

2.3 Generative AI in Business Process Management

With the rise of LLMs, BPM researchers have begun exploring how these powerful tools can be integrated into business process management. Several studies [2,4,33] have examined all stages of the BPM lifecycle to identify areas where LLMs can add value. However, much of the current literature focuses on using LLMs from a conversational perspective–for tasks such as process understanding, information retrieval, and decision support–primarily addressing specific phases like *discovery*, *analysis*, and *monitoring* [7]. While these approaches may treat LLMs as agents in a limited sense, they largely automate isolated tasks, leaving final decision-making and broader, goal-based autonomy to human users.

In contrast, long-term visions of generative AI consider AI-based systems not merely as tools for task automation but as intelligent systems capable of adaptability, self-improvement, and bounded autonomy. Recent works [6,17,25] present AI-Augmented BPM Systems (ABPMSs) that enhance process execution, analysis, and optimization within structured workflows by supporting human decision-making rather than replacing it. Notably, ABPMS are envisioned to support BPM holistically [6], while emerging research explores the application of LLM agents to process mining [3]. Similarly, Large Process Models (LPMs) adopt a neuro-symbolic approach by integrating fine-tuned LLMs with traditional symbolic systems for context-aware process design and analysis.

Despite these advances, a critical open question remains: How can BPM best support the management of increasingly autonomous software agents deployed in organizations, and what shifts in perspective are necessary to achieve this?

2.4 Agentic Business Process Management

Building on the research history of agents in BPM, RPA and generative AI in BPM (see Sect. 2), we introduce the concept of *agentic Business Process Management* (see Fig. 1) to address this key question. Figure 1 presents a historical overview that situates agentic BPM within the evolving relationship between agent systems and business process management. Rather than serving merely as a conceptual map, the timeline illustrates how the field has progressed–from

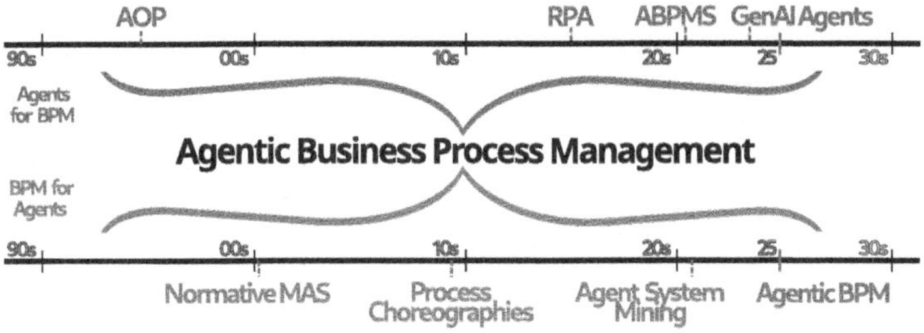

Fig. 1. Emergence of Abstractions and Technologies for agentic Business Process Management over Time: *Agents for BPM* versus *BPM for Agents.*

leveraging agents to support BPM tasks ("Agents for BPM") toward rethinking BPM as a means to govern and coordinate agent-based systems ("BPM for Agents"). This distinction is crucial: while the former reflects the application of agents in traditional BPM contexts, the latter positions BPM as a discipline that adapts and evolves to meet the challenges posed by autonomous, self-adaptive multi-agent systems (MAS). Highlighting this transition helps clarify why a dedicated framework such as ABPM is necessary–not just to harness the capabilities of AI agents but also to ensure their responsible and effective integration into organizations. Our definition of ABPM remains technology-agnostic, draws on established BPM and AI terminology, and is intended to contribute to the broader evolution of BPM as a discipline for agent governance.

Definition 1 (Agentic Business Process Management). Agentic Business Process Management (ABPM) describes *i)* the deployment and execution of autonomous software agents to achieve business process goals and *ii)* the application of agent-based abstractions for the process-oriented design and analysis of autonomous software agents.

Definition 1 makes use of several key concepts, all of which are well-established in the BPM and AI literature.

- An *(autonomous) software agent* is a computer program that "operates autonomously, perceives the environment, persists over a prolonged time period, adapts to change, and creates and pursues goals" [26, pp.21].
- A *process goal* operationalizes the objective that an organization strives to achieve with the corresponding business process, which—in turn–can be defined as "a set of activities that are performed in coordination in an organizational and technical environment", to "jointly realize a business goal" [36].
- An *agent-based abstraction* is a conceptual model explicitly featuring notions of (software or human) agents.
- *Process-oriented design and analysis* describes the modeling business processes, as well as the drawing of inferences from data generated during process execution.

These clarifications build on textbook definitions of agents and business processes. We have deliberately omitted common and nuanced discussions on what constitutes an agent (cf. [1, Chapter 1]). We do not assume that agents are perfectly rational and expect them to always act in a way that achieves the best expected outcome [26, pp.22]. Instead, our only technological commitment is that agents are software-based, supporting self-directed work that extends beyond human capabilities.

Our definition of agentic Business Process Management (ABPM) spans the key phases of the BPM lifecycle by addressing both deployment/execution and design/analysis. Although the discovery phase often overlaps with analysis in practice, our conceptual and technology-neutral definition applies BPM principles to multi-agent systems (MAS).

3 Methodology

To explore how AI agents can be safely adopted in organizations and what the implications are for BPM, we utilized a qualitative research design and conducted in-depth interviews with experienced BPM practitioners. The insights gained reveal both the advantages and challenges associated with deploying AI agents, enabling us to identify key implications and formulate targeted actions to address critical governance and management concerns.

Terminology. Several process management software vendors, including UiPath, Salesforce, IBM, and Workday, use the term *agentic AI* to describe the interplay between genAI and agents. Thus, we leveraged the term during the interviews as a starting point to discuss the impact of agents on process management, expecting it to resonate effectively with the interviewees.

Research Design. Our study leverages a qualitative research design to examine the perspectives of professionals in business process management and automation, focusing on the integration and impact of agentic AI. It investigates the participants' understanding, expectations, and concerns related to agentic AI's autonomy, adaptability, human collaboration, and governance. Alongside discussing the characteristics of agentic AI, the study places greater emphasis on its impact on practitioners' organizations. Therefore, the study also explores to what extent agent governance is required to address management needs that arise from greater software autonomy.

To systematically examine the perceptions and attitudes towards agentic AI, a qualitative content analysis approach [22] was followed. Semi-structured interviews were conducted with professionals across various industries, focusing on their understanding, expectations, and concerns regarding autonomy, adaptability, human collaboration, and governance of agentic AI. The process began with transcription, familiarization with the interview data and analysis using a deductive-inductive coding approach. Initial deductive categories were derived from the research questions, while additional inductive subcategories emerged from the data, allowing themes to derive directly from the participants' responses.

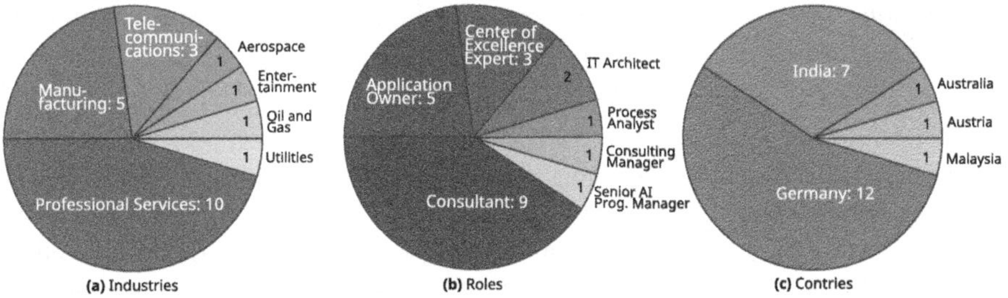

Fig. 2. Demographic Distribution across Participants

We began by reviewing all the interview data to understand the context and identify recurring themes. Next, we applied open coding by labeling segments of text with basic descriptive codes (e.g., benefits, risks, and governance challenges related to agentic AI). We then grouped similar codes into broader categories during axial coding (e.g. "automation potential" and "workflow optimization" to "benefits"), revealing patterns and common themes across the interviews. Finally, we refined and integrated these themes into a coherent framework during selective coding, ensuring that key insights were clearly defined. A total of 22 participants with experience with automation and AI technologies from various industries and roles were interviewed by the first author. In Fig. 2, an overview of the participants demographic characteristics is provided, including their professional roles and the industries and countries in which they are employed. Sessions lasted 60 to 90 min, with some conducted in pairs or groups due to accessibility constraints.

4 Results

The interview results highlight various themes mentioned by the interviewees, which are discussed below.

Understanding. Among the 22 participants, 10 were familiar with the term agentic AI, describing it as a self-learning technology that operates autonomously and adapts to its environment. Some viewed it as an evolution of RPA, overcoming technical limitations with AI. Others associated it with a digital assistant or an orchestration layer that coordinates tasks across specialized agents. In this context, a participant highlights:

> "[Agentic AI is an agent] that can do significantly more than dumb botting, where I always have to tell it everything exactly and have precisely structured it, and it really only helps where I have very clearly structured, stable processes."

Importantly, none of the participants reported to have actual practical experience with agentic AI, i.e., the assessments participants provide is based on expectations, often against the backdrop of existing agent-like automation technologies such as RPA.

Benefits. Participants highlighted several potential benefits of integrating agentic AI into organizational processes (see Fig. 3 (a)). The technology is expected to enhance efficiency by automating routine tasks, streamlining operations, and reducing errors, allowing employees to focus on higher-value work and achieving time and cost savings. It is also anticipated to improve data quality, ensuring consistent and accurate handling of information. Better compliance was another potential benefit, as agentic AI could monitor regulations and enforce standards automatically. Scalability was frequently noted, enabling businesses to handle larger workloads without proportional increases in staffing. Additionally, it is believed to democratize process data, making it more accessible and usable. In this context, a participant highlights:

> "More employees could initiate changes or optimizations if the software supports them directly, reducing dependency on specialized roles."

Risks. Despite its potential, implementing agentic AI carries risks (see Fig. 3 (b)), such as bias from flawed training data, over-reliance leading to diminished human judgment, and lack of transparency in AI-driven decisions. Participants also raised concerns about cybersecurity threats, job displacement, and unauthorized decision-making, stressing the need for adequate human oversight to prevent unintended consequences. In this context, a participant highlights the cultural aspect:

> "It's a cultural thing to be able to accept autonomy and decision making being taken away from a human."

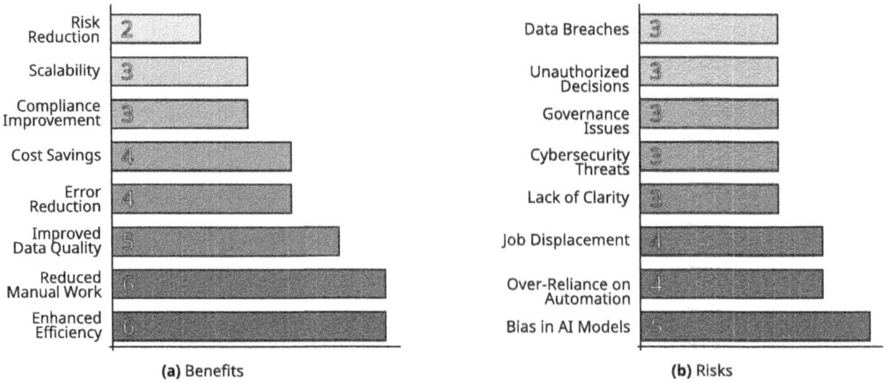

Fig. 3. Agentic AI among Participants: Expected Benefits and Risks.

Use Cases. Participants identified several potential use cases for agentic AI (see Fig. 4 (a)). Key applications include process monitoring to detect inefficiencies and suggest improvements, and predictive analytics to forecast trends and provide actionable insights. Task automation was another major focus, with agentic AI handling routine activities like data entry and document processing. Specific tasks include master data maintenance, user administration, root cause analysis, and decision support through dashboards. A participant highlights:

> "Processes often get stuck due to errors in master data, such as mismatched product codes or pricing issues. [Agentic] AI could analyze and fix these autonomously."

Agentic AI could enhance customer service, while supply chain optimization benefits from agentic AI managing inventory and predicting demand. In finance, agentic AI could aid in fraud detection and transaction monitoring. Additionally, agentic AI could structure unstructured datasets, improving accessibility and analysis.

Requirements. Participants stressed the need for clear rules and guidelines to ensure agentic AI operates ethically and transparently. They highlighted the importance of audit logs, data retention, transparency, robust security, and adherence to corporate policies and regulations. In this context, a participant highlights:

> "[The agent] would basically replace an FTE[3], let's just put it that way; you also have to provide it with the same framework that the employee would be confronted with because what would the employee do if they encounter difficulties?"

Defining roles, responsibilities, and limitations for agentic AI is crucial, along with compliance with data protection laws and risk management frameworks. Seamless integration with existing processes prevents disruptions, while comprehensive employee training ensures effective use of agentic AI. Managing costs, including setup, maintenance, and upgrades, was also highlighted as essential. Preconfigured use cases and applications were recommended to build trust by demonstrating agentic AI's capabilities and delivering clear benefits. Figure 4 (b) gives an overview of requirements that practitioners highlighted.

Autonomy and Adaptability. Participants stressed the importance of managing agentic AI's autonomy to maintain accountability, reliability, and safety. While agentic AI can independently handle low-risk tasks like data preparation, human oversight is essential for critical decisions or significant changes. Configurable autonomy was recommended, granting agentic AI flexibility in low-risk areas while restricting it in high-risk scenarios, such as system changes or external communications. A gradual approach was also suggested, starting with simple tasks and increasing autonomy as trust in the system grows. In this regard, a participant highlights:

---

[3] *Full-Time Equivalent*, i.e., a common unit to measure work in terms of person efforts.

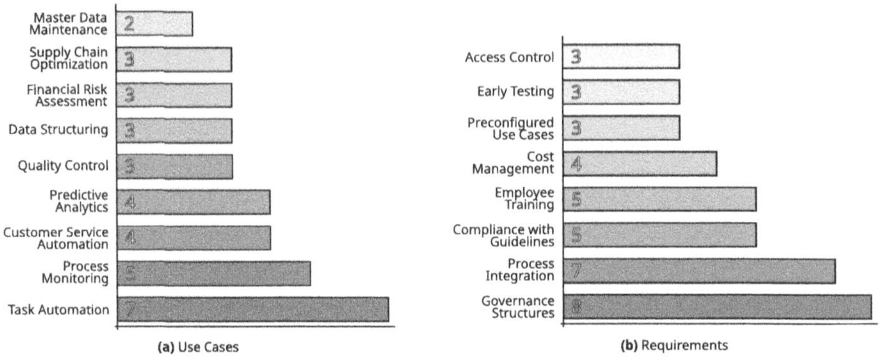

Fig. 4. Agentic AI among Participants: Use Cases and Roll-out Requirements.

> "It shouldn't make changes in source systems or install new apps autonomously. That crosses the line because those areas are managed by different teams and require coordination."

Adaptability is a key strength of agentic AI when operating within clear boundaries, but concerns about trust and consistency highlight the need for human oversight to monitor and guide its adaptations. Transparency was highlighted as essential, with participants calling for clear documentation of all AI decisions and actions. In this context, a participant mentions:

> "If it's routine, the decision itself should be documented. The more complex the task, the more I want to see how the process was developed."

This would include high-level summaries for management and detailed logs for technical teams to ensure traceability and allow for changes to be reverted if necessary. Participants acknowledged agentic AI's potential to enhance workflows and efficiency but emphasized the need to operate within predefined rules to prevent unintended outcomes. Agents should support, not replace, humans in tasks requiring complex judgment. Human validation and oversight remain crucial for critical scenarios.

Human Involvement and Responsibility. Participants agreed that although agentic AI can autonomously handle low-stakes decisions, it should serve as a decision-support tool for complex process scenarios, ensuring human involvement to maintain accuracy and accountability. Its role is to provide clear analyses, suggest actions, and outline impacts, enabling well-informed human decisions. Some proposed methods like briefings or notifications to engage humans in urgent situations, viewing agentic AI as a collaborative assistant offering insights and recommendations. In this regard, a participant underlines:

> "For complex decisions, the [agentic] AI should provide full context; how it arrived at the decision, what data it used, and the potential consequences; just like how humans consult colleagues for advice."

Responsibility for agentic AI failures was seen as a shared effort among organizations, developers, and business leaders. Organizations deploying agentic AI bear ultimate responsibility, particularly by providing oversight in critical tasks to validate agentic AI decisions. Developers play a key role in building reliable systems, while organizations establish robust oversight mechanisms. Key users and application owners were identified as primary contacts for resolving issues, with some suggesting dedicated teams to manage and ensure accountability for agentic AI. Figure 5 summarizes responsibilities of agentic AI failure that practitioners highlighted.

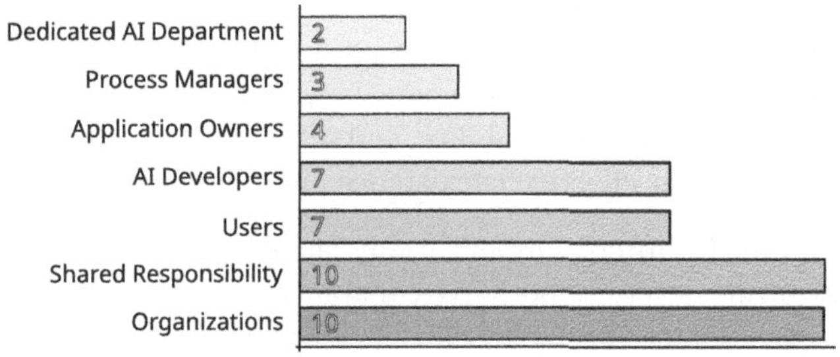

Fig. 5. Agentic AI among Participants: Responsibility for Failures.

5 Recommendations for ABPM Adoption

The findings reveal that organizations require a clear and structured management framework to successfully integrate AI agents into their business processes while ensuring that human actors remain informed, involved, and empowered. Insights from the expert interviews led to a set of practical recommendations and actions that organizations should consider during implementation. These span six critical focus areas: business context, operational guardrails, human-agent collaboration, customization, risk management, and adoption.

Business context. Organizations should define specific goals for the use of AI agents, clarify the expected benefits, and assess the short- and long-term costs associated with their deployment and maintenance.

Guardrails. Organizations must ensure that agents operate within the boundaries of legal, ethical, and organizational rules, while considering the context and environment in which the agents function.

Human-agent collaboration. Clear roles and responsibilities should be established to enable effective collaboration between humans and software agents, including identifying situations where human intervention is necessary in case of agent failure.

Customization. Organizations should tailor agents to meet their specific needs, determining factors such as the level of autonomy granted to agents, how human oversight and review are conducted, the extent of documentation required, and the desired level of reasoning and transparency in agent operations.
Risk management. Organizations must implement safety measures to monitor performance and address potential risks, including preventing undesired adaptability or evolution of the agents.
Adoption. Organizations must ensure the seamless and responsible integration of agents into existing business processes and systems, while implementing fallback mechanisms to enable a safe reversion if necessary.

In addition, the implications for BPM need to be considered. To bridge traditional BPM (focused on structure, control, and standardization) with agentic AI (emphasizing autonomy, adaptability, and self-learning), following actions are proposed to ensure their effective coexistence and alignment.

Human-agent balance. Organizations should strive to design workflows that strategically leverage both AI capabilities and human expertise. This balance is essential for optimizing efficiency while preserving human judgment and creativity.
Human involvement. As intelligent agents become integral to business processes, organizations must clarify whether human actors will collaborate with agents as peers or take on supervisory roles with selective intervention.
Process structures. To fully realize the potential of self-adaptive agents, organizations should consider evolving beyond static process models and enabling more flexible, dynamic structures that can adjust in real time based on agent-driven insights.
Performance metrics. Effective integration of agentic AI requires clear metrics and performance indicators. Organizations should develop governance frameworks that track the impact of agents on process efficiency, decision quality, and broader organizational outcomes.

6 Limitations and Future Work

Several limitations encountered during this study must be acknowledged. First, the sample was relatively small and predominantly drawn from experts at selected multinational organizations, which may not fully represent the diversity of experiences across various industries or among smaller enterprises. Second, the concept of agentic AI remains in its formative stage, as participants employed varied terminology and frameworks, making it difficult to develop a standardized definition. Consequently, the limited adoption and practical experience with this technology constrain the interpretation of the interview results.

Against this background, future research should focus on refining and empirically validating the conceptual frameworks for agentic AI, with particular emphasis on its ethical, legal, and technical implications. Additionally, quantitative studies and broader analyses across different industries are needed to assess the wider impact and evolution of agentic AI within various organizational contexts, and to develop robust governance structures that ensure transparent and accountable decision-making. Finally, future research should also examine how the governance of agentic AI with its core characteristics of autonomy and adaptability should differ from the current frameworks used for established technologies such as Robotic Process Automation.

7 Conclusion

The rise of generative AI has sparked growing interest in its integration with software agents, giving rise to the concept of *agentic AI*. While this emerging technology holds promise, its impact on business processes and the requirements for its effective management remain insufficiently understood. This paper addresses this gap by identifying key recommendations for organizations seeking to implement AI agents responsibly.

Interviews with BPM practitioners reveal that next-generation software agents can enhance efficiency by automating routine tasks and allowing humans to focus on higher-value activities. At the same time, practitioners express concerns about risks such as bias, over-reliance, job displacement, and a lack of transparency. These concerns underscore the need for a clear governance framework to guide agent deployment.

To support informed, involved, and empowered human participation, the study outlines six critical areas for organizational focus: clarifying business objectives, setting legal and ethical guardrails, enabling effective human-agent collaboration, tailoring agent behavior, managing operational risks, and ensuring safe, seamless adoption.

Beyond implementation, the findings have broader implications for Business Process Management. Agentic AI introduces new dynamics–autonomy, adaptability, and self-learning–that challenge traditional BPM principles centered on structure, control, and standardization. To align these approaches, organizations should take strategic steps: design workflows that balance human and agent input, redefine human roles as collaborators or supervisors, evolve static process models to support adaptive agent behavior, and implement robust metrics and governance frameworks to monitor performance and outcomes.

Together, these insights offer a practical roadmap for the responsible integration of AI agents. They help organizations harness the potential of agentic AI while safeguarding process integrity, organizational control, and stakeholder trust.

References

1. Multiagent systems: a modern approach to distributed artificial intelligence. MIT Press, Cambridge, MA, USA (1999)

2. Bennoit, C., Greff, T., Baum, D., Bajwa, I.A.: Identifying use cases for large language models in the business process management lifecycle. In: 2024 26th International Conference on Business Informatics (CBI), pp. 256–263. IEEE (2024)
3. Berti, A., Maatallah, M., Jessen, U., Sroka, M., Ghannouchi, S.A.: Re-thinking process mining in the ai-based agents era (2024). https://arxiv.org/abs/2408.07720
4. Buss, A., Kratsch, W., Schmid, S.J., Wang, H.: Processllm: A llm specialized in the interpretation, analysis, and optimization of business processes. In: Business Process Management Workshops. Springer Nature Switzerland, Cham (2025). https://doi.org/10.1007/978-3-031-78666-2_17
5. Decker, G.: Design and analysis of process choreographies. Ph.D. thesis, Universität Potsdam (2009)
6. Dumas, M., et al.: Ai-augmented business process management systems: a research manifesto. ACM Trans. Manag. Inf. Syst. **14**(1), 11:1–11:19 (2023)
7. Estrada-Torres, B., del-Río-Ortega, A., Resinas, M.: Mapping the landscape: exploring llm applications in business process management. In: Enterprise, Business-Process and Information Systems Modeling, pp. 22–31 (2024)
8. Eulerich, M., Waddoups, N., Wagener, M., Wood, D.A.: The dark side of robotic process automation (rpa): understanding risks and challenges with rpa. Account. Horiz. **38**(2), 143–152 (2024)
9. Flechsig, C., Lohmer, J., Lasch, R.: Realizing the full potential of robotic process automation through a combination with bpm. Logistics Management (2019)
10. Fleischmann, A., Schmidt, W., Stary, C., Obermeier, S., Börger, E.: Subject-oriented business process management. Springer Nature (2012)
11. Grundspenkis, J., Pozdnyakov, D.: An overview of the agent based systems for the business process management (2006)
12. Haase, J., Kremser, W., Leopold, H., Mendling, J., Onnasch, L., Plattfaut, R.: Interdisciplinary directions for researching the effects of robotic process automation and large language models on business processes. Commun. Assoc. Inf. Syst. **54**, 579–604 (2024)
13. Hindle, J., Lacity, M., Willcocks, L., Khan, S.: Robotic process automation. Tech. rep, Interim Executive Research Report (2017)
14. Jennings, N.R., Faratin, P., Johnson, M.J., Norman, T.J., O'Brien, P.D., Wiegand, M.E.: Agent-based business process management. Int. J. Cooperative Inf. Syst. **5**(2&3), 105–130 (1996)
15. Jennings, N.R., Norman, T.J., Faratin, P.: ADEPT: an agent-based approach to business process management. SIGMOD Rec. **27**(4), 32–39 (1998)
16. Jennings, N.R., Norman, T.J., Faratin, P., O'Brien, P.D., Odgers, B.: Autonomous agents for business process management. Appl. Artif. Intell. **14**(2), 145–189 (2000)
17. Kampik, T., et al.: Large process models: a vision for business process management in the age of generative ai. KI - Künstliche Intelligenz (2024)
18. Kirchdorfer, L., Blümel, R., Kampik, T., Van der Aa, H., Stuckenschmidt, H.: AgentSimulator: an agent-based approach for data-driven business process simulation. In: 2024 6th Int. Conf. on Proc. Mining (ICPM), pp. 97–104. IEEE (2024)
19. Kurz, A.F., Kampik, T., Pufahl, L., Weber, I.: Business process improvement with ab testing and reinforcement learning: grounded theory-based industry perspectives. Softw. Syst. Model. (2024)
20. König, M., Bein, L., Nikaj, A., Weske, M.: Integrating Robotic Process Automation into Business Process Management, pp. 132–146 (2020)
21. Ladleif, J., Weske, M., Weber, I.: Modeling and enforcing blockchain-based choreographies. In: Business Process Management, pp. 69–85 (2019)

22. Mayring, P., Fenzl, T.: Qualitative Inhaltsanalyse, pp. 633–648. Springer Fachmedien Wiesbaden, Wiesbaden (2019)
23. Mendling, J., et al.: Blockchains for business process management - challenges and opportunities. ACM Trans. Manag. Inf. Syst. **9**(1), 4:1–4:16 (2018)
24. OMG: Business Process Model and Notation (BPMN), Version 2.0 (January 2011). https://www.omg.org/spec/BPMN/2.0
25. Rosemann, M., Brocke, J.v., Van Looy, A., Santoro, F.: Business process management in the age of ai – three essential drifts. Inform. Syst. e-Bus. Manag. **22**, 1–15 (2024)
26. Russell, S.J., Norvig, P.: Artificial Intelligence: A Modern Approach (4th Edition). Pearson (2020). http://aima.cs.berkeley.edu/
27. Satyal, S., Weber, I., Paik, H., Ciccio, C.D., Mendling, J.: Business process improvement with the AB-BPM methodology. Inf. Syst. **84**, 283–298 (2019)
28. Shen, Q., Polyvyanyy, A., Lipovetzky, N., Kampik, T.: Agent system event data: Concepts, dimensions, applications. In: Conceptual Modeling, pp. 56–72 (2024)
29. Sulis, E., Taveter, K.: Agent-Based Business Process Simulation. Springer (2022)
30. Szimanski, F., Ralha, C.G., Wagner, G., Ferreira, D.R.: Improving business process models with agent-based simulation and process mining. In: Nurcan, S., Proper, H.A., Soffer, P., Krogstie, J., Schmidt, R., Halpin, T., Bider, I. (eds.) BPMDS/EMMSAD -2013. LNBIP, vol. 147, pp. 124–138. Springer, Heidelberg (2013). https://doi.org/10.1007/978-3-642-38484-4_10
31. Tour, A., Polyvyanyy, A., Kalenkova, A.A.: Agent system mining: vision, benefits, and challenges. IEEE Access **9**, 99480–99494 (2021)
32. Tour, A., Polyvyanyy, A., Kalenkova, A.A., Senderovich, A.: Agent miner: an algorithm for discovering agent systems from event data. In: Business Process Management, pp. 284–302 (2023)
33. Vidgof, M., Bachhofner, S., Mendling, J.: Large language models for business process management: opportunities and challenges (2023)
34. Vu, H., Haase, J., Leopold, H., Mendling, J.: Towards a theory on process automation effects. In: International Conference on Business Process Management. pp. 285–301. Springer (2023). https://doi.org/10.1007/978-3-031-41623-1_17
35. Vu, H., Leopold, H., van der Aa, H.: What is business process automation anyway? In: Hawaii International Conference on System Sciences, pp. 5462–5471 (2023)
36. Weske, M.: Introduction. Springer, Berlin Heidelberg, Berlin, Heidelberg (2019)

Motivational Process Mining: A Conceptualization by Adding Self-determination Theory

Amy Van Looy[1](✉) and Peter Vlerick[2]

[1] Faculty of Economics and Business Administration, Department of Business Informatics and Operations Management, Ghent University, Tweekerkenstraat 2, 9000 Ghent, Belgium
Amy.VanLooy@UGent.be

[2] Faculty of Psychology and Educational Sciences, Department of Work, Organization and Society, Ghent University, Henri Dunantlaan 2, 9000 Ghent, Belgium
Peter.Vlerick@UGent.be

Abstract. Since less attention has been paid to the human side of process mining, this study offers a unique psychological lens of self-determination theory for introducing the concept of "motivational process mining". We explore this concept based on a concept analysis approach to call for responsible process mining use from a social sustainability angle. In line with self-determination theory, we conceptualize how work behavior monitoring could affect the fulfillment or deprivation of employees' basic psychological needs (i.e., need for autonomy, relatedness, and competence). Expected outcomes (e.g., performance, well-being) of work behavior monitoring are also considered. We propose a conceptual model and derive propositions for future research.

Keywords: Process Mining · Motivation · Well-being · Self-determination

1 Introduction to Process Mining's Need for Data

Process mining is increasingly used in organizations by collecting event logs and applying data-driven techniques [5]. Also the business process management (BPM) field is increasingly focusing on process mining, but mainly on the technical aspects (e.g., algorithms, event logs) [16]. Given process mining's growth, scholars now call for a more in-depth investigation of various factors affecting its adoption and usage across organizations, business processes and people involved [18].

Gradually, more attention is given to the process mining conditions and success factors at an organizational level (e.g., process mining adoption and maturity) [3, 8]. However, it remains unclear how process mining will be translated at the employee level. Consequently, the human resource management (HRM) implications of process mining at large, and even the perceptions and interpretations of individual employees (as process participants) regarding process mining applications and performance control in their work remain vague [1, 9]. This is especially true when data gathered with

M. ER et al. (Eds.): BPM 2025, LNBIP 565, pp. 44–50, 2026.
https://doi.org/10.1007/978-3-032-02936-2_4

other techniques (e.g., timesheet- or software-based) would be added to event logs, enabling process mining to analyze work behavior on a more fine-grained or instance level. Nevertheless, most organizations implement process mining to detect process inefficiencies, thus only indirectly for improving individual performance and well-being.

In answer of this insufficient knowledge about the human side of process mining and the potential side effects of monitoring and measuring work in event logs [2], this study adds a psychological theory to the BPM field. Building upon self-determination theory [4, 12], we introduce "motivational process mining" as a novel concept. We disentangle the concept to demonstrate that motivational dynamics can affect how process mining is perceived. As a working definition, "motivational process mining" encourages business practices in which process mining efforts not only stimulate process performance but also employee performance and well-being by encouraging the intrinsic motivation of those affected by process mining. "Motivational process mining" foregrounds social sustainability by concentrating on the human-centered and psychological aspects of employees who might feel "surveilled" for the sake of doing a process mining analysis (e.g., on a fine-grained level), affecting their motivation at work. The concept's aim is to better understand how and under which circumstances employees would become more intrinsically motivated by work behavior monitoring. The potential of self-determination theory (i.e., as a core motivational theory) is explored as follows: **"What does employee motivation mean for process mining, especially during data collection? And how can the process mining field contribute to the fulfillment of employees' basic psychological needs and foster their performance and well-being?"**

This concept analysis paper follows the method of [19] to conceptualize "motivational process mining" as a sensitizing (not definitive) concept for further research [11]. It triggers the debate around a responsible and ethical use of process mining for raising awareness. Practically, we interpret how organizations can minimize employee resistance in their attempts to design processes and facilitate the process mining data collection. Accordingly, vendors can advance the development of process mining tools.

2 Gap in Studies on Process Mining Adoption and Acceptance

Since the process mining field is gaining attention from managers and vendors [5, 18], research has started to highlight adoption issues [14]. Scholars have examined the success factors, opportunities, and challenges [20]. Moreover, [3] have derived a maturity model for improving organizational adoption. Other studies have explored how organizations generate business value from process mining [8]. Alternatively, [18] propose a 5-layered framework to unfold process mining research on not only the technical and organizational level but also on the ecosystem, group and individual level. The present study delves into this individual level.

3 Concept Analysis Research Approach

For exploring the individual level, our concept analysis framed "motivational process mining" as a sensitizing concept. According to [11], a sensitizing concept guides in interpreting social phenomena by highlighting themes as a sense of direction, without prescribing a fixed measurement. Similarly, the findings were analyzed to serve future elaborations of concept revision and theory development. This study therefore followed the concept analysis method of [19], guided by the steps in Table 1.

Table 1. Applying the concept analysis steps of [19] to our research.

Concept analysis steps	Current research
1. Select a concept	Motivational process mining (= process mining efforts also fostering employees' intrinsic motivation)
2. Determine the purpose of analysis	Conceptualization as a sensitizing concept
3. Identify concept uses	Where? Organizations using process mining (e.g., for process design, optimization, process performance) What? Focus on data collection efforts for event logs Who? Focus on perceptions of individual employees
4. Determine the defining attributes	Intrinsic motivation (= psychological need fulfilment)
5. Identify a model case	Process mining data capturing, triggering Fig. 1
6. Identify borderline or contrary cases	Strategic motivation for using fictive data in process simulations, or data capturing for extrinsic rewards
7. Identify antecedents and consequences	Antecedents: process mining initiatives (e.g., corporate-related data capturing characteristics) Mediators: basic psychological needs (i.e., autonomy, relatedness, competence) Outcomes: work behavior, well-being
8. Define empirical referents	Ongoing; future research

4 Preliminary Findings for "Motivational Process Mining"

For underpinning our concept of "motivational process mining", we rely on self-determination theory. This is an established theory, recognized and applied in adjacent fields such as human resource management and marketing [6], as well as information systems [10]. To our knowledge, prior studies have not yet investigated the mechanism that underlies the relationship between employees' motivation and performance in light of process mining efforts, though important for success. We contribute to filling this theoretical gap by relying on the self-determination conceptualization of basic psychological need satisfaction, which is the theoretical basis for differentiating between intrinsic and extrinsic motivation [17]. More specifically, self-determination theory is a macro theory of human motivation that explains and predicts the mediating role of a small set of basic psychological needs regarding the impact of organizational factors on employees' motivations and experiences [4, 12]. The idea underlying this theory is that the more an employee's basic psychological needs of autonomy, relatedness and competence are fulfilled or satisfied (i.e., need satisfaction), the more this employee is intrinsically motivated to work well. Proven is that experiencing need satisfaction results in more desirable work performance as well as better well-being [15], or "*thwarting on any of these three basic needs is seen as damaging to motivation and wellness*" [13].

- In this theory, **autonomy** is defined as "*a sense of initiative and ownership in one's actions*" [13] and a "*natural or inherent need to experience (...) behavior as freely chosen and volitional*" [17].
- Next, **relatedness** is defined as "*a sense of belonging and connection*" [13] and "*to care for others and to feel cared for by them*" [17].
- Finally, **competence** is defined as "*the feeling of mastery, a sense that one can succeed and grow*" [13] "*to feel effective and skillful in the activities*" [17].

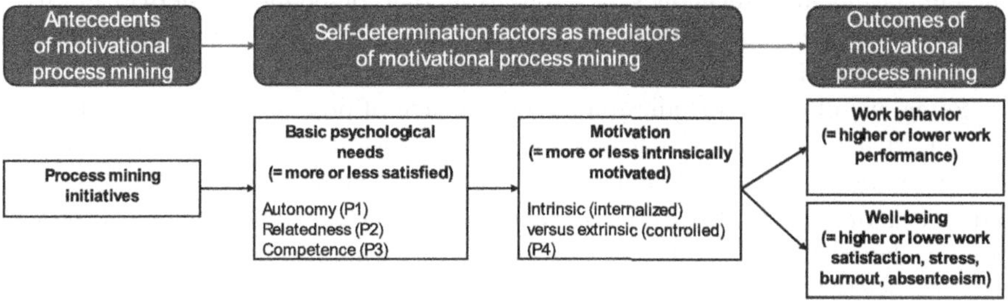

Fig. 1. Proposed conceptual model for "motivational process mining".

Figure 1 applies self-determination theory to process mining. If true for process mining, it is to be expected that not only the advancements in technical or organizational process mining aspects count to realize successful process mining projects (Sect. 2.1). Instead, the notion of "motivational process mining" emphasizes the need to pay equal attention to how individual employees perceive those process mining efforts in terms of their autonomy, relatedness, and competence. Thus, process mining is likely to be more successfully applied when also enhancing employees' sense of autonomy, relatedness, and competence.

Based on self-determination theory, future research can evaluate four propositions.

- **Proposition 1 (P1).** If organizations apply process mining while also enhancing employees' basic psychological need for autonomy, then they are more likely to improve employees' work behavior and well-being.
- **Proposition 2 (P2).** If organizations apply process mining while also enhancing employees' basic psychological need for relatedness, then they are more likely to improve employees' work behavior and well-being.

- **Proposition 3 (P3).** If organizations apply process mining while also enhancing employees' basic psychological need for competence, then they are more likely to improve employees' work behavior and well-being.
- **Proposition 4 (P4).** In organizations with process mining holds that the positive link between employees' need satisfaction and their work behavior/well-being, can be explained by higher intrinsic motivation towards process mining.

5 Discussion of "Motivational Process Mining"

Our conceptualization suggests that satisfying employees' basic psychological needs (i.e., for autonomy, relatedness, competence), and consequently their intrinsic motivation towards process mining initiatives, is of strategic importance for organizations as it also affects employees' work behavior, performance, and well-being. Therefore, process mining as a top-down monitoring technique risks inducing or reinforcing motivational harms. Alternatively, process mining might also be used as a bottom-up technique that supports employees' basic psychological needs for self-improvement.

E.g., work behavior monitoring might decrease one's feeling of autonomy (e.g., as micromanagement) or relatedness (e.g., due to internal competition). Alternatively, it can improve one's perceived competence when acquiring efficiency gains and specialization. Hence, the individual advantages of process mining need to be communicated.

Being one of the first studies on the human side of process mining, we extend the state-of-art with a focus on the individual framework level of process mining [18]. We agree with [2] that evaluating work-related well-being is essential and deserves further examination, while we complement their work with another theoretical lens than resource theory. By promoting a "motivational process mining" concept, our viewpoint helps the BPM field evolve towards a "theory for explaining and predicting" process mining success, namely a theory type 4 according to [7].

Practically, this study has differentiated between basic psychological needs to better understand process mining outcomes, while explaining why they can individually differ. Our study helps derive actions that organizations can take to improve their employees' intrinsic motivation in terms of process mining efforts. For privacy concerns, the captured data are typically anonymized or aggregated into a team or process level. Organizations can adopt a transparent approach to work behavior monitoring and communicate the purpose (why) and ways (how) of data capturing. Process mining training also helps raise awareness without technical details.

This study offers promising avenues on "motivational process mining". In future research, Fig. 1 can be applied to process mining end user organizations, and data may be collected from experts in process mining initiatives. Secondly, qualitative findings can be complemented through a survey with end user organizations to test our propositions. Also, different data capturing techniques (e.g., using dedicated software versus timesheet-based monitoring) can be compared. Finally, research can develop and validate a measurement scale for "motivational process mining" to assess and quantify the concept. More generally, the value of "motivational BPM" can be explored as well.

6 Concluding Remarks

This study touches upon how and why process mining might affect employees' work behavior and well-being. For instance, the data capturing phase may face employee resistance, while being a crucial starting point for process mining analyses. This study has added the psychological self-determination theory for introducing "motivational process mining" (i.e., for advancing process mining while keeping employees intrinsically motivated). In sum, we plead to consider employees' basic psychological needs during process mining initiatives in organizations. Our proposed model and derived propositions contribute to a desired theorization of the BPM field.

References

1. Beerepoot, I., Sinik, T., Reijers, H.A.: Capturing and analysing employee behaviour. Data Knowl. Eng. **154**(102350), 3–16 (2024)
2. Braakman, M.A., et al.: Mining for well-being. In: Gdowska, K., Gomez-Lopez, M.T., Rehse, J.R. (eds.) BPM Workshops. LNBIP, vol. 534, pp. 229–245. Springer, Cham (2024)
3. Brock, J., Brennig, K., Löhr, B., Bartelheimer, C., von Enzberg, S., Dumitrescu, R.: Improving process mining maturity. Bus. Inf. Syst. Eng. **66**, 585–605 (2024)
4. Deci, E.L., Olafsen, A.H., Ryan, R.M.: Self-determination theory in work organizations. Annu. Rev. Organ. Psychol. Organ. Behav. **4**, 19–43 (2017)
5. Gartner. Market share (2024). https://www.gartner.com/
6. Gilal, F.G., Zhang, J., Paul, J., Gilal, N.G.: The role of self-determination theory in marketing science. Eur. Manag. J. **37**, 29–44 (2019)
7. Gregor, S.: The nature of theory in information systems. MIS Q. **30**(3), 611–642 (2006)
8. Mamudu, A., Bandara, W., Wynn, M.T., Leemans, S.J.: Process mining success factors and their interrelationships. Bus. Inf. Syst. Eng. (2024)
9. Martin, N., Beerepoot, I.: Unveiling use cases for human resource mining. Bus. Inf. Syst. Eng. (2024)
10. Mitchell, J.I., Gagné, M., Beaudry, A., Dyer, L.: The role of perceived organizational support, distributive justice and motivation in reactions to new information technology. Comput. Hum. Behav. **28**(2), 729–738 (2012)
11. Podsakoff, P.M., MacKenzie, S.B., Podsakoff, N.P.: Recommendations for creating better concept definitions in the organizational, behavioral, and social sciences. Organ. Res. Methods **19**(2), 159–203 (2016)
12. Ryan, R.M., Deci, E.L.: Self-determination theory and the facilitation of intrinsic motivation, social development and well-being. Am. Psychol. **55**, 68–78 (2000)
13. Ryan, R.M., Deci, E.L.: Intrinsic and extrinsic motivation from a self-determination perspective. Contemp. Educ. Psychol. **61**, 1–11 (2020)
14. Syed, R., Leemans, S.J., Eden, R., Buijs, J.A.: Process mining adoption. In: Fahland, D., Ghidini, C., Becker, J., Dumas, M. (eds.) BPM Forum. LNBIP, vol. 392, pp. 229–245. Springer, Cham (2020). https://doi.org/10.1007/978-3-030-58638-6_14
15. Van Den Broeck, A., Rivkin, W., Slemp, G.R.: Motivating employees using self-determination theory. In: Peeters, M., de Jonge, J., Taris, T. (eds.) An Introduction to Contemporary Work Psychology, pp. 67–87. Wiley, Hoboken (2024)
16. van der Aalst, W.: Process Mining. Springer, Heidelberg (2016). https://doi.org/10.1007/978-3-662-49851-4

17. Vansteenkiste, M., Neyrinck, B., Niemiec, C.P., Soenens, B., De Witte, H., Van den Broeck, A.: On the relations among work value orientations, psychological need satisfaction and job outcomes. J. Occup. Organ. Psychol. **80**, 251–277 (2007)
18. vom Brocke, J., Jans, M., Mendling, J., Reijers, H.A.: A five-level framework for research on process mining. Bus. Inf. Syst. Eng. **63**(5), 483–490 (2021)
19. Walker, L.O., Avant, K.C.: Strategies for Theory Construction in Nursing. Prentice Hall, Upper Saddle River (2011)
20. Zerbato, F., Soffer, P., Weber, B.: Process mining practices. In: Di Ciccio, C., Dijkman, R., del Rio Ortega, A., Rinderle-Ma, S. (eds.) BPM. LNCS 13420, pp. 268–285. Springer, Cham (2022). https://doi.org/10.1007/978-3-031-16103-2_19

Anchoring ESG Goals in Business Processes: Engagement Patterns in Polish Organizations

Natalia R. Potoczek[1], Adriana Paliwoda-Matiolańska[2], Katarzyna B. Homoncik[3](✉), and Mariusz Łapczyński[2]

[1] Institute of Economics of the Polish Academy of Sciences, Warsaw, Poland
npotoczek@inepan.waw.pl

[2] University of Economics, Krakow, Poland
{paliwoda,lapczynm}@uek.krakow.pl

[3] Wyższa Szkoła Biznesu - National-Louis University, Nowy Sącz, Poland
p_kglab@wsb-nlu.edu.pl

Abstract. Process executors should perceive responsible business process management within an enterprise. It is employees who primarily experience changes in daily operations. If they can connect these changes with their company's purposeful and long-term orientation, we can speak of anchoring certain concepts in everyday practice. This paper aims to determine the nature of Polish enterprises' activities in achieving ESG (Environmental, Social, Governance) goals in business processes, their connection with ESG strategy, and institutional pressure and legislative initiatives at the national and European Union levels. The presented survey research conducted among employees representing 547 Polish organizations indicates strong anchoring of ESG goals in business processes and simultaneously statistically significant relationships between strategic and process management. The perceived institutional pressure for ESG goal implementation and its actual impact on enterprise behavior have been confirmed in statistical tests. At the same time, an important research finding is identifying weak support from business environment institutions for sustainable development. The research results may provide knowledge to managers implementing ESG goals, BPMS software suppliers, and training service providers regarding managerial, resource, and technological support.

Keywords: responsible BPM · ESG goals · ESG strategy · institutional pressure · business environment

1 Introduction

Responsible business process management has resonated in the academic community for years, recognizing various threats within enterprises and their environment. Over the past two decades, researchers dealing with BPM issues have

M. ER et al. (Eds.): BPM 2025, LNBIP 565, pp. 51–65, 2026.
https://doi.org/10.1007/978-3-032-02936-2_5

focused mainly on environmental threats resulting from improperly directed requirements for reducing operational costs. This is confirmed by numerous publications oriented toward developing the Green BPM concept. Continuous analysis of the environmental effects of implemented processes should translate into improving methods, technologies, and above all, proper support through information systems providing knowledge about introduced changes [30]. Alongside environmental threats, social threats are attracting increasing attention. The consequences of unilaterally directed cost reduction are also borne by people, especially employees involved in process implementation. Relegating working time standards to the background has begun to contribute to overload and, consequently, decreased employee engagement in process improvement and the introduction of innovative solutions. We had the impression that pressure to increase responsibility for business processes was insufficient for many years, and changes were occurring too slowly. Current observations allow us to assume that only increased institutional pressure and emerging legislative initiatives (e.g., the EU CSRD directive from 2023) are growing enterprise activity in the ESG (Environmental, Social, Governance) area.

In our article, we want to show the nature of enterprise activity in ESG within business processes, its connections with ESG strategy, and institutional pressure and legislative initiatives at national and EU levels. We assume that engagement at the strategic level will be expressed in a developed ESG strategy and appropriate key performance indicators (KPIs), and should also be reflected throughout the enterprise's value chain (VC). For this purpose, we examine changes in individual VC process groups observable by employees and new activities in processes in two periods: up to 2022 and from 2023. The temporal scope of the research is related to the implementation of new ESG reporting obligations for enterprises resulting from the CSRD Corporate Sustainability Reporting Directive of the European Parliament and Council (EU) 2022/2464 of December 14, 2022. Recognizing the importance of the institutional business environment in supporting ESG activities, we also examine the awareness level of employees in government, local government, and budget institutions, their training, and the cooperation undertaken by their institutions with enterprises engaged in ESG. For the purposes of our established aims objectives, we formulated three research questions that organized our research process and ultimately the structure of the content in the article: (RQ1) Do enterprises respond to ESG challenges and is this visible at the strategic and process levels; (RQ2) Is the nature of enterprise ESG activity related to EU regulations/requirements; (RQ3) Do business environment institutions support sustainable development activities?

Subsequently, the article comprises five main sections. The theoretical background connects our research to numerous scientific and legislative sources. In the methodological section, we present our research hypotheses, methods for their verification, and the research sample used. The following two sections include the results of our research and their discussion in light of the adopted assumptions. In the final section, we introduce our conclusions regarding the established objectives, the value of the obtained research results, possibilities for their utilization, and recommendations for possible research directions.

2 Literature Background

In recent years, ESG issues have become crucial in corporate management, influencing strategic direction and companies' financial performance. The ESG acronym was first used in 2004 in the "Who Cares Wins" report, which emerged as a collaboration initiative between the United Nations and leading financial institutions. The main objective of establishing ESG terminology was to integrate non-financial issues with main investment practices and create a common language for investors and enterprises. ESG is also often perceived as a new designation for CSR (Corporate Social Responsibility), as these concepts partially overlap and complement each other. CSR means adhering to authentic corporate stakeholder values, while ESG constitutes a set of measurements from which conclusions about CSR can be drawn. ESG also serves an instrumental function for sustainable development as a tool for measuring progress by providing measurable indicators of sustainable development.

Research shows that integrating ESG practices with corporate strategy leads to increased operational efficiency, better risk management, and enhanced organizational reputation, which translates into higher profitability and market value of enterprises [4,13]. The literature also emphasizes that individual ESG pillars impact corporate management and performance differently. The greatest importance for improving corporate governance and reducing unethical management practices is attributed to the corporate governance component [23]. On the other hand, social and environmental factors strongly influence how the company is perceived by stakeholders and its credit rating [12]. Including ESG in management strategy requires compliance with legal requirements and consideration of ethical aspects, monitoring and motivating employees for pro-social and pro-environmental actions [1]. ESG practices can also mitigate managerial myopia, promoting long-term development and increasing company value [16]. However, some studies indicate that the risk of instrumental use of ESG by companies in difficult financial situations may increase ESG disclosure to mask earnings management or other undesirable practices. The concept itself has critics identifying ESG myths [25], who emphasize that ESG investment practices differ little from conventional investing [3], and their impact on financial performance is negligible [26]. Interestingly, the effectiveness of ESG implementation depends on external factors such as regulatory pressure, economic development, and investor expectations [24]. It is worth mentioning that the most sophisticated and comprehensive ESG reporting regulations, such as the Corporate Sustainability Reporting Directive (CSRD) and the Corporate Sustainability Due Diligence Directive (CSDDD), are in force in the European Union (EU).

It is essential that enterprises not only declare the implementation of the mentioned concepts but also implement them. Real actions are necessary; however, it should be noted that this is a process requiring certain reorganization within the enterprise, change, and increased awareness about sustainable development [18]. Researchers emphasize that enterprises must focus on all three elements of sustainable development. Only such an approach can guarantee success in the long term. Process environments note that business processes constitute a key platform for operationalizing ESG and SDG strategies. In these circumstances, sustainable process management–Green Business Process Management–gains importance. Green BPM is an approach to implementing business processes considering environmental, social, and market criteria [31]. Green BPM assumptions are such that process efficiency is measured in categories such as costs, quality, and time, but also in terms of CO_2 emissions, resource consumption, energy efficiency, and social impact. Researchers dealing with Green BPM topics emphasize that not only do changing legal regulations motivate enterprises to introduce changes with positive effects, including for the environment. Greater consumer awareness is also a motivating factor for introducing changes in organizational processes [18].

The Green BPM concept constitutes a vital tool supporting comprehensive organizational transformation, which is essential in integrating business activities with sustainable development assumptions. These activities involve modeling processes using environmental indicators, analyzing the BPM life cycle, and implementing supporting technologies (e.g., digital twins, AI, and process automation [14,18]. However, it is noticeable that Green BPM implementation does not proceed without problems. Most frequently indicated are the lack of standardized methodology, insufficient managerial competencies, limited data, and reluctance to redefine established processes [10]. Pointing to specific barriers related to Green BPM implementation, it is impossible not to also draw attention to factors favoring the realization of this concept's assumptions. Institutional pressure, previously mentioned legislative actions (EU Taxonomy, CSRD), and customer expectations favor Green BPM CSRD [11,17]. According to dynamic capabilities, organizations that can quickly adapt their processes for sustainable development gain a competitive advantage [8].

In summary, Green BPM is an interdisciplinary combination of process management with new technologies and values promoted by sustainable development. However, it should also be remembered that Green BPM is closely related to employee engagement. It is precisely the participation of personnel in Green BPM design, monitoring, and systematic improvement that is essential for achieving success.

3 Methods

3.1 Materials

To obtain answers to the previously posed questions (in Sect. 1), we formulated three research hypotheses: H1: Enterprise engagement in ESG goal implementation has a strategic character and is confirmed by new activities in business processes; H2: Enterprise engagement in ESG goal implementation is related to non-financial information reporting requirements; H3: The business environment (government, local government, budget) indicates a low level of knowledge and engagement in ESG goal implementation. To verify the hypotheses, we used independent variables such as: place of employment, company stock exchange listing, company headquarters in or outside the EU, and managerial or non-managerial position held. The dependent variables used in the study are: having an ESG strategy, using key performance indicators (KPIs) for ESG goals, knowledge about enterprise obligations resulting from ESG regulations, changes in ESG-related processes, ESG principles in cooperation with external entities, new ESG-related process activities before and after introducing reporting obligations, and ESG training for employees before and after introducing reporting obligations.

We obtained research material using the CAWI interview technique. Our sample selection was purposive; our respondents were professionally active people studying in managerial programs (master's, MBA, and EMBA). We wanted to ensure that the content contained in the survey questionnaire would be properly interpreted. The study was conducted in February-April 2025. Our respondents represent 547 Polish organizations, including enterprises and various business environment institutions within and outside the EU. Table 1 presents the research sample structure considering the workplace, which was a key independent variable in our statistical analyses due to the introduction of ESG reporting obligations.

3.2 Measurement

The chi-square test of independence was applied to examine relationships between variables, and Cramér's V coefficient was used to determine the strength of association. The choice of these methods was determined by the measurement scale of the analyzed variables. Additionally, multiple correspondence analysis (MCA) was applied to visualize relationships between variables. MCA is a data exploration method that examines the co-occurrence of categories in multiple qualitative variables. These categories are typically presented as points in two-dimensional Euclidean space, making this method effective and visually attractive for graphically presenting results. This method was developed in several academic centers [5,19] and is widely used in survey data analysis.

4 Results

4.1 Engagement Patterns in Achieving ESG Goals

The research confirmed a significant relationship between the strategic nature of ESG goal implementation and actual activity expressed in business processes. Among enterprises with an ESG strategy (254; 45.6%), only 8.3% of cases were indicated where performance indicators (KPIs) related to ESG have not yet been developed. In 20.5% of cases, respondents could not confirm using ESG-related KPIs. The strategic dimension of ESG in enterprises was confirmed by strong relationships between having an ESG strategy and its measurement ($p=0.00$;

Table 1. Distribution of respondents by place of employment.

Place of Employment	Short Name	Count	Percent
Large manufacturing or manufacturing-service enterprise (employment of 250 or more people)	LM/MS	150	26.93%
Medium-sized manufacturing or manufacturing-service enterprise (employment less than 250 people)	MM/MS	54	9.69%
Small manufacturing or manufacturing-service enterprise (employment less than 50 people)	SM/MS	32	5.75%
Micro manufacturing or manufacturing-service enterprise (employment less than 10 people)	MiM/MS	16	2.87%
Large service enterprise (employment of 250 or more people)	LS	41	7.36%
Medium-sized service enterprise (employment less than 250 people)	MS	25	4.49%
Small service enterprise (employment less than 50 people)	SS	41	7.36%
Micro service enterprise (employment less than 10 people)	MiS	43	7.72%
Government administration institution	GA	29	5.20%
Local government administration institution	LGA	61	10.95%
Public budget organization	PBO	30	5.39%
Other	O	25	6.29%
Total		**547**	**100%**

Cramér's V = 0.40). Activity related to ESG goal implementation in enterprises with an ESG strategy was additionally confirmed in most cases (90.16%) by identified process activities. This is also evidenced by probability tests and the level of dependency (p=0.00; Cramér's V = 0.41). Examining relationships between process categories and the level of dependency allowed observation of statistically significant associations with strategy creation and development processes (p = 0.00; Cramér's V = 0.41), product and service design processes (p = 0.00; Cramér's V = 0.36), risk management, compliance, adaptation, and flexibility processes (p = 0.00; Cramér's V = 0.39), external relationship management processes (p = 0.00; Cramér's V = 0.33), and processes focused on managing the company's business capabilities (p = 0.00; Cramér's V = 0.30).

The correspondence analysis presented in Fig. 1 indicates the relationship between observed ESG-related activities in process categories Q1-Q12 and large manufacturing and service enterprises (LM and LS), medium manufacturing and service enterprises (MM), and, interestingly, micro manufacturing and service enterprises (MM, MiM). This last case allows us to conclude that relationships connect enterprises operating on a micro scale with large enterprises, and certain operating standards may be imposed by larger entities. Figure 1 also shows that the engagement of business environment entities (LGA, GA, PBO) demon-

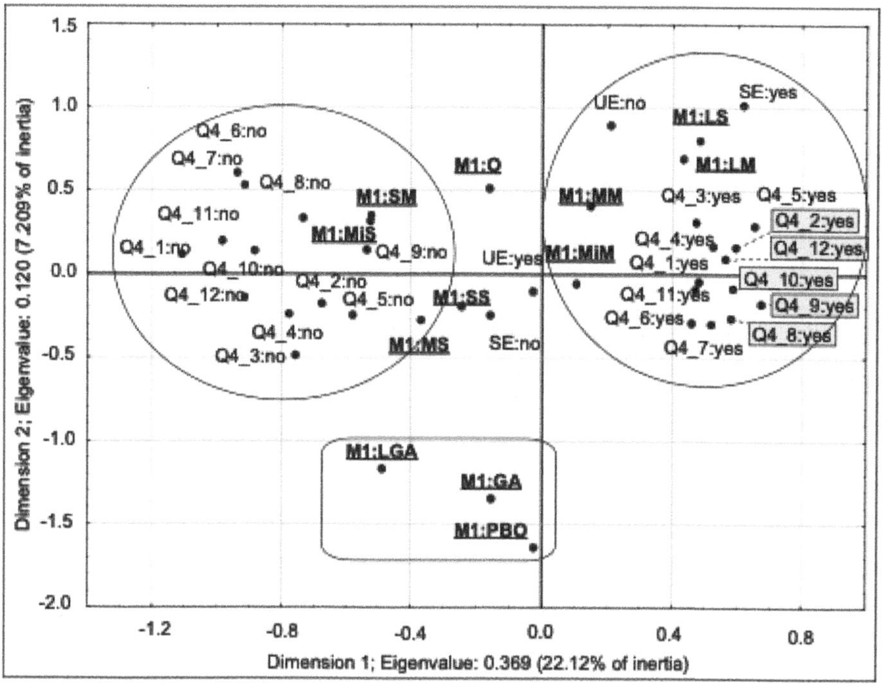

Fig. 1. Engagement in ESG goals in process groups. .Note: The applied process categories are contained in the Process Classification Framework APQC [2]*: Q1 Develop Vision and Strategy; Q2 Develop and Manage Products and Services; Q3 Market and Sell Products and Services; Q4 Deliver Products/Services; Q5 Manage Customer Service; Q6 Develop and Manage Human Capital; Q7 Manage Information Technology (IT); Q8 Manage Financial Resources; Q9 Acquire, Construct, and Manage Assets; Q10 Manage Enterprise Risk, Compliance, Remediation, and Resiliency; Q11 Manage External Relationships; Q12 Develop and Manage Business Capabilities.*

strates minimal activity for ESG goal implementation. The nature of engagement in ESG goal implementation is reflected in new process activities. The research covered two periods: up to 2022 and from 2023, due to the introduction of ESG reporting regulations for companies starting from 2024. Representatives of companies where the ESG strategy is implemented most often notice the emergence of new organizational or process roles related to ESG activity. The level of dependency for these variables is similarly high for the period up to 2022 ($p=0.00$; Cramér's $V=0.30$) (Fig. 2) as for the period from 2023 ($p=0.00$; Cramér's $V=0.37$) (Fig. 3), with attention drawn to the growing dependency over the studied time. Undoubtedly, the strength of association also changed in the period from 2023 (Fig. 3) for activities such as implementing new IT systems supporting ESG in processes ($p=0.00$; Cramér's V increase from 0.17 to 0.30), similarly for new functionalities in existing IT systems ($p=0.00$; Cramér's V increase from 0.23 to 0.30). A similarly high level of dependency in the periods

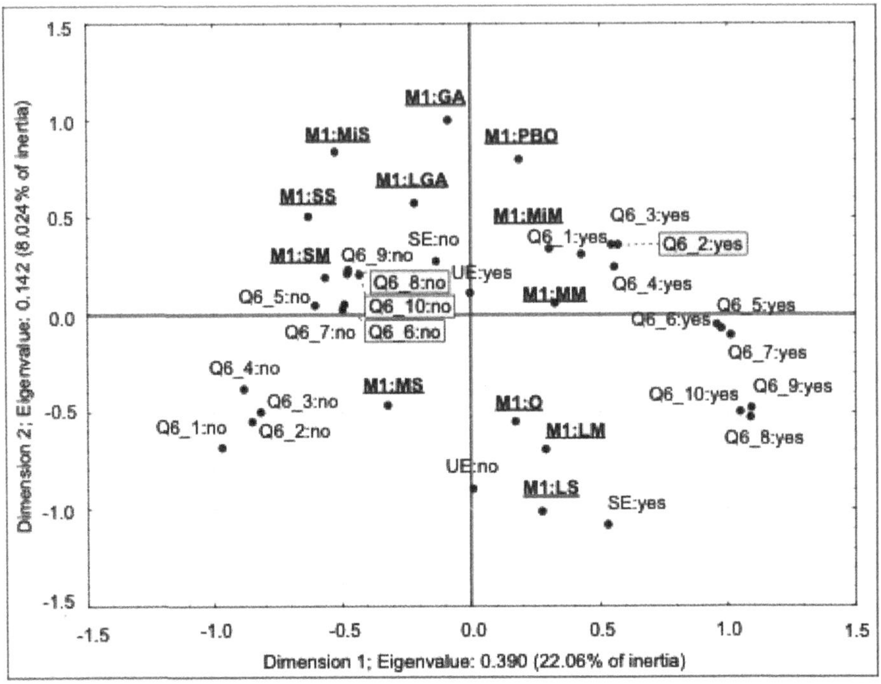

Fig. 2. ESG-related process activities up to 2022.. Note: Categories Q1-Q10 denote new ESG-related activities in business processes: Q1 tasks; Q2 events that must be considered, on which further actions depend; Q3 decisions or decision-making authorities on which further actions depend; Q4 documents or elements in existing documents used in work; Q5 organizational/process roles related to ESG activity; Q6 IT systems supporting ESG activity; Q7 functionalities or tools associated with ESG in used IT systems; Q8 ESG-related reports; Q9 ESG-related elements in existing reporting; obtaining quality certificates related to ESG. Other designations: UE (European Union location); SE (stock exchange presence).

up to 2022 (Fig. 2) and from 2023 (Fig. 3) can be observed for activities related to ESG reporting and obtaining certificates (see Sect. 4.2). The correspondence analysis for the period up to 2022 (Fig. 2) illustrates dependencies occurring between process activities Q1-Q5 and medium and micro manufacturing and service enterprises (MM, MiM) located within the EU (including Poland), while activities Q6-Q10 show greater associations with large enterprises (LM and LS) located outside the EU and, interestingly, with other entities classified as other organizations constituting the business environment (Others) also located outside the EU. In the period from 2023 (Fig. 3), a similar structure of associations can be observed, with the level of dependency between large enterprises located outside the EU and process activities Q15-Q20 significantly increasing, meaning that ESG goal implementation is more frequently reported and realized with greater support from IT systems supporting business processes. The increased

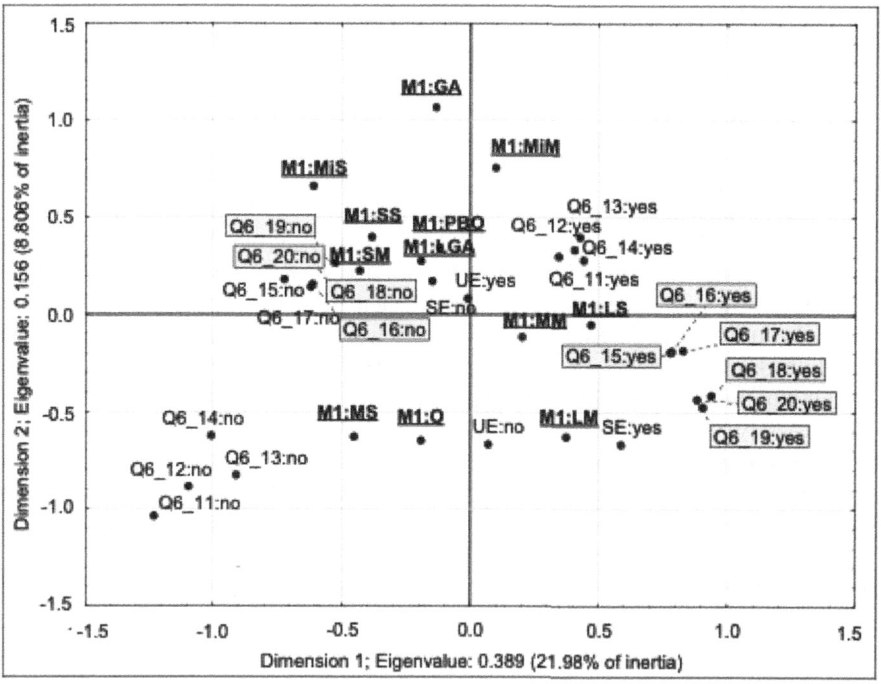

Fig. 3. ESG-related process activities from 2023. Note: Categories Q11-Q20 are coded Q1-Q10, which denote new ESG-related activities in business processes, covering the period from 2023 onwards.

importance of ESG certification may also indicate growing interest in enhancing enterprise competencies and credibility. The nature of engagement in ESG goal implementation also reflects employees' knowledge level about enterprise obligations related to ESG goals implementation, which results directly from external regulations or the adopted ESG strategy. Over 73% of respondents confirmed their companies' activity in informing employees about new requirements. Company activity in this area was confirmed by a high level of variable dependency ($p=0.00$; Cramér's $V=0.49$).

4.2 Regulatory Factors of ESG Engagement Patterns

In the study, respondents were asked to indicate ESG reporting-related actions in two periods: up to 2022 and from 2023. This distinction is based on the fact that most studied enterprises are in the EU, where mandatory sustainability reporting was introduced from 2023. An apparent escalation of new ESG reporting activities within processes in which respondents participated is visible when comparing periods up to 2022 and starting from 2023. Before 2022, 34.01% of respondents indicated the emergence of new ESG-related reports within processes or new elements related to developing existing ESG reports. From 2023, this percentage increased to 44.3%, representing a 10% increase. It should be noted that EU legal frameworks regarding ESG are being implemented gradually. In 2025, these obligations will apply exclusively to large enterprises and public companies (SE). The increased identification of actions between periods before 2022 and after 2023 reflects the emergence of legal requirements related to non-financial information reporting. This phenomenon is particularly visible in the analysis of respondent opinions in groups subject to these obligations. Interestingly, after 2023, over half of large manufacturing enterprises (58%) and service enterprises (53.48%) confirmed the presence of activities related to creating ESG reports in processes or including new elements regarding developing existing ESG reports (LM-59.8%, LS-48.84%). Before 2022, this number was approximately forty percent for new reporting initiatives (LM-45.33%, LS-46.51%) and new actions in existing ESG reporting (LM-42.67%, LS-44.19%). Furthermore, a statistically significant relationship was observed between enterprise size and identification of mandatory process activities, including ESG report creation up to 2022 (p-value=0.00, coefficient 0.33) and after 2023 (p-value=0.00, coefficient 0.26). Statistical analysis revealed analogous correlations regarding respondents' identification of new elements in processes related to developing existing ESG reports up to 2022 (p-value=0.000, coefficient 0.31) and after 2023 (p-value=0.00, coefficient 0.3). This relationship is visually presented in Figs. 2 and 3, illustrating correspondence analysis where the abovementioned activities are presented. We can observe clustering of category points representing enterprise size (LM, LS), stock exchange presence (SE), and identification of new actions resulting from reporting obligations (Q8, Q9 up to 2022; Q18, Q19 from 2023) in the space of main dimensions of correspondence analysis, confirming their functional relationship. Another identified group with a functional relationship consists of medium manufacturing enterprises (MM). The obligation to report non-financial information is internally related to enterprise size, which subsequently translates into identifying new actions in business processes resulting from reporting requirements. Consequently, hypothesis 2 was substantiated.

4.3 Assessment of Knowledge and ESG Engagement Level in Public Sector Organizations

A key element of our analytical framework is examining how organizations constituting the institutional business environment, including government administration bodies, local authorities, and other budget units, position themselves regarding ESG principles. It is worth noting that solving challenges related to achieving sustainable development goals requires a comprehensive and integrated approach. These institutional entities' awareness and scale of engagement positively influence the overall resolution of ecological and social challenges while strengthening enterprise engagement. Survey respondents representing these institutional entities predominantly show a knowledge deficit (35.38%) or cannot determine (27.69%) obligations or recommendations regarding sustainable development for institutions, resulting from legislative frameworks or adopted ESG strategic initiatives. Only 36.92% possess the necessary knowledge, and this percentage remains quite similar across all three types of institutions, although it is highest in budget organizations at 39%. The participation of institutional business environment entities in adhering to ESG principles as a prerequisite for business cooperation shows significantly reduced levels. A nominal 3.09% of institutions implement such practices. Only 20% of institutional business environment entities aspire to maintain cooperative relationships with organizations that adhere to ESG principles, and 20% believe that current circumstances prevent them from choosing cooperating entities that adopt ESG principles. The majority, exceeding half (56.92%), currently shows no ESG orientation in their inter-organizational cooperation frameworks. In the predominant part of these institutions, professional development initiatives related to sustainable development (ESG) were not organized for management staff and employees in both periods: up to 2022 and from 2023. Nevertheless, we have observed an increase in such educational initiatives since 2023. Before 2022, only 26.92% of management personnel in institutional business environment entities received ESG-related professional development, and only 12.31% of non-management employees. From 2023, the frequency of professional development sessions for management personnel increased to 43.08%, and for non-management employees to 23.85%. Statistical analysis did not reveal the occurrence of significant dependencies between variables. In light of the above, hypothesis 3 can be considered confirmed. As a supplement, it is worth referring to the correspondence analysis (Fig. 1), which showed that public sector organizations constitute a separate cluster showing an undefined, passive approach to ESG. They generally do not perceive changes in their organizations in this area. This indicates additional tendencies in the studied phenomenon.

5 Discussion

The obtained research results confirm our first hypothesis, which states that enterprise engagement in ESG goal implementation has a strategic character and is confirmed by new activities in business processes. The significant relationship between having an ESG strategy (in 45.6% of studied enterprises), strategy measurement (KPIs for ESG), and actual activity in processes testifies to relatively high process maturity. In the context of developing ESG regulations, organizations that can implement new tasks in processes increase their capabilities in terms of cost optimization. According to the dynamic capabilities concept, organizations that can quickly adapt their processes for sustainable development also gain a competitive advantage [8]. Of course, there are many barriers to implementing ESG. Many researchers draw attention primarily to resistance from process managers responsible for increasing efficiency, while new ESG-related tasks may initially increase ESG implementation costs [10]. The role of manager and employee engagement in implementing new goals and optimizing processes is emphasized in research conducted within Green BPM by Brajer-Marczak et al. (2024). [9]. Regardless of the nature of new challenges, employee preparation are essential at every stage of the BPM lifecycle [15]. Another critical aspect of the conducted research is its contribution to the academic discussion on the role of institutional pressure and the impact of regulatory instruments on enterprise engagement in environmental and social activities. Legal instruments and regulatory mechanisms motivate organizations to implement pro-ecological practices; however, their effectiveness remains contingent upon additional external pressure factors and the organizational resources of the enterprise [28]. For example, comprehensive environmental management systems, shaped by both coercive and normative pressures, generate higher environmental effectiveness within organizational structures [27]. Moreover, enterprises typically undertake pro-ecological initiatives in response to regulatory requirements, with their attitude characterized by reactivity in many cases, resulting from the necessity to adapt to applicable regulations rather than a proactive organizational strategy [7]. Additionally, introducing new regulatory frameworks initiates increased ecological awareness and transformation of corporate strategies, even when organizations are not initially convinced about the economic benefits of such activities. The implementation of non-financial information reporting requirements necessitated that Polish enterprises adapt their business processes to ensure compliance with established standards, as indicated by both the increase in activity identification between periods up to 2022 and from 2023, and the size of enterprises that most frequently reported ESG activity intensification. These are primarily large enterprises and stock exchange companies that were required to prepare reports in accordance with the CSRD directive for 2024. This study indicates a significant relationship between ESG anchoring in processes and legislative obligation regarding non-financial information reporting. Against this background, the certain passivity of business environment institutions demonstrated by this research is not a factor that positively influences, as institutional theory indicates that the institutional environment has a significant impact on organizational behaviors and practices [20]. The development of ESG regulations at the supranational and ultimately national level responds to the postulates of scientists formulated within the framework of institutional economics. According to this theory, they cannot focus solely on the rational choices of individuals but should take into account the influence of institutions, including law, norms and organizations on human behavior and the functioning of the economy [22]. We can assume that ESG aligns with the mainstream of institutional economics and becomes an integral part of modern management. Research indicates that institutional quality and environmental technologies impact the ecological footprint and support efforts to reduce environmental degradation [21]. Moreover, the institutional environment has a strong influence on managers' positive attitude toward sustainable development, positively affecting the company's overall orientation toward environmental sustainability, particularly in smaller companies [29]. The data reveal a significant gap in ESG awareness among business environment organizations and suggest public sector isolation from ESG practices. This is particularly concerning, considering that government and local government organizations should fulfill the role of leader and initiator in implementing the Sustainable Development Goals (SDGs) [6]. However, a positive signal is the increase in training activity from 2023.

6 Conclusions

The assessment of the ESG goal anchoring level in processes allows us to state that these organizations are transitioning from ESG actions, mainly marketing and legitimizing character for enterprise activities, to a strategic approach toward sustainable development. Enterprises build ESG strategies from the perspective of long-term investments and operationalize ESG in business processes. This confirms organizations' high process maturity while indicating the need for various support from the broadly understood business environment. This is an essential signal for multiple entities, for example, different ESG training services providers or IT systems suppliers supporting business processes in terms of analysis, simulation, monitoring, and reporting activities. It is worth emphasizing that public organizations of the business environment have a unique opportunity to become a catalyst for changes in the ESG area by utilizing their position as regulators, business partners, and organizers of public procurement. Suppose the public sector does not actively support and promote ESG practices. In that case, enterprises may treat these issues as secondary, threatening the implementation of sustainable development goals at national and regional levels. The current state shows untapped potential. Future research should focus on identifying specific organizational, legal, and institutional barriers hampering ESG implementation and comparing solutions in different EU regions to identify best practices for adaptation. Undoubtedly, barriers regarding ESG goal implementation in processes should be analyzed in the context of organizational process maturity, and therefore, a multi-aspect approach to processes in competency, managerial, and technical dimensions.

Disclosure of Interests. The authors have no competing interests to declare that are relevant to the content of this article.

References

1. Almnadheh, Y., Samara, H., AlQudah, M.Z.: Enhancing ESG integration in corporate strategy: a bibliometric study and content analysis. Inter. J. Law Manage. ahead-of-print (2025). https://doi.org/10.1108/IJLMA-10-2024-0385
2. APQC: APQC (n.d.), Process Classification Framework. https://www.apqc.org/process-frameworks (Feb 2025)
3. Atz, U., Tracy, V.H., , Zongyuan Zoe, L., Bruno, C.C.: Does sustainability generate better financial performance? review, meta-analysis, and propositions. J. Sustainable Finance Investment **13**(1), 802–825 (2023). https://doi.org/10.1080/20430795.2022.2106934
4. Bax, K., Bonaccolto, G., Paterlini, S.: Spillovers in Europe: the role of ESG. J. Financial Stability **72**, 101221 (2024). https://doi.org/10.1016/j.jfs.2024.101221

5. Benzécri, J.P.: L'analyse des données: Benzécri. Dunod, J.-P. et al. L'analyse des correspondances (1973)
6. Berniak-Woźny, J., Sliż, P., Siciński, J.: Empowering smart cities through startups: a sustainability framework for incubator-city collaboration. Systems **13**(4), 219 (2025). https://doi.org/10.3390/systems13040219
7. Berry, M.A., Rondinelli, D.A.: Proactive corporate environmental management: a new industrial revolution. Acad. Manag. Perspect. **12**(2), 38–50 (1998). https://doi.org/10.5465/ame.1998.650515
8. Bocken, N., Geradts, T.: Barriers and drivers to sustainable business model innovation: organization design and dynamic capabilities. Long Range Planning **53**(4) (2020). https://doi.org/10.1016/j.lrp.2019.101950
9. Brajer-Marczak, R., Seruga, P., Jarossova, M., Krzywonos, M.: Green business process management in a Polish municipal waste treatment plant-regional case study. J. Mater. Cycles Waste Manage. **26**(5), 3262–3279 (2024). https://doi.org/10.1007/s10163-024-02025-2
10. Cantele, S., Zardini, A.: What drives small and medium enterprises towards sustainability? Role of interactions between pressures, barriers, and benefits. Corp. Soc. Responsib. Environ. Manag. **27**(1), 126–136 (2020). https://doi.org/10.1002/csr.1778
11. Carmagnac, L., Silva, M., Fritz, M.: Exploring sustainable development goals adoption in supply chain management: a typology of coexisting institutional logics. Bus. Strateg. Environ. **33**(4), 3463–3479 (2024). https://doi.org/10.1002/bse.3661
12. Carnini Pulino, S., Ciaburri, M., Magnanelli, B.S., Nasta, L.: Does esg disclosure influence firm performance? Sustainability **14**(13), 7595 (2022). https://doi.org/10.3390/su14137595
13. Di Tommaso, C., Mazzuca, M.: The stock price of European insurance companies: what is the role of ESG factors? Finance Res. Lett. **56**, 104071 (2023). https://doi.org/10.1016/j.frl.2023.104071
14. Di Vaio, A., Palladino, R., Hassan, R., Escobar, O.: Artificial intelligence and business models in the sustainable development goals perspective: a systematic literature review. J. Bus. Res. **121**, 283–314 (2020). https://doi.org/10.1016/j.jbusres.2020.08.019
15. Dumas, M., La Rosa, M., Mendling, J., Reijers, H.A.: Fundamentals of Business Process Management. Springer, Berlin, Heidelberg (2018). https://doi.org/10.1007/978-3-662-56509-4
16. Fan, Z., Chen, Y., Mo, Y.: Management myopia and corporate ESG performance. Inte. Rev. Financial Analy. **92**, 103071 (2024). https://doi.org/10.1016/j.irfa.2024.103071
17. García-Sánchez, I.M., Rodríguez-Ariza, L., Aibar-Guzmán, B., Aibar-Guzmán, C.: Do institutional investors drive corporate transparency regarding business contribution to the sustainable development goals? Bus. Strateg. Environ. **29**(5), 2019–2036 (2020). https://doi.org/10.1002/bse.2485
18. Gohar, S., Indulska, M.: Environmental sustainability through green business process management. Australasian J. Inform. Sys. **24**, 1–30 (2020). https://doi.org/10.3127/AJIS.V24I0.2057
19. Greenacre, M.J. (ed.): Theory and Applications of Correspondence Analysis. Academic Press, London (1989)
20. Gunarathne, N.: Institutional pressures, environmental management strategy, and organizational performance: The role of environmental management accounting. Bus. Strateg. Environ. **30**(2), 25–839 (2020). https://doi.org/10.1002/BSE.2656

21. Hussain, M., Dogan, E.: The role of institutional quality and environment-related technologies in environmental degradation for BRICS. J. Cleaner Product. **304**, 127059 (2021). https://doi.org/10.1016/j.jclepro.2021.127059
22. Kapp, K.W.: The Foundations of Institutional Economics. Routledge, London (May 2012). https://doi.org/10.4324/9780203827864
23. Kim, S., Li, Z.F.: Understanding the impact of ESG practices in corporate finance. Sustainability **13**(7), 3746 (2021). https://doi.org/10.3390/su13073746
24. Krueger, P., Sautner, Z., Tang, D.Y., Zhong, R.: The effects of mandatory ESG disclosure around the world (2024). https://doi.org/10.2139/ssrn.3832745
25. Larcker, D.F., Tayan, B., Watts, E.M.: Seven myths of ESG. Eur. Financ. Manag. **28**(4), 869–882 (2022). https://doi.org/10.1111/eufm.12378
26. Pedersen, L.H., Fitzgibbons, S., Pomorski, L.: Responsible investing: the ESG-efficient frontier. J. Financ. Econ. **142**(2), 572–597 (2021). https://doi.org/10.1016/j.jfineco.2020.11.001
27. Phan, T.N., Baird, K.: The comprehensiveness of environmental management systems: The influence of institutional pressures and the impact on environmental performance. J. Environ. Mana. **160**, 45–56 (2015). https://doi.org/10.1016/j.jenvman.2015.06.006
28. Redondo Alamillos, R., de Mariz, F.: How can european regulation on eSG impact business globally? J. Risk Financial Manag. **15**(7), 291 (2022). https://doi.org/10.3390/jrfm15070291
29. Roxas, B., Coetzer, A.: Institutional environment, managerial attitudes and environmental sustainability orientation of small firms. J. Bus. Ethics **111**(4), 461–476 (2012). https://doi.org/10.1007/s10551-012-1211-z
30. Seidel, S., Recker, J., Vom Brocke, J.: Green business process management. In: Green Business Process Management: Towards the Sustainable Enterprise, vol. 9783642274886, pp. 3–13 (2012). https://doi.org/10.1007/978-3-642-27488-6_1
31. Sohns, T., Aysolmaz, B., Figge, L., Joshi, A.: Green business process management for business sustainability: A case study of manufacturing small and medium-sized enterprises (SMEs) from Germany. J. Cleaner Product. **401** (2023). https://doi.org/10.1016/j.jclepro.2023.136667

Bridging BPM and Sustainability Journeys in Navigating Sustainability Paradoxical Tensions

Ika Nurkasanah[1,2](✉) and Amy Van Looy[1]

[1] Faculty of Economics and Business Administration, Department of Business Informatics and Operations Management, Ghent University, Tweekerkenstraat 2, 9000 Ghent, Belgium
Ika.Nurkasanah@Ugent.be, Amy.VanLooy@UGent.be

[2] Institut Teknologi Sepuluh Nopember, 60111 Surabaya, Indonesia

Abstract. Since Business Process Management (BPM) and sustainability often evolve along separate journeys, few organizations fully recognize the potential of integrating both disciplines to address shared tensions and to optimize sustainability-driven business process transformation. Hence, this study investigates the convergence of BPM and sustainability journeys through multiple case studies, applying paradox theory and BPM capability frameworks to analyze emerging tensions and response strategies. The findings reveal that despite their different starting points, both BPM and sustainability journeys face shared paradoxical tensions. Our research also contextualizes existing BPM capability areas for identifying shared solutions to navigate the tensions. This study offers theoretical contributions by extending BPM capability frameworks to address sustainability paradoxes and by introducing process-level capability considerations to sustainability research. Our research provides practical insight into fostering collaboration between BPM and sustainability functions that enable organizations to more effectively navigate the complexities of managing sustainable processes.

Keywords: Business Process Management · Sustainability · Paradox Theory · BPM Capability

1 Introduction

Given growing pressures to balance all sustainability pillars (i.e., economic, environmental, and social) [1], organizations face challenges to translate their sustainability commitment into actionable business processes. Business Process Management (BPM) and sustainability disciplines are both critical to this endeavor but often operate in silos due to their differing origins. Traditional BPM mainly focuses on the devil's quadrangle (i.e., economic sustainability) [2], whereas the sustainability discipline also requires environmental and social impacts [3]. This separation often leads to fragmented efforts in organizations since few of them recognize the advantages of integrating both disciplines to address shared challenges and to enhance performance outcomes.

M. ER et al. (Eds.): BPM 2025, LNBIP 565, pp. 66–71, 2026.
https://doi.org/10.1007/978-3-032-02936-2_6

Additionally, organizations are challenged by addressing tensions due to the competing business requirements of various stakeholders [4]. Therefore, they need to develop capabilities to manage these tensions effectively. While paradox theory [5] has been widely used to explore tensions particularly, most of such studies focus on the corporate strategic level, thus overlooking the operational process level [6]. In parallel, BPM research has evolved to define capability areas for sustainable process maturity, with efforts being discussed rather fragmentally and by emphasizing either the economic (e.g., traditional BPM) [7], environmental (e.g., green BPM) [8], or social (e.g., benevolent BPM) [9] dimensions. Therefore, our study combines paradox theory and the BPM capability framework [7] views to explore the following research question: *"how can organizations integrate BPM and sustainability over time?"*.

Our study provides theoretical contributions by demonstrating how conventional BPM capabilities can be contextualized to better address sustainability-driven paradoxical tensions when integrating BPM and sustainability efforts. We also introduce a process-oriented lens into sustainability research, highlighting the role of BPM in enabling practical sustainability outcomes. Practically, our findings offer insights into how organizations can foster collaboration between BPM and sustainability functions to navigate the complexities of managing sustainable business processes.

2 Paradox Theory and BPM Frameworks

According to paradox theory, tensions persist when organizations face conflicting yet interrelated demands from diverse stakeholders [5]. These tensions are divided into four categories, namely: (1) **learning tensions**, involving decisions about whether to expand on existing knowledge or pursue new ones, (2) **organizing tensions**, associated with decisions on managing organizational systems (e.g., centralization versus decentralization), (3) **performing tensions**, arising from competing performance objectives (e.g., profit versus ecological versus social outcomes), and (4) **belonging tensions**, based on differing identities or values among individuals or groups or units.

Since paradox theory provides a structured perspective to comprehend and manage the concurrent pursuit of sustainability pillars, it is pertinent to sustainability research. Nonetheless, the majority of research uses paradox theory at the corporate level, thus with limited attention to how it affects the operational level. On the other hand, BPM frameworks (particularly BPM capability areas [7]) are at the process level, but mainly focus on economic outcomes. Combining paradox theory with BPM thus offers a promising path for resolving sustainability tensions by dedicated response strategies.

3 Methodology

Our study adopted a multiple-case research design [10] by examining the phenomenon of integrating BPM and sustainability in real-world organizational contexts. The unit of analysis was holistic for exploring BPM adoption at the organizational level. We selected cases under the condition that organizations (1) had sustainability awards or a formal commitment to sustainable business processes, and (2) had adopted BPM, in which sustainability was also considered in their process initiatives. We then selected two cases with different starting points (i.e. sustainability versus BPM triggers).

Data was collected from multiple sources to ensure triangulation and capture diverse perspectives. First, we conducted six *semi-structured interviews* (of 90–120 min each) with mid-to-senior-level managers of two sub-holding companies. Participants were contacted via LinkedIn and email. Each received an interview guide and consent form beforehand. Company A (or **CA**) included interviews with the Environmental, Social, and Governance (ESG) Manager, Head of BPM & IT, and Human Capital Manager. Company B (or **CB**) included the Director of Sustainability, the Senior Manager of Sustainability, and the Associate BPM Director. Additionally, we analyzed the last 5 years sustainability reports to trace the cases' evolution of sustainability initiatives and to identify embedded process management elements.

Data were analyzed through qualitative coding using MaxQDA. A deductive analysis was performed in two phases: (1) identifying shared paradoxical tensions when integrating BPM and sustainability, using the paradox theory lens, and (2) identifying shared solutions to these tensions using recognized BPM capability areas (i.e., *strategic alignment, method, IT, governance, people, and culture*) [7]. All coding interpretations were refined through iterative researcher discussions until consensus was reached.

4 Preliminary Findings

4.1 Different BPM and Sustainability Starting Points and Journeys

Cases CA and CB were characterized by a notable distinction in their history of integrating BPM with sustainability, particularly in terms of the maturity per discipline. In CA, both disciplines evolved concurrently. The CEO formalized their sustainability commitment by forming a task force in 2020 and an ESG department in 2022 to structure a sustainability roadmap with a broader scope. Simultaneously, a former IT division was transformed into a "BPM & IT" division in 2020 to concretize sustainability and integrate digitalization into this sustainability roadmap. Through this collaboration, process automation could support CA's sustainability goals. E.g., a fatigue management system optimized task allocation based on employees' readiness and reduced accidents.

In contrast, CB had over 30 years of sustainability experience, beginning with waste management and expanding to encompass various sustainability initiatives. BPM was only introduced later, and accelerating since 2022 when some departments had already practiced process excellence and incorporated sustainability considerations. BPM was adopted to standardize business processes across functions, supporting a unified strategy while enhancing efficiency and integrating sustainability into operations. For instance, CB's BPM systems enabled a Source-To-Pay (STP) business process to assess suppliers' sustainability performance, and cut paper usage within the sourcing process.

4.2 Shared Problems: Paradoxical Tensions

Despite their different histories, CA and CB encountered sustainability-related paradoxical tensions while integrating BPM and sustainability (Table 1). First, **learning tensions** emerged as both cases faced resistance in continuing existing process improvements while learning new processes toward sustainability, mainly due to employees' fear of job loss, increased workload, and fatigue from continuous process changes. Second, **organizing tensions** emerged from limited resources despite growing sustainability-driven process initiatives. Thus, both companies were required to dynamically balance between dedicating and sharing resources to support sustainability. Third, **performing tensions** involved balancing economic, environmental, and social goals. E.g., CA's high-risk mining operations required substantial investments in safety to achieve zero accidents, while CB faced high procurement costs when switching to sustainable suppliers. Fourth, **belonging tensions** occurred in CA due to misalignment between sustainability goals and those of its holding company, with employees seeing sustainability as an external mandate. In CB, a tension emerged between local and global requirements, leading to deviations in business processes in some countries that undermined overarching sustainability targets.

Table 1. Shared paradoxical tensions between BPM and sustainability disciplines.

Category	Description
Learning	Continue existing versus learning new sustainability processes [CA,CB]
Organizing	Dedicate versus share resources to support sustainability initiatives [CA,CB]
Performing	Balance economic versus environmental versus social goals [CA,CB]
Belonging	Follow the holding's objectives versus subsidiaries' objectives [CA] Follow the global versus local process practices [CB]

4.3 Shared Solutions: BPM Capability Areas for Embedding Sustainability

Learning Tensions. CA and CB started to address the resistance to relearning by evolving *strategic alignment* in cascading their corporate sustainability strategies into process-level KPIs, helping employees better understand how their work contributed to broader goals. *Methods* for designing sustainability-driven processes were performed differently, for which CA relied on internal learning by mapping existing processes through cross-functional workshops. Then, CA redesigned processes iteratively, starting with a high-level process with the Board of Directors (BoD) and the sustainability team. This approach preserved routine knowledge while enabling documented process transformation. In contrast, CB skipped as-is process mapping and adopted best practices directly, co-designing new processes with external consultants to gain employees buy-in. *IT capabilities* supported these efforts: CA used familiar tools for internal redesign (e.g. BPMN with MS Visio), while CB implemented SAP-based solutions to reinforce new sustainable workflows. Beyond those capabilities, both companies fostered readiness to adopt new practices by pursuing similar *people & cultural* paths: (1) raising awareness through campaigns, (2) building trust through success stories and training, then (3) reinforcing adoption. Both companies also introduced reward systems linked to sustainability KPI's as extrinsic motivation. CA held annual competitions to encourage bottom-up sustainability process innovation.

Organizing Tensions. Both companies addressed the tension between dedicated and shared teams through evolving BPM governance capabilities that clarified roles, responsibilities, and decision rights. CA divided the BPM and IT divisions into three departments: (1) BPM, facilitating the full BPM lifecycle with process owners, (2) IT & Digitalization, developing the supporting tools, and (3) a Corporate Management System department as process governance controller. Governance was structured hierarchically, with authority cascading from the process owners' division heads to assistant managers. CB adopted a shared governance model in which process proposals by process owners across departments were reviewed by BPM Leads and escalated to a governance board, with major changes sent to the design authority. A recent stage of *governance* progress measured the percentage of process adherence.

Performing Tensions. *Methods and IT* capabilities also played an essential role in balancing conflicting performance goals. CA prioritized economic benefits while considering the environmental and social impacts of process improvements. Thus, during process analysis, they combined qualitative (e.g., root cause analysis) and quantitative *methods* (e.g., cost-benefit analysis), with redesign simulations in Bizagi to assess sustainability impacts. With similar prioritization, CB employed advanced technologies such as process mining to detect bottlenecks, ensure process compliance, and simulate sustainability outcomes. Both companies planned to enhance decision-making by using hyper-automation tools for a sustainability pillar's trade-off analysis.

Belonging Tensions. *Strategic alignment* between CA and its holding company was crucial to dispel perceived differences. This enabled communication between BoD, Sustainability, and BPM teams to formulate sustainability policies, goals, and KPIs that met the interests of both CA and the holding company. CB addressed process deviations by enforcing global best practices while allowing local adaptations when necessary, balancing consistency with responsiveness to local business needs.

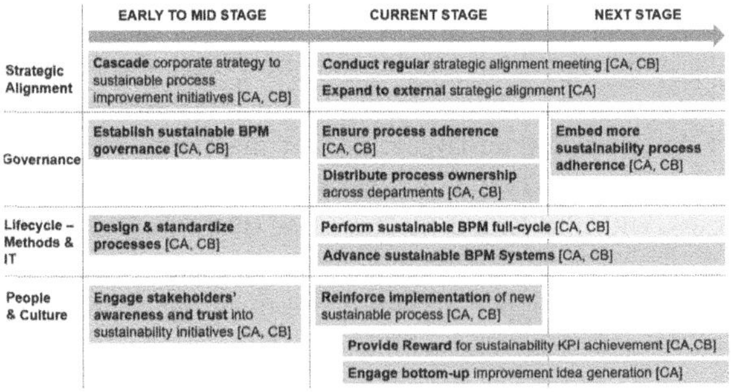

Fig. 1. BPM capability areas progression towards sustainability-driven business process

Figure 1 summarizes the subsequent stages of CA and CB's progression in BPM capability areas toward sustainability, starting with a strategic alignment evolution from internal process improvement coherence to alignment with corporate sustainability. Governance evolved from structures to ensure process adherence and shared ownership.

Methods and IT progressed toward advanced BPM system integration, while people and culture reinforcements followed employee engagement programs.

5 Discussion and Conclusion

This study has shown that the paradox theory lens can help to structure the identification of sustainability tensions, while BPM capability areas have been extended by mapping them with potential response strategies. Despite the differing maturity levels of BPM and sustainability at the start of the integration journey, both cases have experienced similar tensions, except for the belonging tensions. They have both started with responding to the tensions through *strategic alignment* efforts. Other capability areas (e.g., *governance, people, method,* and *culture*) tend to evolve comparably. Since our study is limited to two cases, future research will expand to diverse industries, and validate capability measures through expert panels and surveys. This will enable more generalizable insights for advancing sustainability-driven BPM maturity.

References

1. Elkington, J., John, E.: Cannibals with forks: the triple bottom line of 21st century business. Capstone, Oxford, U.K. (1997)
2. Dumas, M., La Rosa, M., Mendling, J., Reijers, H.A.: Fundamentals of business process management. Springer, Berlin Heidelberg, Berlin, Heidelberg (2018)
3. Karwowski, M., Raulinajtys-Grzybek, M.: The application of corporate social responsibility (CSR) actions for mitigation of environmental, social, corporate governance (ESG) and reputational risk in integrated reports. Corp. Soc. Responsibil. Env. **28**, 1270–1284 (2021)
4. PwC: PwC Voice of the Consumer Survey. PwC (2024)
5. Smith, W.K., Lewis, M.W.: Toward a theory of paradox: a dynamic equilibrium model of organizing. Acad. Manage. Rev. (2011)
6. Zehendner, A.G., Sauer, P.C., Schöpflin, P., Kähkönen, A.-K., Seuring, S.: Paradoxical tensions in sustainable supply chain management: insights from the electronics multi-tier supply chain context. IJOPM **41**, 882–907 (2021)
7. Rosemann, M., vom Brocke, J.: The six core elements of business process management. In: Vom Brocke, J., Rosemann, M. (eds.) Handbook on Business Process Management 1, pp. 105–122. Springer, Berlin, Heidelberg (2015)
8. Sohns, T.M., Aysolmaz, B., Figge, L., Joshi, A.: Green business process management for business sustainability: a case study of manufacturing small and medium-sized enterprises (SMEs) from Germany. J. Clean. Prod. **401**, 136667 (2023)
9. Rosemann, M., Ostern, N., Voss, M., Bandara, W.: Benevolent business processes-design guidelines beyond transactional value. Presented at the International Conference on Business Process Management (2023)
10. Yin, R.K.: Case study research and applications: design and methods. SAGE, Los Angeles London New Delhi Singapore Washington DC Melbourne (2018)

Operationalizing Responsible BPM: A Method for Value-Based Process Redesign

Djordje Djurica[1(✉)], Sandro Franzoi[2], and Sarah Spiekermann-Hoff[1]

[1] Wirtschaftsuniversität Wien, Vienna, Austria
{djordje.djuricahoff,sarah.spiekermann-hoff}@wu.ac.at
[2] University of Münster, Münster, Germany
sandro.franzoi@uni-muenster.de

Abstract. Traditional BPM approaches prioritize operational performance, often overlooking the ethical and societal consequences of how processes are designed and executed. While Responsible BPM (RBPM) calls for the integration of ethical, social, and environmental values into process design, it lacks concrete methods to operationalize these goals. This paper introduces a structured, four-phase method for value-based process redesign, grounded in the principles of Value-Based Engineering and adapted to the BPM context. We position our approach within the BPM redesign landscape and demonstrate its applicability through a fictional case of a hospital's patient discharge process. By bridging normative aspirations with practical modeling guidance, the method contributes a missing methodological foundation to the RBPM agenda.

Keywords: Responsible BPM · Business Process Redesign · Value-Based Engineering · Ethical Design · Stakeholder Values

1 Introduction

Business Process Management (BPM) has long served as a managerial approach for optimizing organizational performance. By structuring work into analyzable and repeatable processes, BPM has enabled organizations to reduce costs, increase efficiency, and improve service quality. These benefits, measured through indicators like cycle time and resource utilization, reflect a techno-economic logic shaping BPM's development [7].

Yet, the very features that make BPM effective (e.g., standardization, automation, and metric-driven control), can also obscure the ethical and societal dimensions of business processes. As organizations increasingly integrate algorithmic systems, artificial intelligence (AI)-based decision-making, and data-intensive automation into core processes, BPM's design decisions begin to influence not just efficiency, but many other values, such as privacy, user control, or fairness [24]. Therefore, the ethical value consequences of business processes are

M. ER et al. (Eds.): BPM 2025, LNBIP 565, pp. 72–86, 2026.
https://doi.org/10.1007/978-3-032-02936-2_7

no longer incidental. Instead, they are built into the way work is structured, how decisions are taken, and how results or consequences are shared across stakeholders.

In this context, scholars have begun to call for a rethinking of information system design as a more value-sensitive activity [9,24]. Research in responsible information systems and responsible AI highlights the risks of embedding opaque or biased decision rules into digital infrastructures and stresses the importance of designing technologies with ethical foresight [27]. Within BPM, similar concerns are gaining attention under the label of Responsible Business Process Management (RBPM). RBPM challenges the assumption that business processes are value-neutral and calls for an explicit integration of ethical, social, and environmental considerations into the design and governance of organizational processes [13,29].

However, while RBPM provides conceptual framing, it remains underdeveloped at the methodological level. Specifically, a recent literature review revealed a gap: BPM lacks structured tools to elicit stakeholder values, resolve ethical trade-offs, or embed normative goals into process models [13]. Unlike traditional design criteria such as cost or speed, values like fairness, dignity, or inclusion are difficult to formalize and often absent from standard BPM toolkits. As a result, organizations seeking to operationalize RBPM principles face a lack of procedural guidance on how to do so in practice.

To address this gap, this paper draws on Value-Based Engineering (VBE) to propose a method for responsible, value-based process redesign. VBE originates in the systems and software engineering domain, where it has been developed to integrate stakeholder values into the design of IT systems [24]. Grounded in ethical theory and formalized in the ISO/IEC/IEEE 24748-7000 standard [1], VBE provides a structured process for eliciting values, prioritizing them, and then translating them into system requirements leading to design artifacts. We argue that adopting the methodological approach of VBE in BPM can translate RBPM's goals into concrete, repeatable design practices during the process redesign phase. Our method consists of four steps: (1) stakeholder value elicitation and clustering, (2) value prioritization and conceptualization, (3) ethical value requirements identification and value-aligned process design, and (4) validation and impact assessment. To demonstrate the method's relevance and application, we illustrate its use in a fictional redesign scenario of a hospital's patient discharge process, guided by the values of patient well-being, employee empowerment, and sustainability. Our work contributes to BPM research and practice by providing a structured and ethically grounded method that enables the systematic integration of stakeholder values into process redesign. It offers a novel perspective on how normative considerations can be operationalized alongside traditional performance goals and advances the methodological foundations for RBPM.

The remainder of this paper is structured as follows. Section 2 builds the theoretical foundation for our method by connecting value-based engineering with responsible business process redesign. It introduces the ethical grounding of design-relevant values, positions our work within the RBPM discourse, and reviews existing redesign methods and value-related approaches. Section 3 presents our four-phase method. Section 4 applies it to a fictionalized case of a hospital's patient discharge process. Section 5 discusses implications and limitations, before Sect. 6 concludes the paper.

2 Theoretical Background

2.1 From Ethical Foundations to Design-Relevant Values

Moral philosophy and material value ethics are powerful theories able to provide the foundation for reasoning about what is right or wrong, just, and appropriate in human action. Unlike technical reasoning, ethical inquiry elicits normative considerations and guides decisions based on principles and context [2]. As Dewey [8] argued, ethics is a reflective practice through which communities determine how to act. Singer [23] reinforces this view, noting that ethics is not merely a matter of personal opinion or cultural convention, but a reasoned investigation into how we should act and why.

Theoretical traditions from moral philosophy are adapted through applied ethics, which enables normative reasoning in fields like medicine, law, and business. In business, applied ethics shapes governance and decision-making, reinforcing corporate responsibility and stakeholder trust [5,21].

Within these applied frameworks, values play a critical role as normative orientations that guide behavior and justify choices. Values such as fairness, privacy, user control, dignity, and transparency are often shared within professions, cultures, or institutions and function as benchmarks for evaluating action. Philosophers like Max Scheler [12,19] have proposed a moral hierarchy of values, treating values not as subjective preferences but as objective, felt *principles of the ought-to-be* that influence how individuals and societies evaluate and prioritize actions.

Spiekermann [24] extends this idea by emphasizing that values are distinct from norms or needs. Norms reflect social conventions, and needs may be unspoken or subjective, but values offer an overarching ethical grammar that can orient both individual and collective reasoning. Material value ethics, as a school of thought that treats values as real and experiential, rather than socially constructed, posits that values can be positive (e.g., privacy) or negative (e.g., violation of privacy), and that they exert emotional force since we experience them as moral attraction or repulsion [12]. Importantly, values are not fully contingent on individual recognition. A CEO might personally dismiss the importance of privacy, but this does not erase its normative significance or the moral obligations it imposes.

This understanding of values has led to their increasing importance in technology and design fields. In BPM, this has sparked calls for frameworks that advocate for the intentional design of processes to promote societal benefit and ethical responsibility, such as Benevolent BPM [18]. In response to such normative ambitions, VBE has emerged as a structured approach for identifying, prioritizing, and embedding values into socio-technical system design. We explore it in more detail in Sect. 2.5.

Despite these advances, integrating values into BPM remains a challenge. As Kern et al. [13] note, although ethical values are increasingly discussed in BPM, they often remain conceptually vague and disconnected from formal modeling practices. Next, we show how RBPM reframes the goals of process design and creates a space where Value-Based Engineering can be meaningfully applied.

2.2 From BPM to Responsible BPM

Since its inception, BPM has emphasized efficiency, standardization, and control, an orientation that has delivered substantial gains in productivity and operational excellence [7]. However, this optimization logic has been critiqued for neglecting the broader societal and human implications of process design [29]. As organizations increasingly rely on algorithmic automation, AI-supported decision-making, and data-centric operations, the limitations of this narrow focus become more consequential. Business processes do not merely execute work; they shape experiences, distribute access and risk, and reinforce or disrupt power relations [8]. To understand these dynamics, it is useful to view business processes as value bearers within socio-technical systems. Rather than neutral pipelines, processes embed and enact values (implicitly or explicitly) through decisions, rules, and interactions. This perspective aligns with socio-technical systems theory, which emphasizes the co-constitution of technical and social subsystems [28], as well as with Simon's [22] notion of design as *changing existing situations into preferred ones*. This reconceptualization is central to the idea of RBPM. RBPM expands the goals of process design to include ethical, social, and environmental values [13,20]. It challenges the assumption that traditional performance metrics alone are sufficient, prioritizing principles such as employee well-being [14] and ecological sustainability [15]. Viewed through this lens, a loan approval process optimized for speed and consistency might obscure algorithmic bias, produce discriminatory outcomes, or exclude mechanisms for appeal. An RBPM approach would instead ask: who benefits, who is left out, and how can the process be redesigned to balance efficiency with fairness while ensuring that decisions are based on transparent and interpretable criteria.

Despite its conceptual promise, RBPM remains difficult to operationalize. Ethical values are often qualitative, context-dependent, and hard to represent in modeling languages or optimization frameworks [13]. Subfields such as Green BPM and Sustainable BPM have begun to address specific concerns (e.g., emissions, resource use, and social accountability) [4,29], but they often treat values as external constraints rather core elements of the process design rationale.

What remains missing are structured ways to identify, prioritize, and resolve value priorities throughout the BPM lifecycle. Although the literature increasingly acknowledges the importance of values, methodologies for embedding them into core practices, such as modeling, redesign, and monitoring, are still scarce [13]. Bridging this gap requires methods that not only recognize values but integrate them systematically into the logic of process design.

2.3 Business Process Redesign Methods

Process redesign plays a key role in BPM and involves analyzing and reconfiguring processes that operate within or across organizational boundaries [6,7]. Its primary aim is to change how work is performed and how value is generated [7]. A wide range of redesign techniques has been developed and adapted over time [7,11]. These techniques differ in their nature (analytical vs. creative), perspective (internal vs. external), and ambition (incremental vs. transformational) [7]. Prior research shows that most approaches remain analytical and incremental [11]. Widely used methods include Lean, which eliminates non-value-adding activities, and Six Sigma, which reduces process variation and defects. These approaches are valued for their structured procedures and measurable performance outcomes [7]. Creative methods such as Design Thinking, brainstorming, or customer journey mapping are employed earlier in the redesign process to encourage innovation and stakeholder engagement [7].

Other structured techniques include heuristic-based redesign [17] and product-based design approaches [16], which use principles like task elimination, parallelism, and integration to improve process efficiency. However, despite these methods being effective for optimizing conventional performance metrics such as cost, time, and quality, they fall short in addressing the normative concerns raised by RBPM.

2.4 Value-Related Design Methods

While the BPM community has created a number of methods for process redesign over the years, these techniques rarely engage with the ethical dimensions of design or offer formal mechanisms for eliciting and prioritizing stakeholder values. To explore how such concerns have been addressed elsewhere, we turn to value-related design methods developed in other domains, most notably human-computer interaction and software engineering, where ethical considerations have been more explicitly integrated into design practices.

Among the well-established of these approaches is Value-Sensitive Design (VSD), a framework developed in the field of human-computer interaction to account for human values in technology design [10]. VSD emphasizes stakeholder engagement, conceptual analysis, and empirical investigation as complementary methods for identifying and addressing values. However, while VSD offers many case studies on how to use values in system design and methodological flexibility, it remains loosely structured and lacks formal integration with engineering or organizational design lifecycles, making it difficult to apply in complex socio-technical settings like BPM.

A second related approach is Value-based Software Engineering (VBSE), which introduces stakeholder prioritization and value trade-off analysis into the software development process [3]. VBSE is primarily concerned with aligning design decisions with business value, aiming to maximize stakeholder satisfaction under conditions of limited resources. Although it brings quantitative rigor and practical relevance, its interpretation of value is largely utility-driven, leaving little room for ethical, social, or sustainability considerations.

A third perspective is offered by the framework of Responsible Innovation, which calls for anticipatory, inclusive, and reflexive design processes to ensure that emerging technologies align with societal needs and ethical principles [26]. Responsible Innovation emphasizes stakeholder involvement and long-term thinking but often remains abstract and lacks concrete tools for integrating values into processes.

2.5 Value-Based Engineering

Given the limitations of both BPM redesign methods and existing value-related design approaches, we adopt VBE in this paper, as it offers a systematic methodology for embedding values into the design of socio-technical systems and combines academic theorization with practical insights grounded in ISO/IEC/IEEE 24748-7000 [1,24].

Unlike conventional approaches that begin with predefined goals or technical features, VBE starts with a stakeholder-centric inquiry into what ought to be achieved in the system-of-interest and the wider system-of-system that serves it. It is grounded in a value ontology informed by philosophical ethics, especially Scheler's material value ethics [19], and draws on the normative ethical theories of deontology, virtue ethics, and utilitarianism to guide stakeholder engagement and the formulation of core values and value qualities that then determine system requirements.

A key distinction of VBE is its broader scope: it addresses not only a system's technical performance but also the organizational and cultural conditions that support human and organizational flourishing. It advocates a shift toward more reflective design practices that prioritize stakeholder concerns and long-term social outcomes over short-term delivery and efficiency [25].

VBE also emphasizes traceability, the systematic documentation of how values shape design decisions, and how value trade-offs are resolved. This supports transparency, auditability, and ethical accountability across the system lifecycle. Rather than prescribing universal principles, VBE provides a methodological means for identifying societal values and translating them into actionable design elements in the respective system context. This makes it particularly suitable for BPM, where diverse stakeholders, iterative modeling, and societal impact converge.

3 A Method for Value-Based Process Redesign

This section introduces a structured method for value-based process redesign, grounded in the logic of VBE. Furthermore, we tailor VBE's logic of ethical system requirement engineering to the domain of BPM. The method is designed for BPM projects in which ethical concerns might be salient and where value tensions may arise between stakeholders or between values and performance goals. It can be applied by an interdisciplinary project team that includes process analysts, domain experts, and trained facilitators. Facilitators should be skilled in participatory sessions, value-laden discussions, and ethical reasoning.

While our method builds on Value-Based Engineering, it also complements existing BPM redesign and value-oriented design approaches. Compared to traditional redesign methods such as Six Sigma or heuristic-based techniques, it explicitly integrates ethical value elicitation and traceability into process modeling. Unlike Value-Sensitive Design and Responsible Innovation, which focus on product-level or strategic innovation processes and often remain conceptual, our method operates at a lower level of abstraction. It embeds ethical deliberation directly into BPM lifecycle activities, provides structured guidance for process modeling, and emphasizes traceability through the ethical value register. In this way, it bridges the gap between normative reflection and operational process redesign.

The method consists of four phases that support analysts in eliciting, prioritizing, translating, and validating values. The phases and their outcomes are described below and depicted in Fig. 1.

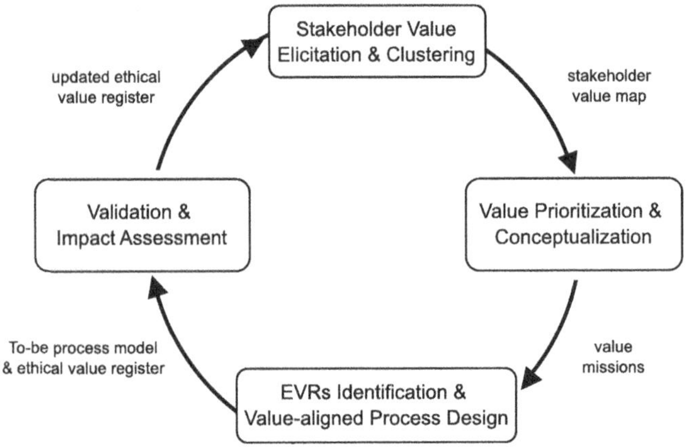

Fig. 1. Method for Value-Based Process Redesign in BPM

3.1 Stakeholder Value Elicitation and Clustering

In BPM, process redesign often begins with understanding functional inefficiencies of the observed business process. Our method extends this early analysis by explicitly eliciting the values that stakeholders associate with the process that is undergoing the redesign initiative. While most BPM methods observe the process from the point of view of the organization or direct customers, our value-based method is not limited to these stakeholders, but can also include indirectly affected stakeholders, oversight bodies, and broader societal actors that may be impacted in the longer term if the system was used at scale. Here, a careful selection of stakeholders, and, where appropriate, weighting their influence, is essential to ensure a balanced and contextually grounded analysis. The object of study is the concrete system-of-interest as well as its system-of-systems: the process itself and its interfaces with external service providers, where many ethical trade-offs may emerge. Using methods such as ethically guided interviews, persona walkthroughs, and participatory workshops, we systematically surface concerns that may otherwise be marginalized in conventional process redesign. To ensure ethical coverage, elicitation is based on three complementary normative theories utilitarianism, deontology, and virtue ethics that provide different lenses through which value questions (e.g., about fairness, autonomy, or responsibility) can be framed [2]. Once collected, values are clustered into themes to reduce redundancy and highlight core values. The output of this phase is a *stakeholder value map* of core values and their qualities, which complements the traditional as-is model with insights about ethical expectations and concerns.

3.2 Value Prioritization and Conceptualization

Traditional BPM methods often prioritize goals like cycle time or resource efficiency. In contrast, the second phase of our method centers on carefully prioritizing core values that will guide the to-be process model design. Facilitating sessions help surface not only concrete ethical value prioritization alternatives (e.g., such as privacy versus transparency or inclusiveness versus speed), but also trade-offs between these values and conventional performance criteria. These deliberations explicitly consider where ethical imperatives should take precedence or where compromises are necessary, depending on the context. To support this, participatory techniques such as deliberative workshops and informal ranking exercises can be employed to guide stakeholders through value trade-offs. This results in the creation of a *value mission* for the redesigned process, similar to a design objective in BPM projects, but oriented around values. We then apply VBE's value ontology in a structured manner, organizing prioritized values into three layers: (1) *core values*, (2) *value qualities*, and (3) *concrete process features* that drive these qualities. For example, the core value of fairness (1) can be operationalized through bias mitigation (2), which is enabled by concrete process features such as excluding sensitive attributes during data preprocessing (3). This structured conceptualization of each value "goal" ensures coherence and traceability when values are later embedded into the redesigned business process.

3.3 Ethical Value Requirements Identification and Value-Aligned Process Design

In conventional BPM, design requirements are typically derived from performance goals, compliance rules, or operational bottlenecks. Our method takes the concept of *Ethical Value Requirements (EVRs)* from the VBE standard as complementary requirements to the traditional design criteria [1]. These requirements capture what a redesigned process should ideally achieve from a stakeholder perspective to respect values such as fairness, autonomy, and transparency.

EVRs are translated into both technical elements (e.g., user interface constraints, algorithmic decision logic, data minimization protocols) and organizational elements (e.g., escalation paths, human-in-the-loop checkpoints, communication channels). To ensure traceability, all EVRs are recorded in an *ethical value register*, which enables traceability by linking each requirement to its underlying value and associated stakeholder concern. The register complements process modeling artifacts by serving as a normative design log used for documenting why certain design choices were made, what values they serve, and what value prioritization alternatives were accepted. Once identified, EVRs are mapped onto concrete process model elements using BPM notations such as BPMN. This includes (1) assigning EVRs to specific activities or subprocesses, (2) embedding decision logic that reflects value-driven rules (e.g., fairness checks, opt-out branches), (3) annotating roles or resources with ethical constraints (e.g., oversight responsibilities), and (4) integrating compensatory mechanisms for value-sensitive edge cases (e.g., appeal workflows).

To support this modeling work, process analysts may use dedicated templates or overlays that make EVRs explicit within the visual representation of the process. Where EVRs conflict with functional objectives (e.g., speed vs. inclusiveness), value priorities are revisited in consultation with stakeholders to find acceptable compromises or safeguards.

3.4 Validation and Impact Assessment

While BPM commonly includes evaluation through performance indicators and process monitoring, our method introduces value-centered validation as an additional step. This involves assessing whether the to-be process design meets its value objectives and how it is perceived by stakeholders. To this end, stakeholder review workshops, value walkthroughs, and surveys can be used to collect feedback on perceived fairness, transparency, and inclusion. In addition to gathering feedback, this phase involves identifying unintended side effects that may have emerged during redesign. These are analyzed in relation to the EVRs, ensuring that ethical concerns are not only addressed in principle but also realized in practice. To support accountability and continuous improvement, traceability is maintained through the *ethical value register*, linking each design decision back to its originating value. This ensures that value-based decisions can be reviewed and justified over time, supporting ethical accountability and guiding future redesigns throughout the BPM lifecycle.

4 Demonstration: Value-Based Redesign of a Hospital's Patient Discharge Process

To demonstrate the application and benefits of our proposed value-based process redesign method, this section presents a fictional redesign scenario of a hospital's patient discharge process. This process covers the steps taken to prepare a patient to leave the hospital and continue their recovery at home or in another care setting. Traditionally, the focus has been on making this process as fast and efficient as possible to free up hospital beds. However, this scenario demonstrates how process redesign can be grounded in explicitly prioritized values such as patient well-being, employee empowerment, and sustainability, shifting the focus toward a more ethically aware and socially responsible form of BPM.

Phase 1: Stakeholder Value Elicitation and Clustering. Stakeholders in the hospital's patient discharge process include not only direct actors, such as patients, nurses, physicians, and case managers, but also administrative staff and organizational stakeholders concerned with compliance and resource efficiency. The engagement process can be guided by a facilitator in a role analogous to the *Value Lead* in Value-Based Engineering, and it starts by reviewing how the process flows right now, which stakeholders then discuss and reflect on. For example, patients may express concerns about emotional readiness, clarity of post-discharge instructions, or feelings of vulnerability. Clinical staff might point to stress caused by fragmented communication and unclear responsibilities, while administrators could emphasize concerns around re-admission rates and ecological sustainability. In this phase, stakeholder dialogue is guided by the three ethical reasoning frameworks of utilitarianism, virtue ethics, and deontology all of which provide a spotlight to unique challenges associated with the process [2]. Based on this analysis, the *stakeholder value map* highlights core value themes such as patient well-being (reflecting care, empathy, and safety), employee empowerment (focused on shared decision-making and professional agency), and sustainability (centered on resource-conscious operations and environmental responsibility).

Phase 2: Value Prioritization and Conceptualization. The second phase helps prioritizing which values should guide the redesign of the hospital's patient discharge process. Stakeholders are invited to reflect on trade-offs, such as throughput time versus patient well-being or speed of decisions versus employee empowerment. In this scenario, the process leads to the prioritization of three core value clusters: patient well-being, employee empowerment, and sustainability. These are further structured using VBE's value ontology, linking core values to value qualities (e.g., emotional safety, shared responsibility, responsible resource use). The result is a *value mission* that provides normative direction for the subsequent process redesign.

Phase 3: EVR Identification and Value-Aligned Process Design. In the third phase, the prioritized value clusters are translated into EVRs that specify what the redesigned discharge process must achieve to reflect the selected values. Each EVR is linked to the underlying value through traceable justifications, aligning with VBE's value ontology (core values → value qualities → system dispositions) and stored in the *ethical value register*. In the hospital scenario, three exemplary EVRs with their corresponding value qualities and core values are illustrated in Table 1.

Table 1. Example Mapping from Core Values to Ethical Value Requirements in Value-Based Process Redesign

Core Value	Materializing in/ Harmed by	value Quality	Ethical Value Requirement
Patient well-being	materializing in	Emotional safety	Include a patient-centered discharge briefing that assesses emotional readiness
Employee empowerment	materializing in	Shared responsibility	Introduce daily team-based discharge planning sessions allowing nurses, doctors, and case managers to collaborate with shared responsibility
Sustainability	materializing in	Responsible resource use	Replace printed discharge materials with digital packets integrated with outpatient care systems

Each core value can have multiple value qualities, which again can have multiple EVRs. These EVRs can then be operationalized using BPM modeling practices. For example, EVR 1 can be assigned to a dedicated activity that is added in the BPMN model, which includes an emotional readiness check before discharge. EVR 2 may be reflected in role annotations that enable equal participation in a redesigned planning activity, with decision logic distributed among stakeholder roles. EVR 3 can be modeled by integrating automated document generation steps and digital handover points with external care systems. To illustrate this concretely, we provide both the as-is and to-be BPMN models in the online appendix.[1] The to-be model incorporates the EVRs as follows: (1) the emotional readiness check is modeled as a new decision gateway and activity, (2) a team-based planning session is added as a collaborative task, and (3) digital discharge paperwork replaces a physical document activity. These additions demonstrate how ethical values can be embedded into process logic in a traceable and operational form.

Phase 4: Validation and Impact Analysis. The final phase focuses on evaluating whether the redesigned discharge process fulfills its ethical value requirements and is perceived positively by stakeholders. In the hospital scenario, patient well-being can be assessed via post-discharge surveys on emotional readiness and clarity of instructions. Staff engagement may be measured through pulse surveys on collaboration and autonomy. Sustainability impacts could be evaluated by tracking reductions in paper use and the adoption of digital discharge materials. Feedback loops help surface any unintended effects, which can inform iterative refinements. All validation activities are documented in the *ethical value register*, maintaining traceability and supporting ongoing accountability for value-based decisions throughout the process lifecycle.

This demonstration shows how value considerations can be systematically elicited, prioritized, implemented, and validated. By incorporating VBE's approach and traceability, the method supports processes aligned with societal and ethical commitments.

[1] https://tinyurl.com/293bcvts.

5 Discussion

5.1 Implications for Research and Practice

Our proposed method offers several important implications for BPM research: First and foremost, our work contributes to the ongoing development of RBPM by translating its normative aspirations into a structured and actionable method. While prior literature has emphasized the importance of integrating ethical and societal values into BPM [13,18,29], few efforts have addressed how to do so systematically during the redesign of processes. By drawing on the logic of VBE [24], our method provides a systematic and repeatable approach for eliciting stakeholder values, prioritizing them, and embedding them as design requirements for process redesign. This expands the scope of BPM research by proposing ethical value requirements as a novel category of design inputs, alongside traditional performance metrics such as cost and efficiency. Moreover, our method invites empirical work to examine how different values (e.g., fairness, sustainability) can be operationalized and how they interact with contextual constraints. In this regard, our work aligns with and advances recent BPM research that foregrounds the importance of embedding normative goals, such as employee well-being [14], ecological sustainability [15], or fairness and inclusion [13] into BPM practices. By offering systematic guidance, our method provides the infrastructure to enact these ambitions in practice. Finally, the method invites further conceptual development on the nature of value in BPM. While BPM has traditionally emphasized operational value in terms of performance metrics, our work highlights the need for a more holistic understanding of value that includes ethical, experiential, and long-term societal dimensions. This opens new directions for BPM research to explore how different types of value can be represented, balanced, and modeled in the process lifecycle. This work also lays the groundwork for more ethically-aware, stakeholder-centric approaches to process redesign.

For practitioners, the method provides a structured yet adaptable approach to ensure that business process redesign initiatives are not only efficient but also based on a holistic value evaluation. It guides organizations through a transparent process of stakeholder engagement and value elicitation, helping surface concerns that might otherwise remain unaddressed. This can be particularly helpful in contexts involving automation, AI-supported decision-making, or sensitive human interactions. Furthermore, By documenting values transparently, the method enhances accountability and supports compliance with ethical and regulatory standards. Additionally, the method encourages dialogue between different organizational actors (e.g., IT, operations, compliance, human resources), supporting inclusive process development, especially in complex environments with broad social or ethical implications and multiple stakeholder groups.

Taken together, these implications highlight the transformative potential of a value-based approach to business process redesign. By placing ethical, social, and experiential values at the core of redesign efforts, our method not only enriches existing BPM methodologies but also equips researchers and practitioners with the tools to address pressing societal challenges through process innovation. As such, it marks a step toward a more reflective, inclusive, accountable, and responsible BPM discipline.

5.2 Limitations and Future Research Directions

While the method structured approach for embedding ethical values into process redesign, this method also has several limitations that should be acknowledged.

First, the method assumes a willingness and capacity within organizations to engage in normative reflection and stakeholder dialogue. In practice, this requires a cultural shift, away from performance-driven logic and toward more value-conscious forms of process management. Such a shift may be difficult to realize in environments dominated by short-term goals, tight resource constraints, or hierarchical decision-making structures. Therefore, the transformational ambition of this method, while powerful, poses challenges for adoption at scale.

Second, the success of the method is dependent on the skill of facilitators and the specific organizational context. Eliciting, prioritizing, and translating values requires nuanced facilitation, ethical sensitivity, and the ability to navigate power dynamics among stakeholders. This reliance on expert facilitation may limit the method's replicability in settings lacking ethical design expertise or stakeholder alignment.

Third, balancing competing values (e.g., privacy versus accountability) remains an open challenge. The method does not offer definitive resolution strategies, as these will often be context-specific and politically charged. Future work could explore more formal mechanisms for value prioritization analysis and stakeholder mediation.

Fourth, while our method is grounded in established theoretical frameworks and structured around repeatable steps, it requires further empirical validation. Future research could include experiments to evaluate its impact on process quality and stakeholder satisfaction. Such studies are needed to refine the phases, assess practicality, and test scalability. Finally, integrating EVRs into existing BPM tools and modeling standards (e.g., BPMN) poses practical challenges. Further work is needed to develop notation, templates, or plugins that support seamless translation of values into process models.

6 Conclusion

As business processes increasingly intersect with sensitive human, societal, and environmental concerns, there is growing need for the field to move beyond traditional efficiency-driven redesign. While RBPM articulates this expanded vision, it remains underdeveloped in terms of concrete, actionable methods. In response to the lack of concrete methods for operationalizing RBPM, we proposed a structured approach that adapts Value-Based Engineering to process redesign. By combining ethical value elicitation, prioritization, and traceable implementation, the method helps embed normative principles into BPM practice. Future work can explore its application across contexts and its impact on process outcomes.

References

1. ISO/IEC/IEEE 24748-7000:2022: Standard Model Process for Addressing Ethical Concerns During System Design. International Standard ISO/IEC/IEEE 24748-7000:2022, ISO/IEC/IEEE JTC 1/SC 7 Software and Systems Engineering, Piscataway (2022), systems and Software Engineering – Life Cycle Management – Part 7000
2. Bednar, K., Spiekermann, S.: Eliciting values for technology design with moral philosophy: an empirical exploration of effects and shortcomings. Sci. Technol. Hum. Values **49**(3), 611–645 (2024)
3. Boehm, B.: Value-based software engineering: reinventing. ACM SIGSOFT Softw. Eng. Notes **28**(2), 3 (2003)
4. Couckuyt, D., Van Looy, A.: A systematic review of green business process management. Bus. Process. Manag. J. **26**(2), 421–446 (2020)
5. Crane, A., Matten, D., Glozer, S., Spence, L.J.: Business Ethics: Managing Corporate Citizenship and Sustainability in the Age of Globalization. Oxford University Press (2019)
6. Davenport, T.H., Short, J.E.: The New Industrial Engineering: Information Technology and Business Process Redesign (1990)
7. Dumas, M., La Rosa, M., Mendling, J., Reijers, H.A., et al.: Fundamentals of Business Process Management, vol. 1. Springer, Heidelberg (2018). https://doi.org/10.1007/978-3-662-56509-4
8. Eubanks, V.: Automating Inequality: How High-Tech Tools Profile, Police, and Punish the Poor. St. Martin's Press (2018)
9. Friedman, B., Hendry, D.G.: Value Sensitive Design: Shaping Technology with Moral Imagination. MIT Press (2019)
10. Friedman, B., Kahn, P., Borning, A.: Value Sensitive Design: Theory and Methods. University of Washington Technical report, 02–12 (2015)
11. Gross, S., Malinova, M., Mendling, J.: Navigating through the maze of business process change methods (2019)
12. Kelly, E.: Material Ethics of Value: Max Scheler and Nicolai Hartmann, vol. 203. Springer, Dordrecht (2011). https://doi.org/10.1007/978-94-007-1845-6
13. Kern, C.J., Poss, L., Kroenung, J., Schönig, S.: Navigating the moral maze: a literature review of ethical values in business process management. Bus. Process. Manag. J. **30**(8), 343–370 (2024)
14. Klessascheck, F., Stiehle, F.: Business value and worker wellbeing? Foregrounding workers in process management. In: 15th International Workshop on Enterprise Modelling and Information Systems Architectures (EMISA) 2025. Gesellschaft für Informatik, Bonn (2025)
15. Klessascheck, F., Weber, I., Pufahl, L.: SOPA: a framework for sustainability-oriented process analysis and re-design in business process management. Inf. Syst. e-Bus. Manag. (2025)
16. Reijers, H.A., Limam, S., Van Der Aalst, W.M.: Product-based workflow design. J. Manag. Inf. Syst. **20**(1), 229–262 (2003)
17. Reijers, H.A., Mansar, S.L.: Best practices in business process redesign: an overview and qualitative evaluation of successful redesign heuristics. Omega **33**(4), 283–306 (2005)
18. Rosemann, M., Ostern, N., Voss, M., Bandara, W.: Benevolent business processes-design guidelines beyond transactional value. In: Di Francescomarino, C., Burattin, A., Janiesch, C., Sadiq, S. (eds.) BPM 2023. LNCS, vol. 14159, pp. 447–464. Springer, Cham (2023). https://doi.org/10.1007/978-3-031-41620-0_26

19. Scheler, M.: Formalism in Ethics and Non-Formal Ethics of Values: A New Attempt Toward the Foundation of an Ethical Personalism. Northwestern University Press (1973)
20. Schmiedel, T., vom Brocke, J., Recker, J.: Culture in business process management: how cultural values determine BPM success. In: vom Brocke, J., Rosemann, M. (eds.) Handbook on Business Process Management 2. IHIS, pp. 649–663. Springer, Heidelberg (2015). https://doi.org/10.1007/978-3-642-45103-4_27
21. Schwartz, M.S.: Business Ethics: An Ethical Decision-Making Approach. Wiley (2017)
22. Simon, H.A.: The Sciences of the Artificial, Reissue of the Third Edition with a New Introduction by John Laird. MIT Press (2019)
23. Singer, P.: Practical Ethics. Cambridge University Press (2011)
24. Spiekermann, S.: Value-based engineering: a guide to building ethical technology for humanity. In: Value-Based Engineering. De Gruyter (2023)
25. Spiekermann, S., Winkler, T.: Value-based engineering for ethics by design. arXiv preprint arXiv:2004.13676 (2020)
26. Stilgoe, J., Owen, R., Macnaghten, P.: Developing a framework for responsible innovation. In: The Ethics of Nanotechnology, Geoengineering, and Clean Energy, pp. 347–359. Routledge (2020)
27. Trier, M., et al.: Digital responsibility: a multilevel framework for responsible digitalization. Bus. Inf. Syst. Eng. **65**(4), 463–474 (2023)
28. Trist, E.L.: Towards A Social Ecology: Contextual Appreciation of the Future in the Present. Plenum Press, London; New York (1973)
29. Vom Brocke, J., Seidel, S., Recker, J.: Green Business Process Management: Towards the Sustainable Enterprise. Springer, Heidelberg (2012). https://doi.org/10.1007/978-3-642-27488-6

Designing Business Processes for Good: A Canvas-Based Framework for Routinized Benevolence

Thamali Chandrasiri[1](✉), Wasana Bandara[1], Michael Rosemann[1], Nadine Ostern[1], and Marleen Voss[2]

[1] Queensland University of Technology, Brisbane, QLD 4000, Australia
thamali.don@hdr.qut.edu.au
[2] University of Münster, 48149 Münster, Germany

Abstract. As organizations confront increasing expectations for responsible practices, embedding benevolence into business operations has become both an ethical intent and a pathway to sustained success. *Routinized Benevolence* (RB) refers to voluntary actions embedded into routine operations of for-profit organizations to enhance customer well-being while pursuing mutualistic gains. Although evidence of RB is observable in practice and closely aligns with the principles of Responsible BPM, its conceptual underpinnings and operational mechanisms remain underdeveloped. Addressing this gap, we introduce the RB Canvas, a structured design artefact that captures key elements of RB, including its drivers, action patterns, characteristics, impacts, and risks. Developed using a Design Science Research approach and grounded in empirical case evidence, the canvas serves as both a conceptual framework and a practical tool to support implementation. This study contributes to the emerging agenda of Responsible BPM by offering a systematic approach to understanding and enabling benevolence within business processes.

Keywords: Canvas · Routinized Benevolence · Responsible Process Design · Mutualistic Value Creation

1 Introduction

In recent years, the long-standing paradigm of shareholder primacy has come under increasing scrutiny. A growing chorus of scholars, practitioners, and policymakers now advocate for a more inclusive and balanced approach, one that accounts for both economic performance and broader societal well-being [1]. This evolving perspective has brought renewed attention to Business Process Management (BPM) and the rising debates about Responsible BPM (RBPM), with calls for embedding ethical, sustainable, and stakeholder-oriented values into the very fabric of business processes. By strategically embedding responsible practices into their core operations, organizations can pursue long-term success without compromising profitability [2]. However, in practice, responsible initiatives often remain peripheral, treated as standalone efforts rather than being systematically integrated into routine business operations [3]. This isolation limits their scalability, consistency, and overall impact.

M. ER et al. (Eds.): BPM 2025, LNBIP 565, pp. 87–101, 2026.
https://doi.org/10.1007/978-3-032-02936-2_8

Evolving approaches to responsible enterprise, exemplified by concepts such as shared value [4] advocate for a rethinking of organizational priorities. These approaches urge businesses to go beyond compliance or philanthropic add-ons and embed pro-social practices, such as benevolence, directly into the core of business processes. Benevolence, in this context, reflects the principle of "doing good." While benevolence has been well explored within psychology and not-for-profit domains, its application in for-profit settings remains underdeveloped and conceptually immature. In corporate environments, benevolence within routine operations is still seen as a foreign concept. Only sporadic and isolated instances of benevolent practices are evident, lacking systematic integration or conceptual clarity. One possible reason for this limited uptake is the enduring tension between benevolence and profitability. Organizations often grapple with the paradox of investing in long-term, benevolent practices versus the pressure to deliver short-term financial results [4].

Despite the expanding discourse on shared value and responsible business practices, traditional business processes continue to prioritize immediate economic efficiency. They often do so without considering the longer-term benefits of embedding stakeholder-centric values such as benevolence. This narrow focus risks rendering business practices obsolete amid growing societal demands for transparency, fairness, and meaningful stakeholder engagement [3]. The current body of knowledge offers limited actionable guidance on how to systematically embed benevolence into routine business processes, particularly those that interface directly with customers - who represent both the most immediate economic stakeholders and the primary targets of sustained relational engagement in for-profit settings.

Rosemann et al. [5] argue that organizations can enhance long-term success by embedding benevolence into their everyday business processes; a practise referred as "Routinized Benevolence" (RB). Real-world cases further demonstrate the practical relevance and feasibility of this concept. For instance, Tesla has embedded a customer-centric contingency practice: when natural disasters such as hurricanes or bushfires threaten regions, the company remotely and temporarily extends the battery capacity of its electric vehicles via over-the-air updates, allowing drivers to evacuate safely. Similarly, Amazon alerts customers when they attempt to repurchase books they have previously bought, helping to prevent accidental duplicate purchases and demonstrating a commitment to customers' financial well-being over short-term revenue.

Although compelling real-world examples demonstrate how benevolence can be embedded into everyday operations to align commercial practices with customer well-being, many organizations continue to operate at the level of aspirational intent. While benevolent values are often articulated in mission statements, CSR initiatives, and sustainability reports, few firms succeed in translating these into systematic, process-level routines. This gap aligns with a core purpose of responsible BPM [6] - the need to move beyond high-level declarations and toward the deliberate design of processes that embed ethical, sustainable, and stakeholder-oriented values.

A key barrier to advancing routinized benevolence is the lack of conceptual clarity, particularly around what it entails and how it can be operationalized in diverse for-profit settings. To address this gap, this paper poses the research question: *'How can routinized benevolence be conceptualized via a canvas?'*.

Using a Design Science Research approach, we develop a canvas-based framework for Routinized Benevolence that systematically identifies its core elements, including key characteristics, triggering conditions, operational design features, expected impacts, and associated risks. This canvas serves as both a conceptual lens and a practical design artefact that guides organizations in the integration of customer-focused benevolent practices into their routine operations. In doing so, this study contributes meaningfully to the Responsible BPM agenda by providing actionable design knowledge that enables organizations to "design business processes for good" embedding ethical, customer-centric values into their process architectures, while maintaining alignment with strategic and economic objectives.

2 Related Work

2.1 Benevolence in Business

Benevolence, in general, is defined as the act of preserving and improving the well-being of individuals with whom one has frequent personal interactions [7]. It is an ethical principle which revolves around a person's care, concern, and the act of "doing good" [8]. Researchers have introduced various forms of benevolence depending on different criteria. A key distinction is drawn between altruistic and mutualistic benevolence. Altruistic benevolence refers to the extent to which a trustee is perceived to act in the trustor's best interests, driven by genuine concern rather than self-interest. In contrast, mutual benevolence involves the trustor demonstrating authentic concern for the trustee's well-being and intentionally seeking outcomes that offer shared benefit [9]. This study adopts the mutualistic form of benevolence, particularly given the for-profit context the study is focused on.

Though benevolence was extensively studied in psychological and not-for-profit organizational contexts, only a few researchers have studied benevolence in the for-profit context. Beveridge and Höllerer [10] define organizational benevolence as "a company's commitment to prioritize the well-being of an external stakeholder group as a primary objective, where advancing the interests of the "other" is the ultimate aim of actions taken" [10], thus, presenting benevolence as an *ultimate* goal, achieved as an end in itself. On the contrary, a few other researchers have examined benevolence as an *intermediate* goal, that enables long-term outcomes. For instance, they have revealed that benevolence fosters customer loyalty [11] satisfaction [12], and stronger commitments [13]. In contrast to Beveridge and Höllerer's [10] view, and in line with these other scholars, who highlight the long-term benefits of benevolence [9–11], we argue that for-profit organizations can view benevolence; in particular routinized benevolence, not as a final objective but as a strategy that ultimately supports profit generation.

2.2 Routinized Benevolence

Based on existing literature on benevolence and insights derived from empirical cases, we define Routinized Benevolence (RB) as "***the voluntary acts embedded into routine business operations of for-profit organizations, which prioritize the well-being of customers - and undertaken as an intermediate goal, with the expectation of longer-term***

gains.". Benevolence becomes truly embedded when it is treated as an integral component of day-to-day operations, part of "business as usual" rather than as a peripheral or ad hoc initiative [5]. This entails translating benevolent intent into repeatable patterns of behaviour, underpinned by appropriate resources, enabling technologies, and clearly assigned responsibilities. Rosemann et al. [5] underscore the value of embedding benevolence directly into a firm's business processes and offer eight actionable design patterns to support the intentional development and institutionalization of benevolent practices.

Several key criteria underpin the definition of RB. First, the intention must explicitly be focused on **prioritizing customer well-being**, positioning the needs and interests of existing or potential customers. Second, RB is inherently **mutualistic**; it delivers immediate or long-term value to the customer while involving a cost to the organization, with the expectation of achieving long-term gains such as loyalty, trust, or reputational benefits. Third, RB requires **operational embedding** - benevolent actions must be integrated into the organization's routine business processes rather than existing as peripheral, ad hoc, or one-off initiatives. Fourth, the act should be **voluntary**, not necessarily driven by regulations or external reporting requirements.

While we acknowledge that conceptual overlaps may exist, RB differs from related concepts such as corporate social responsibility (CSR), shared value, corporate philanthropy, and compassion in several important ways. While CSR seeks social good and is driven by responsible objectives, it is often influenced by external regulations, reporting obligations, and is typically seen as a discrete initiative aimed at gaining visibility or prestige [14]. In contrast, RB is not shaped by regulations or external mandates, and is embedded into a firm's regular operations, not a discrete or one-off act. Additionally, RB specifically targets customers as primary and immediate economic stakeholders, whereas CSR initiatives typically involve a broader range of stakeholders, including communities and the environment. Shared value also seeks to address societal challenges while generating economic benefits for the firm [15]. This broader societal focus distinguishes shared value from RB, which is more specifically centred on enhancing customer well-being. Although customer experience (CX) design focuses on managing point of contacts for a successful interaction [16], RB goes further by prioritizing customer well-being over short-term gains, embedding moral intent into routines, and emphasizing long-term trust over transactional benefits. Finally, while corporate compassion and philanthropy are inherently altruistic and carried out without the expectation of return [17], RB, as examined in this study, aims to deliver value to customers while concurrently advancing the firm's strategic objectives and long-term profitability.

3 Research Methodology

This study adopts Design Science Research (DSR), which is recognized not merely as a methodology, but as a distinct research paradigm [18]. In DSR, the artefact itself is the principal outcome of a project and can take the form of a construct, model, method, instantiation, and/or a refined design theory [19]. Peffers et al. [20] proposed a multi-stage DSR process (DSRP) model which is used both to guide and present design science research in Information Systems. Figure 1 illustrates how Peffers et al.'s [20] DSR process has been adapted and applied in this study. While the stages are presented sequentially for clarity, the research followed an iterative approach with repeated cycles of reflection, design refinement, and feedback integration across requirements definition, and artefact development. This allows continuous learning and refinements through iterative movements across design echelons [21].

As part of a broader research program, this paper presents a preliminary version of a novel artefact: the 'RB Canvas' which encapsulates the outcomes of the initial design and evaluation cycle and reflects the current state of development within the study. The phases completed to date are highlighted in green, while the remaining stages are outlined as future research work in Sect. 5.

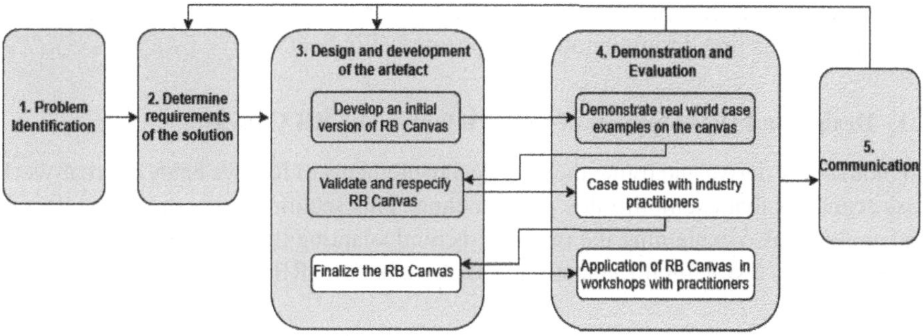

Fig. 1. Adapted DSR approach applied in this study, highlighting current research progress

Stage 1, Problem Identification - as outlined in Sect. 1, drew on both literature and practice cases, to frame the core question: *How can routinized benevolence be conceptualized via a canvas?*

In ***Stage 2*** – Requirements Definition – the design requirements for the targeted artefact were established by synthesizing identified gaps in the existing literature with inferred insights from the analysis of observed practice cases, with a focus on distilling its essential elements. These requirements were:

R1: The artefact should define the core characteristics of RB - This requirement ensures that the artefact provides a clear and distinctive conceptualization of RB by articulating its unique attributes. Doing so enables differentiation from related concepts, thereby establishing a robust conceptual foundation.

R2: The artefact should identify drivers of RB - This requirement ensures the artefact can guide organizations in recognizing the underlying motivations that initiate RB practices within for-profit settings. Understanding these drivers is essential for aligning benevolent actions with strategic intent and operational priorities.

R3: The artefact should demonstrate how RB is operationalized within for-profit organizations - This includes detailing the process design of RB acts by identifying the specific activities performed by various actors (e.g., individuals, systems), the resources mobilized, and the contextual constraints under which these actions occur. This requirement ensures the artefact provides a practical blueprint for embedding benevolence into routine business operations.

R4: The artefact should identify and structure the impacts of RB- This requirement ensures that the artefact captures the multi-level effects of RB. Understanding these impacts is essential for justifying the adoption of RB as a long-term value-creation strategy and for guiding the design, implementation, and evaluation of RB practices.

R5: The artefact should identify and categorize the risks associated with implementing Routinized Benevolence RB- This requirement ensures that the artefact offers a balanced perspective by highlighting potential challenges that organizations must consider when embedding RB practices. Recognizing these risks is crucial to ensuring the resilience, sustainability, and contextual fit of RB initiatives.

Stage 3 involves the design and development of the artefact, while ***Stage 4*** focuses on its demonstration and evaluation, as detailed below. ***Stage 5*** supports continuous communication of results and reflection to guide refinement and learning. These stages occur in tandem as the artefact evolves through feedback and adaptation. We present the outcomes of this first design iteration of the artefact in Sect. 4.

3.1 Design and Development of the Artefact – The RB Canvas

To inductively derive and refine the core design elements of RB, we analyzed real-world case scenarios that exemplify this phenomenon. This section details the case selection and analysis, also explaining the rationale behind adapting the Business Model Canvas (BMC) [22] as a structural foundation of the artefact, the RB Canvas.

3.1.1 The Case Selection, Evidence Collection and Analysis

Drawing on secondary data to theorize from real-world practice is a widely recognized method in organizational and information systems research [23]. It enables researchers to generate conceptual insights grounded in practical realities by leveraging rich, timely information from industry reports, media coverage, and public communications [23]. To this end, we systematically collected and analyzed real-world examples of RB in the for-profit sector. These cases were drawn from both the existing published research on benevolence (e.g. [24, 25]) and publicly available resources documenting benevolence in the for-profit sector such as business news sites (e.g., Forbes, Harvard Business Review, McKinsey Insights), company websites and social networking platforms where customers share their experience (e.g., Reddit, Twitter, LinkedIn).

These cases were evaluated against the predefined criteria for RB, as outlined in Sect. 2. Some cases demonstrated strong alignment with all criteria and were therefore selected for in-depth analysis. The criteria mapping for the selected cases is presented in part A of Appendix. Upon identifying an RB example, we gathered additional details from industry reports, corporate documents, and social media to build the case repository (see Appendix part B). A total of ten qualified cases were identified for this study. Of these, five were selected to inform the development of the artefact based on the richness of available data and their representation across diverse industry sectors. The remaining five cases are reserved for the subsequent demonstration and evaluation phase. The five cases used for artefact development are detailed below (additional details are provided in Appendix parts A and B).

Bolt - The e-scooter provider integrated a cognitive reaction test into its mobile app to prevent drunk driving. If users fail or skip the test, the scooter remains locked and alternative transport options are suggested, proactively enhancing safety.

Tesla - - As part of its customer safety enhancement efforts, Tesla integrated over-the-air updates to automatically extend battery capacity, ensuring drivers have sufficient range to evacuate safely during emergencies.

Commonwealth Bank - Implemented a Compassionate Care Program - a complimentary protection program for eligible owner-occupied home loans which provides financial support if the customer, their spouse, or a dependent pass away or is medically diagnosed with a terminal illness.

Supercheap Auto - The Australian retailer introduced a price-promise program aimed at enhancing customer satisfaction by offering a simplified refund process when the price of a purchased item drops within two weeks.

AT & T - A major US telecommunications provider established a crisis response process that waives overage fees for data, calls, and texts, ensuring customers can stay connected during emergencies.

The case analysis consists of a three-cycle coding approach as detailed in Appendix Part C. The coding process is aligned with the approach proposed by Saldaña and Omasta [26]. NVivo was used as a tool for analysis. In the *first cycle*, Open coding [26] was conducted through line-by-line analysis of case content, aiming to capture all statements and micro themes describing the benevolent act.

In the *second cycle*, the research team engaged in a collaborative process to refine, reorganize, and synthesize the initial codes generated during the first cycle. This involved comparing codes across cases, grouping similar codes, resolving redundancies, and identifying relationships among them. Through iterative discussions, the team consolidated these codes into more abstract, higher-level themes – following axial coding [26] that captured recurring patterns and relationships across cases.

In the *third cycle*, a series of interpretive workshops with the research team were conducted to further synthesize and sensitize the identified themes. While this step extends beyond the two-cycled coding approach outlined by Saldaña & Omasta [26], it aligns with hermeneutic and interpretive traditions [27], where meaning is developed through iterative dialogue, reflection, and contextual understanding. During this stage, the research team also identified the need for a structured way to represent the emerging themes. This led to the consideration of the Business Model Canvas (BMC) [23] as a brainstorming tool for representing themes.

3.1.2 Deriving the RB Canvas: Inspired by the Business Model Canvas

This iterative process led to the emergence of a canvas-based framework; drawing inspiration from the BMC [22], to systematically represent the multifaceted nature of RB practices. BMC is defined as "a shared language that allows organizations to easily describe and manipulate business models to create new strategic alternatives" [22]. It acts as a blueprint for a strategy to be implemented through organizational structures, processes, and systems and describes the rationale of how an organization creates, delivers, and captures value [22]. The original BMC consists of 9 building blocks including key partners, activities, key resources, value propositions, customer relationships, customer segments, channels, cost structure, and revenue streams.

Given the strong alignment between the inductively derived themes from our case analysis and the architecture of BMC, we were drawn to adopt the BMC as a *sensitizing concept* [28]. It served as a flexible reference framework rather than a rigid template, ensuring that the data shaped the final structure rather than being forced to fit predefined elements. A closer examination of prior studies [29, 30] reinforced the adaptability of the BMC as an effective tool for synthesizing and visualizing the relationships between different organizational components. This convergence provided a robust justification for adapting BMC as a foundation for representing RB.

The RB canvas along with the definitions of its key elements (indicative of their origins) are presented in Sect. 4.

3.2 Demonstration/Evaluation of the of the Artefact – The RB Canvas

According to Peffers et al.'s [20] DSRP model, the demonstration phase plays a critical role in establishing the efficacy of the artefact. It refers to showing how well the artefact can solve the real-world problems it was designed to address. It can be achieved through various methods such as experiments, case studies, simulations, or providing concrete proofs that illustrate the artefact's functionality and value in practical settings. In this study, the demonstration phase is exemplified by mapping five real-world case examples into the RB canvas. It assesses how well RB canvas delivers its intended outcomes. One illustrative application - the Apple case canvas is presented in the next section (see Appendix part D for all the case canvases).

These additional cases revealed more nuanced variations in how RB is operationalized, prompting iterative refinement to the canvas. Some of the additional sub-sections were introduced - particularly different types of conditions, and risks. Furthermore, some of the categories (e.g. "resources") were more precisely defined, and the specific characteristics of benevolence such as level of frequency, were further added to enhance conceptual clarity.

To evaluate the validity of RB Canvas, this study will first conduct case studies with organizations, including several featured in the initial case mapping, by assessing its practical value and applicability in real-world settings. Insights from this phase will inform further refinements. The revised version will then be evaluated through a series of workshops with practitioners. Following an iterative approach, the canvas will be continuously improved based on feedback from these workshops. Finally, the refined canvas will undergo additional validation through a survey administered to workshop participants, assessing its usability, usefulness, and acceptance.

4 Routinized Benevolence Canvas

RB Canvas serves as both a conceptual framework and a practical tool that helps organizations systematically analyse, design, and embed benevolent practices into routine operations. By visualizing the key components of RB, the canvas facilitates structured thinking, comparison across cases, and informed decision-making for integrating human-centric, ethically grounded practices within profit-driven environments. This section introduces the RB Canvas as depicted in Fig. 2, in its current form and defines its key components. The RB canvass is made of 5 key parts - each outlined in detail below.

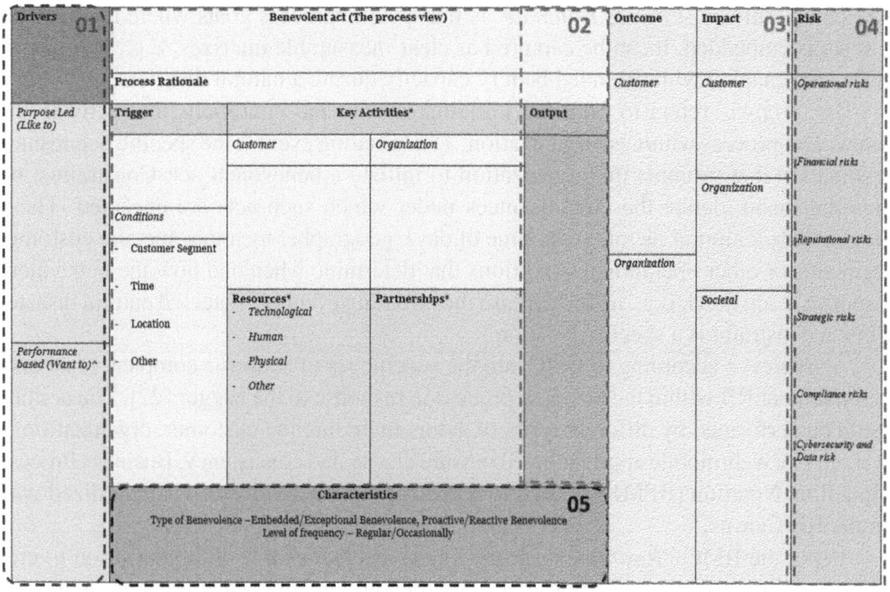

* - Directly relates with BMC components
^ - Indirectly relates with BMC components

Fig. 2. The Routinized Benevolence Canvas

The canvas begins by outlining the drivers of RB (see area denoted by #01 in Fig. 2), which are essential for helping organizations to recognize underlying reason or motivation for carrying out an RB act, providing a rationale that aligns benevolent intent with strategic and economic drivers. Drivers are the key factors that motivate and influence a firm's decision-making process and strategic direction [31].

A driver of RB can be either *purpose-led*, or *performance-based*. Purpose was defined as "a concrete goal or objective for the firm that reaches beyond profit maximization" [32]. Purpose–led in this context refers to the reason for implementing a benevolent act beyond monetary expectations. E.g., Ensure customer safety (Tesla).

In this study, *performance-based* refers to the financial motivations behind the benevolent act, encompassing both the revenue generated and the costs incurred through its implementation. It reflects the organization's ability to achieve desired outcomes efficiently and effectively and includes aspects such as growth, profitability, and competitiveness [33]. This dimension indirectly aligns with the 'revenue stream' and 'cost structure' components of the original BMC [22]. E.g., Increase customer retention/enhance reputation (Tesla).

The center of the canvas (see the area denoted by # 02 in Fig. 2) illustrates the RB act, how RB is operationalized within routine business operations, detailing the initiating event, constraints, actions by various actors, resources utilized, and any relevant partnerships. This operational view is critical as it translates benevolent intent into actionable process elements, enabling organizations to embed benevolence seamlessly into their routine operations. '*Process rationale*' is the specific process goals where the benevolent act is embedded. It can be captured as clear measurable matrixes. E.g., Number of Tesla cars enabled with extended battery capacity during a natural disaster.

The '*trigger*' refers to both the initiating events and conditions that activate the benevolent process within an organization. The initiating event is the specific occurrence or situation that prompts the organization to initiate a benevolent act. Constraints, on the other hand, define the circumstances under which such acts are executed. These may include temporal factors (e.g., time of day), geographic location, specific customer segments, or other operational conditions that determine when and how the benevolent response is activated. E.g., in Tesla's case the initiating event is a detected natural disaster while a constraint is a specific location.

'*Activities*' – according to BMC, are the specific set of actions a company must take to implement RB within the business process in response to the trigger [22]. The actions could be performed by different types of actors including the customer, organization, a system, or a web/mobile application. To ensure clarity and consistency, Business Process Modelling Notation (BPMN) is used to represent these activities in a standardized way on the RB Canvas.

As per the BMC, '*Resources*' are the key assets that enable an organization to create and offer a value proposition, reach its markets, maintain relationships with customers and generate revenues [22]. Similarly, the RB canvas considers the resources required to implement the benevolent act across different stages encompassing technological, human, financial, intellectual, and other operational assets. E.g., Wireless software Update (Tesla). '*Partnerships*' represents "the network of suppliers and partners that make the business model work" [22]. Accordingly, this study recognizes it as the network of stakeholders in the process eco-system that helps to make the benevolent act work. E.g., CommBank – Compassionate Care program, partnering with AIA Insurance.

Output by definition are the immediate results of an action, and it could be a product or service received by the customer as a deliverable of the process [34]. In this study, *output* refers to the immediate result of the benevolent act – when the process is executed. E.g., with Telsa's case the output is 'increased battery capacity'.

The areas indicated with # 3 in Fig. 2 represents the outcomes and impacts of the RB act. Adapting from Koontz [35], '*outcomes*' refer to the short to medium-term effects of a benevolent act, typically resulting in changes to human perceptions/behaviors. For example, improved customer safety as a result of Tesla's RB initiative. In contrast, '*impact*' denotes the long-term effects of such acts, which may be positive or negative, and can be primary or secondary in nature [34]. These impacts represent actual or highly probable changes in the broader environment, often emerging as a consequence of the outcomes achieved [41]. We define impact across three levels: customer, organizational, and societal. For instance, with the Tesla case - enhanced individual safety at the customer level and reduced public healthcare costs at the societal level.

Risk (see area #4 in Fig. 2) refers to the uncertainty and potential for disappointing outcomes resulting from a benevolent act, aligning with the definition by Bottazzi et al., [36] - the likelihood of significant and/or unfavorable consequences of decisions. Following Frase et al. [37] and Lam [38], we adapted a range of RB-related risks within organizational settings, including: *operational* (e.g., internal failures or external disruptions), *strategic* (poor execution of strategy), *financial* (e.g., credit or liquidity risks), *compliance* (legal/regulatory breaches), *reputational* (loss of trust or image), and *cybersecurity/data risks* (e.g., data breaches or mismanagement).

The final element of the RB canvas (see area #5 in Fig. 2) captures the '*meta characteristics*' of RB acts, that is the distinguishing features that define the nature of benevolent behaviour. These characteristics emerged inductively from our case analysis and reflect recurring patterns that differentiate how benevolence is routinized in practice. For example, it indicates whether an act is performed regularly (as in Bolt's case) or occasionally but embedded into routines (such as Tesla's). By explicitly outlining these characteristics, the canvas supports users in specifying and comparing the distinctive features of different RB initiatives.

One of the case-specific RB canvases (The Apple case – which features the detection of car crashes and automatic contacting of emergency services by the latest iPhones and Apple watches) used for the demonstration is presented below, with the remaining canvases included in part D of the appendix. In these canvases, elements highlighted in blue represent themes directly derived from the inductive analysis of secondary data, while those in red indicate propositions introduced by the research team to address gaps where such details were not explicitly mentioned in the source materials. This practice is consistent with established qualitative research methodologies, which allows using inferred propositions to address gaps in empirical data and strengthen analytic generalization [39] (Fig. 3).

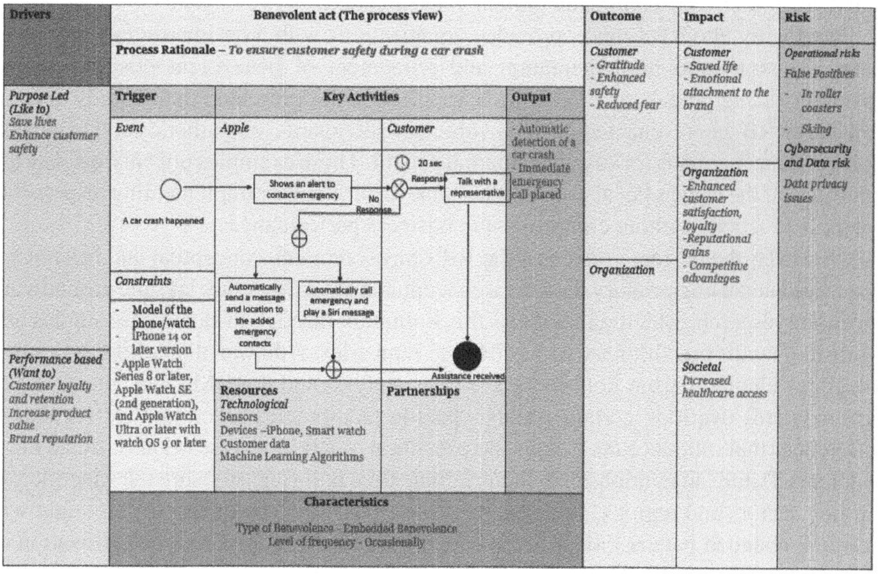

Fig. 3. The Apple Case Canvas

5 Discussion and Future Work

Canvas-based frameworks have gained traction in business and information systems research due to their ability to structure complex constructs into clear, visual, and actionable formats. Widely adapted examples such as the Business Model Canvas [22], demonstrate their value in enhancing stakeholder communication, support iterative refinement, and foster shared understanding across disciplines [40]. Recognized as valid design artefacts, canvases bridge theory and practice by providing structured guidance for real-world application [41].

In line with this approach, the RB Canvas serves not only as a conceptual lens but also as a practical blueprint for embedding benevolence into routine operations, thereby contributing to both theory and practice. Developed through an abductive approach, the canvas draws from inductive insights gained through case analysis while integrating elements adapted from the BMC, including Key Activities, Key Resources, Key Partners, Cost Structure and Revenue Streams. This hybrid composition ensures both conceptual rigor and practical usability, offering a structured tool for organizations to analyze, design, and embed benevolent acts within their business processes.

From an academic perspective, this study advances a formal conceptualization of Routinized Benevolence, an emerging but underexplored area in business and IS literature by articulating its underlying logic and defining characteristics. RB Canvas serves as a novel artefact which reframes benevolence from a discretionary, one-off gesture into a processual, repeatable activity embedded within the operational fabric of organizations. It contributes to the RBPM agenda by introducing a perspective that emphasizes ethical intent, customer well-being, and mutual value creation.

Practically, the RB Canvas provides organizations with a design artefact that facilitates the structured implementation and refinement of benevolent practices across customer-facing processes. Real-world case illustrations presented in this study demonstrate how RB can be enacted through voluntary safeguards, empathetic service adaptations, and mechanisms for sustained mutual benefit. These examples not only validate the relevance of the canvas but also offer actionable inspiration for implementing responsible business practices without compromising business performance.

However, the current version of the RB Canvas remains conceptual and has not yet been validated using primary data. As a conceptual paper, this study lays the groundwork for future research, which will address this limitation through in-depth case studies and workshops with practitioners across diverse industries. Additionally, while the canvas integrates elements from both inductive case analysis and the BMC, its current structure may still overlook certain context-specific factors such as cultural or regulatory differences that influence the design and execution of RB in diverse organizational environments. Future adaptations may require contextual tailoring to ensure relevance across varying sectors and regions. Building on these efforts, the next phase of this study will examine nuanced patterns identified through the canvas to develop a categorization of RB, followed by procedural guidelines tailored to each category, by providing more actionable guidelines for practitioners.

Another promising avenue involves integrating technology-enabled capabilities, including automation, personalization, and real-time decision-making, to support the seamless and scalable execution of benevolent processes, which is already being explored as part of the broader research program. Additionally, future research could investigate the impact of customer involvement in the co-design and co-evaluation of RB. Understanding how such participation influences the relational outcomes of RB could provide valuable insights into the effectiveness and perceived authenticity of these initiatives.

Collectively, these directions underscore the growing importance of ethical and stakeholder-oriented process design. As emphasized in recent Responsible BPM discourse [42], ethics must move from being a peripheral consideration to a foundational element of process thinking. This study contributes to that agenda by offering both a conceptual and practical mechanism to embed "human-centricity" within BPM, advancing transparency, customer well-being, and responsible innovation through the systematic enactment of routinized benevolence, while sustaining economic gains.

Appendix A

This appendix available at https://tinyurl.com/479c66mr is structured into four parts: Part A outlines how each case aligns with the criteria of RB; Part B provides the case repository along with source references; Part C details the coding protocol; and Part D presents the case-specific RB canvases developed during the demonstration phase.

References

1. Boudreau, M.-C.: Corporate social responsibility in the IT Industry: a stakeholder perspective. AMCIS 2023 (2023)
2. Serafeim, G.: Purpose and profit: how business can lift up the world. HarperCollins Leadership (2022)
3. Aguinis, H., Glavas, A.: Embedded versus peripheral corporate social responsibility: psychological foundations. Ind. Organ. Psychol. **6**, 314–332 (2013)
4. Kramer, M.R., Porter, M.: Creating shared value. FSG Boston, MA, USA (2011)
5. Rosemann, M., Ostern, N., Voss, M., Bandara, W.: Benevolent business processes-design guidelines beyond transactional value. In: International Conference on Business Process Management, pp. 447–464. Springer (2023)
6. Mahendrawathi, E., Avigdor, G., Thomas, G., Flavia, S.: Call for responsible BPM forum (2025)
7. Schwartz, S.H., Bardi, A.: Value hierarchies across cultures: Taking a similarities perspective. J. Cross Cult. Psychol. **32**, 268–290 (2001)
8. Livnat, Y.: On the nature of benevolence. J. Soc. Philos. **35**, 304–317 (2004)
9. Mayer, R.C., Davis, J.H.: The effect of the performance appraisal system on trust for management: a field quasi-experiment. J. Appl. Psychol. **84**, 718 (1999)
10. Beveridge, A.J., Höllerer, M.A.: Theorizing organizational benevolence. Organ. Sci. **34**, 1864–1886 (2023)
11. Zhang, C.-B., Li, Y.-N.: How social media usage influences B2B customer loyalty: roles of trust and purchase risk. J. Bus. Indust. Mark. **34**, 1420–1433 (2019)
12. Selnes, F., Gønhaug, K.: Effects of supplier reliability and benevolence in business marketing. J. Bus. Res. **49**, 259–271 (2000)
13. Bove, L.L., Pervan, S.J., Beatty, S.E., Shiu, E.: Service worker role in encouraging customer organizational citizenship behaviors. J. Bus. Res. **62**, 698–705 (2009)

14. Nikolaeva, R., Bicho, M.: The role of institutional and reputational factors in the voluntary adoption of corporate social responsibility reporting standards. J. Acad. Mark. Sci. **39**, 136–157 (2011)
15. Menghwar, P.S., Daood, A.: Creating shared value: a systematic review, synthesis and integrative perspective. Int. J. Manag. Rev. **23**, 466–485 (2021)
16. Tussyadiah, I.P.: Toward a theoretical foundation for experience design in tourism. J. Travel Res. **53**, 543–564 (2014)
17. Vergne, J.-P., Wernicke, G., Brenner, S.: Signal incongruence and its consequences: a study of media disapproval and CEO overcompensation. Organ. Sci. **29**, 796–817 (2018)
18. Baskerville, R., Pries-Heje, J., Venable, J.: Soft design science methodology. In: Proceedings of the 4th International Conference on Design Science Research in Information Systems and Technology, pp. 1–11 (2009)
19. Hevner, A., Chatterjee, S.: Design research in information systems: theory and practice. Springer Science & Business Media (2010)
20. Peffers, K., Tuunanen, T., Gengler, C.E., Rossi, M., Hui, W., Virtanen, V., Bragge, J.: The design science research process: a model for producing and presenting information system research. In: 1st International Conference on Design Science in Information Systems and Technology (DESRIST), vol. 24, pp. 83–106 (2006)
21. Tuunanen, T., Winter, R., vom Brocke, J.: Dealing with complexity in design science research: a methodology using design echelons. MIS Quar. **48** (2024)
22. Osterwalder, A., Pigneur, Y.: Business model generation. Business Model (2010)
23. Das, R., Jain, K.K., Mishra, S.K.: Archival research: a neglected method in organization studies. Benchmark.: Int. J. **25**, 138–155 (2018)
24. Rosemann, M., Bandara, W., Ostern, N., Voss, M.: The benevolent enterprise: how to operationalise and scale doing good (2023)
25. Peppers, D., Rogers, M.: Extreme trust: the competitive necessity of proactive trustworthiness. Marketing Wisdom, pp. 1–16. Springer (2018)
26. Saldaña, J., Omasta, M.: Qualitative research: analyzing life. Sage Publications (2016)
27. Boell, S.K., Cecez-Kecmanovic, D.: A hermeneutic approach for conducting literature reviews and literature searches. Commun. Assoc. Inf. Syst. **34**, 12 (2014)
28. Bowen, G.A.: Grounded theory and sensitizing concepts. Int. J. Qual. Methods **5**, 12–23 (2006)
29. França, C.L., Broman, G., Robert, K.-H., Basile, G., Trygg, L.: An approach to business model innovation and design for strategic sustainable development. J. Clean. Prod. **140**, 155–166 (2017)
30. Jin, Y., Ji, S., Liu, L., Wang, W.: Business model innovation canvas: a visual business model innovation model. Eur. J. Innov. Manag. **25**, 1469–1493 (2022)
31. Ammirato, S., Linzalone, R., Felicetti, A.M.: Business model innovation drivers as antecedents of performance. Meas. Bus. Excell. **26**, 6–22 (2022)
32. Henderson, R., Steen, E.V.D.: Why do firms have "purpose"? The firm's role as a carrier of identity and reputation. Am. Econ. Rev. **105**, 326–330 (2015)
33. Cherrington, D.J.: Organizational behavior: the management of individual and organizational performance. Allyn & Bacon (1989)
34. Intrac: Outputs, Outcomes and Impact (2017)
35. Koontz, T.M., Jager, N.W., Newig, J.: Assessing collaborative conservation: a case survey of output, outcome, and impact measures used in the empirical literature. Soc. Nat. Resour. **33**, 442–461 (2020)
36. Bottazzi, G., Secchi, A., Tamagni, F.: Productivity, profitability and financial performance. Ind. Corp. Chang. **17**, 711–751 (2008)
37. Fraser, J.R., Simkins, B., Narvaez, K.: Implementing enterprise risk management: case studies and best practices. Wiley (2014)

38. Lam, J.: Enterprise risk management: from incentives to controls. Wiley (2014)
39. Timmermans, S., Tavory, I.: Theory construction in qualitative research: from grounded theory to abductive analysis. Sociol Theory **30**, 167–186 (2012)
40. Gassmann, O., Frankenberger, K.: The business model navigator ePub eBook: the business model navigator: 55 models that will revolutionise your business. Pearson UK (2014)
41. Gregor, S., Hevner, A.R.: Positioning and presenting design science research for maximum impact. MIS Quar. 337–355 (2013)
42. Santoro, F.: Ethics at the core: the future of BPM. In: s22nd International Conference, BPM 2024, pp. 24–29. Springer (2024)

ReDPT-CT: Designing Responsible Digital Process Twins for Clinical Trials

Luiz Ricardo Brito Ribeiro[1](✉), Silvia Dallavalle[1], Gerald Kremer[2], and Rainer Stark[2]

[1] University of São Paulo, Av. Bandeirantes, 3900 - Vila Monte Alegre, Ribeirão Preto - SP 14040-900, Brazil
luizrbr@usp.br

[2] Technische Universität Berlin, Pascalstraße 8-9, 10587 Berlin, Germany
kremer@tu-berlin.de

Abstract. This study explores the intersection of ethics, Business Process Management (BPM), and digital process twins in clinical trials. It investigates the research question: What responsible aspects within business processes can a Clinical Research Center manager consider when assessing the adoption of Digital Process Twins? To address this, a multi-phase methodological approach was employed. First, a structured literature review mapped the ethical landscape of Responsible BPM aspects related to clinical trials and DPTs. Based on these findings, the Responsible Digital Process Twin for Clinical Trial (ReDPT-CT) framework was developed, linking 15 ethical responsibility categories to CRC business process groups—core, support, and management. The study highlights transparency, informed consent, and scientific integrity as the most frequently cited ethical aspects, while also emphasizing emerging areas such as innovation governance and ethical capacity-building. The framework offers a novel contribution by translating ethical principles into actionable digital process design for CRC operations. However, the research is limited by necessity of empirical validation, which is recommended for future studies across diverse research settings. Practically, the framework equips CRC managers with structured guidance for ethical DPTs adoption, while socially, it contributes to building a more trustworthy and participant-centered clinical trial ecosystem.

Keyword: Clinical Research Center · Digital Twin · BPM · Ethics

1 Introduction

Clinical trials are a mandatory process for a therapeutic innovation, providing the safety and efficacy evidence required for the approval and commercialization of new medicines and medical devices. However, as research scales up and sponsors and Clinical Research Centers (CRCs) face increasing pressure for financial returns, the conduct of these studies exhibits ever more complex ethical tensions [1]. Conflicts of interest fueled by economic incentives, transparency gaps in data disclosure, opaque financial ties between industry and investigators, weaknesses in the informed-consent process, and premature trial interruptions motivated by commercial considerations are repeatedly highlighted in recent

M. ER et al. (Eds.): BPM 2025, LNBIP 565, pp. 102–116, 2026.
https://doi.org/10.1007/978-3-032-02936-2_9

literature [2–5]. These problems not only compromise scientific integrity but also erode the social trust that underpins biomedical research [6].

In parallel, Business Process Management (BPM) approaches has provided tools to rationalize and standardize organizational processes, including health care. Yet BPM practices traditionally centered on efficiency and cost have proved insufficient to address the multiple ethical and regulatory dimensions that permeate clinical trials [7]. The emerging movement of Responsible BPM (RBPM) therefore advocates the systemic incorporation of values such as safety, equity, transparency and sustainability throughout the entire process life-cycle [8, 9]. In the clinical trial domain, this translates into principles that include, among others: (i) participant safety and well-being; (ii) data integrity and traceability; (iii) transparency and explicability of decision flows; (iv) privacy and personal-data protection; and (v) multistakeholder governance. [10].

While these ethical agendas gain momentum, Digital Process Twins (DPTs) have emerged as a new paradigm for representing, simulating and monitoring complex clinical processes in real time. By combining IoT sensors, artificial intelligence (AI) and formal modelling languages (BPMN/DMN), DPTs has a potential to detect protocol deviations, anticipate adverse events and optimize resources across multiple sites simultaneously [11]. Yet the data-intensive nature of such solutions amplifies risks of privacy breaches, algorithmic bias, opaque decision-making and legal accountability [12, 13]. In other words, the ethical dilemmas that already challenge traditional clinical trials are being reinvented in the digital realm.

Literature therefore lacks a framework that coherently combines DPTs and RBPM principles, offering to CRCs organizations a roadmap for the responsible adoption of digital twins in clinical trial research. Although isolated efforts exist—such as using blockchain to reinforce data traceability [14] or corporate transparency frameworks [15] —there is no integrative model that articulates ethical, regulatory, technological and managerial requirements across all phases of a clinical trial.

Drawing on perspectives from biomedical ethics, process management, and systems engineering, this study addresses the following research question: what responsible aspects within business processes can a CRC manager consider when assessing the adoption of DPTs? To advance on this question, the paper sets out to achieve the following objectives: (i) map the Responsible-ethical landscape, (ii) develop the Responsible Digital Process Twin framework and (iii) set a Forward-looking research agenda.

2 Literature Review

This chapter aims to present the foundational concepts of RBPM, Clinical Trials, and DPTs. Considering the research question established, it is adequate to begin by understanding the concept of responsibility, particularly when it relates to business processes and data management. Being responsible during the implementation of a new system involves adhering to the FACT principles [16]: Fairness, meaning that outputs should avoid producing unjust outcomes, even when technically accurate; Accuracy, ensuring that results are based on reliable data rather than speculation; Confidentiality, whereby insights are provided without compromising sensitive information; and Transparency, which requires that outcomes are accompanied by clear, understandable explanations of how decisions were reached. These principles must be embedded within any technological transformation of business processes [17], especially in complex and ethically sensitive environments such as clinical research.

To understand how responsibility can be embedded in organizational practices, it is important to consider the classification of business processes. These can be classified into Core, Support, and Management categories [8] been: (i) core processes representing the value-generating activities of an organization—in the case of CRCs, this includes direct interactions with trial participants and sponsors; (ii) support processes ensuring the operational infrastructure needed to sustain core activities, such as staff management, facilities, and IT systems; (iii) management processes providing strategic direction and governance, covering activities like protocol portfolio management, strategic planning, compliance, etc. This structure helps identify where responsible practices can be integrated across different levels of CRC operations.

Building on this process-based view of CRCs, it becomes evident that ethical considerations must be integrated across all categories of business processes to ensure responsible and trustworthy trial conduct. Ethics in the execution of clinical trials is a cornerstone of scientific integrity and plays a critical role in maintaining public trust in evidence-based medicine [10, 18, 19]. Among the most prominent concerns are conflicts of interest, particularly in trials financed by pharmaceutical companies, where financial incentives may compromise researcher neutrality [20]. Such influences can affect various stages of the research process, from hypothesis formulation to selective reporting of outcomes. Notably, practices like *p-hacking*—the manipulation of statistical analyses to achieve false significance—have been documented in phase III trials, especially those funded by smaller sponsors [2].

Conflicts of interest in clinical trials critically undermine the ability of Clinical Research Centers (CRCs) to uphold responsible practices across core, support, and management processes. Financial ties—often undisclosed, as seen in 92% of oncology trial authors in Japan receiving payments with only 10% disclosing them [21]—erode transparency and violate key FACT principles like Fairness and Transparency. Industry-funded trials on antidepressants disproportionately report positive outcomes [22], while 93% of meta-analyses omit funding disclosures [23], revealing systemic weaknesses in protocol design, monitoring, and reporting. Declarations alone are insufficient; mitigation must occur throughout the trial lifecycle [24]. The CREATE-X trial exemplifies how unethical financial arrangements can distort CRC management [4], while even minor influences, like academic sponsorships, can impair researcher autonomy and institutional ethics [25]. Embedding responsibility in CRCs thus requires transparent sponsor relations, independent oversight, and financial accountability.

These challenges extend to ethical vulnerabilities within core processes, such as informed consent. When participants are not fully aware of risks or commercial incentives—especially in contexts where clinical trial participation is the only access to care—their autonomy is compromised [26]. Upholding the principles of Transparency and Autonomy under the FACT framework demands not only clear communication but also unrestricted access to study goals and raw data [27]. Ethical dilemmas such as Geron Corporation's abrupt trial termination for financial reasons [5], or trials involving organ sourcing from executed prisoners in China [28], demonstrate the tension between localized practices and universal ethical standards. Confronting such dilemmas requires institutional transformation. Strengthening ethical governance—through compliance mechanisms and empowered ethics committees [29]—is essential for aligning CRC operations with responsible digitalization, promoting transparency, and fostering trust in clinical research.

The concept of a Digital Twin (DT) refers to a real-time digital replica of a physical system, integrating sensor data, computational models, and simulations to represent and predict the behavior of its physical counterpart [21]. This technology has been adopted across various domains—such as engineering, healthcare, and logistics—to anticipate failures, optimize performance, and support personalized decision-making. A Process Twin, unlike a physical Digital Twin, models the dynamic flow of operations within a system—such as patient trajectories in hospitals or goods movement in supply chains—providing critical insights into timing, interdependencies, and workflow efficiency [22],

In healthcare, Digital Twins has been applied to ICU patient monitoring, where they integrate real-time data from life-support equipment and electronic health records to forecast clinical deterioration [23]. In hospital logistics, Process Twins have shown promise in medication supply chain optimization, helping reduce delays and minimize waste in cold-chain management [24]. Another emerging field involves clinical recruitment prediction, using digital models to simulate patient eligibility in real time. However, these systems face significant technical risks, including data latency, model drift (when performance deteriorates over time), and high dependency on physical sensors. These challenges not only compromise operational accuracy but also pose serious ethical concerns, especially if clinical decisions are made based on outdated or biased data [21].

3 Method

This study adopts a multi-phase methodological design. Phase I presents a structured literature review aimed at mapping the ethical–technological landscape surrounding Clinical Trials, RBPM and DPTs. The objective is to identify key responsibility issues and application domains that shape the integration of DPTs in healthcare research settings. Phase II builds on these insights to develop the ReDPT-CT framework. Finally, Phase III consolidates the findings into a forward-looking research agenda, identifying open questions and future directions for scholarship at the intersection of clinical trial ethics, digital innovation, and process management (Fig. 1).

Fig. 1. Research method phases

3.1 Phase I – Responsible-Ethical Landscape

The structured literature review followed papers published that address Clinical-Trial ethics, Responsible BPM, Digital Process Twins, and Ethics issues running in Scopus database using the string: TITLE-ABS-KEY ("clinical trial*" AND "digital twin*" OR "process twin*") AND ("ethic*" OR "privacy" OR "governance"). After retrieval, researchers screened the papers and for each eligible study, was extracted keywords on the principal responsible and ethical issue, regulatory context, digital-twin application domain, and challenges reported by the authors.

3.2 Phase II – Framework Development

Building on the findings from the previously phase, Phase II focused on the development of the ReDPT-CT framework. This phase aimed to translate the ethical challenges and opportunities identified in the literature into a structured model that could guide the ethical-by-design implementation of DPTs in CRCs. The framework was constructed as a three-layer architecture, integrating (i) Responsible Aspects, (ii) CRC business processes, and (iii) DPT implementation guidelines.

- Layer 1 – Responsible Aspects: A set of 15 ethical categories was established through thematic analysis of the literature, capturing critical dimensions such as transparency and disclosure, informed consent and autonomy, conflict of interest, data privacy, regulatory compliance, and stakeholder engagement. These categories represent normative anchors for responsible conduct in clinical trials and digital transformation.
- Layer 2 – CRC Business Processes: CRC operations were mapped into three macro-process categories, following BPM conventions: Core Processes (e.g., protocol startup, execution, monitoring, and closure), Support Processes (e.g., procurement, HR, IT infrastructure), and Management Processes (e.g., strategic planning, portfolio management, budget oversight). Each process was examined to identify its ethical relevance and vulnerabilities, leading to the allocation of applicable responsible aspects from Layer 1.
- Layer 3 – Digital Process Twin Guidelines: For each CRC process, DPT-oriented recommendations were developed to translate ethical imperatives into DPTs functionalities. These included the use of simulation for consent validation, automated alerts for protocol deviations, AI ethics evaluation modules, traceable data flows for transparency, and contextual adaptation for global ethics harmonization. Each guideline was explicitly mapped to connect with the respective process.

3.3 Phase III – Research-Agenda Formulation

In Phase III, the insights generated in the preceding stages are consolidated during a dedicated research retreat. This phase is designed to use the emerging framework as a springboard for future inquiry, enabling scholars and practitioners to anticipate the technological, regulatory and societal trends that shape responsible digital-process twin adoption in the CRCs environment. Through discussion, the team identifies unresolved questions according to their relevance to the for RBPM principles. The resulting research agenda maps specific knowledge gaps, methodological opportunities that could guide the next wave of investigations at the intersection of RBPM, DPTs and Clinical Trials.

4 Results

This chapter presents and analyzes the resulting ReDPT-CT framework, which was developed to integrate RBPM aspects into the design and implementation of DPTs within CRCs processes.

4.1 Responsible-Ethical Landscape

The construction of the responsible aspects layer in the ReDPT-CT framework was grounded in an analysis of keywords extracted from a curated set of academic articles. These articles, cited throughout this study [25–49, 50–52], span the domains of clinical trial ethics, RBPM and DPTs. Through a process of inductive thematic clustering, the keywords were synthesized into 15 ethical categories, which encapsulate the principal dimensions of responsibility in clinical trials. Among these, Transparency and Disclosure (A) emerged with the largest number of keywords—over 18 unique terms—highlighting its cross-cutting importance across all trial stages and processes. This category was extensively discussed in articles focusing on publication bias, selective reporting, and trial registration [25–37], reinforcing transparency as a fundamental concern in both ethical theory and practical compliance. Similarly, Informed Consent and Autonomy (B) and Scientific Integrity and Methodological Rigor (E) stood out with a high density of references and keywords, emphasizing participant protection and reliability of results as pillars of responsible digitalization in clinical trials.

Other categories such as Emerging Technologies and Innovation (M) and Training, Culture, and Ethical Capacity Building (O), while supported by fewer articles [29, 48, 44, 50], bring attention to evolving and often under-addressed areas in clinical research ethics. These dimensions point to the need for proactive governance of AI, blockchain, and data ethics, as well as for fostering institutional cultures of responsibility and continuous education. Likewise, categories such as Justice and Social Inclusion (C) and Global and Contextual Ethical Standards (L) underscore ethical imperatives related to equity and the contextualization of ethical practices across diverse populations and regions—elements especially relevant for multinational trials. Interestingly, while Data Privacy and Security (H) and Regulation and Compliance (K) had moderate keyword counts, they displayed high relevance in process alignment, particularly with support and management layers of CRCs. The distribution of keyword intensity across categories demonstrates both the maturity of certain ethical discussions (e.g., transparency, consent, scientific rigor) and the emerging nature of others (e.g., innovation governance, cultural ethics), which together shape a balanced and forward-looking ethical foundation for DPTs in clinical trial environments. The Table 1 presents all the Responsible aspects and respective keywords.

Table 1. Responsible aspects and respective keywords

Responsible aspects	Keywords
A. Transparency and Disclosure: Includes aspects related to clarity, open communication, and accountability in clinical trials. [18, 25–36]	*transparency, full transparency, reporting transparency, data transparency, trial registration, protocol registration, result reporting, results disclosure, public outcome reporting, publication bias, selective reporting, improved reporting standards, transparency guidelines, transparency in RCT design, document accessibility, results posting, complete reporting, seeding trial avoidance*
B. Informed Consent and Autonomy: Focuses on participants' self-determination and ethical communication of risks and benefits.[18, 20, 37–42]	*informed consent, participant consent, randomization disclosure, consent simplification when risks are low, waived consent implications, participant, communication, autonomy, patient autonomy, participant dignity, cognitively impaired patients*
C. Justice and Social Inclusion: Involves fair distribution of risks and benefits, protection of vulnerable populations, and equity. [20, 32, 38, 41]	*justice, distributive justice, equitable access, equitable selection, participant rights, participant protection, participant welfare, protecting vulnerable populations, vulnerable populations, vulnerable patient protections, recruitment equity, social value, exploitation prevention, anti-exploitation, resource inequality, Western-centric ethics*
D. Ethical Governance and Committees: Refers to the role, structure, and responsibility of ethics committees. [33, 35, 38, 43, 44]	*ethics committees, ethics committee approvals, IRB oversight, IRB compliance, ethics committee recommendation, ethics committee evolution, ethics committee role, ethics committee strength, ethical review, ethical oversight, ethical compliance monitoring, mutual acceptance, review timelines, tailored ethical oversight*
E. Scientific Integrity and Methodological Rigor: Pertains to scientific validity, transparency, and reliability.[25, 27, 31, 32, 45, 46]	*methodological rigor, scientific rigor, clinical trial transparency, CONSORT adherence, scientific integrity, credible results, scientific relevance, bias minimization, bias mitigation, trial integrity, peer review, research integrity*

(*continued*)

Table 1. (*continued*)

Responsible aspects	Keywords
F. Conflict of Interest and Independence: Covers the separation of commercial and scientific interests.[20, 29, 34, 47]	*conflict of interest, conflict of interest disclosure, funding disclosure, independent research, independent oversight, third-party oversight, firewall from industry, separation of funding and conduct, contract clauses*
G. Compensation and Incentives: Addresses fair participant compensation and avoidance of economic exploitation. [19, 39, 41]	*compensation, fair compensation, participant compensation, payment ethics, undue influence, EU compliance*
H. Data Privacy and Security: Related to protection of personal data, confidentiality, and ethical information use. [29, 33, 48]	*data privacy, patient confidentiality, privacy under GDPR, secure data sharing, participant-level data access, cybersecurity, GDPR compliance, responsible data sharing, privacy*
I. Risk and Benefit Assessment: Related to the principle of beneficence and participant safety. [18, 20, 39, 46, 49]	*risk-benefit analysis, risk-benefit balance, risk/benefit assessment, beneficence, minimize risk, safety, no patient harm, risk management, safety evaluations, trial deprioritization*
J. Stakeholder Relationships: Involves communication and participation of interested parties. [15, 30, 33, 40]	*stakeholder engagement, stakeholder trust, stakeholder inclusion, public trust, community partnerships, patient advocacy, institutional involvement, public perception, clinician-patient integrity*
K. Regulation and Compliance: Refers to adherence to regulatory guidelines and legal frameworks.[18, 26, 35, 38, 43, 44]	*regulatory compliance, GCP adherence, FDA/EMA compliance, EU Regulation 536/2014, Clinical Trials Act, WHO best practices, Nuremberg Code, Declaration of Helsinki, GCP inspection gaps*
L. Global and Contextual Ethical Standards: Explores geographic and cultural variations in ethical standards. [20, 38, 43]	*global ethics variation, national alignment, local health needs, international guidelines, local healthcare access*
M. Emerging Technologies and Innovation: Explores the role of new technologies and their ethical governance. [29, 46]	*AI and blockchain transparency, data ethics, digital tools review, explainability, innovation balance, personalized trials ethics*

(*continued*)

Table 1. (*continued*)

Responsible aspects	Keywords
N. Operational Management and Corporate Ethics: Focuses on aligning management practices with ethical principles.[15, 19, 26, 33]	*codes of conduct, business-aligned ethics, organizational ethics, project management integration, transparency in management, quality management, business integration, ethical R&D balance, trial management ethics*
O. Training, Culture, and Ethical Capacity Building: Covers ethics education, institutional culture, and capacity building. [19, 42]	*bioethics training, staff training, moral leadership, professional standards, ethical responsibility, researcher integrity*

4.2 Framework Development

Compiling all these information's, the framework is structured in three interconnected layers: (i) Responsible Aspects, (ii) CRC Business Processes, and (iii) Digital Process Twin guidelines. Each layer builds upon insights gathered in the previous methodological phases and is interconnected to promote an ethics-by-design approach to digital transformation in clinical research operations.

The Responsible Aspects layer was derived through a systematic consolidation of keywords extracted from the literature review conducted in Phase I. Through a thematic clustering, these terms were synthesized into 15 distinct ethical responsibility categories. These categories serve as ethical "lenses" through which CRC processes and DPT functionalities are assessed and aligned.

The second layer of the framework organizes CRC operations into three conventional BPM process groups: Core Processes, Support Processes, and Management Processes. Each group is analyzed for its ethical exposure and mapped to the most relevant responsible aspects as explained:

- Core Processes—comprising *Protocol Startup*, *Conduct Protocol*, *Monitor Protocol*, and *Close Protocol*—are directly associated with research participants and sponsors, making them highly sensitive to ethical dimensions such as *Informed Consent and Autonomy (B)*, *Scientific Integrity and Methodological Rigor (E)*, *Data Privacy and Security (H)*, and *Transparency and Disclosure (A)*. These processes are where the ethical stakes are highest, as they involve human subjects and directly influence trial outcomes. Notably, all core processes are linked with at least three responsible aspects, reflecting their central role in protecting human subjects and ensuring regulatory integrity.
- Support Processes—including *Financial Management*, *Procurement*, *HR Management*, *Infrastructure Management*, and *Communication*—enable core functions but are often overlooked in ethical analyses. The framework reveals that these processes bear ethical implications related to *Compensation and Incentives (G)*, *Operational Management and Corporate Ethics (N)*, and *Training, Culture, and Ethical Capacity Building (O)*. For example, *HR Management* is linked to issues of equity and inclusion (N, O), while *Procurement* connects to fair supplier practices and conflict

of interest prevention (F, G, K). These links highlight the importance of embedding ethical safeguards even in "back-office" operations.

- Management Processes—such as *Budget Management*, *Strategic Planning*, *Management Processes*, and *Protocol Portfolio Management*—are responsible for organizational direction and governance. Consequently, these processes were aligned with aspects like *Regulation and Compliance (K)*, *Global and Contextual Ethical Standards (L)*, and *Stakeholder Relationships (J)*. Particularly, *Protocol Portfolio Management* integrates broader ethical coordination across multiple trials, ensuring that strategy is consistent with both local and international standards. The strong presence of *Transparency in Strategy (A, D, K)* further suggests the increasing expectation that ethical considerations must be embedded into high-level planning and decision-making.

The allocation of responsible aspects across CRC process groups allows for a few analytical resolutions:

1. Ethical density is highest in Core Processes: The core operations of the CRC are linked to a broader and more intense set of ethical categories, especially concerning *informed consent*, *privacy*, and *scientific rigor*. This finding is consistent with clinical trial ethics literature, which places participant protection at the heart of responsible research conduct.
2. Support Processes address organizational integrity and fairness: Though less visible to external stakeholders, support processes are critical in sustaining ethical culture and avoiding structural vulnerabilities. The presence of categories such as *Training*, *Compensation*, and *Corporate Ethics* underscores the need for systemic integrity beyond participant-facing activities.
3. Management Processes reflect ethical governance and strategic accountability: Ethical aspects such as *global standards*, *regulatory compliance*, and *conflict of interest* become more prominent at the managerial layer. This reflects a shift from participant-centered ethics to institutional responsibility, especially in multi-site or multinational research environments.
4. Cross-cutting Aspects suggest the need for holistic DPT Integration: certain ethical aspects—such as *Transparency and Disclosure (A)* and *Regulation and Compliance (K)*—cut across all process types, indicating that their operationalization must be addressed through organization-wide digital strategies. This reinforces the need for DPTs that are not only process-specific but also cross-functional.

Figure 2 presents the ReDPT-CT framework.

By aligning DPT guidelines with ethical-process maps, the framework embeds responsibility into design—translating ethical principles into actionable features like audit logs and AI governance, ensuring ethics are proactively built in, not retroactively applied.

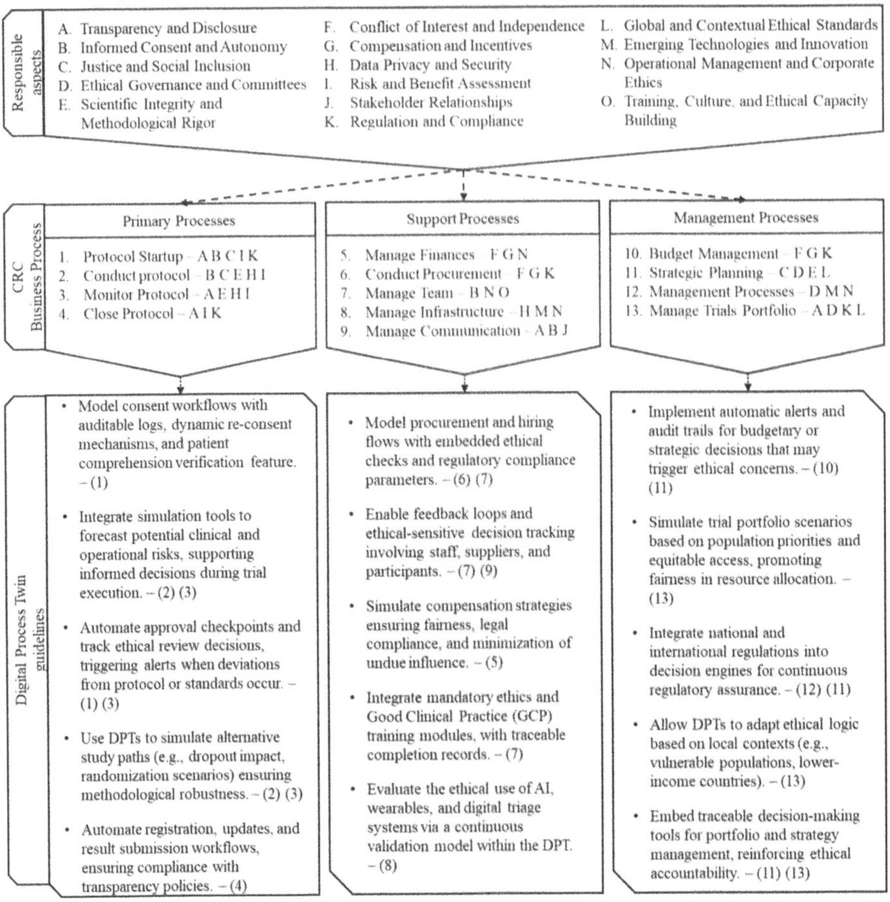

Fig. 2. ReDPT-CT framework

4.3 Research-Agenda Proposition

The implementation of the ReDPT-CT framework in real-world environments requires empirical testing and iterative feedback from CRCs to evaluate its usability, contextual adaptability, and effectiveness in addressing operational challenges. Conducting case studies across a range of CRC settings—including university-affiliated centers, privately managed facilities, and multicenter research networks—will be critical to understanding how institutional characteristics shape the ethical and digital integration of the framework. Such applications will generate comparative insights and highlight important nuances in ethical decision-making and process governance across different organizational contexts.

Furthermore, methodological advancements are needed to quantify the ethical and operational impacts of DPT adoption. Developing indicators and longitudinal study designs would enable researchers to measure how the integration of responsible DPTs influences compliance, transparency, patient autonomy, and process efficiency. Investigating the maturity of digital and ethical practices in CRCs over time can illuminate progress and persistent gaps, especially when comparing pre- and post-DPT implementation phases.

An important direction also lies in the technical integration of DPTs with existing clinical trial management systems such as eTMF, EDC, and CTMS platforms. This integration must be guided not only by interoperability standards but also by ethical safeguards—ensuring that data flows, traceability, and automation features align with principles of transparency, privacy, and autonomy. Finally, research is urgently needed to identify and analyze cultural and institutional barriers that may hinder the responsible deployment of DPTs. These include resistance to change, lack of digital literacy, insufficient ethical training, or misalignment between regulatory expectations and organizational culture. Addressing these challenges requires interdisciplinary efforts that bridge BPM, bioethics, digital health, and organizational studies—laying the foundation for a sustainable, ethically grounded digital transformation in clinical research.

5 Conclusions

The growing complexity of clinical trial operations, combined with rapid digital transformation in health research environments, has raised questions about how to align technological innovation with ethical responsibility. The adoption of DPTs in CRCs presents not only opportunities for operational efficiency but also challenges related to transparency, fairness, autonomy, and accountability. In this context, CRC managers would require structured guidance to understand which ethical dimensions are embedded in their processes and how these can inform responsible DPT implementation.

This study addressed the research question: what responsible aspects within business processes can a CRC manager consider when assessing the adoption of DPTs? The answer is provided through the development of the ReDPT-CT framework, a three-layer model that integrates (i) fifteen categories of responsible aspects derived from a structured literature review, (ii) CRC business process architecture classified into core, support, and management processes, and (iii) practical suggestions of DPT design guidelines mapped to each process. This framework enables a process-aware view of ethical responsibility, offering actionable insights into how each CRC workflow may trigger, reflect, or be enhanced by ethical principles in DPT implementation.

Key highlights from the results include the identification of ethical categories most frequently associated with core processes—such as informed consent and transparency—as well newer concerns in management processes, including portfolio fairness, global ethical variability, and AI governance. The allocation of DPT guidelines, such as automated approval checkpoints, traceable decision-making flows, and scenario-based simulations, demonstrates how digital twin functionalities can be ethically grounded within daily clinical operations. The framework's structure allows it to be used not only as an analytical tool but also as a design and training resource for CRC stakeholders navigating digital transitions.

Nevertheless, the study presents some limitations. First, the structured literature review, while rigorous, may benefit from further expansion given the breadth and interdisciplinarity of topics such as ethics, governance, business process and digital health.

Future iterations could include stakeholder interviews or additional sources such as policy documents. Second, the framework has not yet undergone empirical validation in real-world CRC environments. Feedback from key actors—such as principal investigators, research coordinators, and study sponsors—would be essential to assess the framework's practical applicability, perceived relevance, and usability across different institutional contexts. Despite these limitations, this study establishes a conceptual foundation for responsibly integrating DPTs into Clinical Trials, and sets a future agenda for methodological, empirical, and strategic advancements at the intersection of ethics, digital transformation, and BPM in health research.

References

1. Peyro-Saint-Paul, L., et al.: Compensating patients in trials: perspectives from an ethical committee versus sponsor. Eur. J. Clin. Invest. **53**(10) (2023)
2. Adda, J., Decker, C., Ottaviani, M.: P-hacking in clinical trials and how incentives shape the distribution of results across phases. Proc. Natl. Acad. Sci. **117**(24), 13386–13392 (2020)
3. Miller, J., Ross, J.S., Wilenzick, M., Mello, M.M.: Sharing of clinical trial data and results reporting practices among large pharmaceutical companies: cross sectional descriptive study and pilot of a tool to improve company practices. BMJ l4217 (2019)
4. Ozaki, A.: Conflict of interest and the CREATE-X trial in the New England journal of medicine. Sci. Eng. Ethics **24**(6), 1809–1811 (2018)
5. Eaton, M.L., Kwon, B.K., Scott, C.T.: Money and morals, pp. 297–315 (2014)
6. Coutellec, L.: Ethics and scientific integrity in biomedical research. In: Handbook of Research Ethics and Scientific Integrity, pp. 1–14. Springer International Publishing, Cham
7. Pascalev, A.: 191 ethical considerations of decentralized clinical trials. J. Clin. Transl. Sci. **6**(s1), 26–27 (2022)
8. Dumas, M., La Rosa, M., Mendling, J., Reijers, H.A; Fundamentals of Business Process Management. Springer Berlin Heidelberg (2018)
9. Ahmad, T., Van Looy, A.: Business process management and digital innovations: a systematic literature review. Sustainability **12**(17), 6827 (2020)
10. Tremaine, W.J., Carlson, M.R., Isaacs, K.L., Motil, K.J., Robuck, P.R., Wurzelmann, J.I.: Ethical issues, safety, and data integrity in clinical trials. Inflamm. Bowel Dis. **11**, S17–S21 (2005)
11. Kremer, G., Ribeiro, L.R.B., Blüher, T., Dallavalle, S., Stark, R.: Towards a unified approach: developing a reference model for digital twins and business process management in clinical trials. In: Gdowska, K., Gómez-López, M.T., Rehse, JR. (eds.) Business Process Management Workshops. BPM 2024. Lecture Notes in Business Information Processing, vol. 534. Springer, Cham (2025). https://doi.org/10.1007/978-3-031-78666-2_30
12. Damjanovic-Behrendt, V.: A digital twin-based privacy enhancement mechanism for the automotive industry. In: 2018 International Conference on Intelligent Systems (IS), pp. 272–279. IEEE (2018)
13. Kuštelega, M., Mekovec, R., Shareef, A.: Privacy and security challenges of the digital twin: systematic literature review. JUCS – J. Univ. Comput. Sci. **30**(13), 1782–1806 (2024)
14. Alja'afreh, M., Ahmad, O.B., Karime, A.: Enhancing privacy and security in digital twin and metaverse technologies: mathematical models, blockchain, and future directions. In: 2024 2nd International Conference on Intelligent Metaverse Technologies and Applications (iMETA), pp. 070–077. IEEE (2024)
15. Parris, D.L., Dapko, J.L., Arnold, R.W., Arnold, D.: Exploring transparency: a new framework for responsible business management. Manag. Decis. **54**(1), 222–247 (2016)

16. van der Aalst, W.M.P., Bichler, M., Heinzl, A.: Responsible data science. Bus. Inf. Syst. Eng. **59**(5), 311–313 (2017)
17. Pisoni, G., Moloney, M.: Responsible AI-based business process management and improvement. Dig. Soc. **3**(2), 23 (2024). https://doi.org/10.1007/s44206-024-00105-2
18. Scheinberg, L.C., Goldsmith, L.S.: Ethical and legal aspects of clinical trials. Arch. Neurol. **40**(11), 702 (1983)
19. Shamoo, A.E., Resnik, D.B.: Ethical issues for clinical research managers. Drug Inf. J. **40**(4), 371–383 (2006)
20. Iltis, A.S.: Stopping trials early for commercial reasons: the risk–benefit relationship as a moral compass: Table 1. J. Med. Ethics **31**(7), 410–414 (2005). https://doi.org/10.1136/jme.2004.008771
21. Jyothula, C.J., Sithole, T.: Digital twins in healthcare, pp. 48–68 (2024)
22. Zuhr, P., Rissmann, L., Meißner, S.: Framework for planning and implementation of Digital Process Twins in the field of internal logistics. IFAC-PapersOnLine **55**(10), 2221–2227 (2022)
23. Pop, E., Iliuţă, M.-E., Moisescu, M. A.: Digital twin types and applications in healthcare (2024)
24. Ashok, P., Pillai, S.: Transforming cold chain management: the role of digital twins in post-harvest logistics. In: 2024 2nd World Conference on Communication ; Computing (WCONF), pp. 1–6. IEEE (2024)
25. Altman, D.G., et al.: The revised CONSORT statement for reporting randomized trials: explanation and elaboration. Ann. Intern. Med. **134**(8), 663–694 (2001)
26. Vian, T.: Anti-corruption, transparency and accountability in health: concepts, frameworks, and approaches. Glob. Health Action **13**(sup1), 1694744 (2020)
27. Ross, J.S., Gross, C.P., Krumholz, H.M.: Promoting transparency in pharmaceutical industry-sponsored research. Am. J. Public Health **102**(1), 72–80 (2012)
28. Miller, J.E., Wilenzick, M., Ritcey, N., Ross, J.S., Mello, M.M.: Measuring clinical trial transparency: an empirical analysis of newly approved drugs and large pharmaceutical companies. BMJ Open **7**(12), e017917 (2017)
29. Zarin, D.A.: Participant-level data and the new frontier in trial transparency. N. Engl. J. Med. **369**(5), 468–469 (2013)
30. Nicholls, S.G., Langan, S.M., Benchimol, E.I., Moher, D.: Reporting transparency: making the ethical mandate explicit. BMC Med. **14**(1), 44 (2016)
31. Evans, S.: Societal responsibilities of clinical trial sponsors. BMJ **322**(7286), 569–570 (2001)
32. Matheson, A.: Marketing trials, marketing tricks — how to spot them and how to stop them. Trials **18**(1), 105 (2017)
33. Shabani, M., Obasa, M.: Transparency and objectivity in governance of clinical trials data sharing: Current practices and approaches. Clin. Trials **16**(5), 547–551 (2019)
34. Logvinov, I.: Clinical trials transparency and the Trial and Experimental Studies Transparency (TEST) act. Contemp. Clin. Trials **37**(2), 219–224 (2014)
35. DeVito, N.J., Morley, J., Goldacre, B.: Barriers and best practices to improving clinical trials transparency at UK public research institutions: a qualitative interview study. Health Policy (New York) **142**, 104991 (2024)
36. Abdul Razack, H.I., Aranjani, J.M., Mathew, S.T.: Clinical trial transparency regulations: implications to various scholarly publishing stakeholders. Sci. Public Policy **49**(6), 951–961 (2022)
37. Berger, V.W., Exner, D.V.: Detecting selection bias in randomized clinical trials. Control. Clin. Trials **20**(4), 319–327 (1999)
38. Goldstein, C.E., et al.: Ethical issues in pragmatic randomized controlled trials: a review of the recent literature identifies gaps in ethical argumentation. BMC Med. Ethics **19**(1), 14 (2018)

39. Lublin, F.D., Reingold, S.C.: Placebo-controlled clinical trials in multiple sclerosis: ethical considerations. Ann. Neurol. **49**(5), 677–681 (2001)
40. Lebacqz, K.: Controlled clinical trials: some ethical issues. Control. Clin. Trials **1**(1), 29–36 (1980)
41. Liddell, K., et al.: Recommendations in relation to the EU Clinical Trials Directive and Medical Research Involving Incapacitated Adults. Wien. Klin. Wochenschr. **118**(5–6), 183–191 (2006)
42. Kass, N.E., Faden, R.R., Angus, D.C., Morain, S.R.: Making the ethical oversight of all clinical trials fit for purpose. JAMA **333**(1), 75 (2025)
43. Califf, R.M., Sugarman, J.: Exploring the ethical and regulatory issues in pragmatic clinical trials. Clin. Trials **12**(5), 436–441 (2015)
44. Crosby, S., Malavisi, A., Huang, L., Jan, S., Holden, R., Neal, B.: Factors influencing the time to ethics and governance approvals for clinical trials: a retrospective cross-sectional survey. Trials **24**(1), 779 (2023)
45. Ashcroft, C., Clark, E., Frith, H.: Implications of socio-cultural contexts for the ethics of clinical trials.. Health Technol. Assess (Rockv) **1**(9) (1997)
46. Kadane, J.B.: Progress toward a more ethical method for clinical trials. J. Med. Philos. **11**(4), 385–404 (1986)
47. Ozaki, A., Takita, M., Tanimoto, T.: A call for improved transparency in financial aspects of clinical trials: a case study of the CREATE-X trial in the New England Journal of Medicine. Invest. New Drugs **36**(3), 517–522 (2018)
48. Gudi, N., et al.: Regulatory frameworks for clinical trial data sharing: scoping review. J. Med. Internet Res. **24**(5), e33591 (2022)
49. Rosenblatt, M.: Is it ethical to conduct placebo-controlled clinical trials in the development of new agents for osteoporosis? An industry perspective. J. Bone Miner. Res. **18**(6), 1142–1145 (2003)

Measuring Moral Performance: A Framework for Ethical Business Process Management

Leo Poss[1(✉)], Christopher Julian Kern[2], Christoph Stoiber[1,3], Julia Kroenung[2], and Stefan Schönig[1]

[1] University of Regensburg, Regensburg, Germany
leo.poss@ur.de
[2] University of Hagen, Hagen, Germany
[3] Freudenberg Flow Technologies, Wolfratshausen, Germany

Abstract. This paper proposes a novel approach to integrating ethical considerations into process management. The research endeavor is timely and relevant, as ethics in Business Process Management (BPM) have been scarcely researched, and practice increasingly focuses on ethical principles. Our framework, developed based on the fundamentals of design science research (DSR), provides a structured method for integrating ethical assessments into BPM. It combines established ethical perspective questionnaires with customizable evaluations of specific ethical values, enabling both high-level and granular assessments. This framework supports the modeling phase by allowing annotation of ethical aspects, the execution phase by facilitating real-time ethical assessment by users, and the monitoring phase by visualizing collected ethical performance data. It enables customizable ethical value tracking, enhancing ethical awareness in business operations. The concept was implemented as a prototype and evaluated with domain experts, demonstrating its utility in operationalizing ethical considerations in BPM. This research contributes to ethics-aware BPM by providing a practical framework to make ethical considerations explicit, measurable, and manageable within daily business operations and expanding the scope of research on quantifying and enhancing ethical process performance.

Keywords: BPM · Ethical Values · Sociotechnical System · Benevolent BPM · Ethics-aware BPM

1 Introduction

Business Process Management (BPM) is central to how modern organizations operate, enhancing the efficiency and effectiveness of business operations [6]. While ethical considerations are often implicit in process design and execution, current BPM practices primarily focus on traditional performance metrics, overlooking explicit mechanisms for the systematic assessment and management

M. ER et al. (Eds.): BPM 2025, LNBIP 565, pp. 117–132, 2026.
https://doi.org/10.1007/978-3-032-02936-2_10

of ethical values [17,26]. This gap highlights a critical need: to move beyond efficiency gains and develop methods that systematically assess, incorporate, and ensure transparency of ethical values during process execution.

The existing literature on ethical issues and organizational culture highlights gaps in broader ethical considerations, while security and privacy concerns have received some attention [3,17]. Adopting emerging digital technologies, such as machine learning and robotic process automation, needs to be designed to avoid compromising fairness, reducing biases, or eroding trust [19]. Therefore, it is relevant to align ethics and processes to mitigate the ethical pitfalls of digital technologies while maintaining efficiency by directly focusing on the underlying socio-technical system [15]. Novel research streams, such as benevolent business process management, emphasize the importance of designing business processes that incorporate ethical, environmental, and social considerations beyond operational efficiency and transactional value [26]. The imperative for ethical considerations is further underscored by the emerging concept of Digital Responsibility [29]. This concept emphasizes promoting responsible behavior and integrating ethical considerations directly into digital technologies. While existing research articulates these crucial conceptual approaches and calls to action, a notable gap remains: practical implementations are rare, and their systematic adoption within organizations is even scarcer.

To address these gaps, we introduce a framework, theoretically developed and realized as a prototypical implementation, for making ethical considerations tangible in BPM. This framework facilitates the assessment of ethical perspectives and specific values during task execution, enabling the collection of data for monitoring and improvement, designed for integration with BPMS. Our approach aims to answer the following key research question: *How can ethical considerations be effectively integrated (from modeling through execution to monitoring) within the BPM lifecycle to enable ongoing ethical assessment and improvement?*

Our research provides a novel framework for measuring moral performance by systematically integrating ethical perspectives and values into the generic BPM lifecycle and everyday process execution. This IT artifact is designed to be system- and model-language-agnostic, providing functionalities for collecting, measuring, and visualizing ethical values associated with processes. By enabling the collection and analysis of ethical performance data, our framework provides a basis for more thorough process monitoring and data-driven discussions on ethical improvement, thereby contributing to a more sustainable and responsible digital future.

2 Theoretical Background

2.1 Business Process Management (BPM)

Business Process Management (BPM) is an important factor in driving business value creation by enabling the digitization, automation, and strategic planning of operational processes [6]. While traditional BPM research predominantly focuses on the design and execution of processes, encompassing methodologies, languages, and tools for modeling, simulation, and execution, contemporary research has expanded this scope. This broadened perspective includes integrating business processes with critical organizational assets and resources, such as information systems, human resources, their interplay, and collaborative networks including socio-technical interplay and acting benevolent [6,26]. A business process, in its essence, is a structured set of activities designed to achieve a specific organizational objective [6,31]. These processes can be represented through one or more process models that then serve as blueprints for real-world scenarios, providing a structured and analyzable representation of the process flow [31]. In this way, a business process model serves a dual purpose: it allows for a visual representation of stakeholder communication and understanding, and acts as an executable specification for automated process execution within a BPMS.

2.2 Ethics

Ethics, often used synonymously with morality [9], is the branch of philosophy that investigates the principles guiding right and wrong human conduct [27]. It systematically examines ethical concepts, values, reasoning, and justifications [2], aiming to understand and guide human actions towards achieving a good life. This includes *normative ethical theories*, which provide frameworks for evaluating moral actions and character, *applied ethics*, which adapts these theories to specific domains [32], *descriptive ethics*, which empirically analyze existing moral beliefs [13] and *meta-ethics*, which explore the nature of ethical judgments themselves. Ethical theories, which provide systematic approaches to moral reasoning, can broadly be categorized into schools such as *teleological ethics* (focusing on outcomes or consequences, e.g., utilitarianism), *deontological ethics* (emphasizing duties, rules, and rights), and virtue ethics (focusing on moral character). While our framework does not rigidly enforce a single school, the chosen ethics scale (see below) incorporates elements reflecting both deontological (e.g., justice, contractualism) and relativistic considerations, offering a multifaceted lens for assessment.

Ethics scales are measurement instruments used in psychology, sociology, and philosophy to evaluate individuals or specific problems. In psychology, ethics scales are designed to assess ethical decision-making and behavior [5,25]. They often include dimensions such as moral equity, relativism, contractualism, and other ethical philosophies [18]. In contrast, sociology may use ethics scales to evaluate ethical behavior at different educational stages or organizational contexts [30]. Ethics scales are grounded in various ethical theories, including virtue, deontology, utilitarianism, and justice principles. The development of scales is influenced by moral philosophy, including the application of moral principles to analyze ethical issues [20]. One widely used and developed ethics scale and questionnaire is the 2013 scale developed by [25], as presented in Fig. 1. It is a more straightforward, shorter version of the scale developed in 1988 [24] and consists of three dimensions or constructs that are each evaluated on a 7-item

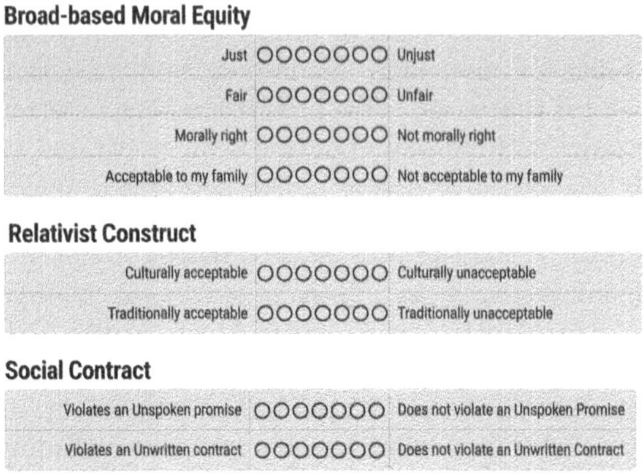

Fig. 1. Multidimensional ethics scale (MES) [25]

Likert-scale: *(i) Broad-based Moral Equity*, which describes the moral philosophy of justice (i.e., fairness and justness) with a deontological and relativistic concept for "broader-based notion of good and bad," *(ii)* a *Relativistic Dimension* that includes questions on guidelines, requirements, and parameters inherent in social/cultural systems and *(iii)* a *Contractualism Dimension* which is a deontological construct that further evaluates the "social contract" between businesses and society. Explicitly absent are egoistic and utilitarian considerations, as they did not contribute strongly to explainability and were hard to understand. Additionally, relativism can be seen as both a source and a standard for ethical evaluation and was deemed difficult to apply (cf. [25]). This multidimensional approach enables a nuanced assessment that encompasses concepts central to deontological thought (e.g., fairness, unspoken promises inherent in social contracts) and relativistic considerations tied to cultural and traditional acceptability.

2.3 Related Work

Ethics and BPM. The integration of ethics into process management is increasingly recognized as essential for aligning technological advancements with ethical and societal values – values such as transparency, accountability, and human well-being are essential aspects in the design and execution of processes, particularly as data-driven systems become more prevalent [15,17,19]. [8] highlight the significance of individual contributions in the formation of ethical organizational cultures. Shared values and ethical conduct play a pivotal role in implementing business processes. Ethical considerations in workflow automation extend to the displacement of jobs and the potential dehumanization of the workforce. Organizations are expected to proactively address these concerns, as this aligns with their ethical responsibility to mitigate conflicting impacts on the workforce [7].

Business Process Evaluation and Gathering Direct Feedback During Process Execution. The BPM lifecycle includes monitoring and improvement based on fundamental metrics. These metrics are collected during process execution in addition to classical performance metrics [1]. Our work builds on the idea of expanding these metrics, similar to approaches like [12], explicitly generating additional metrics and capturing implicit feedback, such as user interactions and behavioral patterns, to enrich process understanding. Furthermore, recommender systems and personalization research rely on information collected during process execution. By analyzing user preferences, behavior, and feedback, these systems can tailor recommendations and personalize experiences, leading to increased user engagement and satisfaction (e.g., [16]).

3 Methodology

Our research endeavor aims to enhance the representation of ethical considerations within business processes, which involves the explicit articulation, systematic assessment, and visual feedback of ethical values and perspectives as integral components of process models and their execution instances. In line with existing research on this topic, we position this contribution within the *design science research (DSR)* paradigm, as it enables the development and evaluation of practical artifacts [23]. To adopt established methods and draw on specific methodological guidance, we relied on the well-established seven guidelines for DSR as defined by [11]:

- *Design as an Artifact.* This research develops an executable artifact for integrating ethical considerations into business process execution. The artifact provides functionalities for: *(a)* annotating process tasks with predefined ethical perspective questionnaires (e.g., [25]) and user-defined specific ethical values during the modeling phase; *(b)* enabling process participants to provide assessments against these during task execution; and *(c)* collecting this data and visualizing ethical performance indicators for monitoring and discussion.
- *Problem Relevance.* The rapid adoption of technology in business outpaces ethical integration in BPM. Existing frameworks prioritize performance, overlooking moral implications in automation. This research addresses the prevalent lack of systematic and transparent approaches for embedding and managing ethics in BPM, a gap that can impact organizational integrity, stakeholder trust, and societal well-being, especially with increasing automation.
- *Design Evaluation.* The artifact's utility and effectiveness are assessed through integration into a BPMS and user feedback. Effectiveness in supporting ethical decision-making is evaluated using empirical data and cognitive load analysis via NASA-TLX [10], following the DSRM evaluation principle.
- *Research Contributions.* This research advances ethics-aware BPM by providing a practical framework and IT artifact for explicitly embedding, assessing, and visualizing ethical values in business processes, offering a new perspective on BPMS as a system capable of actively supporting ethical process execution

and monitoring, and contributing initial empirical insights into the feasibility and user perception of integrating ethics assessments in BPM.
- *Research Rigor.* The study applies rigorous methods from ethics, BPM, and IS, ensuring empirical validity. Data collection adheres to ethical guidelines, ensuring anonymity and maintaining research rigor principles in DSR.
- *Design as a Search Process.* The artifact undergoes iterative refinement based on stakeholder feedback, ensuring continuous improvement. The iterative approach aligns with DSRM's emphasis on design as an adaptive search process.
- *Communication of Research.* Findings are disseminated through conference publication, covering problem definition, design, evaluation, and demonstration. The artifact remains available for future refinements (Section Sect. 4.2).

To address our research question and the identified problem relevance, we derive the following design objectives to guide the artifact's development: ❶ *Perspective assessment* where the artifact must provide a way to assess broad-based ethical perspectives using an existing questionnaire, ❷ *Value assessment*, which requires the consideration and assessment of singular ethical values for a more granular view next to the general evaluation using a predefined, widely acknowledged measurement scale, that provide the base for the artifact, ❸ *Consider all phases of the BPM lifecycle* for a holistic approach similar to related approaches; the artifact must be supported and usable during the modeling, executing, and monitoring phases of BPM. And for general reusability and best-practices: ❹ *Independence of BPMS and modeling language* to ensure that the conceptual artifact can work independently from both the modeling language or paradigm and the chosen BPMS implementation that executes the process model, and ❺ *Modularity* which means, along the phases, the artifact must be as modular as possible to be as easily adaptable as the process modeling itself. This includes providing support for monitoring and visualizing different aspects of the data.

4 Design and Development

Based on DSR, the following sections outline the steps involved in designing and developing the artifact, including collecting, categorizing, and evaluating existing scientific questionnaires for ethical assessment, conceptualizing the framework based on these findings, and implementing this framework as a software prototype.

4.1 Concept – Ethics-Aware BPM Through Questionnaire and Ethical Values

Our framework's ability to collect ethical assessment data during task execution is designed to be independent of specific process modeling languages or BPMS. It leverages fundamental concepts of activity lifecycles common to most BPMS, specifically requiring at least an '*enabled*' (task is active or being worked on) and a '*completed*' (task is finished) state [31]. The ethical assessment interface is presented to the user while the task is '*enabled*' (i.e., during or immediately after completing the primary task work), and the collected assessment data is transmitted to a data store once the task instance transitions to '*completed*' and thus terminates. To contextualize the results, we need to provide a means to connect the values to a specific task type or process instance, along with a timestamp for temporal ordering. Finally, to fulfill our Design Objectives, all phases and parts must be modular and easily adjustable, from the possibility of manually modeling and defining which values to ask to the final visualization and data analysis of results. The first opportunity to assess ethical aspects is by using existing questionnaires that address ethical perspectives in various contexts. We provide the possibility for comparing high-level information based on the previously introduced questionnaire by [25], while explicitly leaving room for choosing different Likert-based ethical perspective questionnaires.

Binary Assessment of Singular Ethical Values. In addition to this approach for a broader, perspectival overview, we iteratively identified the need for a more specific survey that would enable the binary evaluation of specific ethical values. Ethical values, sometimes referred to as "common values," are shared by a particular group of people, such as those within a company or a country. The authority of these values typically originates from established agreements [4]. Several lists combine different sets of shared ethical values that can serve as a base catalog for selecting values to include for specific tasks. Two examples from related fields are [33], which describe values for sustainable information systems design, and [14], describing values for ethics in artificial intelligence. Both authors combine even more fine-grained values into overarching principles, including *Accessibility, Accountability, Efficiency, Environment, Privacy, Reliability, Transparency,* and *Trust.*

Nevertheless, using specific, singular ethical values comes with specific shortcomings: *(i)* their definition may be subjective and vary from person to person, and *(ii)* choosing one value from many may be challenging. Lastly, assessing large numbers of values requires additional effort during and after task execution (cf. Sect. 5). On the other hand, it enhances granularity, allowing for more precise monitoring and analysis.

4.2 Implementation – Evaluating Ethical Aspects in Process Execution

The following sections cover the development of a prototypical implementation for assessing ethical considerations during task execution[1]. Here, we implemented the questionnaire [25] (discussed in Sect. 2.2) for assessing broad ethical perspectives, along with functionality that allows users to evaluate predefined singular ethical values relevant to specific tasks. Still, depending on the execution context

[1] Prototype source: https://github.com/LeoPoss/assessingEthicsInBPM, Demo server: http://pcinf00057.ur.de/, Visualization: http://pcinf00057.ur.de/kibana/app/dashboards.

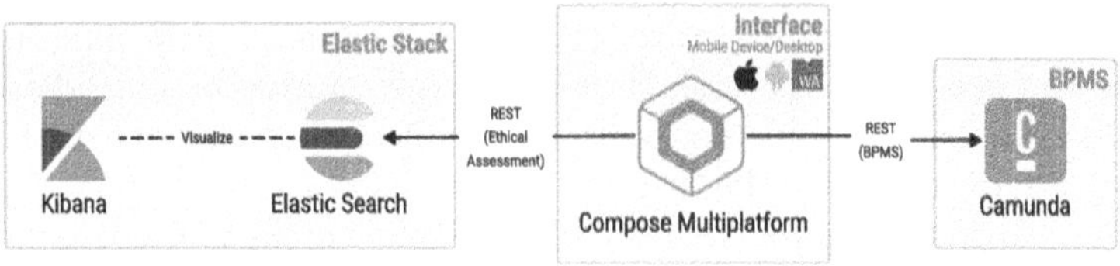

Fig. 2. Architecture of prototypical implementation

and tasks involved, the questionnaire can easily be adapted to another context if it is not relevant or applicable. The overall approach supports all phases of the BPM lifecycle, and the following sections follow this pattern, beginning with a brief introduction to the general system architecture.

The selected BPMS is *Camunda*[2], which is widely used within the research community. The system is readily adaptable and offers well-structured and documented access and the functionality for modeling and executing Business Process Model and Notation (BPMN) process models (cf. [21]). The interface is implemented using *Compose Multiplatform*[3], which provides an abstraction layer for cross-platform native mobile applications based on *Jetpack Compose*[4]. This decision enables broad native usability on mobile devices, desktop operating systems that use the Java Virtual Machine (JVM), and browsers through Kotlin WebAssembly (WASM). Finally, we rely on the *Elastic Stack*[5], which combines *Elasticsearch* to store information and uses *Kibana* as a tool for discovery and visualization, allowing easy adaptation based on necessary parameters and configurations.

Architecture. Figure 2 shows how the prototype for assessing values is implemented in our framework. At a high level, it consists of three main parts: Starting with the BPMS that provides the execution logic for the process model and well-defined endpoints to interact with it; second, the main application running on mobile devices, browsers, or other devices providing an interface to the BPMS and implementing the concept from above; and lastly the existing integration of Elastic Search and Kibana for data collection, monitoring, and visualization.

Functionality. For a concise overview of all functionalities, the process used for explanation and visualization represents a basic HR process modeled using BPMN with three consecutive User Tasks (cf. [21]): *Screen Application* followed by *Conduct Interview* and *Evaluate Candidate.* For simplicity, no additional steps or decisions are required; the process proceeds directly, and it concludes after evaluating the candidate.

Modeling. The first value assessment step is providing additional information during process modeling. Integration into the modeling phase is achieved by annotating User Tasks within the BPMN model. In our Camunda-based pro-

[2] https://camunda.com.
[3] https://www.jetbrains.com/compose-multiplatform.
[4] https://developer.androind.com/compose.
[5] https://www.elastic.co/elastic-stack.

(a) Start screen with task list.

(b) Task details including singular value assessment.

(c) Submission sheet with questionnaire grouped in perspectives.

Fig. 3. Prototypical implementation running on a mobile device.

totype, this utilizes the existing capability to associate forms or metadata (via form keys) with tasks. We define a specific JSON structure for this form key, allowing process designers to specify: *(a)* whether the full ethical perspectives questionnaire (e.g., Reidenbach & Robin [25]) should be presented for a task (via a binary `perspectives` flag), and *(b)* a list of specific, singular `values` (e.g., 'Fairness', 'Transparency') to be assessed for that particular task. Our interface then queries this configuration during task execution to present the appropriate assessment options dynamically.

Task Execution and Data Collection. The application interface allows access to the BPMS and enables the user to assess value fulfillment and perspectival evaluation of tasks. The first screen in Fig. 3 displays a list of all currently available tasks and their current assignee, a search bar, and a filter for assigned tasks. Additionally, it provides the possibility to start new process instances. The following two screens include the added functionality for assessment: The detailed view for the task contains general information like the title and more detailed information like duration, due date, description, and the possibility to claim and return tasks followed by *(i)* a stack of swipeable cards that each contain one ethical value defined in the process model if any that can be swiped right or left based on the fulfillment during task execution, and *(ii)* when the task is completed and is submitted by the user, an overlay will show the questionnaire including all perspectives and sliders for assessing their satisfaction level following the seven-item scale defined in [25]. The task is finally completed after assessing binary values and perspectives. The application triggers task completion in Camunda, which advances the process and sends the data to Elasticsearch in a structured manner.

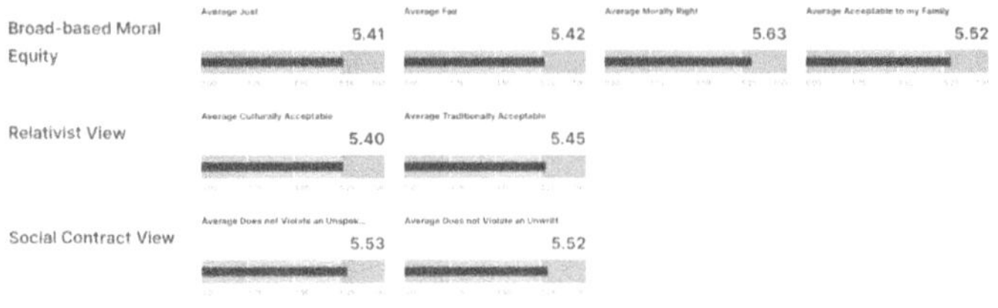

(a) Average assessment based on perspectives and broad views.

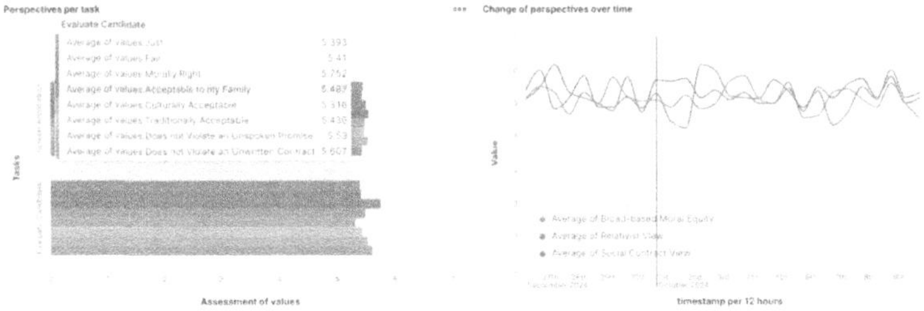

(b) Average perspectives per task and aggregated change of views over time.

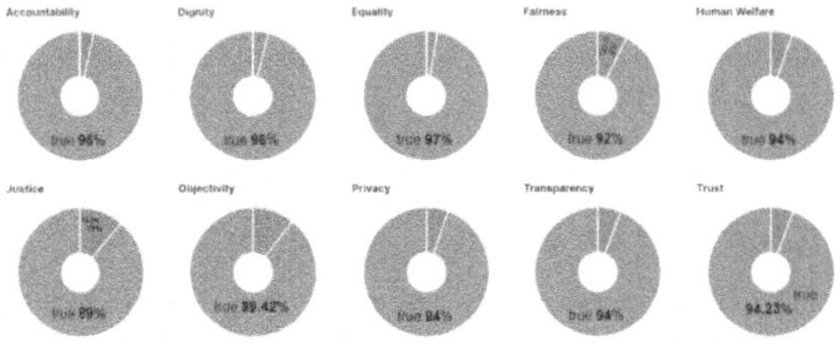

(c) Visualization of binary fulfillment of specific ethical values

Fig. 4. Exemplary visualizations based on the collected data.

Data Evaluation and Visualization. The last component collects, stores, discovers, and visualizes the data to gain deeper insights into the ethical performance of the executed process. The combination of Elasticsearch and Kibana is a widely adopted industry standard for observability, and the structured approach and format of data handling enable seamless integration with other metrics from various sources. Figure 4 shows examples of dashboards created in Kibana using data collected via the prototype. These visualizations include: The first shows the current average satisfaction for each perspective, grouped by the broader views, while the second provides a more detailed view at the task level, along with the temporal evolution of the averages for the three principal perspectives. The latter provides an overview of the binary assessment of specific ethical values, allowing for drill-down analysis to individual tasks. All parameters and visualizations are easily customizable, allowing for filtering by specific tasks and different periods, as well as visualizing changes over time using built-in functionality. This can be achieved by selecting all tasks where the value of privacy is false and then immediately displaying all visualizations for only those process

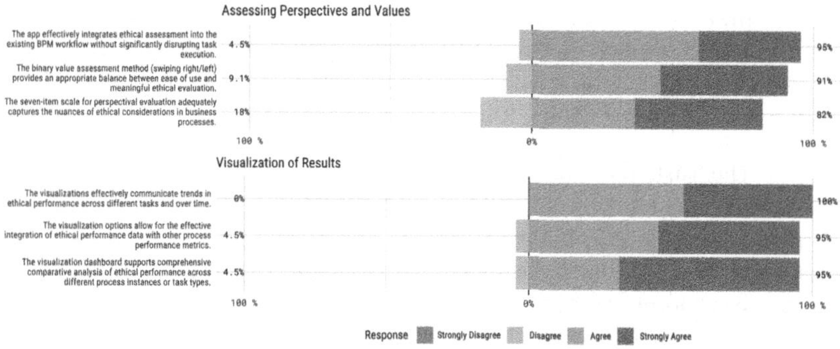

Fig. 5. Evaluation of Assessment and Visualization.

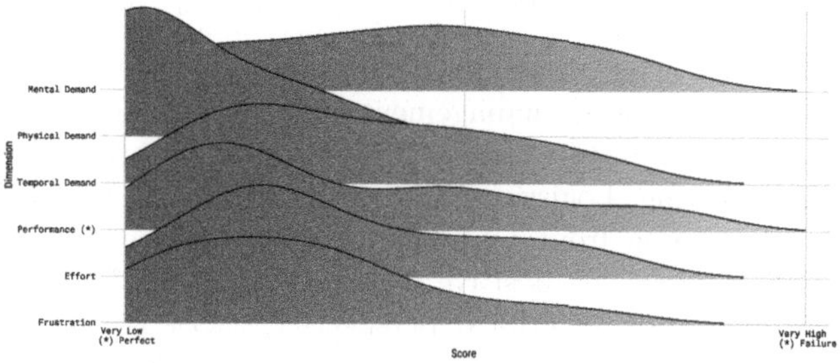

Fig. 6. Evaluation NASA-TLX Cognitive Load Assessment

instances. Another idea could be to explicitly select a single task to examine its performance further and gain insight into possible correlations between ethical values, perspectives, and other indicators. This visualization capability is crucial for the monitoring phase of the BPM lifecycle, enabling stakeholders to identify trends, potential areas of ethical concern, and facilitating discussions around ethical process improvement.

5 Evaluation

The DSR methodology encompasses a range of techniques for assessing the quality of the resulting artifact. The above approach aligns with the prototype method category. Consequently, it can be regarded as an "implementation of an artifact aimed at demonstrating the utility or suitability of the artifact" [22]. Additionally, to gather initial feedback on the artifact's utility and usability, we conducted an evaluation with $n = 22$ process experts and practitioners. Participants were provided with a document briefly introducing the research context and the artifact's purpose, a description of the artifact's functionalities (modeling annotation, task execution assessment, visualization dashboard), and access to the fully functional prototypical implementation, including the mobile/web interface and the Kibana dashboard, pre-populated with sample data from the

HR hiring process scenario (*Screen Application, Conduct Interview, Evaluate Candidate* – as described in Sect. 4.2). This scenario was chosen for its relatability, allowing participants to focus on the artifact's mechanics rather than complex domain-specific ethics. Participants were asked to:

- Explore the task list and complete a few instances of the predefined HR process tasks using the ethical assessment features (both singular values and the perspective questionnaire).
- Interact with the Kibana dashboard to view and filter the (sample and their own newly generated) ethical assessment data.
- Subsequently, they completed an online questionnaire comprising specific questions about the assessment interface and the monitoring dashboard (results shown in Fig. 5), as well as the NASA-TLX questionnaire to assess the cognitive load (results shown in Fig. 6).

Participants range in age from 23 to 48 (median: 32) and have, on average, 5.3 years of experience in process management and business processes. Participating companies are based in Germany and vary widely in terms of economic sectors (e.g., Manufacturing, Logistics, Automotive, Financial Services, and Technology) and size, ranging from a minimum of 5 employees to a maximum of 320,000 (median: 3,500). The positive responses shown in Fig. 5 regarding the app's effective integration of ethical perspectives (95% agree/strongly agree) and the usefulness of the seven-item scale (82%) provide initial validation for our *Perspective Assessment* and *Value Assessment* design objectives. Similarly, high agreement on the utility of the visualizations for understanding ethical performance (100% and 95%) supports the *monitoring* aspect of our framework. The artifact has been well-received overall by all participants. However, minor adjustments are still needed to facilitate its active use during everyday work (cf. Sect. 6). Finally, we evaluated the artifact using the NASA-TLX questionnaire to measure the cognitive task load (e.g., [10]), assessed ethical behavior during process execution using existing ethical questionnaires and specific ethical values, and presented the results in Fig. 6. NASA-TLX was chosen to evaluate the cognitive workload imposed by the artifact, as it provides a validated, multidimensional assessment of mental effort, frustration, and key factors in determining usability and efficiency. Given that ethical decision-making involves complex cognitive processing and high mental demand [28], this approach ensures that the artifact supports users without overburdening their cognitive capacities. The NASA-TLX results indicate that while participants were generally successful in using the artifact with moderately low frustration, the assessment process itself was perceived as having a strong mental and temporal demand, and requiring effort. This suggests that while the interface is usable, the cognitive task of ethical evaluation is inherently demanding [28]. This finding highlights the importance of features such as customizable questionnaires and sampling strategies (see Sect. 6) in managing this load in practice.

Considering the objectives defined, we can assess the evaluation of the implementation of ❶ and ❷. The other objectives ❸, ❹, and ❺ (consideration of the complete lifecycle, independence of paradigm and system, and modularity) are conceptual requirements, where the implementation of the concept provides a structural and functional evaluation, as described in [11].

6 Discussion

Based on the broader evaluation with practitioners, we can again reiterate the importance of ethical values and their support in process management. The artifact created in our paper is agnostic to business specifics and can be easily adapted to any business process and culture. To effectively integrate ethical considerations into the BPM lifecycle, our approach embeds ongoing ethical assessment as a seamless and adaptable component of BPM methodologies, thereby effectively answering our research question. While the proposed approach offers a novel method for integrating ethical considerations into business process execution, it is essential to acknowledge potential limitations identified during the evaluation. One challenge is the approach's scalability and usability for process sizes and task volumes that span a wide range. Evaluating every value and every perspective each time a task is finished can be time-consuming. However, simple sampling techniques can mitigate this issue, where only subsets of values and ethical perspectives are collected (e.g., ask to assess one value for every tenth task or one perspective every hour, depending on the number of actors and process instances), directly influencing temporal demand and effort. Another inherited concern is the potential for self-reporting bias in ethical assessments, which can be partly mitigated by using standardized and well-researched questionnaires and clear definitions of ethical values, thereby promoting consistency and comparability. Furthermore, while the adaptability of the framework is a strength, allowing for customization, it also requires careful consideration and adjustments based on specific organizational needs and cultural contexts when implementing (e.g., selecting or refining the set of ethical values or modifying questionnaires). Beyond these concept-specific limitations, it is important to recognize the inherent subjectivity in interpreting ethical values. Finally, collecting and analyzing data on individuals' ethical behavior, even if aggregated, requires careful consideration of data privacy and potential misuse. The proposed solution suggests anonymizing data where possible (e.g., not linking specific assessments to individual actors for performance review purposes unless explicitly designed for transparent coaching and with their consent) and securing storage to ensure compliance with privacy regulations, such as the GDPR.

7 Conclusion

This paper presents a novel framework and a prototypical implementation, representing the first step towards systematically integrating explicit ethical considerations into the daily execution and management of business processes. We combined ethics scales with specific value evaluations to create a multifaceted assessment in the BPM lifecycle. This customizable and agnostic approach enables the tracking of ethical values and performance monitoring from both conceptual and prototypical implementations, which we evaluated with the assistance of experts. By doing so, we also introduce mechanisms for real-time feedback on ethical performance during process execution, enabling immediate adjustments and promoting continuous ethical learning. While our current approach has its limitations, as previously outlined, it is essential to consider these limitations when formulating future strategies. To ensure the integrity of the results, it is critical to minimize self-reporting bias and subjectivity by promoting ethical processes. This includes future work in a few directions, like a quantitative study of the artifact at scale (e.g., gather quantitative data on ethical performance over a more extended period, including correlation with other process KPIs), further research on specific ethical values in process management based on interviews (e.g., interviews with stakeholders across different roles to uncover a richer understanding of the values relevant to specific process contexts), and leveraging *Value-Driven Process Improvement*, i.e., aligning business objectives with ethical considerations and foster ethics as a first-class citizen in BPM.

It is essential to reiterate the gap in BPM research regarding the explicit, manageable, and measurable integration of ethics, particularly in light of the growing role of automation and its impact on human actors [17,19]. By providing this practical framework and its prototype, we aim to stimulate further interdisciplinary research into quantifying the performance of ethical processes, developing methods for ethics-driven process improvement, and integrating ethical assessments into routine business operations.

References

1. van der Aalst, W.M.P.: Process Mining: Data Science in Action. Springer, Heidelberg (2018). https://doi.org/10.1007/978-3-662-49851-4
2. Becker, L.C.: Encyclopedia of Ethics. Routledge (2001)
3. vom Brocke, J., Zelt, S., Schmiedel, T.: On the role of context in business process management. Int. J. Inf. Manage. **36**(3), 486–495 (2016)
4. Carr, D.: Common and personal values in moral education. Stud. Philos. Educ. **17**(4), 303–312 (1998)
5. Dufrene, R.L., Glosoff, H.L.: The ethical decision-making scale-revised. Meas. Eval. Couns. Dev. **37**(1), 2–14 (2017)
6. Dumas, M., La Rosa, M., Mendling, J., Reijers, H.A.: Process-Aware Information Systems. In: Fundamentals of Business Process Management, pp. 341–369. Springer, Heidelberg (2018). https://doi.org/10.1007/978-3-662-56509-4_9
7. Dykman, C.A., Davis, C.K.: Addressing resistance to workflow automation. J. Leadersh. Account. Ethics **9**(3), 115–123 (2012)
8. Ferreira, A., Otley, D.: The design and use of performance management systems: an extended framework for analysis. Manag. Account. Res. **20**(4), 263–282 (2009)
9. Garnett, A.C.: Distinctions and definitions in ethics. Philos. Phenomen. Res. **12**(1), 69–82 (1951)

10. Hart, S., Staveland, L.: Development of NASA-TLX (Task Load Index): Results of Empirical and Theoretical Research, pp. 139–183. Elsevier (1988)
11. Hevner, A.R., March, S.T., Park, J., Ram, S.: Design science in information systems research. MIS Q. **28**(1), 75–105 (2004)
12. Hu, Y., Koren, Y., Volinsky, C.: Collaborative filtering for implicit feedback datasets. In: 2008 Eighth IEEE International Conference on Data Mining. IEEE (2008)
13. Hämäläinen, N.: Descriptive Ethics. Palgrave Macmillan US (2016)
14. Jobin, A., Ienca, M., Vayena, E.: The global landscape of AI ethics guidelines. Nat. Mach. Intell. **1**(9), 389–399 (2019)
15. Jonny, J., Kriswanto, K., Toshio, M.: Building an implementation model of IoT and big data and its improvement. Int. J. Technol. **12**(5), 1000 (2021)
16. Kaptein, M., Parvinen, P.: Advancing e-commerce personalization: Process framework and case study. Int. J. Electron. Comm. **19**(3), 7–33 (2015)
17. Kern, C.J., Poss, L., Kroenung, J., et al.: Navigating the moral maze: a literature review of ethical values in business process management. Bus. Process. Manag. J. **30**(8), 343–370 (2024)
18. McMahon, J.M., Harvey, R.J.: Psychometric properties of the Reidenbach–Robin multidimensional ethics scale. J. Bus. Ethics **72**(1), 27–39 (2006)
19. Mendling, J., Decker, G., Hull, R., et al.: How do machine learning, robotic process automation, and blockchains affect the human factor in business process management? Commun. Assoc. Inf. Syst., 297–320 (2018)
20. O'Fallon, M.J., Butterfield, K.D.: A review of the empirical ethical decision-making literature: 1996–2003. J. Bus. Ethics **59**(4), 375–413 (2005)
21. OMG: Business Process Model and Notation (BPMN), Version 2.0.2 (2011). http://www.omg.org/spec/BPMN/2.0.2/
22. Peffers, K., Rothenberger, M., Tuunanen, T., Vaezi, R.: Design science research evaluation. In: Peffers, K., Rothenberger, M., Kuechler, B. (eds.) DESRIST 2012. LNCS, vol. 7286, pp. 398–410. Springer, Heidelberg (2012). https://doi.org/10.1007/978-3-642-29863-9_29
23. Peffers, K., Tuunanen, T., Rothenberger, M.A., et al.: A design science research methodology for information systems research. J. Manage. Inform. Syst. **24**(3), 45–77 (2007)
24. Reidenbach, R.E., Robin, D.P.: Some initial steps toward improving the measurement of ethical evaluations of marketing activities. J. Bus. Ethics **7**(11), 871–879 (1988)
25. Reidenbach, R.E., Robin, D.P.: Toward the development of a multidimensional scale for improving evaluations of business ethics. In: Michalos, A., Poff, D. (eds.) Citation Classics from the Journal of Business Ethics. Advances in Business Ethics Research, vol 2, pp. 45–67. Springer, Dordrecht (2013). https://doi.org/10.1007/978-94-007-4126-3_3
26. Rosemann, M.: Benevolent business processes - design guidelines beyond transactional value. In: BPM 2023. LNCS, vol. 14159, pp. 447–464. Springer, Cham (2023). https://doi.org/10.1007/978-3-031-41620-0_26
27. Roth, J.K.: Ethics: Ready Reference. Salem Press (1994)
28. Street, M.D., Douglas, S., Geiger, S.W., et al.: The impact of cognitive expenditure on the ethical decision-making process: the cognitive elaboration model. Organ. Behav. Hum. Decis. Process. **86**, 256–277 (2001)
29. Trier, M., Kundisch, D., Beverungen, D., et al.: Digital responsibility. Bus. Inform. Syst. Eng. **65**(4), 463–474 (2023)

30. Valentine, S., Barnett, T.: Ethics codes and sales professionals' perceptions of their organizations' ethical values. J. Bus. Ethics **40**(3), 191–200 (2002)
31. Weske, M.: Business Process Management, 3rd edn. Springer, Heidelberg (2019)
32. Winkler, E.R. (ed.): Applied Ethics. Blackwell (1993)
33. Winkler, T., Spiekermann, S.: Human values as the basis for sustainable information system design. IEEE Technol. Soc. Mag. **38**(3), 34–43 (2019)

Process Technology Forum

Preface of the Process Technology Forum

Process systems combine advanced computer science methods and techniques with real-world applications that impact nearly every industry. They sit at the heart of modern applications, helping enterprises to streamline their operations, enhance process efficiency, and respond to the needs of ever-changing markets.

Process system engineers face complex challenges such as developing flexible modeling frameworks that can capture diverse processes and complex process dependencies in business operations. They also develop simulation tools to test these processes under different scenarios, providing valuable insights into resource allocation, bottlenecks, and process optimization. Process systems often incorporate advanced analytics to track performance, allowing for the real-time monitoring of processes, the detection of inefficiencies, and data-driven improvements. The enactment component needs to be able to handle large numbers of concurrently executed processes efficiently, securely, and reliably. It further requires seamless integration with various enterprise software systems, ensuring that processes align across all organizational functions. As organizations increasingly adopt AI-driven automation, process systems are evolving to support automated decision-making, predictive analytics, and self-optimization. Additionally, process systems are increasingly designed for cross-organizational scenarios, requiring engineers to address scalability, security, privacy, and fault tolerance across multiple process systems, ensuring high availability.

The Process Technology Forum provides a platform for discussion and introduction of new research ideas related to the technical and engineering aspects of process systems. We invited submissions on technical and engineering aspects related to the study, design, development, and maintenance of software systems that support the modeling, simulation, analysis, enactment and monitoring of business processes, "process systems" in short.

The forum received 20 paper submissions for evaluation, out of which 10 were selected for presentation. Each paper was reviewed by at least three reviewers, after which the program chairs made a decision. The papers presented cover a broad range of different areas of process technology, including resource allocation in business processes, blockchain and smart contracts, robotic process automation, Internet of things, privacy, and process mining. To kick off the Process Technology Forum, Stefanie Rinderle-Ma gave a keynote on Process Systems in the Era of Artificial Intelligence.

We hope that the reader finds the selection of papers and presentations useful to get an insight into advances of the state of the art in process technology.

Organization

Program Chairs

Mathias Weske	HPI, University of Potsdam, Germany
Remco M. Dijkman	Eindhoven University of Technology, The Netherlands
Dimka Karastoyanova	University of Groningen, The Netherlands

Program Committee

Andreas Metzger	Universität Duisburg Essen, Germany
Andrés Jiménez Ramírez	Universidad de Sevilla, Spain
Barbara Weber	Universität St. Gallen, Switzerland
Barbara Re	University of Camerino, Italy
Bernhard Rumpe	RWTH Aachen, Germany
Carl Corea	Universität Koblenz, Germany
Chiara Di Francescomarino	University of Trento, Italy
Cristina Cabanillas	University of Seville, Spain
Dimitris Karagiannis	Universität Wien, Austria
Dominik Bork	TU Wien, Austria
Gregor Engels	Universität Paderborn, Germany
Ingo Weber	Technische Universität München, Germany
Irene Vanderfeesten	KU Leuven, Belgium
Jochen Küster	Hochschule Bielefeld, Germany
Julius Köpke	Universität Klagenfurt, Austria
Luise Pufahl	Technische Universität München, Germany
Manfred Reichert	Universität Ulm, Germany
Matthias Weidlich	HU Berlin, Germany
Mirela Riveni	University of Groningen, Netherlands
Monica Vitali	Politecnico di Milano, Italy
Orlenys López-Pintado	University of Tartu, Estonia
Patrick Delfmann	Universität Koblenz, Germany
Peter Fettke	Universität des Saarlandes, DFKI, Germany
Peter Loos	Universität des Saarlandes, DFKI, Germany
Pierluigi Plebani	Politecnico di Milano, Italy
Raimundas Matulevicius	University of Tartu, Estonia
Rik Eshuis	Eindhoven University of Technology, Netherlands

Robin Bergenthum	Fernuniversität Hagen, Germany
Ronny Seiger	Universität St. Gallen, Switzerland
Sander Leemans	RWTH Aachen, Germany
Shazia Sadiq	University of Queensland, Australia
Simon Hacks	Stockholm University, Sweden
Stefan Schönig	Universität Regensburg, Germany
Stefanie Rinderle-Ma	Technische Universität München, Germany
Chun Ouyang	Queensland University of Technology, Australia

KRAFT – A Knowledge-Graph-Based Resource Allocation Framework

Leon Bein[1(✉)], Niels Martin[2], and Luise Pufahl[1]

[1] School of Computation, Information, and Technology, Technical University of Munich, Heilbronn, Germany
{leon.bein,luise.pufahl}@tum.de

[2] Hasselt University, Hasselt, Belgium
niels.martin@uhasselt.be

Abstract. Resource allocation in business process management involves assigning resources to open tasks while considering factors such as individual roles, aptitudes, case-specific characteristics, and regulatory constraints. Current information systems for resource allocation often require extensive manual effort to specify and maintain allocation rules, making them rigid and challenging to adapt. In contrast, fully automated approaches provide limited explainability, making it difficult to understand and justify allocation decisions. Knowledge graphs, which represent real-world entities and their relationships, offer a promising solution by capturing complex dependencies and enabling dynamic, context-aware resource allocation. This paper introduces KRAFT, a novel framework that leverages knowledge graphs and reasoning techniques to support resource allocation decisions. We demonstrate that integrating knowledge graphs into resource allocation software allows for adaptable and transparent decision-making based on an evolving knowledge base.

Keywords: Business Process Management · Resource Allocation · Knowledge Graphs

1 Introduction

Organizations execute business processes to deliver products or services to their customers. Effective and efficient process execution relies on various resources, including staff, machines, vehicles, and software that perform process activities [13]. These resources are diverse and have varying capabilities and functionalities. At the same time, they are valuable, often costly, and scarce.

During process execution, each newly enabled task, i.e., activity instance, requires allocating one or more suitable resources, a decision that is inherently complex [22]. For instance for human resources, factors such as individual expertise, past experience, and team membership, but also their availability, task-specific characteristics, other competing tasks, etc. can influence the speed, outcome, and quality of task execution [2]. Additionally, dependencies between

M. ER et al. (Eds.): BPM 2025, LNBIP 565, pp. 137–152, 2026.
https://doi.org/10.1007/978-3-032-02936-2_11

preceding and subsequent tasks within the process further complicate resource allocation [30]. We refer to such factors as *knowledge* that needs to be considered during resource allocation.

While resource allocation can be performed manually, information systems and business process management systems (BPMS) typically provide capabilities to support this complex decision [13,22]. In many cases, process experts define resource allocation knowledge as extensions to a process model, e.g., by specifying rules and constraints for process activities [7,26], formulating allocation problems and solution techniques [16], or incorporating additional visual elements [6]. Furthermore, process experts must define the set of available resources along with their attributes and, in some cases, the organizational structure, such as hierarchical relationships [32]. This approach often requires *significant manual effort* for specification and raises *maintenance challenges* as business processes are regularly adapting to changing environments [23], which also necessitates changes in resource allocation. Approaches have been developed to extract resource allocation knowledge from past process execution data and use this knowledge for real-time resource allocations [22]. However, such approaches are *limited in the concepts* for which knowledge can be extracted [24]. Further, *explainability* is key for complex process-decision-making systems [12] as it builds trust in the system's decisions, which is currently missing in resource allocation support [22]. System-provided reasoning allows human decision-makers to make informed decisions or adapt system-made decisions while saving the time to look up the relevant information [27].

Methods from the fields of knowledge graph reasoning and ontology-based business process modeling appear promising to bridge this gap of easily adaptable approaches that provide comprehensive reasoning based on wide allocation knowledge. Knowledge graph reasoning aims to derive new knowledge from rich graph representations of the real world [8], providing tools that can consider all that knowledge in an explainable way [11,27]. In turn, ontology-based business process modeling is concerned with extending classical process models with instances of ontologies that represent additional background process knowledge [10], resulting in rich graph structures of process knowledge.

Research Objectives. While existing works, e.g., propose to consult taxonomies of resource criteria [2,20] or propose the usage of graph reasoning to recommend next activities [4], the use of knowledge graph reasoning in combination with ontological modeling has not yet been investigated for resource allocation in business processes. To this end, this paper investigates this use, considering the following objectives derived from the gaps explained above:

- O1 It should support capturing and encoding a broad set of allocation knowledge
- O2 It should support an easy adaptation of the encoded allocation knowledge
- O3 It should determine (the most) suitable resources based on that knowledge and comprehensively explain the decisions made

To realize these objectives, this paper proposes *KRAFT, a Knowledge-graph-based Resource Allocation Framework for Tasks* in business processes. It allows populating, updating, and reasoning on a knowledge graph that captures resource allocation knowledge for explainable resource allocation decision support. Further, we present a prototypical implementation and demonstration of it on a loan application process, provided as a real-world data set in [28]. In Sect. 2, we discuss preliminaries and related work on resource allocation and knowledge graphs. Then, in Sect. 3, KRAFT is presented in detail, and demonstrated using a proof of concept implementation in Sect. 4. Finally, we discuss our findings in Sect. 5 and conclude in Sect. 6.

2 Background and Related Work

2.1 Resource Allocation

In the business process management (BPM) field, resource allocation describes the decision of assigning concrete process tasks (i.e., enabled activity instances) to resources during process execution, adhering to run-time information (e.g., the resource availability) and previously defined allocation goals and constraints [7]. In this work, we mainly focus on allocating tasks to human resources. Resource allocation can have various *(optimization) goals*, such as balancing workloads, or minimizing cost or cycle time. Further, resource allocation must follow *hard constraints* (e.g., the separation of concerns principle or specific permission requirements) and *soft constraints* (e.g., preference for resources with a particular experience), where non-adherence may be associated with a penalty [22].

To avoid lengthy manual decision-making, resource allocation can be supported by information systems, which either fully automate the decision or provide recommendations for human decision-makers [26,32]. These systems need to be aware of knowledge for performing suitable resource allocations [22]. We define *allocation knowledge* as all this (potentially implicit) knowledge relevant for making an allocation decision. In literature, Pufahl et al. [22] describe how goals (e.g., w.r.t. process performance or balancing workload) and constraints (e.g., for complying with specific compliance rules or allowing a certain process quality) have to be considered for resource allocation. Each activity might be associated with its own subgoals and specific constraints [16]. Further, allocation mechanisms [32] are part of the allocation knowledge and can be selected to support decisions. They include allocation rules as collected by Russel et al. [26], such as *role-based allocation*, i.e., requiring assigned resources to possess specific roles, or assigning based on previous work on the case. Beyond this, Ouyang et al. [20] and Arias et al. [2] discuss knowledge about individual resources such as their roles, their expertise, workload, social context, and the organizational setting as well as their previous performance. Additionally, resource behavior, such as prioritization, handovers, or batching is summarized by Rubensson et al. [24]. Finally, Pufahl et al. [22] highlight that knowledge about the tasks to assign is also relevant, such as resource requirements, priority, expected performance, and its history. Based on these related works, we can observe that existing research selectively considers certain allocation knowledge, but no overall knowledge categorization for resource allocation exists.

Many state-of-the-art resource allocation approaches use black-box models such as reinforcement learning algorithms (e.g., [19]), hindering explainability and discarding knowledge not encoded in event logs. Conversely, related to our work, Havur et al. [14] use answer set programming, a logical reasoning approach, for resource allocation based on a knowledge graph encoding role and availability constraints. However, their work does not provide a framework for adapting the knowledge graph and applying explainable reasoning techniques.

When information systems or even BPMSs support the execution of business processes, process execution data is stored. Using this data in the form of event logs, *process mining* offers data-driven techniques to discover the process, check its conformance, and enhance it with additional information [12]. Process mining offers a set of techniques to derive knowledge on behavioral and structural aspects of resources in business processes as reviewed by Martin and Beerepoot [18] and Rubensson et al. [24]. These techniques provide opportunities to derive structural knowledge for resource allocation from process execution data, but are limited to the information stored in event logs.

2.2 Knowledge Graph Reasoning

Knowledge graphs (KG) can be defined as data graphs whose nodes and edges represent real-world entities of interest and their relations, and that are extended by context, constraints, and inference rules in the form of ontologies [15]. This context can be used to formally and informally define the domain, e.g., in terms of types of entities and properties of relations [15].

Knowledge graphs allow encoding a wide range of knowledge and can thus be used in diverse cases, such as for product recommendations, using the graph to encode product and user relations [15], or for personalized medicine, using the graph to encode biochemical interactions [25].

Formally, KGs can be expressed as sets of triples of the form *(subject, predicate, object)*, where the subject and object are unique ids representing entities, and the predicate identifies a usually non-unique relation between them. In this paper, we use graph visualizations as exemplified by Fig. 1, where the tail, label, and head of an edge represent a subject, predicate, and object, respectively. For simplification, we further use node labels in the form "id:typename" which imply the relation $(id, \texttt{type}, typename)$.

Fig. 1. Knowledge graph example. Node labels include types for simplification.

Working with knowledge graphs can be categorized into knowledge extraction, ontology engineering, and knowledge graph reasoning [15]. *Knowledge extraction* refers to the act of identifying the relevant entities and their relations based on unstructured or structured data. When integrating extracted knowledge into an existing graph, knowledge fusion needs to be performed. *Ontology engineering* helps to structure and add meaning to the extracted knowledge. It encompasses methods ranging from manual elicitation to (semi-)automated learning approaches. *Knowledge graph reasoning* (KGR) aims to derive new knowledge from existing graphs, i.e., mainly new entities and especially new relations [8,11]. AI advances with subsymbolic methods have yielded new neuro-symbolic approaches for KGR, which integrate symbolic logic-based reasoning techniques with subsymbolic neural-net-based methods [11]. Consequently, many of these approaches provide explainability [8,27].

2.3 Ontology-Based Business Process Modeling

Ontology-based Business Process Modeling (OBPM), also called *Semantic Business Process Modeling*, encompasses approaches to extend classical process models with instances of ontologies that represent additional background process knowledge [10]. This semantic underpinning promises to provide missing context to model users and to allow for applying advanced reasoning techniques, e.g., for validation and compliance checking of process models [1,3,9]. Existing works show how ontology languages such as OWL can be used to express a subset of the BPMN language [1,3] as well as to express event-log-like process execution traces [1]. While some of the works already apply ontology-based reasoning methods to instances of the modeled ontologies (cp. [9]), these works are limited to reasoning at process design-time and the major focus lays on modeling and integrating with existing process modeling approaches rather than reasoning [10]. In contrast, we consider process run-time, when allocation decisions happen, and put an emphasis on automated reasoning.

3 A Knowledge-Graph-Based Resource Allocation Framework

This section presents *KRAFT, a Knowledge-graph-based Resource Allocation Framework for Tasks*, that integrates organizational process mining, ontology-based allocation knowledge modeling, and knowledge graph reasoning, to deliver suitable and explainable resource allocations in processes. Figure 2 provides an overview of KRAFT.

At the heart of KRAFT, there is the *resource allocation knowledge graph* that encodes the relevant entities and their relations, as well as, at an abstracted level, the organization's allocation concepts and mechanisms. This graph is populated in two ways. First, the *process expert*, i.e. the person responsible for defining the organization's allocation mechanisms, can update the graph both conceptually, e.g., by introducing new allocation mechanisms, and at an instance level, e.g., by adding new employees or changing roles. To this end, the process expert can leverage a combination of process mining insights and domain knowledge. Second, the (potentially also KG-based [4]) *BPMS* can update the graph with execution information such as running tasks. Using the resource allocation knowledge graph, *reasoning approaches* are applied to infer potential allocations for

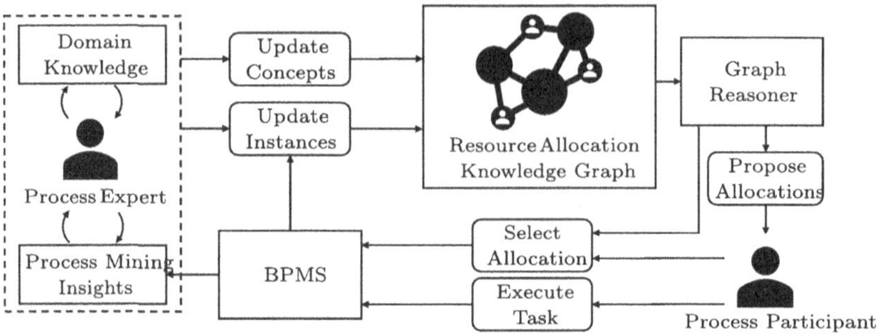

Fig. 2. Conceptual overview of KRAFT, a Knowledge-graph-based Resource Allocation Framework for Tasks

currently open tasks, together with an assessment and reasoning for each potential allocation. The *best proposals* can either be (i) automatically allocated or (ii) reviewed by the process participants to make an allocation decision. By executing the allocated tasks, the process state gets updated, which the BPMS forwards to the graph and reasoner by updating the respective information.

3.1 Resource Allocation Knowledge Graph

The resource allocation knowledge graph is aimed at making all necessary allocation knowledge available for reasoning whilst supporting easy adaptability. In the following, we provide our classification of the *allocation knowledge* with a focus on human resources to be encoded in the graph, leveraging and extending the literature presented in Sect. 2.1. Figure 3 visualizes this classification, distinguishing resource, task, case and allocation strategy knowledge.

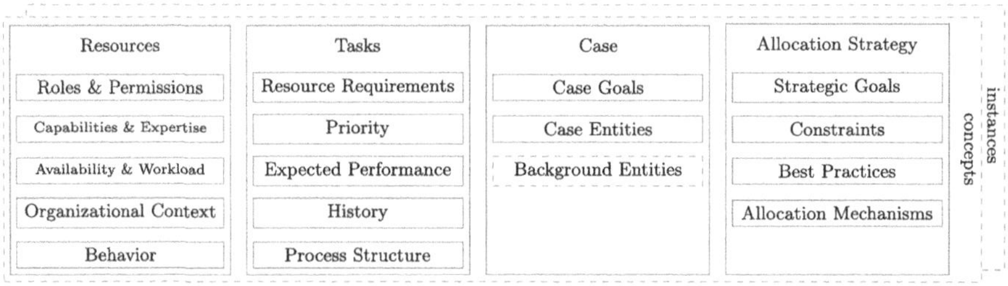

Fig. 3. Overview of resource allocation knowledge, extended from [2,20,22]

Resource knowledge covers knowledge about the resources as main entities for allocation. This includes knowledge regarding the individual roles and permissions (determining which tasks a resource is responsible for and permitted to do), the individual capabilities and expertise (determining the resource's aptitude for tasks and consequently its performance), the resource's availability (e.g., working hours) and workload (e.g., general amount of tasks being handled), the organizational context (i.e., where resources are based, which teams they belong to, etc.), and – often implicit – behavior (e.g., prioritization).

Task knowledge pertains to knowledge about single tasks, i.e., instances of activities. This groups resource requirements related to a particular task, e.g., defined based on the resource's knowledge and experience, as well as other task-related factors such as the tasks's priority and expected performance. Further, it includes the aggregated history of previously executed and currently running tasks. Lastly, we count the structure of the process, which encodes the relations between the activities, in this category.

Case knowledge refers to the knowledge about cases, i.e., instances of processes, including individual case goals and entities. Notably, we introduce the notion of *background entities* and concepts, which has only sparsely been touched in literature. Under this subcategory, all the relevant domain-specific entities and concepts related to a case are collected. For example, in a medical process, this would relate to, e.g., medical concepts and their interrelations, connected to the patient case. In a loan application process, this would relate to, e.g., risk classes and loan goal types attached to a loan application case.

Lastly, we introduce the notion of *allocation strategy knowledge*. Strategy is influenced by organizational goals exceeding individual case goals, e.g., "the workload should be distributed equally". Further, *constraints*, stemming from regulations or organizational *best practices*, are collected here. Finally, *allocation mechanisms* refers to knowledge about the methods applied for making allocation decisions, such as allocation rules (e.g., separation of concerns).

It is further useful to divide allocation knowledge along another axis into knowledge about *concepts* and *instances*. Under concept knowledge, analogously to KG ontologies, we group knowledge about what kind of entities can exist, how they relate, what they mean, and how allocation reasoning can be performed on them. For example, in a loan application process, concept knowledge includes that there are activities, resources, loan goals, and risk classes, as well as that tasks are instances of activities, or that certain risk classes imply the need for more senior clerks. On the other hand, we consider the knowledge that there is a specific activity "Check for Fraud", a task "t576" that is an instance of this activity, and an encompassing loan application case "c56" referencing the loan goal of "Financing a Car", as knowledge about instances of these concepts.

To implement the resource allocation knowledge graph, common knowledge graph technologies can be utilized, such as standardized ontology languages like OWL. Further, following guidelines from ontology-based process modeling [10], a respective system implementing the resource allocation knowledge graph should provide an initial set of standard concepts to organizations, which captures the most common allocation knowledge as described above. Organizations can build upon what they need from this knowledge and extend it with their individual concepts. Existing ontologies of organizational knowledge, such as the *W3C*

Organization Ontology[1] or the *BPMN-Based Ontology*[2] can be taken as foundation to build this standard concept graph. However, such existing ontologies conflict with each other, e.g., in how to represent roles, and cover the dimensions described above only insufficiently, either focusing on resources or task and process knowledge and mostly ignoring allocation strategy knowledge.

3.2 Knowledge Extraction

Relevant information must be extracted as a first step in populating or updating the knowledge in the graph. We identify two main sources that organizations can leverage: (1) process mining insights and (2) domain knowledge. *Process mining techniques* can extract resource allocation knowledge from historical process data captured in event logs. These techniques are particularly effective in identifying knowledge related to resources, e.g., roles, expertise, and behavior [18,24]. Additionally, they can extract task-specific knowledge, such as process structures [13], resource requirements, and case-specific knowledge, including case entity discovery [24]. Furthermore, historic allocation mechanisms [22] can be inferred. We assume that a process expert utilizes a process mining tool supporting such techniques, e.g., PM4Py [5]. Based on the insights obtained from process mining analysis, the expert selects and integrates relevant knowledge into the graph.

While process mining excels at extracting resource- and task-related knowledge, *domain knowledge* is essential for aspects that cannot be directly observed in execution data, such as allocation strategies, strategic goals, constraints, and best practices. Domain knowledge can take different forms: it may exist as implicit expert knowledge, which can be elicited through a conversational interface, or as explicit data. This data may be unstructured (e.g., regulatory documents) or structured, originating from existing systems managing allocation knowledge.

Different extraction techniques are applicable based on the knowledge source and format. For textual sources, including human input, Natural-Language-Processing-based information extraction is effective for knowledge graph construction [15]. Recent advances in large language models further enhance these capabilities, making them a promising tool for structuring domain knowledge into graphs [31]. Extracted knowledge then needs to be translated into updates to be applied to the graph, which should be manually checked to ensure validity. However, deriving valid updates while incorporating human feedback is complex and beyond the scope of this paper. Instead, we focus on the overall framework for knowledge-graph-based resource allocation, while future work may explore conversational process knowledge elicitation and knowledge fusion techniques to enhance this integration.

[1] https://www.w3.org/TR/vocab-org/.

[2] https://www.irit.fr/recherches/MELODI/ontologies/BBO, see also [1].

3.3 Allocation Reasoning

Goal of the allocation reasoning is to consider the knowledge encoded in the graph, assess potential allocations, and provide explanations for the assessments. Following the different degrees of automation, the component must both allow automatically selecting an allocation and letting a human select from a set of best options with explanations. In case human decision-making strongly deviates from system proposals, self-improvement using a conversational interface similar to knowledge extraction appears sensible.

The internal workings of the reasoning strongly depend on the technologies utilized. Knowledge graph reasoning (KGR) and adjacent fields provide a wide toolbox of approaches to achieve explainable recommendations (cp. [11,15]). Specifically, we can interpret allocation as the problem of *predicting links* between resource and task entities, one of the most common use cases of KGR. Notably, KGR approaches often combine multiple technologies, borrowing from both symbolic and subsymbolic methods to varying degrees, placing them as neuro-symbolic [11]. When selecting appropriate methods, a tradeoff has to be made. Subsymbolic approaches tend to be more efficient, generalizable, and more suitable for continuous target variables. Conversely, symbolic approaches are more comprehensible, more stable to change, and allow for better explainability [8,15,27]. For instance, one family of neuro-symbolic approaches learns and considers weighted path or graph pattern rules for link prediction, providing the deciding rules as fairly interpretable output along with the edge [8,11,15].

Specifically for resource allocation, on the symbolic side, description logic ontologies in combination with answer set programming have already been utilized for compliance checking [9] and scheduling under role and availability constraints [14]. On the subsymbolic side, neural nets have already been used to predict execution times based on process execution traces [29], which should be also applicable to graph-based inputs. As an exemplary combination, these run-time estimations could be utilized in rule-based reasoning.

4 Proof of Concept

This section outlines our prototypical implementation of KRAFT and demonstrates the general feasibility and usefulness of knowledge-graph-based resource allocation support.

4.1 Illustrative Setting

As a synthetic demonstration case for our prototype, we assume a loan application process in a financial institution. This process is based on an adapted form of the Business Process Intelligence Challenge 2017 dataset [28], as provided by the Business Process Optimization Challenge 2023[3] (BPOC). The dataset has been cleaned and aggregated w.r.t. the activity lifecycle. Figure 4 represents the resulting main process flows as a BPMN diagram.

[3] See https://sites.google.com/view/bpo2023/competition.

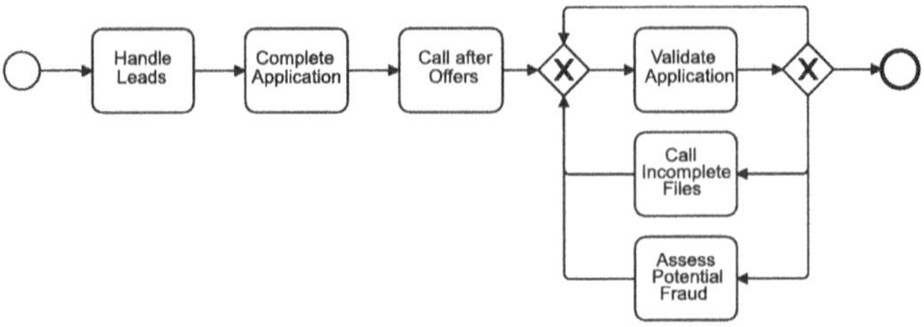

Fig. 4. Main flows for the example loan application process

Cases of this process have two categorical attributes, *Application Type* and *Loan Goal*, as well as one numerical attribute, the *Requested Amount*. In addition to the original log, we define further allocation knowledge and associated mechanisms. This includes *Risk Class*es attached to loan goals and *Seniority* attached to resources, where higher risk classes require higher seniority when working on a case, as well as *Separation of Concerns* (ensuring different resources handle a specific set of activities). Further, we added individual *Expertise* for specific case attributes. To substitute a real-world BPMS, a business process simulation frame adapted from the BPOC is used, which logs relevant events and triggers necessary allocation decisions. The simulation frame requires one-to-one allocation based on provided direct resource-to-activity permissions.

4.2 Prototype Design

To demonstrate KRAFT, we have implemented a simple application that showcases the framework's concepts on an elementary level. As an interface to the prototype, we provide a Jupyter notebook, which is prefilled with a demonstration that walks users through, i.a., importing resource expertise from process mining and utilizing the prototype to make allocation decisions for the illustrative setting[4].

The prototype's resource allocation graph is based on common knowledge graph technologies: all nodes and edges are encoded in the W3C Resource Description Framework (RDF) format. Concept knowledge is stored using the W3C standards Web Ontology Language 2.0 (OWL) and SHACL, a W3C standard language for expressing constraints on RDF graphs.

Knowledge extraction has been performed on the BPIC 2017 event log, applying Pika et al.'s [21] Resource Behavior Indicators to determine resource seniority and expertise. To input knowledge, RDF triples can be imported from respective files or directly created with code, and then added to the respective subgraphs.

Reasoning in the prototype is based on the above-mentioned SHACL, which allows to define rules in the form of graph patterns, as well as dynamic messages and scores for pattern matches. During "run-time", i.e., when running the

[4] The prototype code and the demonstration notebook are available at github.com/INSM-TUM/knowledge-graph-resource-allocation/tree/bpm25_paper.

simulation, it updates the graph based on incoming events, checks these rules, and determines an aggregated score and assessment for potential assignments, which can be considered in two modes. In *automatic* mode, the resource with the highest score is automatically selected. In *human-in-the-loop* mode, users are prompted to select the appropriate resource. The prototype supports pausing the simulation to switch modes or input knowledge at run-time. The detailed workings of the prototype are described in the following section using demonstrative examples.

4.3 Demonstration

We can use this prototype to integrate allocation standard concepts as well as the allocation knowledge described in the example case of Sect. 4.1. Following the guidance from Sect. 3.1, tasks, activities, cases, resources, and roles, as well as direct and role-based allocation are part of the base concept graph of the prototype. In contrast, the domain-specific concepts of the example case, such as resource seniority, application types, and risk-class-based allocation, are use-case-specific extensions of this schema. For an interactive visualization of the resulting graph structure, we refer to the repository.

Considering the example case, let us assume that at some point the company added this non-standard knowledge in the form of domain-specific entities and allocation rules. In the prototype, this would be expressed in the form of OWL and SHACL structures to be added. For instance, adding loan goals, attached risk types, and the respective allocation rule encompasses adding classes `LoanGoal` and `RiskType` and their respective relations `hasLoanGoal` and `hasRiskType` to the ontology. Further, a SHACL rule defining the graph pattern for the allocation rule has to be added. Figure 5 visualizes the pattern as a graph[5].

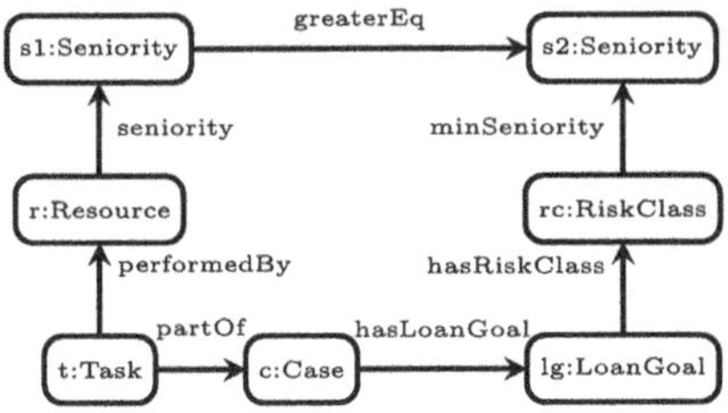

Fig. 5. Risk class based seniority requirement expressed as graph pattern

The pattern, forming a simple circle, can be interpreted as follows: *There is a resource* `r` *performing a task* `t` *for a case* `c` *with a specific loan goal* `lg` *and attached risk class* `rc`. *The seniority* `s1` *of the resource must be greater or equal than the minimum seniority* `s2` *of the risk class.* In addition to the pattern, the rule formulates a parameterizable message for matches, such as "Seniority '`s1`' is sufficient for risk class '`rc`' of loan goal '`lg`' ". Note that, analogous to this positive pattern that matches adherences to the rule, we also define a negative pattern matching violations, which is then used to penalize violating allocations.

[5] For ease of reading, we have refrained from putting SHACL code into the paper. We refer interested readers to the code repository.

Now consider the case of a concrete application for raising the credit limit for a car: handling leads, completing the application, calling after offers, and validating the application have already been performed, and the respective events have been logged in the graph. Now, potential fraud should be assessed and the proper resource needs to be selected. Resources `User_26`, `User_55`, and `User_83` are available and all possess roles to be allowed to perform the activity. `User_55` has been responsible for the case so far. Figure 6 shows an excerpt of the prototype's resource allocation knowledge graph representing relevant entities and relations for this example.

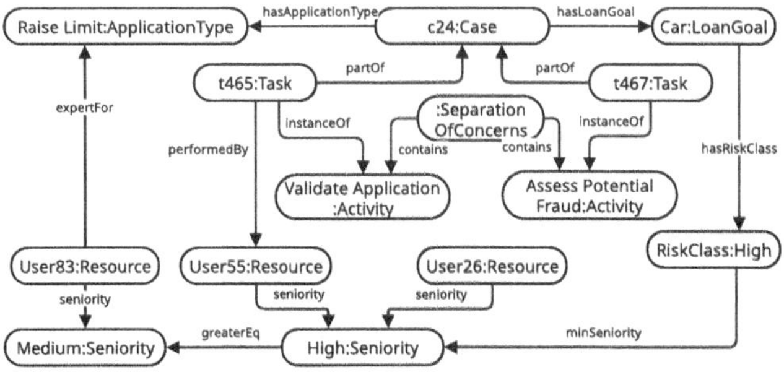

Fig. 6. Allocation knowledge graph excerpt with key nodes and edges for example allocation decision

The system now tries out different assignments and considers its known rules. `User_55` has performed `Validate application` in the same case, which is connected to `Assess Potential Fraud` via a separation of concerns node, and thus is not eligible to perform the task. Assigning `User_83` would not break the separation on concerns rule. The person is additionally an expert on the application type `Raise Limit`. However, their seniority is lower than the high seniority required for high-risk car loans. In contrast `User_26` does not have special expertise on limit raises, but their high seniority matches the requested one. Assuming that the defined scores weigh the seniority higher than the expertise, in automatic mode, the system selects `User_26` and logs the reasoning as shown in listing 1. In human-in-the-loop mode, a human could overrule this weighting.

```
case-1 task-7: W_Assess potential fraud
Resources Available: {'User_26', 'User_55', 'User_83'}
Assigning: User_26 to task-7 considering the following:
    Assignment conforms separation of concerns with activity 'W_Validate application'
    Seniority 'High' is sufficient for risk class 'High' of loan goal 'Car'
```

Listing 1: Example output for on allocation decision

5 Discussion and Future Work

Based on the presented framework details and the prototypical demonstration, this section discusses KRAFT w.r.t. the objectives defined in Sect. 1 and elaborates on avenues for future research.

5.1 Discussion

O1 Holistic Integration. The prototype shows how common base allocation knowledge and domain-specific knowledge can be integrated and jointly considered for reasoning. The demonstration integrates various kinds of allocation knowledge (cp. Fig. 3), such as resources, tasks, and cases as classes, direct allocation in the form of `canBeExecutedBy` edges between activities and resources, expertise as `expertFor` edges from resources, task history through attributes such as `startedAt` and `directlyFollowedBy` edges, and domain-specific case attributes and entities in the form of the application type, loan goal, and risk type classes. Further, allocation mechanisms in the form of rules, such as separation of concerns and risk-attached seniority, have been demonstrated.

O2 Easy Adaptability. The demonstration showed how the concept of risk-class-based allocation can be introduced with simple additions to the allocation knowledge graph. We further want to highlight that, for instance, changing risk classes for a specific loan goal can be easily implemented by changing one edge. In contrast, in a system that does not explicitly consider the "hidden" dependencies of allocation decisions, one would explicitly have to know to adapt the allocation mechanisms when risk classes are adjusted. We argue that other allocation knowledge aspects can be as easily integrated and adapted in a similar fashion as demonstrated. However, for extracting some of these aspects, more complex techniques might be necessary, such as predictive process monitoring techniques for expected performance. Moreover, a more user-friendly interface is necessary to demonstrate the ease of adaptation.

O3 Explainability. The prototype shows a simple, rule-based implementation of explaining its allocation decisions. For automated decisions, reasonings are logged which allows users to understand why the system has made a decision and to make adjustments to the allocation knowledge in the graph. Further, in manual mode, the system provides reasoning that humans can interpret for semi-automated decision-making. Consequently, users do not have to blindly accept system decisions or manually reproduce all the reasoning, while being able to understand where wrong conclusions were made and override them.

5.2 Limitations and Future Work

While the prototype demonstrates the general feasibility and usefulness of the KRAFT framework, its complexity and scope are rather limited, yielding interesting challenges for future research. First, on the knowledge input side, the realization of a conversational knowledge input interface can be researched. Further, integration with existing organizational mining techniques and the development of knowledge-graph-based ones can be investigated. Second, more sophisticated reasoning methods need to be inspected. Conceptually, approaches that look ahead, e.g., into the expected process continuation and future resource (un)availability, consider a broader range of allocation problems such as many-to-many allocation, or proactively learn, e.g., from user decisions and occurrences, can provide more optimal allocations. Technologically, integrating subsymbolic neural approaches is promising to achieve this, e.g., by utilizing remaining-time predictions as a decision input. Lastly, an investigation into run-time complexity needs to be launched, as decision-making during process execution needs close-to-real-time system reasoning.

Apart from examining how to implement the framework, further evaluation and validation are needed. We want to especially highlight the human factor here. As a major part of allocation knowledge currently exists in the form of implicit human knowledge, the interplay between these humans and systems w.r.t. to knowledge transfer, contradictory knowledge, but also impact on an emotional level needs to be further investigated [17]. Further, this paper only used the KRAFT framework in one reality-inspired use case. An evaluation with further use cases would help to strengthen the validity of the framework and potentially provide valuable pointers for improvement.

6 Conclusion

In this paper, we presented KRAFT, a framework for resource allocation based on knowledge graph reasoning. The framework provides guidance on how to integrate domain knowledge, process mining insights, and knowledge graph reasoning with business process management systems for resource allocation. Using a prototypical proof of concept, we demonstrated that such an approach has potential to support resource allocation in three major dimensions. First, to capture a wide range of allocation knowledge from various sources and integrate them into one data structure. Second, to easily adapt the stored knowledge on the relevant entities and mechanisms for allocation. Third, to get automated, comprehensive reasoning for selecting the best suitable resources for tasks. We argue that existing methods and technologies can serve as a basis for implementing the framework and highlight interesting opportunities for further research.

References

1. Annane, A., Aussenac-Gilles, N., Kamel, M.: BBO: BPMN 2.0 based ontology for business process representation. In: 20th European Conference on Knowledge Management (ECKM 2019), vol. 1 (2019)
2. Arias, M., Munoz-Gama, J., Sepúlveda, M.: Towards a taxonomy of human resource allocation criteria. In: Teniente, E., Weidlich, M. (eds.) BPM 2017. LNBIP, vol. 308, pp. 475–483. Springer, Cham (2018). https://doi.org/10.1007/978-3-319-74030-0_37
3. Bachhofner, S., Kiesling, E., Revoredo, K., Waibel, P., Polleres, A.: Automated process knowledge graph construction from BPMN models. In: Database and Expert Systems Applications, vol. 13426. Springer, Heidelberg (2022). https://doi.org/10.1007/978-3-031-12423-5_3
4. Bein, L., Pufahl, L.: Knowledge graphs: a key technology for explainable knowledge-aware process automation? In: BPM Workshops 2024. Springer, Heidelberg (2025). https://doi.org/10.1007/978-3-031-78666-2_2

5. Berti, A., van Zelst, S., Schuster, D.: Pm4py: a process mining library for python. Softw. Impacts **17** (2023)
6. Cabanillas, C., Knuplesch, D., Resinas, M., Reichert, M., Mendling, J., Ruiz-Cortés, A.: RALph: a graphical notation for resource assignments in business processes. In: Zdravkovic, J., Kirikova, M., Johannesson, P. (eds.) CAiSE 2015. LNCS, vol. 9097, pp. 53–68. Springer, Cham (2015). https://doi.org/10.1007/978-3-319-19069-3_4
7. Cabanillas, C., Resinas, M., del-Río-Ortega, A., Ruiz-Cortés, A.: Specification and automated design-time analysis of the business process human resource perspective. Inf. Syst. **52** (2015)
8. Chen, Y., et al.: An overview of knowledge graph reasoning: key technologies and applications. J. Sensor Actu. Netw. **11**(4) (2022)
9. Corea, C., Delfmann, P.: Detecting compliance with business rules in ontology-based process modeling. In: Wirtschaftsinformatik 2017 Proceedings (2017)
10. Corea, C., Fellmann, M., Delfmann, P.: Ontology-based process modelling - will we live to see it? In: Ghose, A., Horkoff, J., Silva Souza, V.E., Parsons, J., Evermann, J. (eds.) ER 2021. LNCS, vol. 13011, pp. 36–46. Springer, Cham (2021). https://doi.org/10.1007/978-3-030-89022-3_4
11. DeLong, L.N., Mir, R.F., Fleuriot, J.D.: Neurosymbolic AI for reasoning over knowledge graphs: a survey. IEEE Trans. Neural Netw. Learn. Syst. (2024)
12. Dumas, M., et al.: AI-augmented business process management systems: a research manifesto. ACM Trans. Manag. Inf. Syst. **14**(1) (2023)
13. Dumas, M., La Rosa, M., Mendling, J., Reijers, H.A.: Fundamentals of Business Process Management. Springer, Heidelberg (2018)
14. Havur, G., Cabanillas, C., Mendling, J., Polleres, A.: Resource allocation with dependencies in business process management systems. In: La Rosa, M., Loos, P., Pastor, O. (eds.) BPM 2016. LNBIP, vol. 260, pp. 3–19. Springer, Cham (2016). https://doi.org/10.1007/978-3-319-45468-9_1
15. Hogan, A., et al.: Knowledge graphs. ACM Comput. Surv. **54**(4) (2022)
16. Ihde, S., Pufahl, L., Völker, M., Goel, A., Weske, M.: A framework for modeling and executing task-specific resource allocations in business processes. Computing **104**(11) (2022)
17. Klessascheck, F., Bein, L., Haase, J., Pufahl, L.: A critical investigation of rationalities in automation with BPM. In: 2024 26th International Conference on Business Informatics (CBI). IEEE (2024)
18. Martin, N., Beerepoot, I.: Unveiling use cases for human resource mining: a framework of past and future research. BISE (2024)
19. Middelhuis, J., Bianco, R.L., Sherzer, E., Bukhsh, Z., Adan, I., Dijkman, R.: Learning policies for resource allocation in business processes. Inf. Syst. **128** (2025)
20. Ouyang, C., Wynn, M.T., Fidge, C.: Modelling complex resource requirements in business process management systems. In: ACIS 2010 Proceedings (2010)
21. Pika, A., Leyer, M., Wynn, M.T., Fidge, C.J., Hofstede, A.H.M.T., Aalst, W.M.P.V.D.: Mining resource profiles from event logs. ACM Trans. Manag. Inf. Syst. **8**(1) (2017)
22. Pufahl, L., Stiehle, F., Ihde, S., Weske, M., Weber, I.: Resource allocation in business process executions—a systematic literature study. Inf. Syst. **132** (2025)
23. Reichert, M., Weber, B.: Enabling Flexibility in Process-Aware Information Systems: Challenges, Methods, Technologies, vol. 54. Springer, Heidelberg (2012). https://doi.org/10.1007/978-3-642-30409-5

24. Rubensson, C., Pufahl, L., Mendling, J.: A conceptual framework for resource analysis in process mining. In: EDOC 2024 Workshops, vol. 537. Springer, Heidelberg (2025). https://doi.org/10.1007/978-3-031-79059-1_12
25. Ruiz, C., Zitnik, M., Leskovec, J.: Identification of disease treatment mechanisms through the multiscale interactome. Nat. Commun. **12**(1) (2021)
26. Russell, N., Hofstedeter, A., Edmond, D., Aalst, van der, W.: Workflow Resource Patterns. BETA Publicatie: Working Papers, TU Eindhoven (2004)
27. Tiddi, I., Schlobach, S.: Knowledge graphs as tools for explainable machine learning: a survey. Artif. Intell. **302** (2022)
28. van Dongen, B.: BPI challenge 2017 (2017)
29. Verenich, I., Dumas, M., Rosa, M.L., Maggi, F.M., Teinemaa, I.: Survey and cross-benchmark comparison of remaining time prediction methods in business process monitoring. ACM TIST **10**(4) (2019)
30. Xu, J., Liu, C., Zhao, X., Ding, Z.: Incorporating structural improvement into resource allocation for business process execution planning. Concurr. Comput. Pract. Exp. **25**(3) (2013)
31. Zhu, Y., et al.: LLMs for knowledge graph construction and reasoning: recent capabilities and future opportunities. World Wide Web **27**(5) (2024)
32. Zur Muehlen, M.: Organizational management in workflow applications – issues and perspectives. Inf. Technol. Manag. **5**(3/4) (2004)

An RPA-BPMS Integration Framework Leveraging RPA as a Resource

Lingtong Zhang[1], Roy Jing Yang[2], Chun Ouyang[2], and Yang Yu[3](✉)

[1] School of Computer Science and Engineering, Sun Yat-sen University, Guangzhou, China
[2] School of Information Systems, Queensland University of Technology, Brisbane, Australia
[3] School of Software Engineering, Sun Yat-sen University, Guangzhou, China
yuy@mail.sysu.edu.cn

Abstract. Robotic Process Automation (RPA) is a technology that creates software robots emulating human interactions with computer software to automate low-level, routine tasks in business processes. The integration of RPA technology within Business Process Management Systems (BPMS) improves process automation at operational level and accelerates digital transformation in modern organizations. Existing research has proposed various solutions to streamline a low-cost integration. But they require modification of the existing process models and do not fully support the use of RPA bots as active resources in process execution. In this paper, we present an integration framework for RPA within BPMS that address those gaps. The framework is designed following the established Workflow Reference Model (which underpins BPMS) and supports two ways of having RPA bots as active process participants: acting as "digital employees" that can independently complete process tasks as human resources, or acting as "human assistants" that collaborate with human resources in completing tasks. We implemented the proposed framework as an open-source software prototype and evaluated its feasibility through a use case of bank account opening process.

Keywords: Robotic Process Automation · Business Process Management System · Integration · Resource

1 Introduction

Modern organizations are constantly seeking efficient and flexible process automation solutions to accelerate digital transformation. Robotic Process Automation (RPA) and Business Process Automation (BPA) are two major types of process automation technologies. BPA, typically supported by Business Process Management Systems (BPMS), focuses on *end-to-end process* automation, emphasizing the orchestration and coordination of tasks and resources

M. ER et al. (Eds.): BPM 2025, LNBIP 565, pp. 153–167, 2026.
https://doi.org/10.1007/978-3-032-02936-2_12

across the entire process. In contrast, RPA operates at *the task level*, automating rule-based operations that are often performed by humans [4]. As such, RPA functions more as an automated worker than a process manager, forming a new resource category of RPA bots within business processes. Together with human actors, RPA bots can be orchestrated by BPMS, highlighting the technologies' complementary roles and the need for their integration to enable end-to-end automation.

Existing research highlights the value of integrating RPA with BPMS, suggesting that both technologies benefit from such integration [1,4]. At the same time, RPA has been seen as participating in business processes in two different ways: RPA bot as a "digital employee", capable of autonomously performing tasks like a human actor [5,8] (see left side of Fig. 1); RPA bot as a "human assistant", assisting human actors in completing tasks under their coordination [7,9] (see right side of Fig. 1). To this end, integration solutions proposed by König et al. [6] and Flechsig et al. [2] represent state-of-the-art efforts. However, both approaches require modifications to the process engines within BPMS and updates to the process models to explicitly define RPA-specific tasks. As a result, they face challenges in enabling direct task assignment to RPA bots via standard process engine interfaces, according to the Workflow Reference Model [3].

In this paper, we build on the concept of *RPA as a resource* that can function both as a digital employee and as a human assistant in business processes. Guided by this design rationale, we propose a novel integration framework that includes specification of the task execution process of RPA bots using sequence diagrams. Our integration solution follows the engine interface protocol in the established Workflow Reference Model [3] to enable task assignment directly from the workflow engine to RPA bots. The engine does not need to distinguish whether a task is preformed by an RPA bot or a human. As a result, no modification to the workflow engine or process models is necessary, enhancing compatibility and scalability while reducing implementation overhead. To validate the proposed integration framework, we developed an open-source pro-

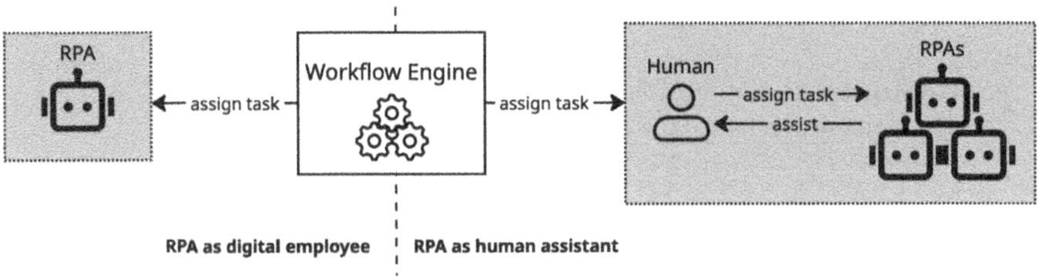

Fig. 1. An illustration showing two methods for the integration of RPA within BPM. Left: RPA bots act as digital employees that can complete a process task on its own like a human resource. Right: RPA bots act as human assistants and collaborate with the human resource to complete a process task decomposed into sub-tasks. The gray boxes represent resources from the perspective of the workflow engine.

totype and evaluated its feasibility through a use case of bank account opening process.

The remainder of this paper is organized as follows. Section 2 reviews related work and identifies gaps in existing solutions. Section 3 presents the design rationale behind the integration. Section 4 and Sect. 5 present an integration framework leveraging RPA as a resource acting as a digital employee and as a human assistant, without the need to modify process engine and process model. Section 6 validates the feasibility of the framework through a prototype implementation and a use case scenario. Section 7 concludes the work and explores future research.

2 Related Work

Guided by the evaluation framework by Venable et al. [11], we investigate the literature and evaluate the existing solutions for integrating RPA within BPMS. We consider the following requirements for a solution to achieve low-cost integration. (1) An integration solution requires *minimal modification* to existing business process models and workflow engine, so that the organization avoids incurring extra cost due to the change of process model or its current BPMS. (2) An integration solution can be deployed with *minimal coupling* with specific RPA or workflow technologies, such that the organization retains the software adaptability in case of future business changes. (3) An integration solution should maintain *task assignment directly from the workflow engine to resources* (see Fig. 1); in other words, there is minimal change to task assignment rules resulted from the deployment of RPA bots alongside human resources.

König et al. [6] analyzed the lifecycle of BPM and proposed an integration framework based on the use of adapters that bridges Robotic Process Automation System (RPAS) and BPMS. In their integration framework, the workflow engine sends the RPA tasks to the RPA software through a protocol, and the RPA software completes it (similar to the left side of Fig. 1). Note that the workflow engine needs to be modified to identify RPA tasks. At the same time, the existing business process needs to be modified to replace the original elements with the RPA task elements. Using this integration framework, different workflow engines need to be able to identify RPA tasks and implement protocols with different RPA software, that is, to adapt to different workflow engines and RPA software. Specifically, from an interface perspective (cf. Workflow Reference Model [3]), the workflow engine in this framework would treat RPA as applications instead of resources that can be directly assigned tasks. Also, this framework does not consider RPA assisting human resources in completing tasks.

Flechsig et al. [2] explored the integration of RPA and BPM from the perspective of using a standardized language for process modeling and orchestration. They proposed to use Business Process Model and Notation (BPMN) to uniformly describe RPA processes and business processes and introduce two key components: RPA parsers and RPA orchestrators. When a workflow engine parses RPA tasks, it uses RPA parsers to convert RPA task elements into a specific format. Then, when the workflow engine assigns RPA tasks to RPA bots, it uses an RPA orchestrator to decide how to assign them to RPA (see left side of Fig. 1). Therefore, this integrated architecture enables RPA to act as digital employees. It is also capable of describing the collaboration between RPA bots and humans using BPMN to realize RPA as a human assistant (see right side of Fig. 1). The use of RPA parser and orchestrator enables adaptation to different RPA software and thus remain decoupled from specific technology. However, note that the solution requires changes to an existing process model to include tasks performed by RPA bots and changes to the workflow engine to identify those RPA tasks and use the parsers and orchestrator. Also, with this solution, the task assignment from the workflow engine to resources is no longer direct but requires an RPA orchestrator to decide.

In addition, Ruiz et al. [10] proposed a method for collaboration between humans and RPA bots. The method uses RPA software to orchestrate the human-RPA collaboration and does not involve the use of BPMS. Therefore, we do not consider it as a solution that integrates RPA into BPMS.

Table 1. A summary of evaluating existing work, based on the requirements for low-cost integration and support for RPA acting as digital employees or human assistants.

Requirement & Support	König et al. [6]	Flechsig et al. [2]
Minimal process model modification	✘	✘
Minimal engine modification	✘	✘
Minimal coupling to specific technologies	✔	✔
Direct engine-to-resource task assignment	✘	✘
Supporting RPA as digital employees?	✔	✔
Supporting RPA as human assistants?	✘	✔

Table 1 shows a summary of the above evaluation of the existing integration solutions. It can be seen that gaps remain in the need for modification of existing process and workflow engine and engine-to-resource task assignment, motivating us to research on a new solution.

In doing this, we refer to the established workflow reference model [3], specifically concerned with the engine assigning tasks to the work lists of resource through Interface II. In the remainder of this paper, we elaborate on our new solution for the integration of RPA with BPMS and explain how it may address all the requirements and enables the support for RPA to act as both digital employees and human assistants.

3 Design Rationale

This section discusses the two methods for integrating RPA with BPMS: RPA bots as digital employees and as human assistants, and describes how RPA bots participates in business process execution, respectively.

3.1 RPA Bot as Digital Employee

When an RPA bot acts as a digital employee in an organization, it needs to be able to complete tasks alone. Like a human resource, an RPA bot is assigned a role associated with some task [3] in the process definition, and the bot can be considered an alternative to human resources in the same role when performing the task. The workflow engine can assign tasks directly to RPA bots.

An RPA bot performs a task according to a predefined RPA process, without human intervention. From a process architecture perspective, such an RPA process is a lower-level, *Individual Operation Process (IOP)* [12], i.e., it uses operations as logical steps and describes the procedure of how a task may be performed by an individual resource through the operations. When an RPA bot acts as a digital employee in an organization, it can be added to the organizational model, assigned roles, and execute a task decomposed into a sequence of sub-tasks in accordance with lower-level process definitions specified by IOPs.

3.2 RPA Bot as Human Assistant

RPA bots cannot assume responsibility. Therefore, tasks associated with responsibility must conventionally be performed by a human. However, if most of the steps in a task are simple and repetitive, human involvement is only required for decision-making and acceptance of responsibility. In this case, a human can use multiple RPA bots to assist with carrying out simple and repetitive steps and confirm the work results of RPA bots. The RPA bots collaborate under the decision and coordination of the human actor and become human assistants.

Figure 2 illustrates an example. A bank employee receives the work to review a bank account opening application. To do so, the employee needs to verify the applicant's identity, check credit, and request for a criminal record check. The final task of confirming the review outcome is dependent on the information

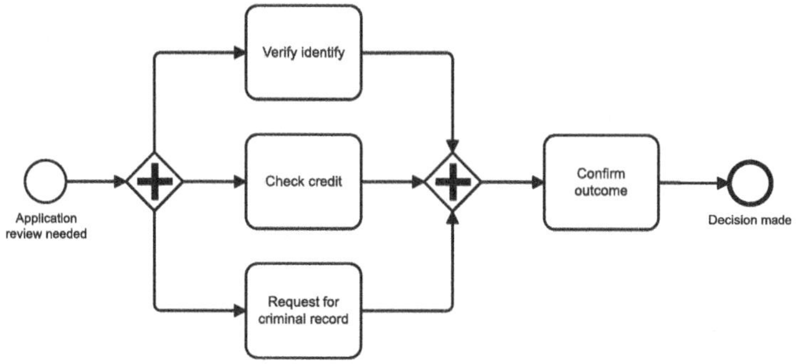

Fig. 2. An example illustrating the idea of RPA bot as human assistant. A human employee may delegate the first three tasks to RPA bots performing the tasks in parallel, and the human needs to only perform the last task to confirm the outcome.

gathered in the first three tasks, and the employee takes the responsibility of the decision on the application. Here, three RPA bots may be used to handle the first three tasks of information gathering and basic check, leaving only the final confirmation task to be performed by a human. These bots may execute tasks in parallel, improving overall process efficiency.

When an RPA bot acts as a human assistant in an organization, it needs to collaborate with a human actor to complete tasks under human control. The workflow engine assigns tasks directly to a human, and the human decides how to assign tasks to RPA bots. RPA bots indirectly participate in business process execution and requires the human to control their collaboration, instead of the workflow engine.

From a process architecture perspective, such a process may be seen as an *Individual Business Process* (IBP), which involves an individual human resource assigned by the workflow engine to complete a task or a subprocess with several tasks. It is up to the human resource to further decompose the task or the subprocess to enable automation by RPA bots. As such, the human gains enhanced working abilities beyond an ordinary human resource—this group of a human and multiple RPA bots can be viewed as a "superman" resource. In the aforementioned example (Fig. 2), the four tasks may be viewed as an IBP.

To summarize, the two proposed methods for RPA-BPMS integration correspond to the following view on process architecture:

- Business Process: A typical (end-to-end) process managed by a BPMS. It consists of tasks, and its execution involves the participation of multiple human actors and RPA bots (using one of the proposed integration methods).
- Individual Business Process (IBP): Part of a business process performed by a single human actor, which may be a task or a subprocess. The execution of an IBP involves a human actor managing and collaborating with one or more RPA bots as her assistants.
- Individual Operational Process (IOP): A lower-level process that consists of operations, describing how a task in a business process is performed. An IOP can be fully automated by having an RPA bot performing the operations as predefined, without the need for humans participation.

4 Framework

This section describes the framework corresponding to the two integration methods and provides a technical foundation for seamless integration of RPA into BPMS.

Whether RPA bots acts as an alternative to humans through IOPs or assist humans through IBPs, RPA bots and "superman" teams of human and bots are resources executing tasks within business processes. The integration of RPA into BPMS is done through Interface II [3]: A workflow engine parses business process definitions, and tasks to be performed by RPA bots or "superman" are directly allocated to worklists.

4.1 RPA Bot as Digital Employee: IOP-Based Integration

The framework for IOP-based integration is shown in Fig. 3. An RPA bot assumes tasks originally performed by a human, liberating human from repetitive operations. Existing workflow engine technologies enable task allocation to clients without modifying the engine. An RPA bot is added as a resource to the organizational model, eliminating the need to alter existing business processes. For workflow engines, task allocation follows predefined rules and conditions, with no requirement to distinguish between a human and an RPA bot.

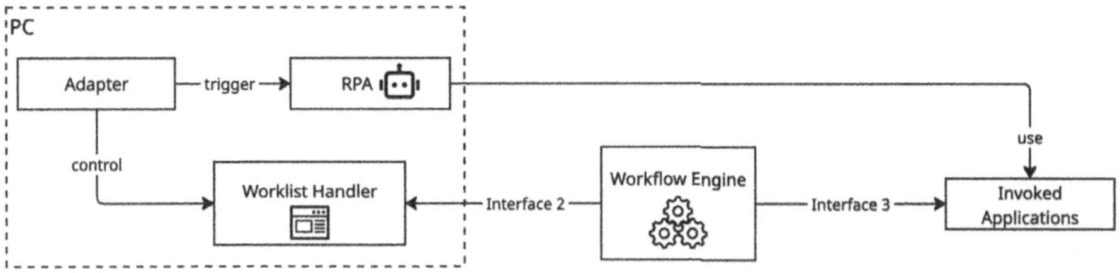

Fig. 3. IOP-based Integration Framework Diagram.

The integration framework achieves fully automated control over workflow engine clients via an adapter, enabling an RPA bot to simulate a human using invoked applications and directly participate in business process execution. Business processes are deployed onto the workflow engine, while IOPs are executed on PC.

Workflow Engine. The workflow engine parses deployed business processes and allocates tasks to the corresponding client through Interface II. The client receives tasks from the workflow engine, where the worklist stores all tasks that need to be completed by RPA bot. The client implements all the functions specified in the Interface II protocol and constitutes the basic mechanism for the operation of the framework.

Adapter. The adapter is the core module in the integrated architecture. The main function of the adapter is to control the engine client and to trigger the correct IOPs. It utilizes Interface II protocol functions, such as querying worklist, retrieving task details, and completing task—to identify required IOPs and trigger the correct IOPs. During the execution of an IOP, the adapter sleeps until the IOP is completed, and then controls client to report the result to the workflow engine. The adapter can be implemented as an RPA bot itself or as a software program with a listener feature. Either way, the implementation must ensure full automation and compliance with the basic functions outlined in the subsequent section.

RPA Bot. An RPA bot is assigned a role in the organizational model, and the engine client logs in to the account of the role. After being triggered, RPA bot simulates a human using the invoked application to finish the task, thereby participating in the business process.

This IOP-based integration framework enables an RPA bot to autonomously handle assigned tasks without human intervention, positioning an RPA bot as a resource in business processes to replace human. An RPA bot acts as a digital employee.

4.2 RBA Bot as Human Assistant: IBP-Based Integration

The framework for IBP-based integration is shown in Fig. 4. The integration framework describes how a "superman" team completes tasks through Interface II and how an RPA bot assists a human as a human assistant in business process execution.

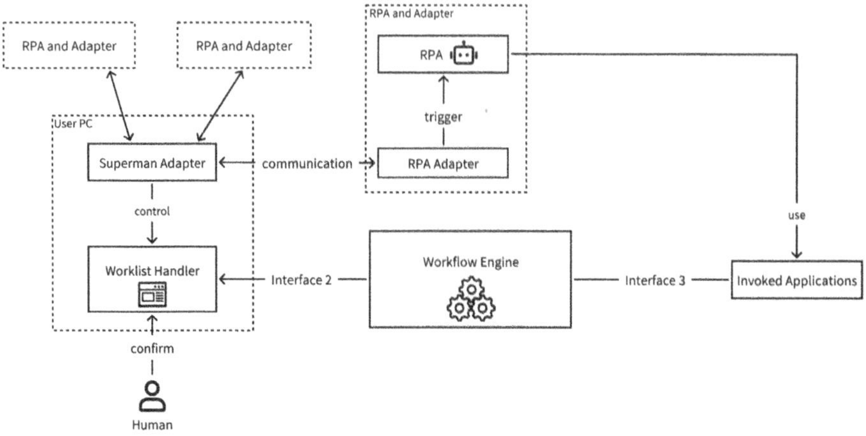

Fig. 4. IBP-based Integration Framework Diagram.

An RPA bot assists a human and indirectly participates through "superman" collaboration. The human in the "superman" team coordinate multiple RPA bots to complete tasks, with IBP deployed to the engine via BPMN notation. Existing business processes remain unchanged. The workflow engine parses IBP and allocates all tasks to the client of the human, who then distributes tasks among multiple RPA bots. From the workflow engine's perspective, the "superman" team appears as a single human resource. Engine operations remain unaffected, as tasks are allocated to the worklist of the human via Interface II, eliminating the need for engine modifications.

In this integration approach, in addition to deploying business processes and IOPs, it is essential to deploy manually designed IBPs to the engine.

Adapters. Modifications are made to the adapter module in the IBP-based integration framework. The adapters are divided into two categories: A Superhuman Adapter and multiple RPA Adapters.

The Superhuman Adapter controls the engine client, while the RPA Adapters trigger the correct IOPs. These adapters communicate via a protocol to collaboratively complete tasks. The framework includes only one Superhuman Adapter but multiple RPA Adapters, each corresponding to an RPA bot. The Superhuman Adapter distributes tasks requiring RPA bot execution to individual RPA Adapters over the network. Upon task completion, RPA Adapters return results to the Superhuman Adapter via network communication. Finally, the Superhuman Adapter controls the client to report task completion to the engine. Additionally, the Superhuman Adapter initiates the client to start IBPs. The basic function of the Superhuman Adapter and RPA Adapters, as well as their communication protocol, will be detailed in the following section.

Human Resource. A human participate in the IBP-based integration approach by verifying the outcomes of RPA task completion and assuming responsibility for the results. Specific tasks are executed by multiple RPA bots. The human design IBPs using BPMN to coordinate multiple RPA bots. If parallel RPA bot execution is required, each RPA bot must operate on a computer.

Workflow Engine. The organization recognizes only a human resource and remains unaware of the "superman" team or the RPA bots managed by the human. The engine allocates tasks to the client of the human, and the Superhuman Adapter subsequently distributes the tasks to RPA Adapters. Once triggered by their respective RPA Adapters, RPA bot simulates a human using the invoked application to complete the task, thereby participating in business processes.

Through this IBP-based integration framework, an RPA bot is positioned as a resource that assists a human.

5 Task Execution in Integration

This section demonstrates how tasks are executed in two integration approaches, accompanied by UML sequence diagrams. Based on the sequence diagrams, we analyze the functional requirements of the adapter modules and the communication protocol between the Superman Adapter and RPA Adapters. Process deployment details are excluded from the diagrammatic representation.

5.1 Sequence Diagram for IOP-Based Integration

The sequence diagram for the IOP-based integration approach is illustrated in Fig. 5. It demonstrates how an RPA bot directly replaces a human in task execution within business processes. Parentheses denote data parameters required for interaction actions.

During the preparation phase, the adapter must first open the worklist on the client. Subsequently, the adapter periodically polls the worklist to check for pending tasks that require RPA bot execution. Tasks assigned to RPA bot by the business process engine automatically appear in the worklist and are detected

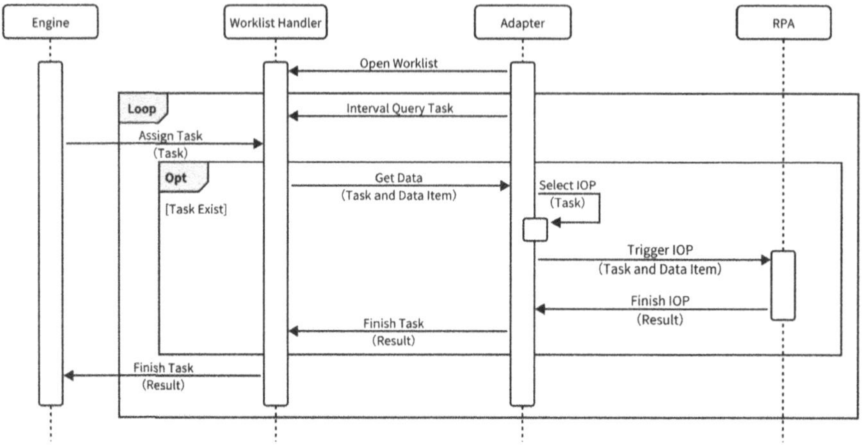

Fig. 5. Sequence for the execution of an RPA task in IOP-based integration.

during the next polling cycle of adapter. The adapter retrieves data from the client, including task metadata (e.g., task name, priority) and business object data items. The data is used to select the appropriate IOPs, trigger the correct IOPs, and complete tasks. After triggering an IOP, the adapter enters the sleep state until RPA bot completes the task, then reports task completion to the engine via client.

Functional Requirements for Adapter. By analyzing interactions between the adapter, engine client, and RPA in the sequence diagram, the essential functions of the adapter can be summarized. The interactions between the adapter and the engine client primarily relies on functions defined in the Interface II protocol, which are already implemented by the client. These functions include opening the worklist, fetching work items, extracting task information, and completing tasks. The adapter controls the client to invoke these protocol-based functions. In contrast, the interactions between the adapter and RPA requires the adapter to independently implement functions such as IOP selection, triggering of the correct IOPs, and waiting for IOP completion. Specifically, after triggering an IOP, the adapter remains in the sleep state until the IOP completion is detected.

Triggering the Correct IOPs. To trigger the correct IOPs, the adapter must first map task information (e.g., task name) to the corresponding IOPs. This mapping can be stored in an Excel file, where task attributes are linked to predefined IOP trigger conditions.

The adapter reads the Excel file to identify the IOPs required to complete the tasks and satisfy its triggering conditions. For example, if an IOP is triggered on file creation, the adapter generates a specific file in the monitored directory. The RPA bot listens to the trigger condition and initiates the IOP. The method of triggering depends on the types of triggers supported by the RPA software, such as file creation, hotkey activation, or email-based trigger. In our implementation, a file creation trigger mechanism is adopted. The adapter continuously monitors the condition status (e.g., file deletion) to confirm IOP completion before notifying the engine.

5.2 Sequence Diagram for IBP-Based Integration

The sequence diagram for the IBP-based integration approach is illustrated in Fig. 6 and Fig. 7.

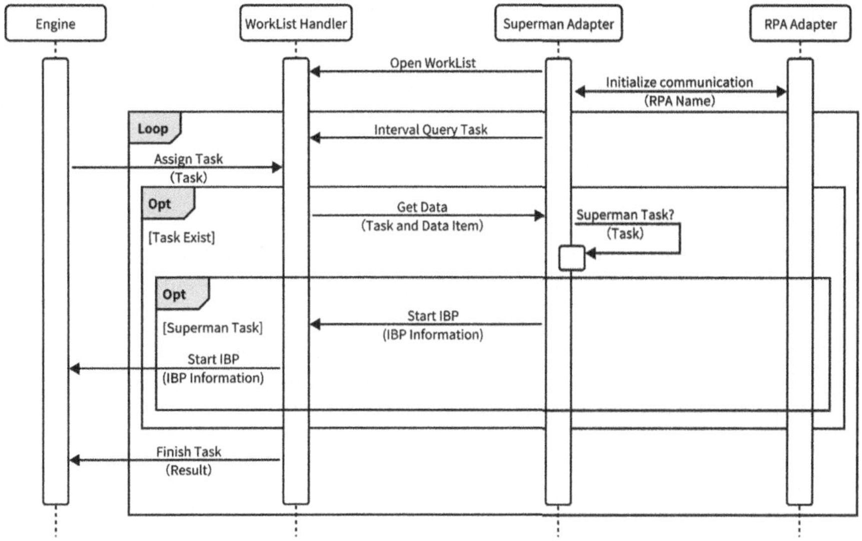

Fig. 6. Sequence for the execution of a Superman task in IBP-based integration.

During the preparation phase, the Superhuman Adapter opens the worklist on the client and establishes communication with all RPA Adapters. When the Superhuman Adapter queries the worklist, tasks may belong to three categories: Superhuman Tasks (requiring the initiation of an IBP), RPA Tasks (to be executed by an RPA bot), or Human Tasks (requiring confirm by a human). The Superhuman Adapter identifies task types by analyzing task metadata, such as the task name. After completing the preparation phase and recognizing a Superhuman Task that starts an IBP, RPA Tasks and Human Tasks subsequently appear in the worklist, as illustrated in Fig. 6.

When a task requiring RPA bot execution is identified, the Superhuman Adapter determines which RPA bot should handle the task, transmits the data to the corresponding RPA Adapter via network communication, and triggers the IOP to complete the task. The specific process by which the RPA Adapter triggers the correct IOP is not elaborated here. Upon IOP completion, the

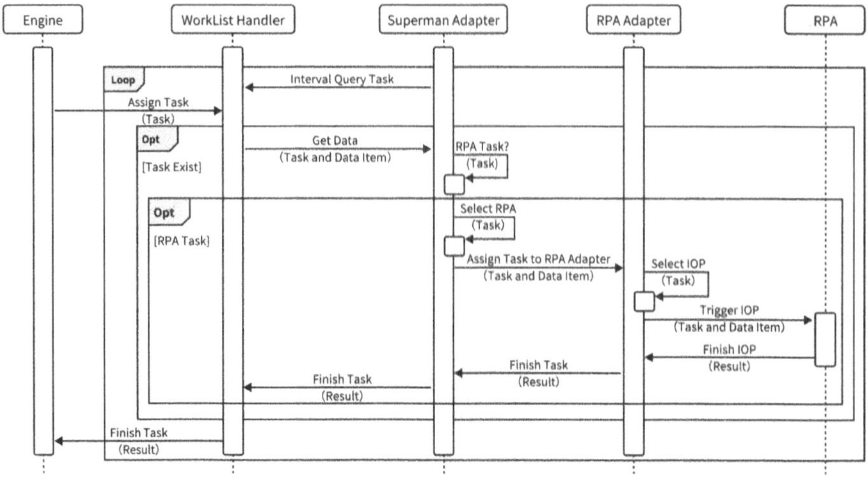

Fig. 7. Sequence for the execution of an RPA task in IBP-based integration.

RPA Adapter reports the results to the Superhuman Adapter through network communication, and the Superhuman Adapter subsequently submits the results to the engine via the client, as illustrated in Fig. 7.

The Superhuman Adapter determines which RPA bot should execute a task based on task information (such as task name). Our implementation specifies that when a task name follows the format $RPA\_\{RPAName\}\_\{TaskName\}$, it is designated as an RPA task. In such cases, the RPA bot named $RPAName$ is responsible for executing the task named $TaskName$. With this method, there is no need to modify the engine to make it identify an RPA task.

When a human task is identified, the adapter takes no action. Instead, the human resource must manually confirm the outcomes and assume responsibility for the results.

6 Evaluation

This section demonstrates the feasibility of the proposed integration framework through software implementation and an application case.

Prototype Implementation. We provide an open-source software prototype that implements the proposed framework[1]. The prototype employs Camunda BPMN Workflow Engine 7[2] as the BPMS and Automation Anywhere[3] as the RPA software. Camunda provides the web-based Tasklist as a client to obtain the worklist of human. The prototype provides the implementation of the adapter

[1] Prototype implementation code repository: https://github.com/0SocialExp/RPA_BPM_Integration.

[2] https://camunda.com/de/platform-7/workflow-engine/.

[3] https://www.automationanywhere.com/.

module. The adapter uses the selenium library[4] in Python to control the web-based Tasklist to use the functions provided by the Interface II.

Assume a bank provides account opening services to customers. Historically, bank employees manually reviewed and approved applications before processing account creation. To improve employee efficiency and streamline these time-consuming and error-prone tasks, the bank has decided to implement process automation through RPA and BPM.

The account opening process is illustrated in Fig. 8. The process begins with receiving customer applications via email. Bank employees extract required information from the applications and conduct approval checks to determine eligibility. If the application is approved, employees proceed with offline card issuance. If the application is rejected, employees notify the customer via email, specifying the reasons for rejection.

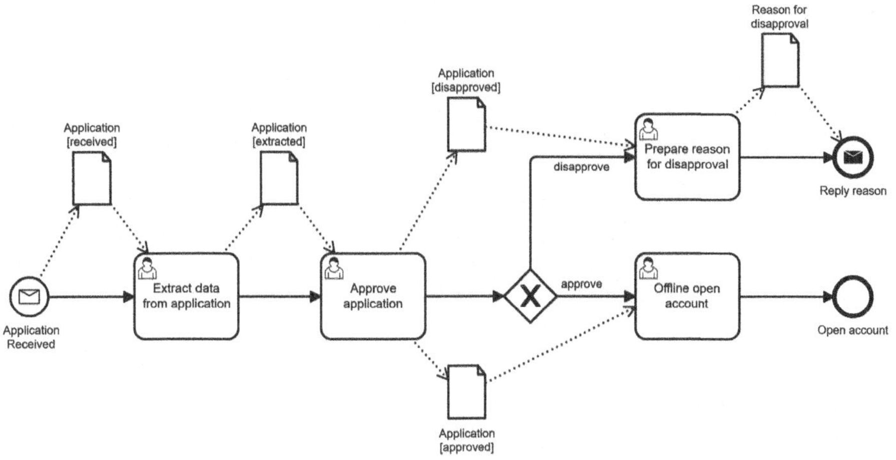

Fig. 8. Business process of bank account opening.

To achieve the integration of RPA and BPM, it is necessary to analyze each task in the process to determine how RPA participates. Tasks such as "Extract data" and "Prepare reason" can be fully automated by RPA and integrated through the IOP approach, where RPA bot acts as a digital employee, directly replacing humans. However, the "Approve application" task requires manual approval and responsibility, which is not suitable for RPA bot alone. Nonetheless, RPA bot can assist in completing the steps in this task through IBP, forming a "superman" team of RPA bots and a human. The "Open account" task must be manually executed by a human. Human is responsible for designing the IBP as depicted in Fig. 9.

Bank employees have defined three distinct RPA roles, each responsible for specific tasks: Identity Checker (verifies the authenticity of customer identity

[4] https://robotframework.org/SeleniumLibrary/.

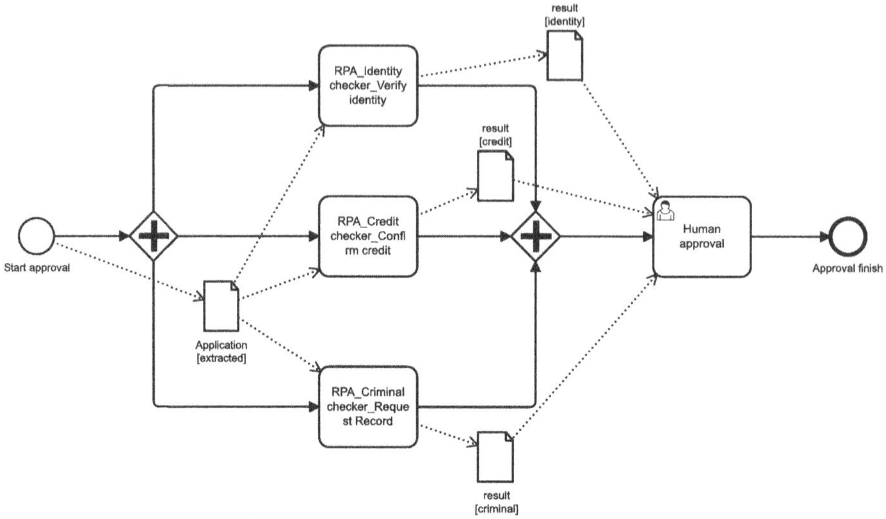

Fig. 9. IBP of bank account opening approval.

information), Credit Checker (ensures no adverse credit history), and Criminal Checker (queries criminal records). The three RPA bots perform tasks in parallel to improve work efficiency. Ultimately, bank staff make decisions based on the work results of the three RPA bots to decide whether to approve the user's account opening application and are responsible for the decision results.

7 Conclusion

This paper proposes an integration framework that follows the engine interface protocol from the Workflow Reference Model to integrate RPA into BPMS without requiring modifications to the process model or workflow engine. RPA is leveraged as a resource, functioning as both digital employee and human assistant within the organisation. We developed an open-source prototype implementation using the Camunda BPMN Workflow Engine as the BPMS platform and Automation Anywhere as the RPA software. The feasibility of the integration was demonstrated through a use case of bank account opening process.

Future research will extend the current framework to more use cases to validate its practical applicability. In parallel, we plan to enhance the software prototype by improving user interface and user experience, as well as extending the RPA trigger functions. In addition, it would also be interesting to explore new paradigms for human-bot collaboration within "superman" team, e.g., leveraging dynamic state machine models to enable more adaptive and context-aware interactions.

Acknowledgements. This work is Supported by the NSFC-Guangdong Joint Fund Project under Grant No.U20A6003, the National Natural Science Foundation of China (NSFC) under Grant No.61972427, and the Research Foundation of Science and Technology Plan Project in Guangdong Province under Grant No.2020A0505100030.

References

1. Flechsig, C., Lohmer, J., Lasch, R.: Realizing the full potential of robotic process automation through a combination with BPM. In: Bierwirth, C., Kirschstein, T., Sackmann, D. (eds.) Logistics Management. LNL, pp. 104–119. Springer, Cham (2019). https://doi.org/10.1007/978-3-030-29821-0_8
2. Flechsig, C., Völker, M., Egger, C., Weske, M.: Towards an integrated platform for business process management systems and robotic process automation. In: Marrella, A., et al. (eds.) BPM 2022. LNBIP, vol. 459, pp. 138–153. Springer, Cham (2022). https://doi.org/10.1007/978-3-031-16168-1_9
3. Hollingsworth, D., Hampshire, U.: Workflow management coalition: the workflow reference model. Document Number TC00-1003 **19**(16), 224 (1995)
4. Kirchmer, M., Franz, P.: Value-driven robotic process automation (RPA). In: Shishkov, B. (ed.) BMSD 2019. LNBIP, vol. 356, pp. 31–46. Springer, Cham (2019). https://doi.org/10.1007/978-3-030-24854-3_3
5. Kokina, J., Blanchette, S.: Early evidence of digital labor in accounting: innovation with robotic process automation. Int. J. Account. Inf. Syst. **35**, 100431 (2019)
6. König, M., Bein, L., Nikaj, A., Weske, M.: Integrating robotic process automation into business process management. In: Asatiani, A., et al. (eds.) BPM 2020. LNBIP, vol. 393, pp. 132–146. Springer, Cham (2020). https://doi.org/10.1007/978-3-030-58779-6_9
7. Lin, M., Zheng, Z.: Design and implementation of intelligent search assistant based on AIGC and RPA. In: 8th International Conference on Electronic Information Technology and Computer Engineering, pp. 534–538 (2024)
8. Nahak, K., Mishra, A., Dash, S.S.: Exploring the future of robotic process automation in the digital workforce. J. Data Acquisition Process. **38**(2), 2446 (2023)
9. Rizk, Y., et al.: A conversational digital assistant for intelligent process automation. In: Asatiani, A., et al. (eds.) BPM 2020. LNBIP, vol. 393, pp. 85–100. Springer, Cham (2020). https://doi.org/10.1007/978-3-030-58779-6_6
10. Ruiz, R.C., Ramírez, A.J., Cuaresma, M.J.E., Enríquez, J.G.: Hybridizing humans and robots: an RPA horizon envisaged from the trenches. Comput. Ind. **138**, 103615 (2022)
11. Venable, J., Pries-Heje, J., Baskerville, R.: A comprehensive framework for evaluation in design science research. In: Peffers, K., Rothenberger, M., Kuechler, B. (eds.) DESRIST 2012. LNCS, vol. 7286, pp. 423–438. Springer, Heidelberg (2012). https://doi.org/10.1007/978-3-642-29863-9_31
12. Zhang, L., Yang, J., Ouyang, C., Yu, Y., Pan, M.: Conceptualizing processes in robotic process automation: a basic meta-model. In: 10th International Conference on Systems and Informatics (ICSAI), pp. 1–8 (2024)

Towards Automated Handling of Object State Exceptions in Smart Contracts

Julius Köpke[1(✉)], Giovanni Meroni[2], and Mattia Salnitri[3]

[1] University of Klagenfurt, Klagenfurt, Austria
julius.koepke@aau.at
[2] Technical University of Denmark, Lyngby, Denmark
giom@dtu.dk
[3] University of Bergamo, Bergamo, Italy
mattia.salnitri@unibg.it

Abstract. Blockchain-based process execution has gained significant attention from the BPM community in the previous years. The ability to guarantee the correct execution of processes is considered an important factor for applications in low-trust environments and for the implementation of smart contracts in general. However, enforcing the correct execution of the control flow is a two-edged sword as it naturally minimizes flexibility. At the same time, the enforcement of the process is only partial since the blockchain cannot control the state of real-world objects: physical objects and off-chain data objects. Consequently, the actual object state may deviate from the one assumed by the smart contract. Smart contract models should allow the automatic handling of exceptions arising from such deviations while still guaranteeing contractual behaviors. This paper introduces a modeling approach for smart contracts that utilizes imperative modeling of processes in combination with declarative modeling of object life-cycles and consensus mechanisms. This combination allows us to agree on exceptions, to automatically derive plans for resolving object state exceptions at runtime within the boundaries of the agreed smart contract, and to agree on a selected plan.

Keywords: Smart Contracts · BPM on Blockchain · Exception Handling · Guard Stage Milestone

1 Introduction

The execution of business processes on blockchains has received substantial interest in research throughout the previous years [18]. Various research prototypes exist for executing business processes on blockchains. However, there is no widespread adoption in the industry yet. While the execution of business processes on blockchains is considered beneficial in low-trust environments, important aspects for practical use - such as privacy [8] and flexibility [4,11] - were not addressed by the initial prototypes. Flexibility is a challenging property, as it contradicts the aim of guaranteeing the correct execution of a process

M. ER et al. (Eds.): BPM 2025, LNBIP 565, pp. 168–183, 2026.
https://doi.org/10.1007/978-3-032-02936-2_13

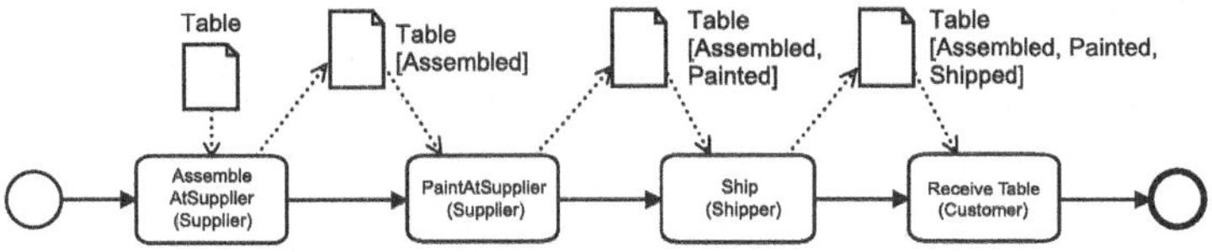

Fig. 1. Running Example (actors in parenthesis)

based on the underlying process model. On the one hand, blockchain technology is considered beneficial for low-trust environments as it guarantees the correct process execution. On the other hand, unforeseen events may always happen, preventing the enforced process from continuing its execution if not addressed.

In an ideal world, exceptions are not likely to happen if the execution of a process is fully enforced by the blockchain. However, the blockchain can only enforce the process based on facts known to the blockchain system (typically stored on-chain). For real-world objects - physical objects or off-chain data objects - no enforcement of the state is possible by the blockchain system since the blockchain does not control the actual state of the objects, but only their on-chain representation. Thus, deviations between the prescribed state in the process model and the actual state are prone to happen. Such deviations can lead to process instances that cannot continue since no alternative behavior of the process flow is permitted, but a continuation as prescribed is also impossible.

To clarify the problem with an example, Fig. 1 shows a fragment of a simplified process, where a table is assembled and then painted by a supplier. Next, the table is shipped by a shipper and finally received by a customer. The table is modeled as a data object using the state notation in square brackets. The blockchain cannot guarantee the actual state of the table; it is possible that the recorded state of the table (e.g., *painted* and *assembled*) does not match the state of the table when the customer receives it. If the customer now refuses to receive the table because it is not adequately painted or assembled, current blockchain-based solutions for process execution would not know how to continue executing this process instance. Thus, the instance would get stuck forever.

In general, we argue that processes executed on blockchains should effectively handle exceptions arising from a mismatch between the actual state of a real-world object and the state reported on the blockchain. However, it is infeasible to model all potential state exceptions and their handling mechanisms a priori. In this paper, we explore how imperative modeling of the expected execution flow of blockchain-based processes can be complemented by declarative object life-cycles to derive exception handlers at runtime. We provide the following contributions: (i) a method for modeling blockchain-based processes supporting the recovery of state exceptions of objects; (ii) algorithms to automatically derive recovery plans in case of state exceptions; (iii) a dispute protocol for handling disputes on object states and agreeing on a repair plan.

The paper is structured as follows: Sect. 2 presents related work, Sect. 3 introduces the proposed modeling approach. In Sect. 4, we provide the details on the design-time, and in Sect. 5 on the runtime phases of the approach. Finally, Sect. 7 concludes the paper.

2 Related Work

The concept of Smart Contracts as the execution of legal contracts via hardware and software was already coined in the last millennium by N. Szabo [19]. However, it recently achieved substantial momentum with the introduction of second-generation blockchain systems such as Ethereum [2]. Notably, this has also led to a different interpretation of the term. A blockchain smart contract is code on a blockchain that may or may not implement an actual contract.

The BPMN community started to contribute various approaches for the model-based implementation of Business Process Models on blockchain. There are two major classes: (1) The process itself is executed on blockchain, e.g. [10,20], (2) the choreography is enforced by the blockchain, e.g. [3,23]. Since the required degree of enforcement depends on the actual security requirements of the collaboration, a BPMN extension to explicitly model smart contract-specific requirements was proposed in [9]. It allows deriving which parts of a process should be secured by the blockchain, and for what parts this strong enforcement is not required.

The strong enforcement of the control flow as provided by blockchain systems is a two-edged sword: Strong enforcement limits flexibility, and flexibility limits enforcement. Consequently, solutions for flexibility in blockchain-based processes need to balance both sides. In [11] a consensus-based mechanism to assign and reassign participants and predefined parts of the control flow at runtime is introduced. The control flow can be changed by selecting alternative sub-processes and by gateways that are based on runtime consensus decisions. Consequently, flexibility is restricted to explicitly modeled cases. An alternative viewpoint is taken by [4], where the blockchain enforces a choreography, but the participants can change the actual rules of the choreography at runtime if a consensus on the change is reached. A similar approach based on DCR graphs is followed in [7], where participants can agree on changing the DCR graphs and partner assignments. However, none of the approaches addresses exceptions.

Flexibility and exception handling have been extensively studied for classical business processes [16]. Similarly, approaches that combine the process view with the lifecycle of the participating artifacts to ensure consistency already exist [1,13,17]. However, the existing approaches do not take into account the specific requirements of low-trust environments, where flexibility is needed but deviations must still comply with the original intentions of the participants and single points of control are not acceptable.

This paper explores the combination of imperative and declarative process modeling to derive contractual repair plans for object state exceptions at runtime. This way, the declarative model sets the frame for the flexibility of the original contract. Automatic planning has been applied in the BPM domain to repair data inconsistencies due to concurrency issues [21], and to support flexibility in business processes [12]. However, our work proposes to model possible adaptations declaratively in an established modeling language, and it covers a consensus protocol for the occurrence of exceptions and for agreeing on a recovery plan. Both are key factors in low-trust environments.

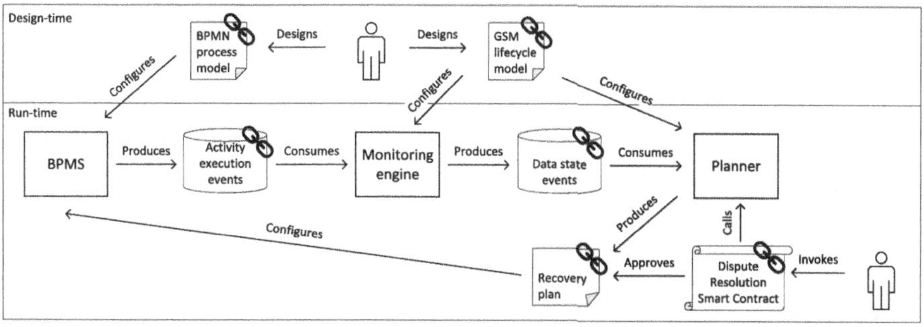

Fig. 2. Proposed approach. The chain symbol indicates elements on a blockchain.

3 Approach

To handle exceptions arising from a misalignment between smart contract object states and real-world object states, we propose the approach outlined in Fig. 2. It is structured in two phases: a design-time phase, where the process is modeled, and a runtime phase, where the actual execution, monitoring, and exception handling take place. During the design-time phase, the modeler is in charge of:

- Designing a BPMN or SecBPMN2BC [9] process model that captures the expected execution flow of the smart contract. Real-world objects are represented as BPMN data objects, including state annotations.
- Designing life-cycle models representing the life-cycle of each data object. The existing implicit state transitions of the BPMN model (input and output states of activities) can be used as a starting point. However, additional state changes and possible activities for reaching these changes are included to define the contractual behaviors in a declarative form. For allowing a dispute on the actual state of data objects, the challengeable states are defined in the life-cycle model.

The BPMN and life-cycle models are then used in the runtime phase to execute the process, monitor the states of the data objects, and identify recovery plans when an exception in the object states occurs. In particular, the BPMN model is used to instruct a **process executor**. This component can be a traditional off-chain BPMS, an on-chain BPMS such as Caterpillar [10], or smart contract code on a blockchain implementing the logic of the process. In all these cases, we assume these components produce an event whenever a task is executed. These events are then consumed by a **monitoring engine** that is configured with the life-cycle models. In this way, the engine is aware of how the process is progressing and can determine the state of data objects. Whenever a data object achieves a new object state, the monitoring engine will produce an event.

At runtime, whenever a participant has to start or end an activity, he/she can challenge the actual achievement of the object state by invoking a **dispute resolution smart contract**. This smart contract determines, through a consensus protocol involving selected participants, if the claim is correct.

If so, the smart contract invokes a **planner**, which is configured with the life-cycle models and takes as input the state being challenged and all events produced by the monitoring engine. Based on this information, the planner derives a set of recovery plans to recover from the exception, that is, to eventually reach the challenged object state without side effects on other milestones and opened stages. Then, the smart contract determines, through a consensus protocol involving selected participants, which plan to approve. Once a plan has been approved, the process executor puts the process defined in the BPMN model on hold until the challenged state is reached. Then, the original process resumes[1].

The envisioned approach stores the BPMN and life-cycle models, as well as the events and the recovery plans, on a blockchain. In this way, each participant can compute all possible plans and check if a selected plan is valid. Thus, we can guarantee that a contractual strategy is applied once consensus is reached. Since the process is executed on a blockchain and all participants can access the process state, each participant can also execute their own instance of the monitoring engine. Consequently, the monitoring component is neither a single point of failure nor does it introduce central control. However, to keep the discussion compact, we treat the monitoring component as a single logical component in the following discussion.

4 Design-Time Modeling

4.1 Preliminaries on GSM

For representing life-cycle models of data objects, we use Guard Stage Milestone [5] (GSM). It is an artifact-centric language, aimed at the automation of knowledge-intensive processes. Instead of treating activities and control flow dependencies as first-class citizens - as it is typically the case for process-centric languages like BPMN - GSM focuses on the possible states in which an artifact - a physical or virtual object - can be during its life-cycle, and on the admissible transitions from one state to another.

In GSM, activities that can be performed to change the state of an artifact are represented by *stages*, which are represented as rounded rectangles. Stages can be atomic if they represent a single activity, or they can be composite if they nest other ones. Stages can be *open*, making it possible to open inner stages and - for atomic stages - to execute the associated activities, or they can be *closed*.

[1] We assume that the component implementing the process executor supports stopping and resuming the process.

To indicate under which conditions activities can be executed, stages are decorated with *guards*, which are represented as diamonds. In particular, a guard is a sentry - an Event-Condition-Action (ECA) rule [5] - that is evaluated when the corresponding stage is closed. If that sentry holds, then the corresponding stage becomes open.

To specify when the artifact has reached a specific state, stages are decorated with *milestones*, which are represented as circles. A milestone can be either *achieved*, indicating that the artifact has reached a state, or *invalidated*, indicating that the artifact is no longer in that state. Sentries are used to specify under which conditions a milestone is achieved or invalidated. In particular, the achieving sentry is evaluated while the corresponding stage is open. Conversely, the invalidating sentry is evaluated while the stage is closed. In addition, opening a closed stage invalidates all its milestones, and achieving a milestone makes the attached stage and all its inner stages closed.

In GSM, sentries are triggered by events which can be internal (e.g., the achievement of a milestone) or external (e.g., a notification from an information system). It is worth noting that, when a sentry holds true, it will change the state of the model (e.g., a stage will become open). Thus, it will generate an internal event that could trigger one or more sentries. GSM is designed to evaluate sentries to avoid an infinite chain of internal events [5]. The chain of events that originates from a single external event is called B-step.

GSM also includes a data structure (which has no visual representation) named *information model.* This component is used to store runtime information on the current configuration for stages (e.g., open or closed) and for milestones (e.g., achieved or invalidated), as well as the external events that the model can accept and their payload.

4.2 Modeling of Admissible Object Behaviors

In our approach, we complement the imperative process-oriented smart contracts with models explicitly defining the contractual state transitions of objects. While there are various modeling methods for defining allowed state transitions of data objects, such as state charts [6] or Petri-nets [15], we have opted for GSM life-cycle models for the following reasons:

1. It must be possible to express that an object can have multiple states simultaneously. In our example, a table should be in the state *assembled* and in the state *painted.* The modeling method should support this without duplications of states.
2. GSM is a formally grounded modeling method supporting declarative and compact life-cycle models. At the same time, the building blocks of GSM are well-known since they serve as the foundation for OMG's case handling standard CMMN [14].

We use life-cycle models as parts of the overall smart contract. Therefore, at build-time, the participants of the smart contract agree on the life-cycles of all data objects in the form of GSM planning models. At runtime, these models have two purposes: they are used to derive plans for recovery actions, and they are used to derive models for synchronizing the state of the imperative process with that of its data objects. We refer to the former as GSM planning models and to the latter as GSM monitoring models.

Example. In the example process in Fig. 1, there is one data object *Table*. From the process, we can identify the following activities leading to state transitions of tables: *AssembleAtSupplier*, *PaintAtSupplier*, and *Ship*. The activity *AssembleAtSupplier* requires a non-assembled table and produces an *assembled* table. The activity *PaintAtSupplier* requires an assembled table and returns a *painted* (and still *assembled*) table as output. Finally, *Ship* requires an *assembled* and *painted* table and has an additionally *shipped* table as output.

Based on this information, we can create a stub of a GSM Planning model with three composite stages: *Assemble*, *Paint*, and *Shipment*. We nest the previously identified activities in the form of atomic stages into the composite stages. For each composite stage, we add a milestone with the label of the newly achieved output state (*Assembled*, *Painted*, *Shipped*). We define these milestones as achieved once the corresponding atomic stage is completed. In addition, we create guards for the new composite stages to comply with the semantics of the input process. For example, a table can only be assembled if it has not yet been assembled. This pattern allows us to easily extend the model with additional activities on tables that are not part of the BPMN model.

The resulting model stub represents the table's assumed behavior during the execution of the imperative process. However, planning models may also contain additional stages for state transitions that are physically possible and contractually allowed. Therefore, the partners agree on the allowed state transitions via the planning models. These atomic actions are not necessarily executed during the standard paths of the process. However, the participants agree that they can potentially be used to handle exceptions.

In the example, we include the additional atomic stages *PaintAtCustomer* and *ReturnService*. Furthermore, we add a composite stage, *Return*, and adopt *Shipment* to allow shipments in any state. Overall, this model now expresses all agreed-upon state transitions of the table and their preconditions. Finally, we mark all milestones that may be challenged by the participants at runtime as *challengeable*. A graphical representation of the resulting planning model is shown in Fig. 3. In the graphical representation, challengeable milestones are shown as grey circles. All composite stages have events in the on-part. These events are called planning events (e.g. *P:Paint*) that are used for plan generation (see details later). However, they can also be derived and do not need to be provided manually by the modeler. In the same way, for all challengeable milestones, an invalidation sentry is needed to invalidate the milestone once the corresponding dispute message arrives.

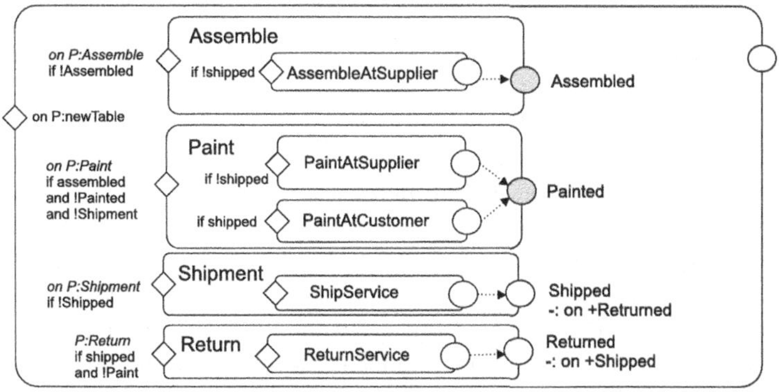

Fig. 3. Example GSM Planning Model for the Table Object

Formalizing Planning Models. After the informal introduction of planning models based on an example, we now formalize the necessary parts of planning models as an extended subset of GSM and refer the user to the definition of GSM in [5] for all the details. We use the typical object-oriented dot notation to access the components of tuples in the following.

Definition 1. (*GSM Planning Model*) A GSM Planning Model is a tuple $Mod = (G, S, M, E, I)$, where

- G is a set of Guards. Each guard g has a property $g.sen$ referring to an expression called Sentry (see Definition 2) defining the condition under which the guard is activated.
- S is the set of Stages. Every stage $s \in S$ has a parent stage $s.p$. For the root stage r, $r.p = r$. The function $s\_type : S \rightarrow \{atomic, composite\}$ maps each stage s to its type such that: $s\_type(s)$ is $\underline{composite}$ if $s \in \{s \in S : \exists s' \in S : s'.p = s\}$ otherwise, the type is $\underline{atomic}$. Each stage s has a non-empty set of guards $s.g$ defining under which conditions a stage is opened. During runtime a stage opens if at least one guard evaluates to $true$.
- M is the set of Milestones. Each milestone m in M has a property $m.s \in S$ referring to the stage of that milestone. Every milestone m has a non-empty set of sentries $m.Sen^+$ defining when m is achieved and a non-empty set of sentries $m.Sen^-$ defining when m is invalidated. If at least one sentry evaluates to true, an activation or deactivation occurs. Whether a Milestone can be challenged is defined by the modeler and expressed by the function $challengable : Mod.M \rightarrow \{true, false\}$.
- E is the set of events. The function $e\_type : E \rightarrow \{atomicStart, atomicEnd, internal, planning, ChallengeSuccessful\}$ maps each event to its type.
  Events of type $\underline{atomicStart}$ and $\underline{atomicEnd}$ are the start and end events of services enclosed in atomic stages. Such events have an additional property $e.s$ referring to the atomic stage.
  Events of type $\underline{atomicStart}$ are in control of the participants, while $atomicEnd$ events can only be controlled indirectly: They will eventually occur after the

corresponding start event is issued.
Events of type *planning* are external events under the control of the participants. They are used for planning purposes.
Events of type *ChallengeSuccessful* are sent to the GSM system once a participant successfully challenges a milestone. Such events e have a payload $e.p$ referring to the challenged Milestone in M.
Events of type *internal* indicate the closing or opening of stages, achieving or deactivation milestones. See the full GSM semantics in [5] for details.

- $I = (S, M)$ is the *information model* containing one boolean attribute for each stage in $I.S$ and one boolean attribute for each milestone in $I.M$. In an instance i of the information model I. A value $true$ for an attribute s in i indicates that stage s is currently open. A value $true$ for an attribute m in $i.M$ indicates that Milestone m is currently achieved. The value of false indicates that a milestone is not achieved and a stage is closed.

Definition 2 (GSM Planning Model: Sentries). A sentry in a GSM planning model $Mod = (G, S, M, E, I)$ is an expression of the form: $on\ e\ if\ condition$ or $on\ e$ or $if\ condition$, where $e \in Mod.E$ and $condition$ is a boolean expression over the information model $Mod.I$

Given the general structure of planning models, we can now define correct planning models. In such models, all events that may trigger the opening or closing of stages must be under the control of the executors of the plan. Therefore, all composite stages are activated by specific planning events if the data condition is evaluated as true. Atomic stages must be activated by the corresponding start events of the enclosed services. Finally, for decentralized planning, we require that B-Steps' results are deterministic.

Definition 3 (Correct GSM Planning Model). A GSM planning model $Mod = (G, S, M, E, I)$ is correct if:

- For every stage $s \in Mod.S$ with $s\_type(s) = composite$ every guard g in $s.g$ has a sentry $g.sen$ with an on part referring to an event in $\{e \in Mod.E : e\_type(e) = planning\}$
- For every stage $s \in Mod.S$ with $s\_type(s) = atomic$ every guard g in $s.g$ has a sentry $g.sen$ with an on part referring to an event in $\{e \in Mod.E : e\_type(e) = atomicStart\}$
- No two different atomic stages have the same start or end events: Let EA be the set of events of atomic tasks: $EA = \{e \in M.E : e\_type(a) \in \{atomicStart, atomicEnd\}\}$ $\forall\ (e1, e2) \in EA \times EA : e1.s = e2.s \wedge e_{type}(e1) = e_{type}(e2) \Rightarrow e1 = e2$
- $ChallangeSuccessful$ events are only permitted for challengeable milestones: $\forall e \in \{e \in Mod.E : e\_type(e) = ChallangeSuccessful\} \Rightarrow challengable(e.p)$
- Each challangeable milestone m has one invalidation sentry in $m.Sen^-$ of the form: $on\ e$, where $e\_type(e) = ChallangeSuccessful\ and\ e.p = m$.

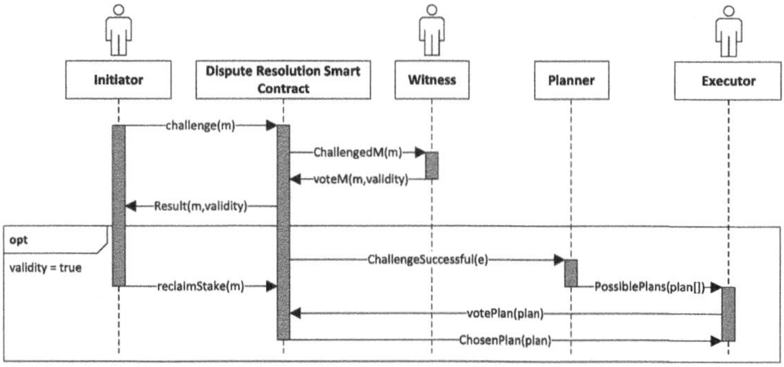

Fig. 4. UML sequence diagram showing the dispute resolution protocol.

- For every $e \in Mod.E$ where $e\_type(e) \neq internal$, and for every instance i of $Mod.I$, the resulting instance of the information model i' derived by the GSM B-Step of e on i is unique.

Please note that the requirement of planning events for each composite stage and specific invalidation sentries for all challengeable milestones does not need to be provided by the modeler. Instead, driving a correct model from a model without these features is straightforward based on the definition. Since the B-Step must be deterministic, we will use a B-Step function *B-Step(i,e,Mod)* in the further discussion.

Deriving Monitoring Models from Planning Models. Since at runtime of the standard flow, no planning actions are issued, and the process is fully driven by atomic services and their events, we need to derive monitoring models for planning models. This is achieved by replacing each guard with a planning event with one guard for every atomic stage that may be called after opening this stage (recursively). Due to space limitations, we cannot provide the full details here.

5 Runtime: Handling of State Exceptions

5.1 Dispute Protocol

As explained in Sect. 3, the dispute resolution smart contract is in charge of handling disputes between object states and the actual conditions of real-world objects, and then taking countermeasures. To do so, it relies on the dispute resolution protocol shown in Fig. 4, which starts from the idea of deferred mediation outlined in [22]. When the smart contract is deployed, the following lists of participants in the process need to be provided[2]:

[2] The participants need to agree on their roles and on the need to involve other subjects, such that the risk of collusion is minimized.

- Initiators: participants who can challenge a specific milestone.
- Witnesses: participants who can verify if a milestone corresponds to the state of a real-world object.
- Executors: participants who are in charge of executing the service represented by an atomic stage in the GSM model.

When one initiator detects a discrepancy, (s)he calls the smart contract by invoking the *challenge* function, passing as input the milestone representing the challenged object state. To prevent a malicious initiators from sending fraudulent claims and from flooding the smart contract with disputes, (s)he is also required to put a stake on the milestone being challenged. When called, the smart contract notifies the witnesses of the milestone being challenged by emitting the *ChallengedM* event. Witnesses then have to verify if a discrepancy actually exists, and to vote accordingly. To do so, they invoke the *voteM* function, passing as input the challenged milestone and a boolean outcome representing their vote. Based on the voting outcome, the smart contract determines whether the claim holds or not. In both cases, the *Result* event is emitted.

If the claim holds, the initiator can reclaim its stake by invoking the *reclaimStake* function, passing the challenged milestone as input. Also, the smart contract notifies the planner that a milestone has been challenged by emitting the event *ChallengeSuccessful*. The planner then determines one or more plans of possible remediation actions and notifies the executors by emitting the *PossiblePlans* event. Note that multiple planners could possibly exist for a process (e.g., each executor may have its own planner), thus avoiding centralizing trust.

Executors then have to vote on which plan to follow, and they do so by invoking the *votePlan* function, passing the plan they want as input. Once consensus has been reached on a plan, the smart contract notifies the executors on the chosen plan by emitting the *ChosenPlan* event. To avoid deadlocks, witnesses and executors may also be asked to put a stake before the process starts, which can be reclaimed only when the process ends. In this way, they are incentivized to take part in the dispute resolution protocol.

5.2 Repair Plan Generation

Generation of a State Transition System. To derive repair plans, we generate a state transition system in the form of a directed multi-graph $DG = (V, A)$, where V is a set of states (nodes) and A is a set of directed, labeled arcs of the form (s, t, e), representing a transition from state s to state t via event e. V is a subset of all possible instantiations of $Mod.I$ of the correct GSM planning model $Mod = (G, S, M, E, I)$ used as input.

Initialization: When an event e of type $ChallengeSuccessful$ is received, we first add the current instance i of $Mod.I$ as the start node i_{target} to our state transition system: $DG.V = \{i_{target}\}$. We then use the function $B - Step(i_{target}, e, Mod)$ to derive the next state $i_{challenged}$ resulting from a GSM B-Step on i_{target} with event e for Mod (see Subsect. 4.2). As a consequence, in $i_{challenged}$ the challenged milestone is invalidated. Since the state $i_{challenged}$ is derived from i_{target} by applying e, we add the edge $(i_{target}, i_{challenged}, e)$ to $G.A$. Accordingly, i_{target} represents the goal state, and $i_{challanged}$ represents the exceptional state as the starting point.

Generation of All States: Since the participants can only control planning and start events of atomic tasks, we can only generate plans based on these events. Given the input GSM model Mod we define the set of controllable events E_c as $E_c = \{e \in Mod.E : e\_type(e) \in \{planning, atomicStart\}\}$. To complete the state transition system, we apply the $B - Step$ function for every event e in E_c on $i_{challenged}$ to derive all reachable states RS from $i_{challenged}$. We add each reachable state rs to $DG.V$ and connect it to $i_{challenged}$ by adding the edge $(i_{challenged}, rs, e)$ to $DG.A$. Here, e is the event that resulted in achieving rs. If the resulting state already exists, the reached node is not further explored. Otherwise, we recursively apply the $B - Step$ function over all events in E_c on the newly derived state.

The approach's core is realized by Algorithm 1. The algorithm is called with the most recent instance of the artifact information model i_{target} and the $challengeSuccessful$ event e as input. This way, in the first call of $ProcessState()$, the successor state $i_{challanged}$ of i_{target} is derived. This deactivates the challenged milestone. Afterward, all states that are reachable by controllable events are derived recursively. For each atomic start event of services, the next state is derived by additionally ingesting the eventual end event of that service. This is required since the controllable start event leads to a not directly controllable but eventually appearing end event.

Plan Generation. Once the state transition system is created, we can use it to derive repair plans by searching for paths starting at the exceptional state $i_{challanged}$ and ending at the state before the exception i_{target}. For deriving actionable and non-redundant plans, we are only interested in paths that do not visit any state twice. For each such path, we project only the start events of

Algorithm 1 ProcessState

```
function ProcessState(i, e, Mod, DG)
    if e_type(e) = atomicStart then               ▷ Special care for start events of services
        i' ← B-step(B-step(i,e,Mod), start2End(e), Mod))
    else
        i' ← B-step(i, e, Mod)
    end if
    if i' ∈ DG.V then                                               ▷ We found a loop
        DG.A = DG.A ∪ {(i,i',e)}
        return
    end if
    DG.V = DG.V ∪ {i'}
    DG.A = DG.A ∪ {(i,i',e)}
    E_c = {e ∈ M.E : e_type(e) ∈ {planning, atomicStart}}.
    for all pe ∈ E_c do                  ▷ Generate next-state for each controllable event.
        ProcessState(i', pe, Mod, DG)
    end for
end function
```

atomic tasks to obtain an execution plan. Deriving all paths between two nodes without loops is straightforward using a depth-first or breadth-first traversal.

6 Illustrative Case Study

We assume the customer should sign to receive the table (last step in the BPMN diagram in Fig. 1), thus acting as initiator. We also assume that, in case of disputes on the state of the table, the shipper would act as witness. For this step, the table must be in the state *assembled*, *painted*, and *shipped*. The customer observes that the paint is faulty. Therefore, he challenges the milestone *painted* by invoking the *challenge* function of the dispute resolution smart contract, passing *painted* as input. The shipper then confirms the challenged milestone by invoking the *voteM* function of the dispute resolution smart contract, passing *painted* and *true* as input. After that, an event e of type *ChallengeSuccessful* with payload $c.p = \{painted\}$ is sent by the dispute resolution smart contract to the planner component. The planner, by replaying the object state events produced by the monitoring engine, reconstructs the current state of the information model i_{target}: $i_{target}.M = \{(Assembled, true), (Painted, true), (Shipped, true)\}$, $i_{target}.S = \{(Assemble, false), (AssembleAtSupplier, false), (Paint, false), (PaintAtSupplier, false), ...\}$. First, $i_{challenged}$ is derived by applying the B-Step of e on i_{target}. In particular, in $i_{challenged}$, milestones painted changes from *true* to *false*: $i_{challanged}.M = \{(Assembled, true), (Painted, false), (Shipped, true)\}$, $i_{challenged}.S = \{...\}$. After this, all reachable next states are derived by examining the applicability of every planning or start event of atomic stages recursively. In particular, issuing the planning event *P:Return* leads to a new state where the *Return* stage is open. Alternatively, issuing the planning event *P:openPaint* leads to a new state where the stage *Paint* is open. For both new states, the processing continues recursively. All other planning events do not lead to a new state because, based on the GSM model, the event will not lead to a new state. An illustration of the entire state transition graph is shown in Fig. 5. For a compact representation, we only show elements of the information model where the value is *true*. We indicate planning events by *P:event* and start events of atomic tasks by *S:event*.

We can derive recovery plans from this state transition system. A plan candidate is any path between $i_{challanged}$ and i_{target}, where no state is visited twice. In the example, there are the plan candidates ⟨*P:Paint, S:PaintAtCustomer*⟩ *and* ⟨*P:Return, S:returnService, P:paint, S:PaintAtSupplier, P:Shipment, S:ShipService*⟩. All other paths between $i_{challanged}$ and i_{target} contain one state multiple times. They are, therefore, discarded. Since only actual actions are needed to execute a plan, we project only the start events of atomic stages and get the following two actionable plans: ⟨*S:PaintAtCustomer*⟩, ⟨*S:ReturnService, S:PaintAtSupplier, S:ShipService*⟩. Potential plan candidates can be computed by each participant individually; however, the participants need to reach a consensus on one repair plan. This is realized through the dispute resolution smart contract. The shipper, the manufacturer, and the customer vote for the plan they want by invoking the *votePlan* function of the smart contract, passing the chosen plan as input. Once the voting is over, the smart contract notifies the plan to the executors and, after the plan is executed, the process can continue as prescribed.

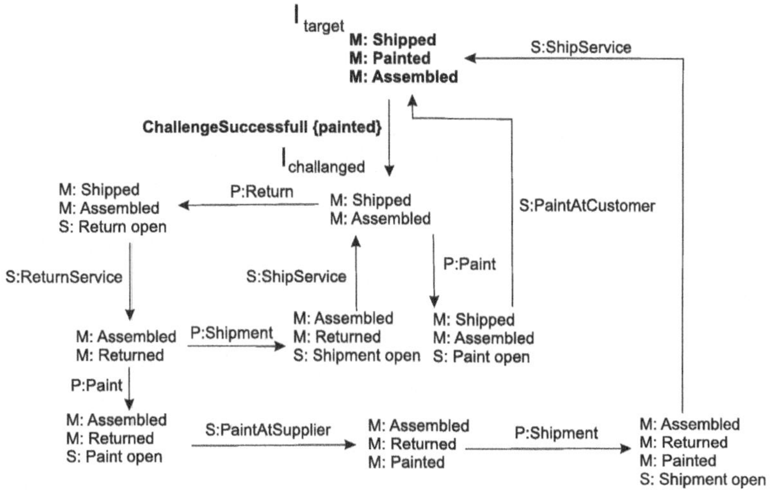

Fig. 5. Example State Transition System for Recovery Plan Generation

7 Conclusion and Future Works

Executing business processes on blockchains is considered highly beneficial for enforcing contractual behaviors in low-trust environments. However, on-chain enforcement does not cover physical or off-chain data objects. This leads to a possible deviation between object states recorded on the chain and actual object states in the real world. Ignoring these concerns may lead to process instances that are unable to proceed. In this paper, we have provided three contributions: (i) A consensus mechanism to agree on such deviations; (ii) a method for modeling contractual state transitions of objects; (iii) a method for deriving recovery plans for handling the resulting state exceptions at runtime. With these techniques, we can move forward to more robust smart contracts dealing with such deviations. For future work, we plan to implement the approach in a state-of-the-art process engine on blockchain. This also includes the potential translation of the planning problem to PDDL for using standard planners. With our proposed modeling, an avenue of interesting questions arises: Is a smart contract transactional complete? E.g., is it always possible to reach an acceptable end-state no matter what state deviations occur? If not, either the life-cycles may be extended at design time, or specific changes to life-cycles may be permitted at runtime. We will also explore how the increased complexity in process modeling and execution will affect the performance and cost of the approach.

Acknowledgements. This work was supported in part by project SERICS (PE00000014) under the NRRP MUR program funded by the EU - NGEU.

References

1. Born, M., Dörr, F., Weber, I.: User-friendly semantic annotation in business process modeling. In: Weske, M., Hacid, M.-S., Godart, C. (eds.) WISE 2007. LNCS, vol. 4832, pp. 260–271. Springer, Heidelberg (2007). https://doi.org/10.1007/978-3-540-77010-7_25
2. Buterin, V.: Ethereum: a next-generation smart contract and decentralized application platform (2014). https://ethereum.org/en/whitepaper/
3. Corradini, F., Marcelletti, A., Morichetta, A., Polini, A., Re, B., Tiezzi, F.: Chorchain: a blockchain-based framework for executing and auditing BPMN choreographies. In: BPM 2022 Demo and Resources, pp. 132–136. Springer, Cham (2022)
4. Corradini, F., Marcelletti, A., Morichetta, A., Polini, A., Re, B., Tiezzi, F.: A flexible approach to multi-party business process execution on blockchain. Future Gener. Comput. Syst. **147**, 219–234 (2023)
5. Damaggio, E., Hull, R., Vaculín, R.: On the equivalence of incremental and fixpoint semantics for business artifacts with guard-stage-milestone lifecycles. Inf. Syst. **38**(4), 561–584 (2013)
6. Estañol, M., et al.: Conformance checking in UML artifact-centric business process models. Softw. Syst. Model. **18**(4), 2531–2555 (2019)
7. Henry, T.: Towards trustworthy, flexible, and privacy-preserving peer-to-peer business process management systems. Ph.D. thesis (2022). https://theses.hal.science/tel-03931346/document
8. Köpke, J., Franceschetti, M., Eder, J.: Balancing privity and enforceability of BPM-based smart contracts on blockchains. In: Di Ciccio, C., et al. (eds.) BPM 2019. LNBIP, vol. 361, pp. 87–102. Springer, Cham (2019). https://doi.org/10.1007/978-3-030-30429-4_7
9. Köpke, J., Meroni, G., Salnitri, M.: Designing secure business processes for blockchains with SecBPMN2BC. Future Gener. Comput. Syst. **141**, 382–398 (2023)
10. López-Pintado, O., García-Bañuelos, L., Dumas, M., Weber, I., Ponomarev, A.: Caterpillar: a business process execution engine on the ethereum blockchain. Softw. Pract. Exp. 1162–1193 (2019)
11. López-Pintado, O., Dumas, M., García-Bañuelos, L., Weber, I.: Controlled flexibility in blockchain-based collaborative business processes. Inf. Syst. **104** (2022)
12. Marrella, A., Lespérance, Y.: A planning approach to the automated synthesis of template-based process models. SOCA **11**(4), 367–392 (2017). https://doi.org/10.1007/s11761-017-0215-z
13. Meyer, A., Pufahl, L., Fahland, D., Weske, M.: Modeling and enacting complex data dependencies in business processes. In: Daniel, F., Wang, J., Weber, B. (eds.) BPM 2013. LNCS, vol. 8094, pp. 171–186. Springer, Heidelberg (2013). https://doi.org/10.1007/978-3-642-40176-3_14
14. OMG: Case Management Model and Notation, Version 1.1. Technical report, Object Management Group (2016). http://www.omg.org/spec/CMMN/1.1

15. Popova, V., Fahland, D., Dumas, M.: Artifact lifecycle discovery. Int. J. Cooperative Inf. Syst. **24**(1), 1550001:1–1550001:44 (2015)
16. Reichert, M., Weber, B.: Enabling Flexibility in Process-Aware Information Systems - Challenges, Methods, Technologies. Springer, Cham (2012)
17. Ryndina, K., Küster, J.M., Gall, H.: Consistency of business process models and object life cycles. In: Kühne, T. (ed.) MODELS 2006. LNCS, vol. 4364, pp. 80–90. Springer, Heidelberg (2007). https://doi.org/10.1007/978-3-540-69489-2_11
18. Stiehle, F., Weber, I.: Blockchain for business process enactment: a taxonomy and systematic literature review. In: Marrella, A., et al. (eds.) BPM 2022. LNBIP, vol. 459, pp. 5–20. Springer, Cham (2022). https://doi.org/10.1007/978-3-031-16168-1_1
19. Szabo, N.: Formalizing and securing relationships on public networks. First Monday **9**(2) (1997)
20. Tran, A.B., Lu, Q., Weber, I.: Lorikeet: a model-driven engineering tool for blockchain-based business process execution and asset management. In: BPM (Dissertation/Demos/Industry), pp. 56–60. Springer, Cham (2018)
21. van Beest, N., Kaldeli, E., Bulanov, P., Wortmann, J., Lazovik, A.: Automated runtime repair of business processes. Inf. Syst. **39**, 45–79 (2014)
22. Wagner, E., Völker, A., Fuhrmann, F., Matzutt, R., Wehrle, K.: Dispute resolution for smart contract-based two-party protocols. In: ICBC 2019, pp. 422–430. IEEE (2019)
23. Weber, I., Xu, X., Riveret, R., Governatori, G., Ponomarev, A., Mendling, J.: Untrusted business process monitoring and execution using blockchain. In: La Rosa, M., Loos, P., Pastor, O. (eds.) BPM 2016. LNCS, vol. 9850, pp. 329–347. Springer, Cham (2016). https://doi.org/10.1007/978-3-319-45348-4_19

HiProX: Highly Efficient Process Execution on IoT and Edge Devices

Dominik Voigt(✉), Lisa Podszun, Juergen Mangler, and Stefanie Rinderle-Ma

Chair of Information Systems and Business Process Management, TUM School of Computation, Information and Technology, Technical University of Munich, 85748 Garching, Germany
{dominik.voigt,lisa.podszun,juergen.mangler,stefanie.rinderle-ma}@tum.de

Abstract. The application of process technologies and systems has shifted from traditional business processes such as loan application to processes in domains such as manufacturing and healthcare. In these domains, Internet of Things (IoT) technology is widely used, requiring an integration of process and IoT technologies. While existing approaches focus on mining and analyzing IoT data in a process context, this work aims at providing the technology for highly efficient process execution on IoT and edge devices. This poses many challenges with respect to available computing resources that cannot be met by existing monolithic process management systems. Hence, this work presents HiProX, an transpilation-based implementation of the cloud process execution engine (CPEE) in RUST. It can be shown that HiProX meets the requirements of resource utilization and speed to run average process instances with full communication capabilities on suitable IoT platforms that provide a memory allocator using less than 8 MiB of memory per instance.

Keywords: Process Execution · IoT Data · Process Monitoring · Process Logging · Industry 4.0

1 Introduction

In recent years, Internet of Things (IoT) technology has found widespread adoption across many domains, such as manufacturing and healthcare [3,20]. IoT offers several benefits such as improved data quantity, quality, and "freshness" for decision-making and analysis processes [16]. However, IoT integration also introduces many architectural and operational challenges, such as the heterogeneous nature of the different interfaces and capabilities of devices, scalability, unreliable power provisioning, and network connectivity [2,20,32].

Business Process Management (BPM) and Business Process Management Systems (BPMS) are nowadays the backbone of many enterprise-scale software [11]. From an architectural point of view, they provide service integration and orchestration, which has become paramount with the rise of service-oriented

M. ER et al. (Eds.): BPM 2025, LNBIP 565, pp. 184–199, 2026.
https://doi.org/10.1007/978-3-032-02936-2_14

and microservice architectures in enterprise and specifically IoT-centric applications [17,22,24].

Hence, due to the symbiotic capabilities of IoT and BPM technologies [20], a wide variety of application scenarios have been discussed such as: BPM based Digital Twins for Smart Cities [27], BPM coordinated Cross-docking of Shipping Containers [13], or vertical integration in manufacturing companies, connecting shop-floor IoT systems to high level planning processes [26].

While progress was made, many challenges remain before all the envisioned advantages of IoT and BPM integration can be realized. In the classical integration approach, the Process Engine acts as an orchestration layer for the different software components and IoT devices [13] and is typically hosted within the cloud [9]. However this leads to several issues. For one, due to the latency between the Process Engine and the IoT devices, even for on-premise deployment, this approach is infeasible for latency critical applications [9,24]. Secondly, this centralized approach can lead to a disconnect between sensor data and event logs [16,20], as sensor data collection is often handled independently from the process enactment and just aggregated through the process engine. This makes the aggregation, preparation, and use of IoT data for decision mining and process mining challenging [16,20]. Lastly, since the number of data sources in IoT scenarios is quite large, e.g., more than 10.000 IoT nodes in case of the Cross-docking example, network related issues due to data volume and velocity can become an issue [29,32].

To address the issues of centralized components in IoT scenarios, decentralized approaches leveraging the Fog and Mist Computing Paradigms, have been proposed [3,9,23]. However none of these approaches moves the process execution out of the cloud, which would allow for a fully process oriented approach enabling easy correlation between IoT and process level events. Figure 1 depicts an example scenario for a fully process oriented approach in manufacturing. Here different process instances are directly deployed on the manufacturing machines. This minimizes latency for control and data access. Furthermore by local processing and filtering, as depicted in the expanded process instance (CNC Milling Machine), data volume and velocity over the network are greatly reduced. In this scenario the process engine routes data between instances and forwards it to standard process monitoring or logging software. Traditional IoT implementations would have a similar architecture, but rely on hard-coded components instead of easy to change process instances.

However executing processes on resource-constrained devices such as the manufacturing machines in our example, requires an efficient approach for process execution [33]. Hence, the research question (**RQ**) tackled in this work is how process instances can be deployed and executed on IoT devices.

Existing process engines are often realized as custom interpreters. This makes their use in resource-constrained IoT fog and edge environments difficult. Hence, in this work, we present HiProX, a fast and resource-efficient general-purpose process engine that enables efficient process execution on IoT and edges devices. In detail, this work contributes to the field of IoT-enhanced BPM by (1) iden-

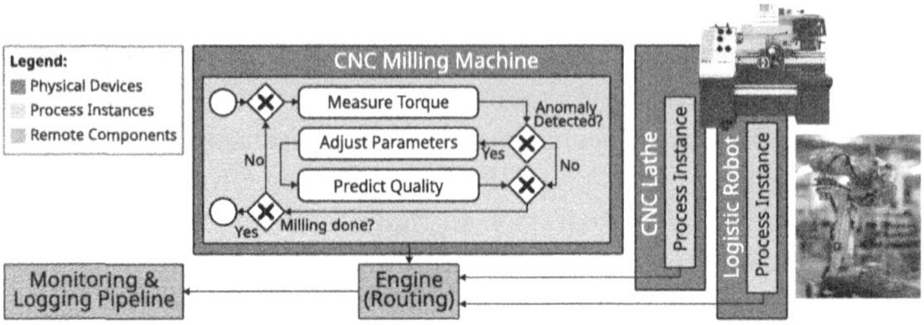

Fig. 1. Example Scenario of a Process Oriented Approach for Manufacturing

tifying existing resource-efficient approaches and deriving requirements for a resource-efficient process engine (2) presenting the HiProX process engine for process execution on resource-constrained devices, and (3) evaluating the engine regarding resource utilization and speed, which allows to identify suitable IoT platforms for instances to run on.

The remainder of the paper is structured as follows. Section 2 discusses execution scenarios and requirements in IoT-enhanced BPM. Section 3 presents HiProX, its architecture, and design decisions. Section 4 evaluates the proposed solution on two compute platforms. Section 5 discusses related work and Sect. 6 concludes the paper.

2 Requirements for Process Execution on IoT Devices

As integration of IoT into processes continues to increase, such as in IoT-enhanced BPM, the need to deploy and execute processes on IoT devices has been identified repeatedly [3,33]. Due to the resource constraints of IoT devices, the following requirements for developing a process engine for process execution on IoT devices, independently of the application domain, arise:

Requirement R1: While modern IoT devices offer increasing amounts of computing capabilities, the overall quantity of computing resources on these platforms is still constrained compared to desktop or server environments, making utilization of these computing resources more challenging (e.g., due to the general lack of an operating system or the restricted feature set of the provided OS) [2]. To enable the execution of business processes on such devices, resources must be utilized efficiently. Therefore, memory consumption and process execution time are used as quantitative measurements. Memory efficiency and execution time should be significantly better than an existing reference implementation, which has no focus on resource-efficiency.

Requirement R2: Logging of the process execution is an essential part of the process-oriented paradigm. This allows the BPMS to monitor, stop, and resume instances and freely modify the instance model, the execution context, and move the execution pointer(s) [34]. The resulting event log, containing classical process events and IoT collected data, furthermore enables meaningful data analysis as the collected IoT data is contextualized by the process events and other meta data [18].

Requirement R3: Although a variety of application protocols are commonly used in IoT scenarios (e.g. MQTT, CoAP, HTTP, ...), we decided to provide HTTPS support including multipart. This decision ensures that the process engine can be used in a wide variety of application scenarios (general-purpose), as HTTP is the de-facto standard application protocol of the Internet [35]. Nevertheless, supporting protocols would be feasible and easily implementable in our extensible architecture. Nonetheless this would required providing a protocol gateway to communicate with classical HTTP-based web services. Finally, while cryptographic protocols are expensive on a resource-constraint device, ensuring secure communication is essential. Thus we decided on supporting HTTPS (TLS).

3 HiProX Architecture

The HiProX architecture is depicted in Fig. 2. The implementation is based on the Cloud Process Execution Engine (CPEE) [19]. New contributions to the architecture are highlighted in green. The adapted architecture is discussed by providing a walk-through based on how a business process would be transpiled, deployed, and executed in Sect. 3.1. Section 3.2 discusses the advantages of a transpilation based approach that was chosen for the proposed solution instead of using model interpretation. Section 3.3 discusses the criteria considered when selecting a host language and why the Rust programming language was selected for this solution. Section 3.4 discusses the process DSL used for the model-to-code approach as an intermediate representation between graphical process model and binary code.

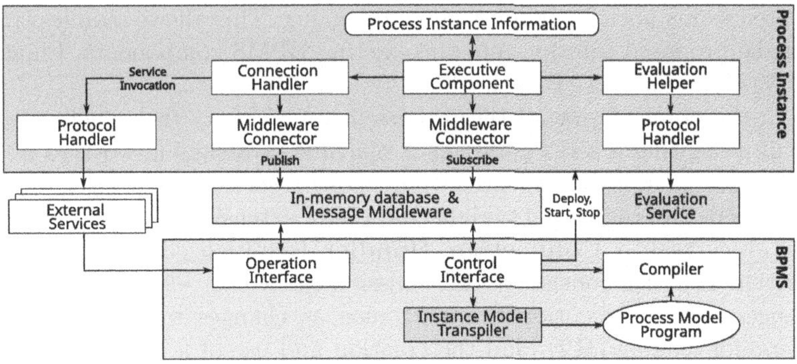

Fig. 2. HiProX Architecture (Contributions of This Paper Highlighted Green) (Color figure online)

3.1 Architecture Walk-Through

HiProX consists of two major components, i.e., the central **BPMS** component and the individual lightweight **Process Instance** component. The BPMS component is responsible for general features like logging, monitoring, and asynchronous callbacks. This component is the one proposed by Mangler [19]. This component was extended by providing a Transpilation Unit. For the enactment of a process model the BPMS spawns a process instance. These process instances are standalone programs that are semantically equivalent from the process model they have been generated from.

Execution of a process is initiated by the **Control Interface** of the BPMS which in turn passes the process model to the newly implemented **Instance Model Transpiler** that transpiles it to an equivalent program in the Process DSL described in Sect. 3.4. Thereafter, the **Control Interface** invokes the compiler to create the executable process instance for the target architecture at hand.

This resulting instance binary contains the instance logic and all the operative code like logging and protocol handlers and can thus be deployed and executed on a different device than the BPMS component. Lastly, the binary, execution context (variables and endpoints), and configuration files are deployed on the target, and the instance is executed.

Instance execution on the target starts by loading the configuration. Then, the **Executive Component** is initialized. This component semantically represents the instance as a whole. All *Process Instance Information* is managed by this component, and the DSL is defined as methods on the **Executive Component** instance. Subsequently, a **Message Middleware Connector** is initiated to listen for events emitted by the management component. This is used for features such as callbacks for asynchronous tasks [19]. The interfacing of process instances with the BPMS component follows the solution proposed by Mangler [19]. Here any events emitted during the execution of the process engine are send to the **Message Middleware**. The **BPMS** then retrieves these events for every process instance for monitoring and logging. This allows seamless integration of the proposed solution into the existing BPMS component[1]. Finally, the execution of the instance code begins.

From the point of view of the process instance, every task, such as sending an email, executing a script, starting a machine or assigning work to a human process participant, is implemented by an external service [19]. This architectural decision keeps the engine lightweight and easily extensible [19].

For every task a **Connection Handler** is initialized. The tasks of the Connection Handler consist of the emission of events, e.g., position changes or changes in the Execution Context, such as changes to variables, through Handlers, for logging (**R2**), and the creation and use of a **Protocol Handler** instance to invoke external services.

For the execution of scripts, e.g., to prepare data for service invocations or changing the execution context based on the result of a service call, an **Eval-**

[1] https://cpee.org/ last access: 2025-05-29.

uation Helper is instantiated. This component handles the invocation of the corresponding **Evaluation Service** via a **Protocol Handler** instance. The Evaluation Service executed the script against the execution context and the Evaluation Helper subsequently handles the result. Such results can include the manipulation of endpoints or data elements, signaling to retry execution or to stop the process instance execution.

The core motivation behind the architecture was to ensure easy extensibility and resource-efficient process execution while providing the features of a general-purpose process engine. Each of the following sections discusses a specific architectural decision and the resulting properties.

The communication between the BPMS and an instance is facilitated completely via the Message Middleware and In-Memory-Database component (excluding deployment, start, and stop of each instance).

This decision also decouples the **Process Instance** from the **BPMS** and thus allows the deployment of the individual components on possibly different tiers of the computing hierarchy [2]. Therefore, deploying instances directly on the IoT devices, while hosting the **Middleware** and the **BPMS** on the middle- or inner-edge tiers is possible. Furthermore this still allows to model, deploy and monitor the process instances via the BPMS component. And thus enabling Mist Computing architectures for IoT-enhanced BPM applications. Additionally this decision allows for horizontal scalability of certain services of the BPMS, such as the logging service by decoupling the event source (instance) and sink (BPMS).

Furthermore utilizing an in-memory database as a middleware component, combined with an appropriate DSL implementation, allows the complete application state of every instance and the corresponding execution logs to be automatically persisted during execution, satisfying the logging requirement (**R2**) (cf. Sect. 2). This feature, combined with the existing **BPMS** implementation in [19], to stop instances during execution and adapting the control flow, execution context, and even the instance model to handle errors occurring during instance execution, e.g. due to divergence of the instance model from reality [16].

The execution of scripts via the Evaluation Helper by invocation of the Evaluation Service differs from how scripts are handled in the architectures proposed by [19,34] that both support direct evaluation by the runtime (Ruby VM). Each approach has certain advantages and disadvantages. While using an external service increases latency and the potential for network-related issues, it provides significant advantages. First and foremost, this allows the formulation of scripts in arbitrary programming languages that do not have to be supported on the deployment target. Simplifying development of processes by Process Designers. Secondly, the possibility for handler code to rely on arbitrary execution environment requirements, e.g., access to a GPU. Lastly, it allows to either link the code directly into the *Instance Binary* or to deploy the execution service on the IoT device by replacing the Protocol Handler component for scenarios where latency and network concerns are predominant. Thus, this approach offers flexibility and ease of use depending on the application scenario.

3.2 Transpilation Vs. Interpretation

The execution of process models can be realized by interpretation or transpilation (model-to-code). The approach used by most process engines in literature and industry, is interpretation of the process model. While interpretation is intuitive, it has performance implications. Wherever an instance is executed an instance of the interpreter needs to be executed. This can require a significant amount of compute resources and is thus not viable for many IoT devices (e.g. on devices with less than 32 MiB RAM). Furthermore, since the model is interpreted, there is no automatic optimization performed, and thus either the model is executed as is, or a separate optimizer needs to be run prior to deployment which needs to be maintained. A transpilation based approach, which we refer to as model-to-code, transforms the process model into semantically and structurally equivalent code in a target programming language. This is the approach used by HiProX to realize process execution on a separate device from the BPMS.

Although this model-to-code approach generates code in a general purpose programming language, it has two advantages over manually writing the application code. First, changes in the business logic can easily be realized by updating the model. Second, this approach, given the actual DSL implementation is tested rigorously, the generated code is less likely to contain errors.

To ensure the maintainability of the produced instance code artifacts [10], an internal DSL within a host programming language is used, as proposed by Stürmer et al. [34]. The DSL, thus, should be easy to understand and extend and be powerful enough to express common workflow patterns. This has several advantages, depending on the programming language which we will discuss further in Sect. 3.3.

3.3 Language Selection Criteria

Selecting an appropriate DSL host programming language is vital to ensure that the instance binary (transpiled process model) is easily maintainable and is efficient to execute. Hence, we formulate the following Criteria C1–C3. The first criterion is whether the language requires a runtime or is compiled natively (**C1**), since runtime environments require additional memory. Furthermore, most popular languages of this category (e.g. Java, C#, Ruby, Python) use a garbage collector mechanism to manage memory. This makes memory-consumption less deterministic and requires additional compute resources. This is the main reason why the original Ruby Process DSL proposed by Stürmer et al. [34] is not suitable for IoT scenarios. The second criterion is the capability of the language to define a DSL that is easy to understand, and the resulting code fragment is easy to modify (**C2**), as an intermediate step between a graphical and a textual representation. Since process execution can be highly concurrent, the third criterion refers to how easy it is to write memory-safe and race-condition free code to ensure stability and maturity of the implementation (**C3**).

Based on these criteria, we select Rust as programming language. Rust provides support for easily readable DSLs via macros and multi-line closures. Furthermore, there are examples of successful Rust DSLs such as SeaQuery[2] for formulating SQL queries in native Rust (C2). Additionally, in contrast to other performance-oriented languages such as C and C++, Rust uses the Ownership[3] memory management model instead of relying on the programmer to manage memory manually. Thus, using safe Rust[4] (and thoroughly tested unsafe blocks where necessary) prevents a large class of memory-related issues such as data races[5] and most memory leaks[6]. Additionally, safe Rust provides a memory safety guarantee[7] (C3).

3.4 Process Domain Specific Language

To ensure that the HiProX engine supports the modeling and execution of general-purpose processes, i.e., independent of any application domain, the process DSL has to be defined such that every component of the control flow has a corresponding DSL element. Since HiProX utilizes the CPEE for modeling, its DSL is based on the existing CPEE DSL in Ruby [21]. This design decision bears the advantage that the CPEE DSL has been already evaluated w.r.t. its expressiveness: it supports all basic control flow patterns and a wide range of more advanced patterns [21], thus satisfying requirement **R3** by design.

4 Evaluation

We evaluate our implementation based on a set of process models. Each process model represents a basic control flow pattern. We compare our novel Rust Process DSL implementation against the original Ruby Process DSL, currently used in the CPEE, based on memory consumption and run time.

4.1 Evaluation Setup

For the evaluation, we created and executed a process model for each of the following basic control flow patterns:

- **Sequence**: A sequence of 3 tasks
- **Parallel**: Parallel execution of 2 tasks

[2] https://github.com/SeaQL/sea-query, last access: 2025-05-29.
[3] https://doc.rust-lang.org/book/ch04-01-what-is-ownership.html, last access: 2025-05-29.
[4] https://doc.rust-lang.org/nomicon/meet-safe-and-unsafe.html last access: 2025-05-29.
[5] https://blog.rust-lang.org/2015/04/10/Fearless-Concurrency.html, last access: 2025-05-29.
[6] https://doc.rust-lang.org/book/ch15-06-reference-cycles.html, last access: 2025-05-29.
[7] https://blog.rust-lang.org/2015/04/10/Fearless-Concurrency.html, last access: 2025-05-29.

- **Choose**: Alternative execution of 2 tasks (only 1 is executed)
- **Loop**: A loop that executes a task a total of 3 times

While these patterns do not cover the full range of features supported by the DSL, they represent a foundational set of process components. We measured the execution time and memory usage of our Rust Process DSL against the original Ruby Process DSL proposed by Stürmer et al. [34] for each model which, in turn, have been evaluated against the van der Aalst Workflow Pattern [1] which compares and summarize the feature set of a multitude of process engines. We decided to limit our comparison to this reference implementation due to the following reasons. (1) Most of the engines we will discuss in Sect. 5 did not provide any code. (2) While some implementations provide memory consumption figures, none documented how these values are measured (Heap allocations vs. RSS). (3) Solutions that provided no network connectivity are not applicable for IoT scenarios. For more details see Sect. 5.

For each model, tasks are implemented as a call to an external service that delays its response for a single second and then responds. For the proposed implementation, a dynamically (to minimize memory and disk usage on nodes with multiple running instances) and a statically linked version were evaluated. Memory and execution time are measured for all patterns. For the memory analysis, residual set size (RSS) and heap memory are considered. RSS is measured for both implementations using the `time`[8]. To measure the heap allocations `heaptrack`[9] is used for the proposed implementation in Rust and `MemoryProfiler`[10] is used for the reference implementation in Ruby. For time measurements `time` is used for both implementations. For all measurements the time delay of the external service was eliminated from the reported time. Evaluation is done on two different platforms. To represent an unconstrained computing environment, a desktop computer with an AMD Ryzen 7 8745H and 32 GB of RAM (furthermore referred to as **Echo**) is used. To represent a constrained computing device an Orange Pi Zero 2W[11] (furthermore called **Orange**) is used. While this device is still a powerful IoT device, we show in the following subsection that the instance binary could be executed on a device with significantly less memory.

Echo uses an evaluation service instance that is hosted on the machine. **Orange** utilizes an external evaluation service hosted within the internet and thus suffers from increased network delays. Echo is connected to the network via an Ethernet cable. Orange is connected via 2.4 GHz Wifi.

4.2 Results

Table 1 shows the execution time measured for both implementations on both platforms. Here we can see that the proposed implementation in Rust performs

[8] https://www.man7.org/linux/man-pages/man1/time.1.html last access: 2025-05-29.
[9] https://github.com/KDE/heaptrack last access: 2025-05-29.
[10] https://github.com/SamSaffron/memory_profiler last access: 2025-05-29.
[11] http://www.Orangepi.org/html/hardWare/computerAndMicrocontrollers/details/Orange-Pi-Zero-2W.html last access: 2025-05-29.

better for every pattern on both platforms. For Echo, the proposed solution is more than 4 times faster for every pattern. This effect is even more pronounced on the low-compute Orange platform where the proposed implementation is faster by a factor of 5–13 times. The execution time measured for the proposed implementation increased by about one second when switching from the Echo to the Orange platform. For the Ruby implementation it increased by more than three seconds. This demonstrates that a compiled approach shows significant performance improvements, especially on resource-constrained platforms.

Table 1. Time Measurements

Pattern	Rust				Ruby	
	Time Echo		Time Orange		Time Echo	Time Orange
	Static	Dynamic	Static	Dynamic		
Sequence	0.25 s	0.26 s	1.12 s	1.42 s	1.35 s	6.27 s
Parallel	0.09 s	0.10 s	0.40 s	0.42 s	1.16 s	5.42 s
Choose	0.09 s	0.10 s	0.73 s	0.51 s	1.32 s	5.59 s
Loop	0.27 s	0.30 s	1.31 s	1.47 s	1.49 s	6.69 s

The results of memory profiling for the experiments is depicted in Tables 2 and 3. Table 2 shows RSS of both implementations across different patterns. Table 3 shows the maximum heap size of both implementations during the execution of the different patterns. These measurements show that the proposed implementation uses significantly less heap, by a factor of about 6–10 depending on the pattern across all platforms.

Table 2. Memory Measurements (RSS)

Pattern	Rust				Ruby	
	Memory Echo		Memory Orange		Memory Echo	Memory Orange
	Static	Dynamic	Static	Dynamic		
Sequence	6.78 MiB	14.00 MiB	5.82 MiB	13.60 MiB	94.5 MiB	85.0 MiB
Parallel	7.86 MiB	15.33 MiB	7.10 MiB	14.82 MiB	94.5 MiB	85.3 MiB
Choose	6.67 MiB	14.25 MiB	6.18 MiB	13.87 MiB	94.0 MiB	85.3 MiB
Loop	6.88 MiB	14.53 MiB	6.42 MiB	14.28 MiB	94.3 MiB	86.4 MiB

RSS of the proposed solution is also significantly smaller than for the reference solution (see Table 2). The statically linked approach has an RSS that is smaller by a factor of more than 10 across both platforms. The dynamically linked approach by a factor of about 5. However, here it is important to mention that these factors are a lower bound if multiple instances of the same program are executed. The reason for this is that for each instance an individual ruby interpreter has to be instantiated and loaded into memory, consuming a significant amount of memory. For the dynamically linked version of the proposed solution this does not happen, the standard library code and the complete application code are linked dynamically, and thus only need to be kept in memory once which makes each process instance very lightweight. Therefore, given that multiple instances are executed on the same platform, significantly less memory is used by the proposed Rust implementation.

When comparing the memory consumption of the proposed solution with the implementations discussed in the related work (see Sect. 5), it becomes clear that the implementation uses comparative amounts of memory to the most resource-efficient approaches [7,8] while providing essential features for general purpose process engines like network connectivity. Thus the proposed solution shows higher memory efficiency compared to related engines and compared to the reference implementation (**R1**).

Table 3. Memory Measurements (Heap)

Pattern	Rust				Ruby	
	Memory Echo		Memory Orange		Memory Echo	Memory Orange
	Static	Dynamic	Static	Dynamic		
Sequence	619.1 KiB	619.2 KiB	598.6 KiB	598.6 KiB	5790.0 KiB	5790.0 KiB
Parallel	966.1 KiB	977.4 KiB	946.5 KiB	799.1 KiB	5800.0 KiB	5810.0 KiB
Choose	949.9 KiB	950.0 KiB	982.3 KiB	982.4 KiB	6850.0 KiB	6850.0 KiB
Loop	998.5 KiB	998.3 KiB	1000.0 KiB	1000.0 KiB	6840.0 KiB	6850.0 KiB

5 Related Work

While there is a lot of work done on IoT-aware process execution, most of these works follow a similar approach by introducing some middleware layer that enables interoperability between the IoT and Process Engine, e.g. [30]. While these approaches certainly work, they utilize the Process Engine mainly as an orchestration layer, losing out on the correlation capabilties between IoT and Process level events a fully integrated approach of Process Engine and IoT could bring. Thus our research question was how process models could be deployed and efficiently executed on IoT devices. Therefor we discuss related work in the area of specialized process engines that are focused on process execution in a mobile or IoT setting as these engines would have the potential to execute processes on IoT devices enabling a fully integrated approach. Through a structured literature analysis [36], we identified 11 approaches that are summarized with their key characteristics in Table 4 and described in the following.

Hackmann et al. [14] implement Java-based BPEL workflow engine Sliver for mobile devices with focus on efficient implementation with SOAP support. While

Table 4. Existing process engines with key characteristics

	Language	Model	Connectivity	Mem.	Comp. or Interpr.	Execution
2006 [14]	Java	BPEL	SOAP	<22 MB	Interpreted	Centralized
2007 [25]	Java [3]	BPEL	SOAP	?	Interpreted	Centralized
2008 [31]	Java	BPEL	SOAP	?	Interpreted	Distributed
2009 [8]	C	SISARL-XPDL	-	1MB	Compiled	Centralized
2011 [7]	C	SISARL-XPDL	-	8 MB	Compiled	Centralized
2011 [28]	.Net	ADEPT	Standalone	?	Interpreted	Distributed
2011 [12]	C++	BPEL	SOAP	1 KB (???)	Compiled	Centralized
2012 [4]	Objective-C	BPMN	HTTP	?	Interpreted	Distributed
2014 [5]	Objective-C	BPEL	HTTP	?	Interpreted	Distributed
2015 [6]	?	BPEL	HTTP	?	Interpreted	Distributed
2016 [15]	Java	ADEPT2	?	?	Interpreted	Centralized

no information on the memory consumption is available, the authors state that their approach outperforms the memory consumption of ActiveBPEL of 22 MB. Pajunen et al. [25] follow a similar approach as [14]. However, they focus less on the resource-efficient implementation of the engine and more on support for local services and different communication protocols such as HTTP, SMS, and email. Sen et al. [31] leverage the Sliver engine for choreographic workflow execution in mobile ad-hoc networks. Both, Chou et al. [8] and Chen et al. [7] develop C-based embedded workflow engines for robotic applications supporting an augmented subset of XPDL called EMWF. In this approach, models are compiled into native code. However, both engines only provide support for local tasks and do not provide any form of network connectivity. While executing a workflow, the engines consume about 1 MB and 8 MB of memory respectively. Pryss et al. [28] propose a distributed interpreting workflow engine for a seemingly custom process definition language using .Net called MARPLE. Communication is facilitated through a custom protocol [3]. No resource utilization measures for MARPLE are available. Glombitza et al. [12] present a model-to-code approach for the execution of BPEL models on sensor nodes. Their approach facilitates network connectivity via LTP. Hence, realizing memory efficiency at the cost of compatibility or additional gateway services, as most of the Web utilizes HTTP-based communication. Additionally, their approach does not include an intermediary internal DSL that simplifies the maintainability of the generated code. Chang et al. [5] develop an application framework for the social private cloud called SPiCa. In their approach, individual devices host a SPiCa instance, including a BPEL workflow component. Conceptually similar works have been proposed by Chang et al. in [4,6]. In all three approaches, resource-efficient process execution was not actively discussed and no resource utilization measurements were reported. Schobel et al. [15] develop a lightweight Java process engine to conduct medical surveys. Surveys are modeled within a graphical modeling language and then transpiled into a process model (ADEPT2). No resource utilization measurements are provided.

Summary and Distinction from Existing Research: Prior works focus on specific aspects of constrained process execution, e.g., Schobel et al. [15] supporting offline execution, Glombitza et al. [12] and Chou et al. [8] focusing on resource-efficient execution using a model-to-code approach. By contrast, HiProX aims at proving comprehensive support by realizing the following three properties: (1) A model-to-code approach that translates the process model to executable code, specifically an internal DSL as in [21,34] that can be compiled and natively executed on devices with low compute capabilities. This contrasts approaches that require a runtime environment to execute the instance which needs to be supported therefore adding a further deployment dependency and leading to an resource overhead (e.g. in Java or C# based solutions [14,15, 25,28]). (2) Full HTTP support to communicate with external web services, including support for HTTP Multipart. This is essential since HTTP is one of the de-facto standard communication protocols for IoT applications and the web [35]. This contrasts this work from prior solutions that provide non-standard or no connectivity, and thus cannot be used as general purpose process engines (e.g. solutions proposed by Glombitza et al. [12] that utilizes LTP or Chou et al. [8] and Chen et al. [7] that provide no network connectivity. (3) Full logging of the process execution [19] via a distributed engine architecture. Full logging of the process instance execution allows for monitoring, stopping, modifying (control flow, execution context, and even the instance model), and resuming the process instance [34] to recover from errors during execution.

The solution proposed in this paper is a full-featured process engine similar to the implementation by Mangler and Stürmer [21,34], while improving the resource utilization by a factor of up to 13, thus making it viable option to use on IoT compute platforms.

6 Conclusion

To advance the integration of IoT and BPM technology, the aim of this paper was to develop general-purpose process engine HiProX that allows the deployment and execution of processes on IoT devices which are generally resource-constrained. For meeting the requirements of resource-constrained execution, the HiProX architecture includes a model-to-code approach for efficient process execution, logging capabilities, and, to complement the model-to-code approach, a powerful internal DSL[12]. For the host language, Rust was chosen due its support fo easily readable DSLs and memory safety guarantee. As shown in the evaluation, HiProX shows significantly more efficient use of computing resources. The evaluation also shows that the engine is more memory efficient than most of the related engines, and consumes significantly less than the threshold described in

[12] https://github.com/DominikVoigt/rusty_weel/blob/main/weel_lib/src/dsl.rs last access: 2025-05-29.

the paper that proposed the Sliver engine [14] that provides comparable features. Future work will tackle the following challenges. While logging can provide fault tolerance, no mechanism was implemented to handle disconnects when talking to external services or the middleware component. Furthermore, the possibility of recovery after a crash, e.g., due to power supply issues [32], common in an IoT environment, is not yet implemented. Lastly, the implementation relies on the Rust Runtime[13] for the `std` crate[14], and thus requires a memory allocator to function.

References

1. van der Aalst, W., ter Hofstede, A., Kiepuszewski, B., Barros, A.: Workflow patterns. Distrib. Parallel Databases **14**(1), 5–51 (2003)
2. Buyya, R., Srirama, S.N.: Internet of Things (IoT) and New Computing Paradigms, pp. 1–23. Wiley (2019)
3. Chang, C., Srirama, S.N., Buyya, R.: Mobile cloud business process management system for the internet of things: a survey. ACM Comput. Surv. **49**(4), 1–42 (2016)
4. Chang, C., Srirama, S.N., Ling, S.: An adaptive mediation framework for mobile P2P social content sharing. In: Liu, C., Ludwig, H., Toumani, F., Yu, Q. (eds.) ICSOC 2012. LNCS, vol. 7636, pp. 374–388. Springer, Heidelberg (2012). https://doi.org/10.1007/978-3-642-34321-6_25
5. Chang, C., Srirama, S.N., Ling, S.: Spica: a social private cloud computing application framework. In: Mobile and Ubiquitous Multimedia, pp. 30–39 (2014)
6. Chang, C., Srirama, S.N., Mass, J.: A middleware for discovering proximity-based service-oriented industrial internet of things. In: Services Computing (2015)
7. Chen, W.C., Shih, C.S.: ERWF: embedded real-time workflow engine for user-centric cyber-physical systems. In: Conference on Parallel and Distributed Systems, pp. 713–720 (2011)
8. Chou, T., et al.: EMWF for flexible automation and assistive devices. In: Real-Time and Embedded Technology and Applications Symposium, pp. 243–252 (2009)
9. Del Gaudio, D., Hirmer, P.: A lightweight messaging engine for decentralized data processing in the internet of things. SICS Softw.-Intensive Cyber-Phys. Syst. **35**(1–2), 39–48 (2019)
10. Do Nascimento, L.M., Viana, D.L., Neto, P., Martins, D., Garcia, V.C., Meira, S.: A systematic mapping study on domain-specific languages. In: Software Engineering Advances (ICSEA 2012), pp. 179–187 (2012)
11. Dumas, M., La Rosa, M., Mendling, J., Reijers, H.A.: Fundamentals of Business Process Management. Springer, Heidelberg (2018)
12. Glombitza, N., Ebers, S., Pfisterer, D., Fischer, S.: Using BPEL to realize business processes for an internet of things. In: Frey, H., Li, X., Ruehrup, S. (eds.) ADHOC-NOW 2011. LNCS, vol. 6811, pp. 294–307. Springer, Heidelberg (2011). https://doi.org/10.1007/978-3-642-22450-8_23
13. Grefen, P., Brouns, N., Ludwig, H., Serral, E.: Co-location specification for IoT-aware collaborative business processes. In: Information Systems Engineering in Responsible Information Systems, pp. 120–132 (2019)

[13] https://doc.rust-lang.org/reference/runtime.html last access: 2025-05-29.
[14] https://doc.rust-lang.org/std/ last access: 2025-05-29.

14. Hackmann, G., Haitjema, M., Gill, C., Roman, G.C.: Sliver: A BPEL workflow process execution engine for mobile devices. In: Service-Oriented Computing, pp. 503–508 (2006)
15. Schobel, J., Pryss, R., Schickler, M., Reichert, M.: A lightweight process engine for enabling advanced mobile applications. In: Cooperative Information Systems, pp. 552–569 (2016)
16. Janiesch, C., et al.: The internet of things meets business process management: a manifesto. IEEE Syst. Man Cybern. Mag. **6**(4), 34–44 (2020)
17. Loke, S.W.: Service-oriented device ecology workflows. In: Service-Oriented Computing, pp. 559–574 (2003)
18. Mangler, J., et al.: Datastream XES extension: embedding IoT sensor data into extensible event stream logs. Future Internet **15**(3), 109 (2023)
19. Mangler, J., Rinderle-Ma, S.: Cloud process execution engine: architecture and interfaces. CoRR abs/2208.12214 (2022)
20. Mangler, J., et al.: From internet of things data to business processes: challenges and a framework. CoRR abs/2405.08528 (2024)
21. Mangler, J., Stuermer, G., Schikuta, E.: Cloud process execution engine - evaluation of the core concepts. CoRR abs/1003.3330 (2010)
22. Mass, J., Chang, C., Srirama, S.N.: Context-aware edge process management for mobile thing-to-fog environment. In: European Conference on Software Architecture: Companion Proceedings (2018)
23. Mass, J., Chang, C., Srirama, S.N.: Edge process management: a case study on adaptive task scheduling in mobile IoT. Internet Things **6**, 100051 (2019)
24. Mass, J., Srirama, S.N., Chang, C.: Step-one: simulated testbed for edge-fog processes based on the opportunistic network environment simulator. J. Syst. Softw. **166**, 110587 (2020)
25. Pajunen, L., Chande, S.: Developing workflow engine for mobile devices. In: Enterprise Distributed Object Computing Conference, p. 279 (2007)
26. Pauker, F., Mangler, J., Rinderle-Ma, S., Pollak, C.: centurio.work - modular secure manufacturing orchestration. In: Business Process Management, pp. 164–171 (2018)
27. Pańkowska, M., Żytniewski, M.: Digital twins for smart city. In: Smart Spaces, pp. 269–286 (2024)
28. Pryss, R., Tiedeken, J., Kreher, U., Reichert, M.: Towards flexible process support on mobile devices. In: Soffer, P., Proper, E. (eds.) CAiSE Forum 2010. LNBIP, vol. 72, pp. 150–165. Springer, Heidelberg (2011). https://doi.org/10.1007/978-3-642-17722-4_11
29. Schönig, S., Ackermann, L., Jablonski, S.: Internet of things meets BPM: a conceptual integration framework. In: Simulation and Modeling Methodologies, Technologies and Applications, pp. 307–314 (2018)
30. Schönig, S., Ackermann, L., Jablonski, S., Ermer, A.: IoT meets BPM: a bidirectional communication architecture for IoT-aware process execution. Softw. Syst. Model. **19**(6), 1443–1459 (2020)
31. Sen, R., Roman, G.C., Gill, C.: Cian: a workflow engine for MANETs. In: Coordination Models and Languages, pp. 280–295 (2008)
32. Sneps-Sneppe, M., Namiot, D.: On web-based domain-specific language for internet of things. In: Ultra Modern Telecommunications and Control Systems and Workshops, pp. 287–292. IEEE (2015)
33. Stoiber, C., Schönig, S.: Event-driven business process management enhancing IoT – a systematic literature review and development of research agenda. In: Innovation Through Information Systems, pp. 645–661 (2021)

34. Stürmer, G., Mangler, J., Schikuta, E.: A domain specific language and workflow execution engine to enable dynamic workflows. In: Parallel and Distributed Processing with Applications, pp. 653–658 (2009)
35. Uckelmann, D., Harrison, M., Michahelles, F.: An architectural approach towards the future internet of things. In: Architecting the Internet of Things, pp. 1–24 (2011)
36. Voigt, D.I.: RUST process model DSL. Master's thesis, Technische Universität München (2025)

Leveraging Counterfactuals for Prescriptive Process Analytics

Ngoc-Diem Le(✉), Alessandro Padella, Francesco Vinci, and Massimiliano de Leoni

University of Padua, Padua, Italy
{ngocdiem.le,francesco.vinci.1}@phd.unipd.it,
alessandro.padella@unipd.it, deleoni@math.unipd.it

Abstract. Prescriptive process analytics aims to provide actionable recommendations for process instances that are predicted to fall short of achieving satisfactory outcomes. A common type of recommendation typically focuses on assigning activities to specific resources, as it is a general task that naturally applies across many domains. Given that processes may involve hundreds of resources, brute-force approaches for evaluating all possible activity-resource combinations are computationally infeasible. Current state-of-the-art techniques, conversely, adopt a sequential approach that selects the most suitable activity and then allocates it to one of the suitable resources: this is inherently sub-optimal. This paper leverages counterfactual generation techniques to formulate recommendations. Counterfactual-based methods offer innovative strategies that efficiently converge to highly effective interventions. Experimental evaluations conducted on several real-life case studies demonstrate that our counterfactual-based technique outperforms a baseline approach that follows a sequential activity-to-resource assignment strategy.

Keywords: Process Mining · Resource Allocation · Process-aware Recommendation Systems · Counterfactual Generation · Process Improvement

1 Introduction

Process mining aims to discover and improve processes based on the analysis of process transactional data that records how single executions have been carried out. Within this field, prescriptive process analytics focuses on recommending actions to mitigate "risky" executions that are predicted to lead to unsatisfactory outcomes, such as high costs or poor customer satisfaction. One of the most common types of recommendations is the real-time suggestion of resources to carry out certain activities [19]. Prescriptive process analytics is currently gaining momentum, with several techniques having been proposed over the years (see Sect. 2). This paper contributes by providing recommendations to optimize process performance, with a particular focus on minimizing the total execution time, aligning with a large share of the related work in the field.

Existing techniques typically leverage a divide-and-conquer approach: they first select the activity based on its potential to enhance the process efficiency, and then assign it to a suitable resource [19]. While intuitive, this strategy can result in sub-optimal outcomes. Conversely, optimizing the activity-resource pair by evaluating all

M. ER et al. (Eds.): BPM 2025, LNBIP 565, pp. 200–215, 2026.
https://doi.org/10.1007/978-3-032-02936-2_15

possible combinations is often intractable in practice, especially in organizations with hundreds of resources.

This paper proposes a technique based on counterfactuals [14,29] to recommend the activity-resource pairs for each running process instance. We use the term *counterfactual* in the context of explainable AI as introduced by [30]: a counterfactual identifies the minimal changes to an input instance that would alter the prediction of the oracle function toward a desired outcome. This extensive employment of counterfactual-based approaches is linked to their nature of defining *what-if* scenarios, such as "*You were denied a loan because your loan duration was 20 years. If your loan duration had been 30 years, you would have been offered a loan*" [30]. This capability makes counterfactual-generation frameworks well-suited for recommending activity-resource pairs. The advantage of prescriptive process analytics is evident: frameworks for counterfactual generation are readily available and can be directly used to generate activity-resource recommendations along with explanations. Also, these frameworks can quickly generate multiple counterfactuals, which translates in our setting to producing several alternative recommendations for the same process instance. Notably, while counterfactuals have been used to explain predictions and recommendations, they have never been used to generate recommendations themselves. This flexibility is beneficial because a rigid resource-to-activity imposition is against the principles of resource-aware recommender systems [9], which emphasize the need to support decision-makers with multiple viable options rather than prescribing fixed assignments.

Our counterfactual-based technique for prescriptive process analytics has been evaluated on several real-life processes and event logs, and compared with a baseline where the best activity is first chosen based on the best predicted next activity, and then it is allocated to one random resource. This baseline is reasonable because, while the number of activities in the process is relatively limited, the number of process resources can be very high, even reaching hundreds (cf. the case studies used in the experiments reported in Sect. 5 and summarized in Table 1). It is indeed intractable to compute the prediction for each of the hundreds of different resources separately. The evaluation results show that the average potential reduction in process-instance total duration is considerably higher than what the baseline can achieve. This is due to the use of counterfactual generation, which can efficiently build effective recommendations in the form of activity-resource pairs.

The remainder of this paper is organized as follows: Sect. 2 discusses related works in the domain of prescriptive process analytics and the use of counterfactuals in these fields. Section 3 introduces the necessary background concepts, while Sect. 4 puts forward our framework that leverages counterfactuals. Section 5 reports on the evaluation setup and results, while discussing limitations in the evaluation. Finally, Sect. 6 concludes the paper, summarizing the contribution of our paper.

2 Related Works

Prescriptive process analytics has recently gained significant attention, with various methods being applied to provide Key Performance Indicator (KPI)-driven recommendations. Numerous data-driven frameworks have been proposed to recommend the optimal next activity for KPI optimization [12], incorporating AI-based approaches such as Reinforcement Learning [6], Generative Adversarial Networks [20], Linear Temporal Logic formulas [10], and explainability techniques [25]. Adopting a different perspective, different authors address the problem of correctly timing the interventions [5], aiming to balance the risks of intervening too early (leading to unnecessary delays/costs) or too late (diminishing effectiveness), also guaranteeing a white-box approach [27]. Beyond next-activity recommendations, researchers have demonstrated that KPI improvements can also result from enhanced resource allocation [3,24,26].

Literature proposes a few counterfactual frameworks [4,8,15–17,22] to explain the predictions, namely the predictive-monitoring outcome. The work by Bhattacharya et al. [4] is particularly noteworthy. They propose a counterfactual-based visual framework designed to support patients and healthcare professionals in making decisions aimed at improving patient outcomes. However, this framework is specifically tailored to the healthcare domain. More critically, it is not process-centric: while it assists in identifying potential next actions, it does not directly generate recommendations, intended as pairs of suggested activity and performing resource. The same limitation also holds for Mothilal et al. [22]. Furthermore, the effectiveness of the approach by Bhattacharya et al. [4] is evaluated through subjective user feedback, rather than through objective metrics that are measured on real(istic) process executions to assess the actual process improvements.

This paper is the first to introduce a recommendation framework that utilizes counterfactuals to explicitly suggest the next activities to be executed for various ongoing process instances, as well as the resources that should carry them out. It is important to emphasize that our problem setting differs fundamentally from job-shop scheduling, where some counterfactual-based approaches have been proposed (see, e.g., [21]). In job-shop scheduling, the sequence of activities is predetermined and cannot be altered, whereas our approach allows for flexibility in recommending alternative next activities.

3 Preliminaries

The starting point for any process mining technique is an *event log*. An event log is a collection of *traces*, where a trace is a sequence of events with their attributes. Each trace describes the life-cycle of a particular *process instance* (i.e., a *case*) regarding the *activities* executed by certain *resources* at specific *timestamps* and the process *attributes* manipulated.

Definition 1 (Events). *Let $\mathcal{A}$ be the set of process activities. Let $\mathcal{T} \subset \mathbb{N}$ be the set of possible timestamps. Let $\mathcal{R}$ be the set of possible resources. Let $\mathcal{V}$ be the set of process attributes. Let $\mathcal{W}_\mathcal{V}$ be a function that assigns a domain $\mathcal{W}_\mathcal{V}(x)$ to each process attribute $x \in \mathcal{V}$. Let $\overline{\mathcal{W}} = \cup_{x \in \mathcal{V}} \mathcal{W}_\mathcal{V}(x)$. An event is a tuple $(a, t, r, v) \in \mathcal{A} \times \mathcal{T} \times \mathcal{R} \times (\mathcal{V} \not\rightarrow \overline{\mathcal{W}})$ where a is the event activity, t is the timestamp associated with the event, r is the resource that performs it, and v is a partial function assigning values to process attributes with $v(x) \in \mathcal{W}_\mathcal{V}(x)$.*

Definition 2 (Traces & Event Logs). *Let $\mathcal{E} = \mathcal{A} \times \mathcal{T} \times \mathcal{R} \times (\mathcal{V} \not\rightarrow \overline{\mathcal{W}})$ be the universe of events. Let a trace σ be a sequence of events, i.e., $\sigma \in \mathcal{E}^*$. An event-log $\mathcal{L}$ is defined as a set of traces, i.e., $\mathcal{L} \subset \mathbb{B}(\mathcal{E}^*)$.*

Without loss of generality, the events in each trace are considered sorted chronologically. Given an event $e = (a, t, r, v)$, the remainder uses the following shortcuts: $activity(e) = a$, $time(e) = t$, $resource(e) = r$ and $variable(e) = v$. Furthermore, given a trace $\sigma = \langle e_1, \ldots, e_n \rangle$, $prefix(\sigma)$ denotes the set of all prefixes of σ, including σ, namely $prefix(\sigma) = \{\langle\rangle, \langle e_1 \rangle, \langle e_1, e_2 \rangle, \ldots, \langle e_1, \ldots, e_n \rangle\}$.

In developing our technique, it is imperative to delineate the recommendation objective. We focus on minimizing the total execution time of a trace, as this is a general goal that naturally fits many real-world settings where process efficiency is a priority. The total execution time can be defined as a function over a given trace $\sigma = \langle e_1, \ldots, e_n \rangle \in \mathcal{E}^*$ that returns the difference between the first and the last timestamp related to σ. Note that the total time is assumed to be computed a posteriori when the execution is completed and leaves a complete trail of the timestamps for a particular trace σ. We can now define the total time prediction problem as follows:

Let $\sigma' = \langle e_1, \ldots, e_k \rangle \in \mathcal{L}$ be a prefix of a running case, which eventually will complete as $\sigma = \langle e_1, \ldots, e_k, e_{k+1}, \ldots, e_n \rangle$. The total execution time prediction problem can be formulated as forecasting the value of $\mathcal{T}_{total}(\sigma') = time(e_n) - time(e_1)$. In the process mining literature, this problem has been faced with different models [13,18,23]. We approach the model by estimating a **Total Time Oracle function** $\Phi : \mathcal{E}^* \times \mathcal{A} \times \mathcal{R} \rightarrow \mathbb{N}$, which, for an incomplete trace σ', an activity a and a resource r, forecasts the total execution time $\mathcal{T}_{total}$ if the next event e_{k+1} satisfies the conditions: $activity(e_{k+1}) = a$ and $resource(e_{k+1}) = r$.

Each prediction technique requires the definition of the domain $X_1 \times \ldots \times X_m$ and a **Trace-to-instance Encoding function** $\rho : \mathcal{E}^* \rightarrow X_1 \times \ldots \times X_m$, which maps each (prefix of a) trace σ in a vector $\rho(\sigma) \in X_1 \times \ldots \times X_m$ of m elements that can be of different nature, such as a process attribute, a timestamp, or the number of executions of an activity in σ.

This paper aims to provide a framework that recommends the most appropriate activity-resource pairs for reducing the execution time of running instances. However, to ensure domain validity, we assume that an activity is considered valid in a given process state, which refers to the current execution context defined by the sequence of past events, if it has been observed in similar past executions. This requires providing a state-representation function, as we will explain in more detail below.

Definition 3 (State-representation Function). *Let σ be a trace, and $\mathcal{S}$ a set of possible state representations, where each representation corresponds to a unique state of the process based on the observed events. The function $l^{state} : \mathcal{E}^* \rightarrow \mathcal{S}$ is called the state-representation function, where it returns the state for each (prefix of a) trace.*

Determining the activities allowed after a sequence of events requires building a transition system where nodes represent the observed states, and arcs represent transitions between states. Each arc corresponds to an event and is labeled with the name of the associated activity [1]. Multiple arcs can bear the same label.

Definition 4 (Transition System). *Let l^{state} be a state-representation function, $\mathcal{L}$ an event log, and $\mathcal{E}$ its set of events over a set $\mathcal{A}$ of activities. A transition system abstracting $\mathcal{L}$ is a tuple $TS_{\mathcal{L}} = (S, T) \subseteq \mathcal{S} \times (\mathcal{S} \times \mathcal{A} \times \mathcal{S})$ where*

- $S = \bigcup_{\sigma \in \mathcal{L}} \bigcup_{\sigma' \in prefix(\sigma)} l^{state}(\sigma')$
- $T = \bigcup_{\sigma \in \mathcal{L}} \bigcup_{\sigma' \oplus \langle e \rangle \in prefix(\sigma)} (l^{state}(\sigma'), activity(e), l^{state}(\sigma' \oplus \langle e \rangle))$

Fig. 1 illustrates an example of a transition system in accordance with Definition 4. It has been built on an event log $\mathcal{L}^{ex} = \{\langle a, b, c, d \rangle, \langle a, c, e \rangle, \langle a, e, b, d \rangle, \langle a, b, d, e \rangle\}$[1] using a sequence-based state-representation $l^{state}_{sequence\_based}(\langle e_1, \ldots, e_n \rangle) = \langle activity(e_1), \ldots, activity(e_n) \rangle$. Through this function, the state of a (prefix of a) trace is identified with its ordered list of activities. In the example with $\mathcal{L}^{ex}$, the set of possible states is thus $\mathcal{S} = \{\langle a, b, c, d \rangle, \langle a, b, d, e \rangle, \langle a, e, b, d \rangle, \langle a, c, e \rangle, \langle a, e, b \rangle, \langle a, b, d \rangle, \langle a, b, c \rangle, \langle a, c \rangle, \langle a, e \rangle, \langle a, b \rangle, \langle a \rangle\}$. Transition systems are built based on historical data, representing which activities are permissible following a given sequence of activities. While these systems can become exceedingly large and unintelligible, this is not a concern as they are used internally and are not exposed to the actors involved in the process.

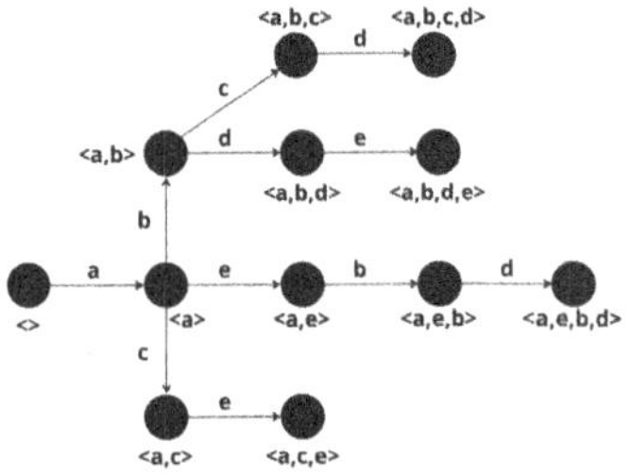

Fig. 1. Transition system for an example log.

The purpose of this work, however, is not to recommend the optimal next activity but rather to identify the most suitable activity-resource pair, ensuring feasibility from both activity and resource perspectives. This requires verifying if a certain resource is eligible to perform a given activity, according to the company and process constraints. The determination of whether a resource r can perform an activity a is based on the information in the event log $\mathcal{L}$; namely, there must exist an event e in the event log $\mathcal{L}$ such that $activity(e) = a$ and $resource(e) = r$. At the end, for each activity $a \in \mathcal{A}_{\sigma'}$, we identify the set of **eligible resources** $\mathcal{R}_a$ capable of performing a, such that $\mathcal{R}_a = \{r \in \mathcal{R} : \exists \sigma \in \mathcal{L}, \exists e \in \sigma$, where $activity(e) = a$ and $resource(e) = r\}$.

4 A Counterfactuals-Based Framework for Generating Recommendations

In this section, we aim to propose a framework that leverages counterfactual generations to recommend the next activities and resources.

Figure 2 depicts our proposed framework. The framework is structured into two phases: Offline and Online. The starting point of the offline phase is a completed event log $\mathcal{L}$. From the event log, we build a **transition system** $\mathcal{TS}_{\mathcal{L}}$ (cf. Definition 4) based on the sequence of activities in $\mathcal{L}$. This transition system captures the sequential relationships between activities, representing how tasks are performed one after another. In parallel, the event log is input into a **Trace-to-instance Encoding function** ρ (cf. Sect. 3). These encoded traces are then used as input to a Predictive Oracle function

[1] For simplicity, events are referred to as activity names, and the event's attributes are abstracted out.

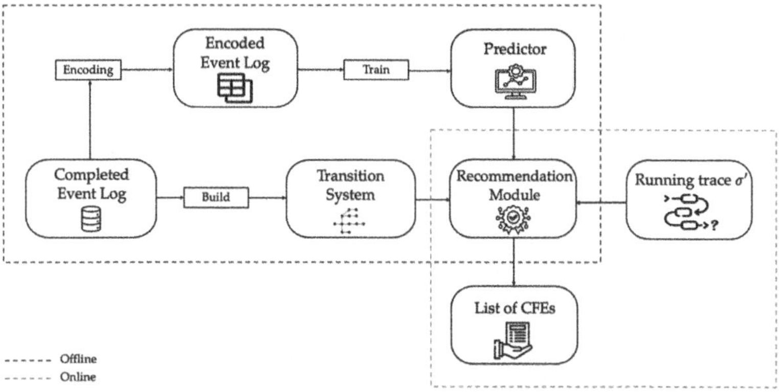

Fig. 2. Schema of our Counterfactual-based Recommender Systems. The schema distinguishes the part executed off-line, namely before activating the system, and that on-line, when recommendations are provided to running cases.

to estimate the relevant process-related outcomes for a given running instance. In this work, we focus specifically on total execution time prediction, where the corresponding **Total Time Oracle function** is defined in Sect. 3.

In the online phase, for every running trace σ', let (a_{next}, r_{next}) be the tuple of the next activity and resource that will be followed if no recommendation is given. We aim to recommend a suitable activity-resource pair that minimizes the total execution time. The proposed framework takes as inputs a Total Time Oracle function Φ, a transition system $\mathcal{TS}_{\mathcal{L}}$, and a running trace σ'. We first generate a set of k **counterfactual examples** (hereafter referred to as CFEs), $C = \{c_1, c_2, \ldots, c_h\}$ (with $h \leq k$), such that $c_i = (a_i, r_i)$ and each example predicts a different outcome from σ' (i.e., $\Phi(\sigma', a_{next}, r_{next}) > \Phi(\sigma', a_i, r_i)\ \forall i \in \{1, \ldots, h\}$).[2]

Assuming that the transition system $\mathcal{TS}_{\mathcal{L}}$ and the Total Time Oracle function Φ are both constructed from an event log $\mathcal{L}$. First, we build the set of **possible next activities** $\mathcal{A}_{\sigma'}$ that are allowed to occur after observing the events in σ', assuming those coincide with what is observed in $\mathcal{L}$ and thus modeled by $\mathcal{TS}_{\mathcal{L}} : \mathcal{A}_{\sigma'} = \{a : \exists(l^{state}(\sigma'), a, l^{state}(\sigma' \oplus \langle e \rangle))\}$.

The **Counterfactuals Recommendation Module** aims to generate a set C of CFEs, by modifying independently the next activity and resource in the running trace σ'. Each CFE $c_i \in C$ represents a hypothetical scenario in which the next activity-resource pair is replaced with (a_i, r_i), where $a_i \in \mathcal{A}_{\sigma'}$ and $r_i \in \mathcal{R}_{\mathcal{A}_{\sigma'}}$. The expected total execution time for each c_i is predicted through the Total Time Oracle function Φ as defined in Sect. 3, evaluating $\Phi(\sigma', a_i, r_i)$. However, there is a possibility that the recommended resource r_i is not allowed to perform the activity a_i (i.e., $r_i \notin \mathcal{R}_{a_i}$). In these cases, the recommended action (a_i, r_i) is discarded. The final recommendation is the activity-resource pair (a_{rec}, r_{rec}) that minimizes the expected total execution time

[2] In contrast to classification-based counterfactuals, where a change in class label is sufficient, our setting requires a numerically improved (i.e., lower) outcome prediction.

Query instance (original outcome : 2634175.0)

	CLOSURE_TYPE	CLOSURE_REASON	ACTIVITY	RESOURCE	ROLE	...	NEXT_ACTIVITY	NEXT_RESOURCE	TOTAL_TIME
0	Bank Recess	1 - Client lost	Service closure Request with BO responsibility	BOC	BACK-OFFICE	...	Pending Request for Reservation Closure	BOC	2634175.0

Diverse Counterfactual set (new outcome: [0, 2370757])

	CLOSURE_TYPE	CLOSURE_REASON	ACTIVITY	RESOURCE	ROLE	...	NEXT_ACTIVITY	NEXT_RESOURCE	TOTAL_TIME
0	-	-	-	-	-	...	Pending Request for Reservation Closure	574	1992700.0
1	-	-	-	-	-	...	Network Adjustment Requested	0	2154634.0
2	-	-	-	-	-	...	Pending Request for Reservation Closure	SB23	2181635.0

Fig. 3. Example output of the recommendation module with 10% of reduction threshold. The chosen counterfactual has been highlighted in the red box. (Color figure online)

(i.e., $(a_{rec}, r_{rec}) = \underset{(a_i, r_i) \in C}{\text{argmin}}\, \Phi(\sigma', a_i, r_i))$. This ensures that the proposed next activity and resource are both efficient and feasible from the organization's perspective.

Our framework is equipped with a parameter, hereafter referred to as **Reduction Threshold**, that indicates the minimum reduction of process-instance execution times that the recommendation can provide when the activity-resource pair is perturbed. For example, if we set this parameter to 5%, only recommendations that lead to a predicted reduction in total execution time of *at least* 5% are allowed. It is worth noting that the larger reduction thresholds would generally produce fewer counterfactuals, possibly even none. While the proposed approach supports automated decision making, in scenarios involving a human decision-maker, the system could present the top-k ranked counterfactuals, allowing the user to choose the most appropriate one.

Figure 3 visualizes the output of the recommendation module for a query instance in which a_{next} = "Pending Request for Reservation Closure" and r_{next} = "BOC". The trace is initially predicted to complete in 30 days (2,634,175 s); the module aims to reduce execution time by at least 10%, bringing it to a maximum of 27 days (2,370,757 s). Therefore, the algorithm modifies the **next_activity** and **next_resource** fields in the original query to explore alternative combinations, generating viable counterfactuals, specifically c_1 = ("Pending Request for Reservation Closure", "574"), c_2 = ("Network Adjustment Requested", "0") and c_3 = ("Pending Request for Reservation Closure", "SB23"). As depicted in the figure, all the counterfactuals satisfy the target range, but c_1 is chosen as it offers the lowest expected execution time.

5 Implementation and Evaluation

This section reports on the results of the proposed counterfactual-based framework for generating recommendations introduced in Sect. 4. The assessment is conducted using four real-life processes that are outlined in Sect. 5.1. Section 5.2 discusses the selection of the predictive model for implementing the Total Time Oracle function and reports on the evaluation of its predictive quality across multiple case studies. Section 5.3 provides an overview of the experimental settings and configurations employed for evaluation,

Table 1. Descriptive Statistics of Event Logs

Event Log	# cases	# activities	# resources	mean case duration
Bank Account Closure	29194	15	654	16.4 days
BPIC13	7554	13	649	12.1 days
BPIC17AO - Before	12207	18	103	20.2 days
BPIC17AO - After	14164	18	119	21.9 days
Consulta	954	18	561	14.9 days

while Sect. 5.4 details the evaluation metrics used. Finally, Sect. 5.5 reports the results to assess the effectiveness of the recommendations.

5.1 Case Studies

The validity of our framework was assessed using four real-life processes, which are described below. The **Bank Account Closure (BAC)** is a process executed at an banking institution.[3] The process deals with the closure of customer's accounts, which may be requested by the customer or the bank for several reasons. **BPIC13** is a dataset provided by Volvo IT Belgium that contains events from the incident management system.[4] **BPIC17AO** refers to a subprocess that includes application-relevant (A) and offer-relevant activities (O) in the 2017 BPI Challenge. This dataset is derived from a loan application process from a Dutch financial institution.[5] **Consulta** is the Academic Credentials Recognition (ACR) process of a Colombian University that was obtained from its BPM system (Bizagi) for the fourth case study.[6] A concept drift was found in the event log BPIC17AO, in which an increase in the workload of resources at week 22 led to a decrease in the service times at week 28 (cf. [2]). Since our framework assumes no concept drift appears in event logs, we partitioned the event log BPIC17AO into two sub-logs: one contains traces before week 22, namely **BPIC17AO - Before**, and the other contains those after week 28, namely **BPIC17AO - After**. This allows us to experiment across five distinct logs. Table 1 shows the descriptive statistics for each case study, where the mean case duration represents the average total execution time of process instances in that event log.

Under the assumption of no concept drift, each event log is temporally split into a training set $\mathcal{L}^{train}$ (first 75% of the traces) and a test set $\mathcal{L}^{test}$. This approach is useful since predictions rely on past information to forecast future outcomes. Then, we create a test log $\mathcal{L}^{run}$ of running traces to evaluate the effectiveness of the recommendation module. These running traces are derived from $\mathcal{L}^{test}$ by truncating traces at a random point to simulate the scenario of ongoing, incomplete executions.

[3] The BAC dataset is confidential and cannot be publicly shared.

[4] https://doi.org/10.4121/uuid:500573e6-accc-4b0c-9576-aa5468b10cee.

[5] https://doi.org/10.4121/uuid:5f3067df-f10b-45da-b98b-86ae4c7a310b.

[6] https://zenodo.org/records/11067569.

5.2 Prediction Model and Quality

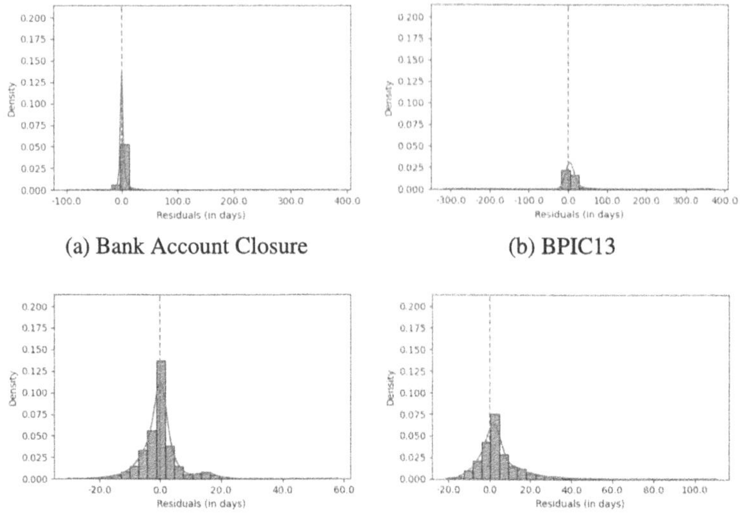

(a) Bank Account Closure (b) BPIC13

(c) BPIC17AO (Before) (d) BPIC17AO (After)

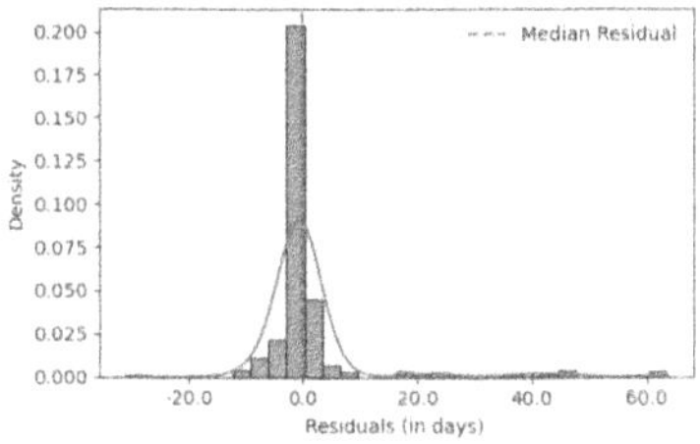

(e) Consulta

Case Study	RMAE
BAC	0.215
BPIC13	0.555
BPIC17AO (Before)	0.145
BPIC17AO (After)	0.265
Consulta	0.753

(f) Relative Mean Absolute Error

Fig. 4. Report on prediction quality across different case studies. Subfigures (a) - (e) display the residual distributions (in days) for five case studies: (a) Bank Account Closure (BAC), (b) BPIC13, (c) BPIC17AO (Before), (d) BPIC17AO (After), and (e) Consulta. The dashed red line represents the median residual. Subfigure (f) shows the Relative Mean Absolute Error (RMAE) for each case study. (Color figure online)

The framework is independent of the specific algorithm to instantiate the predictor of the total execution time of process instances. However, to demonstrate its effectiveness, we instantiated it using CatBoost [11], because of its excellent trade-off between predictive quality and training time [13] and the presence of an open-source implementation. Additionally, we employed a cross-validation approach to optimize the model's hyperparameters, ensuring robust generalization across different process instances. We randomly picked the 20% of traces in $\mathcal{L}^{train}$ and used them as a validation set $\mathcal{L}^{val}$ to apply a cross-validation approach to optimize the following training hyperparameters of

Catboost: *learning_rate, depth, iterations*. The framework has been fully implemented in Python, and the implementation is available on GitHub.[7]

Figures 4a–4e visualize the residual distributions, namely the distribution of the difference between the predicted and actual total execution times for each process instance across the case studies. The nearly symmetric shape of the residuals and the median close to zero highlight the validity of the predictor, indicating that the model is not systematically over- or under-predicting. A model with systematic errors is more predictable, thereby making counterfactual reasoning more effective. The strong performance of the predictors is supported by the relative mean absolute error (RMAE) results as shown in Fig. 4f, this value is computed by dividing the mean absolute error between the predicted and actual values by the mean of the actual values, providing a normalized measure of prediction error across different case studies. A lower RMAE value indicates higher prediction quality.

5.3 Evaluation Methodology

The whole framework is independent of the specific technique to generate counterfactuals. However, we need to leverage a specific framework, and we opted for DiCE (Diverse Counterfactuals Explanations) [22], which has already been used in predictive process monitoring but only for explanation generation purposes (cf. [7, 15]). In our setup, we used DiCE's random sampling method to generate counterfactuals. The framework allows users to specify the maximum number of generated counterfactuals. In our experiments, we set this limit to 100 to balance between diversity and computational efficiency. However, depending on process constraints and resource availability, fewer counterfactuals may be generated in practice. Unlike many counterfactual methods that focus on producing a single optimal solution, DiCE can generate a diverse set of counterfactuals. This diversity is essential in recommendation contexts, offering a variety of actionable next activity-resource pairs, allowing flexibility in operational decisions. By using the data about historical events and features, DiCE can generate a series of *what-if* scenarios by altering features related to the next activity and resource while keeping others constant.

The reduction threshold parameter of our framework directly maps to a corresponding parameter of DiCE. In our experiment, we evaluate thresholds of 5%, 10%, 20%, and 50%. A higher threshold results in greater potential improvements when a counterfactual is found. However, increasing the threshold also reduces the likelihood of generating recommendations, as more pairs of activity-resource may fail to meet the stricter criteria. By varying these thresholds, we assess the framework's ability to generate diverse recommendations under different levels of execution time reduction.

To evaluate the effectiveness of the proposed method, we compare it against a baseline, referred hereafter to as *Random Resource Assignment (RRA)*. In this method, given a running trace σ', we first extract the list of possible next activities using the transition system. For each activity, we use the Total Time Oracle function to estimate the execution duration and select the activity that minimizes the predicted total time as the next

[7] https://github.com/ngocdiemle296/CounterfactualsPrescriptiveAnalytics.

activity. Once the next activity is determined, a resource is randomly chosen from the list of feasible resources to perform that activity.

To assess the effectiveness of our recommendation module, we compare the total execution time of the process instances with and without following recommendations. The best option would be to have an A/B testing with the system running, but that is practically unfeasible. Therefore, building upon prior work in business process simulation [28], this study employs simulation to generate process instances. We developed simulation models for five case studies using established techniques to discover such models [26]. In particular, for each running trace $\sigma' \in \mathcal{L}^{run}$, the simulator is set to the process state (in fact, a Petri-net marking) that would be reached after performing the activities in $\sigma' \oplus \langle(a, t, r, \emptyset)\rangle$ where the recommendation is to have resource r perform activity a at simulation time t.[8] From the reached state, the process instance was simulated to reach the final state for 10 times, thus mitigating the potential consequences of simulation stochasticity.

Note that the sequence $\sigma_r = \sigma' \oplus \langle(a, t, r, \emptyset)\rangle$ is sometimes not possible in the simulation, namely when the sequence of activities in σ_r is not allowed according the simulation model. If the issue is related to σ' not being possible according to the simulation model, we compute the optimal alignment of σ' [1] to determine the state (i.e., the marking) that would be reached by the execution closest to σ': this state is thus used. If $(a, t, r, \emptyset)$ is not allowed by the simulator in the reached state, the recommendation is discarded, and executions are simulated without following the recommendation.

5.4 Evaluation Metrics

The effect of recommending an activity-resource pair for the running trace is computed as the average total execution time of that trace obtained through simulation. In contrast, the effect of not following the recommendation is the actual average total execution time observed in the test set without applying any intervention.

Let $\mathcal{L}^{test} = \{\sigma_1, \ldots, \sigma_n\}$ be the test set, where each trace $\sigma_i \in \mathcal{L}^{test}$ is associated with its actual total execution time $\mathcal{T}_{total}(\sigma_i)$ for $i \in \{1, ..., n\}$ (cf. Sect. 3). Let $\mathcal{L}^{run} = \{\sigma'_1, \ldots \sigma'_n\}$ denote the set of running traces, where $\sigma'_i \in prefix(\sigma_i)$. Then, let $\mathcal{L}^{sim} = \{\sigma_1^{sim}, \sigma_2^{sim}, \ldots, \sigma_n^{sim}\}$ be the set of simulated traces where $\sigma_i^{sim} = \sigma'_i \oplus \langle(a_{rec}, t_{rec}, r_{rec}, \emptyset)\rangle \oplus \sigma''_i$ is generated by following recommendation $(a_{rec}, t_{rec}, r_{rec}, \emptyset)$ after σ'_i, and then simulating the continuation σ''_i to reach the final state.

The effectiveness of the recommendations is evaluated based on the improvement gained by following the activity-resource pairs generated by our proposed technique compared to the case where no recommendations are applied. This requires to compute average execution time of the traces in a log - say $\mathcal{L}$ - which is $avg_{time}(\mathcal{L}) = \frac{1}{|\mathcal{L}|} \sum_{\sigma \in \mathcal{L}} \mathcal{T}_{total}(\sigma)$, where $|\mathcal{L}|$ denotes the number of traces in $\mathcal{L}$.

The performance improvement is thus computed as follows:

$$\Delta_{perf\_imp}(\mathcal{L}^{test}, \mathcal{L}^{sim}) = \left(1 - \frac{avg_{time}(\mathcal{L}^{sim})}{avg_{time}(\mathcal{L}^{test})}\right)$$

[8] Symbol $\emptyset$ indicates a function with an empty domain.

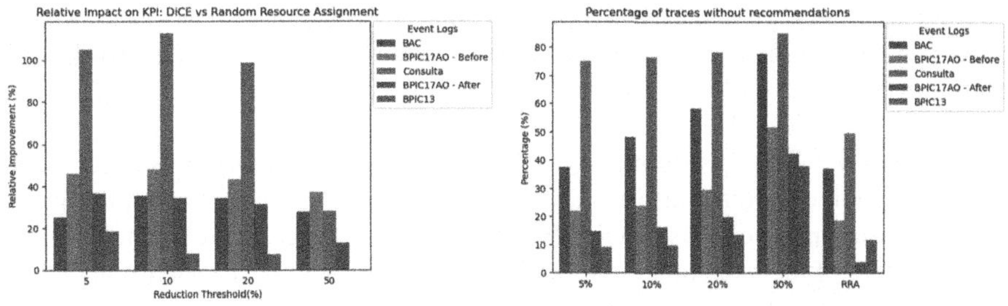

(a) (b)

Fig. 5. (a) Relative improvement related to the average total execution time when applying DiCE compared to the Random Resource Assignment method. On the x-axis, there are the reduction thresholds, while on the y-axis, the relative improvement is computed on the whole $\mathcal{L}^{run}$. (b) Percentage of traces without recommendations when applying DiCE compared to the Random Resource Assignment method. On the x-axis, there are the percentages of reduction thresholds (5%, 10%, 20%, and 50%) when applying DiCE and the RRA column indicating the Random Resource Assignment method.

which quantifies the relative gain in total execution time achieved by applying our method over the ground truth performance, which was observed in the event log when no recommendations are provided. We introduce a metric called **relative improvement**, Δ_{rel_imp}, which measures the effectiveness of our method compared to the RRA approach:

$$\Delta_{rel\_imp}(\mathcal{L}^{test}, \mathcal{L}^{sim}_{counter}, \mathcal{L}^{sim}_{RRA}) = 1 - \frac{\Delta_{perf\_imp}(\mathcal{L}^{test}, \mathcal{L}^{sim}_{counter})}{\Delta_{perf\_imp}(\mathcal{L}^{test}, \mathcal{L}^{sim}_{RRA})}$$

where $\mathcal{L}^{sim}_{counter}$ and $\mathcal{L}^{sim}_{RRA}$ are the event logs obtained via simulation when the recommendations are produced by our method based on counterfactual generations and by the RRA method, respectively.

5.5 Results Analysis

This section reports on the main results of our counterfactuals recommendation module.

Figure 5a presents a comparative visualization of the relative improvement achieved by our framework over the RRA method. Overall, our framework outperforms the RRA approach, with particularly substantial gains observed in the Consulta process, where improvements exceed 100% for the three lowest reduction thresholds. This confirms how a poor choice of a resource to perform a recommendation activity may significantly affect performance and how our framework efficiently generates recommendations in the form of activities and corresponding allocated resources. A noticeable trend is observed where the relative improvement for all methods declines at the 50% reduction threshold. Moreover, when looking at the BPIC13 process, we observe that the proposed method is not able to provide any relative improvement at the 50% total execution time. The underlying reasons for this behavior will be discussed in the following paragraph.

Another aspect we want to take into account is the percentage of traces without recommendations. This is important because the effectiveness of our framework is not solely determined by the performance improvement but also by how many traces can actually be optimized. This may be attributed to two different causes: (i) the model's threshold is set too high, resulting in no recommendations that meet this criterion, or, conversely, (ii) the ongoing trace already represents the optimal execution for the model, making further improvements practically unfeasible. From Fig. 5b, we can notice that if the percentage of the reduction threshold increases, there will be a decrease in the percentage of traces that receive recommendations. This highlights that *if the threshold is pushed too high, several process instances remain without potential recommendations*, which implies that the overall improvement for all running instances is more limited. Since the framework generates recommendations based on patterns and relationships identified in the event log data, setting a high percentage of reduction thresholds may cause the algorithm to focus only on the most extreme cases - those that can achieve significant improvements. As a result, the algorithm becomes more selective, filtering out traces that do not meet the strict total time criteria. This leads to a decrease in the number of eligible traces for recommendations. We are aware that choosing a higher percentage for reduction may seem very optimistic. However, it requires balancing the trade-off between achieving significant execution time improvements and finding a suitable number of recommendations for a substantial number of running cases.

As shown in Fig. 5b, the RRA approach can consistently generate a higher number of recommendations. This is because RRA does not rely on complex optimization techniques. Instead, it selects the next activity based on the smallest predicted total time and assigns a resource randomly based on the chosen best activity. This random allocation often results in suboptimal resource utilization.

In conclusion, our framework can effectively recommend activity-resource pairs that significantly reduce the total execution time, even under strict constraints, resulting in substantial improvements across all five event logs and reduction thresholds. Meanwhile, the random resource assignment prevents the RRA method from achieving equally good results, especially when resource choice critically impacts outcomes. The experiments show that setting overly high reduction thresholds may limit applicability, as many traces cannot receive recommendations. This highlights a key strength of the framework: the explicit tuning of the reduction threshold enables process experts to balance applicability - the proportion of traces that can receive recommendations - and utility - the magnitude of the improvements achieved through these recommendations.

6 Conclusion

This paper has put forward a framework for prescriptive process analytics that utilizes counterfactuals to generate real-time recommendations for the next resource and activity. In particular, DiCE was employed as a framework to generate these counterfactuals. In prescriptive process analytics, recommendations need to be given, usually in the form of suggesting a certain activity to be assigned to a resource for those process instances that are predicted to likely complete with unsatisfactory outcomes. In this paper, we focus on total case duration as the outcome, which is aimed to be minimized.

The idea of using counterfactuals for generating recommendations in prescriptive process analytics has stemmed from the idea that processes usually consist of dozens of potential activities and even hundreds of potential resources to whom these activities can be assigned. While brute-force approaches are not applicable in practice with these numbers, several existing approaches use a divide-and-conquer approach in which, first, the activity is chosen, and then it is assigned to a suitable resource. The leverage of counterfactual generation has the advantage of relying on more innovative approaches to converge to highly effective recommendations quickly.

Experiments were conducted on several real-life processes. These showed that our technique, based on counterfactual generation, outperforms a natural baseline where the best activities are first chosen and then allocated to resources. Experiments also highlight the framework's ability to be fine-tuned, providing a proper balance between applicability, referring to the proportion of traces that receive recommendations, and utility, which reflects the extent of the improvements achieved.

We acknowledge that this is the first work in the literature to provide recommendations based on counterfactual generation, and several avenues for future work exist. Currently, we aim to reduce the execution time since, although simplistic, it is a general indicator desired across various domains. However, extending it to other KPIs is conceptually simple. Furthermore, since the proposed technique does not depend on the method used for generating counterfactuals, we also plan to experiment with alternative counterfactual-generation frameworks, such as large language models, for explaining/producing recommendations about activities and resources.

Acknowledgments. The Ph.D. scholarships of Ngoc-Diem Le and Francesco Vinci are financially supported by MUR (PNRR) and the University of Padua. Authors also acknowledge the support of the project "Future AI Research (FAIR) - Spoke 2 Integrative AI - Symbolic conditioning of Graph Generative Models (SymboliG)" funded by the European Union under the National Recovery and Resilience Plan (NRRP), Mission 4 Component 2 Investment 1.3 - Call for tender No. 341 of March 15, 2022 of Italian Ministry of University and Research âĂŞ NextGenerationEU, Code PE0000013, Concession Decree No. 1555 of October 11, 2022 CUP C63C22000770006. The authors gratefully acknowledge Dr. Massimiliano Ronzani and Mr. Andrei Buliga, from Fondazione Bruno Kessler - Trento, for their valuable feedback provided during the final stages of manuscript preparation.

References

1. van der Aalst, W.M.P.: Process Mining: Data Science in Action. Springer, Heidelberg (2016)
2. Adams, J.N., van Zelst, S.J., Quack, L., Hausmann, K., van der Aalst, W.M.P., Rose, T.: A framework for explainable concept drift detection in process mining. In: Polyvyanyy, A., Wynn, M.T., Van Looy, A., Reichert, M. (eds.) BPM 2021. LNCS, vol. 12875, pp. 400–416. Springer, Cham (2021). https://doi.org/10.1007/978-3-030-85469-0_25
3. Arias, M., Rojas, E., Munoz-Gama, J., Sepúlveda, M.: A framework for recommending resource allocation based on process mining. In: Proceedings of the 13th International Conference on Business Process Management (BPM 2015) (2015)
4. Bhattacharya, A., Ooge, J., Stiglic, G., Verbert, K.: Directive explanations for monitoring the risk of diabetes onset: introducing directive data-centric explanations and combinations

to support what-if explorations. In: Proceedings of the 28th International Conference on Intelligent User Interfaces (IUI '23), IUI '23, pp. 204–219 (2023)
5. Bozorgi, Z.D., Teinemaa, I., Dumas, M., Rosa, M.L., Polyvyanyy, A.: Prescriptive process monitoring for cost-aware cycle time reduction. In: Proceedings of the 3rd International Conference on Process Mining (ICPM 2021), pp. 96–103 (2021)
6. Branchi, S., Di Francescomarino, C., Ghidini, C., Massimo, D., Ricci, F., Ronzani, M.: Learning to act: a reinforcement learning approach to recommend the best next activities. In: Proceedings of the 20th International Conference on Business Process Management (BPM 2022), pp. 137–154. Springer, Heidelberg (2022). https://doi.org/10.1007/978-3-031-16171-1_9
7. Buliga, A., Di Francescomarino, C., Ghidini, C., Maggi, F.M.: Counterfactuals and ways to build them: evaluating approaches in predictive process monitoring. In: Proceedings of the 35th International Conference on Advanced Information Systems Engineering (CAiSE 2023), Pp. 558–574. Springer, Heidelberg (2023). https://doi.org/10.1007/978-3-031-34560-9_33
8. Buliga, A., Di Francescomarino, C., Ghidini, C., Montali, M., Ronzani, M.: Generating counterfactual explanations under temporal constraints. In: Proceedings of the AAAI Conference on Artificial Intelligence, vol. 39, pp. 15622–15631 (2025)
9. Comuzzi, M.: Ant-colony optimisation for path recommendation in business process execution. J. Data Semant. **8**(2), 113–128 (2019)
10. Donadello, I., Di Francescomarino, C., Maggi, F.M., Ricci, F., Shikhizada, A.: Outcome-oriented prescriptive process monitoring based on temporal logic patterns. Eng. Appl. Artif. Intell. **126**, 106899 (2023)
11. Dorogush, A.V., Ershov, V., Gulin, A.: Catboost: gradient boosting with categorical features support. In: Proceedings of the Workshop on ML Systems at NIPS 2017 (2017)
12. Fahrenkrog-Petersen, S., et al.: Fire now, fire later: alarm-based systems for prescriptive process monitoring. Knowl. Inf. Syst. **64** (2022)
13. Galanti, R., et al.: An explainable decision support system for predictive process analytics. Eng. Appl. Artif. Intell. **120**, 105904 (2023)
14. Guidotti, R.: Counterfactual explanations and how to find them: literature review and benchmarking. Data Min. Knowl. Disc. **38**(5), 2770–2824 (2024)
15. Hsieh, C., Moreira, C., Ouyang, C.: DiCE4EL: interpreting process predictions using a milestone-aware counterfactual approach. In: Proceedings of the 3rd International Conference on Process Mining (ICPM 2021), pp. 88–95. IEEE (2021)
16. Huang, T.H., Metzger, A., Pohl, K.: Counterfactual explanations for predictive business process monitoring. In: Proceedings of EMCIS 2021. Springer, Heidelberg (2022). https://doi.org/10.1007/978-3-030-95947-0_28
17. Hundogan, O., Lu, X., Du, Y., Reijers, H.A.: Created: generating viable counterfactual sequences for predictive process analytics. In: Proceedings of the 35th International Conference on Advanced Information Systems Engineering (CAiSE 2023) (2023)
18. Käppel, M., Ackermann, L., Jablonski, S., Härtl, S.: Attention please: what transformer models really learn for process prediction. In: Proceedings of the 22nd International Conference on Business Process Management (BPM 2024), pp. 203–220. Springer, Cham (2024). https://doi.org/10.1007/978-3-031-70396-6_12
19. Kubrak, K., Milani, F., Nolte, A., Dumas, M.: Prescriptive process monitoring: Quo vadis? PeerJ Comput. Sci. **8**, e1097 (2022)
20. Li, J., Ren, Y., Deng, K.: Fairgan: gans-based fairness-aware learning for recommendations with implicit feedback. In: Proceedings of the the Web Conference 2022, pp. 297–307 (2022)
21. Mehdiyev, N., Majlatow, M., Fettke, P.: Counterfactual explanations in the big picture: an approach for process prediction-driven job-shop scheduling optimization. Cogn. Comput. **16**(5), 2674–2700 (2024)

22. Mothilal, R.K., Sharma, A., Tan, C.: Explaining machine learning classifiers through diverse counterfactual explanations. In: Proceedings of the 3rd ACM Conference on Fairness, Accountability, and Transparency (ACM FAT 2020), pp. 607–617 (2020)
23. Navarin, N., Vincenzi, B., Polato, M., Sperduti, A.: LSTM networks for data-aware remaining time prediction of business process instances. In: Proceedings of the IEEE Symposium Series on Computational Intelligence (SSCI 2017) (2017)
24. Padella, A., de Leoni, M.: Resource allocation in recommender systems for global kpi improvement. In: Proceedings of the 21st International Conference on Business Process Management (BPM 2023) (2023)
25. Padella, A., de Leoni, M., Dogan, O., Galanti, R.: Explainable process prescriptive analytics. In: Proceedings of the 4th International Conference on Process Mining (ICPM 2022). IEEE (2022)
26. Padella, A., Mannhardt, F., Vinci, F., de Leoni, M., Vanderfeesten, I.: Experience-based resource allocation for remaining time optimization. In: Proceedings of the 22nd International Conference on Business Process Management (BPM 2024) (2024)
27. Shoush, M., Dumas, M.: White box specification of intervention policies for prescriptive process monitoring. Data Knowl. Eng. **155**, 102379 (2025)
28. Shoush, M., Dumas, M.: Prescriptive process monitoring under resource constraints: a reinforcement learning approach. KI - Künstliche Intelligenz (2024)
29. Verma, S., Dickerson, J., Hines, K.: Counterfactual explanations for machine learning: a review, **2**(1), 1 (2020). arXiv preprint arXiv:2010.10596
30. Wachter, S., Mittelstadt, B., Russell, C.: Counterfactual explanations without opening the black box: automated decisions and the GDPR. Harvard J. Law Technol. **31**, 842–887 (2017)

Privacy and Confidentiality Requirements Engineering for Process Data

Fabian Haertel[1], Juergen Mangler[1], Nataliia Klievtsova[1], Celine Mader[2], Eugen Rigger[2], and Stefanie Rinderle-Ma[1](✉)

[1] Technical University of Munich, TUM School of Computation, Information and Technology, Garching, Germany
{fabian.haertel,juergen.mangler,nataliia.klievtsova, stefanie.rinderle-Ma}@tum.de

[2] Zumtobel Lighting GmbH, Dornbirn, Austria
{celine.mader,eugen.rigger}@zumtobelgroup.com

Abstract. Privacy and confidentiality concerns are major reasons for companies to abstain from sharing their process data. Privacy concerns refer to personal data and have been addressed in existing work by providing privacy-preserving techniques for event logs. However, the concept of confidentiality in process data pertaining to, e.g., sensitive business data, remains unclear. This work addresses confidentiality of process data based on the privacy and confidentiality engineering method (PCRE). PCRE interactively explores privacy and confidentiality requirements with process stakeholders and defines privacy-preserving actions to address possible concerns. We co-construct and evaluate PCRE based on structured interviews with process analysts in two manufacturing companies. The goal is to systematically scrutinize process data and balance the trade-off between privacy and utility loss.

Keywords: Process data · Privacy and Confidentiality Requirements · Requirements Engineering · Privacy-enhancing techniques

1 Introduction

Sharing their process data holds several benefits for companies, e.g., outsourcing advanced analysis tasks or building process collaborations to achieve common business goals. However, companies often hesitate to share information due to privacy and confidentiality concerns [5]. Privacy refers to rights of individuals regarding the collection, usage, and disclosure of their personal data as specified in regulations such as the GDPR[1]. For process mining, privacy has been investigated by various approaches in the last years, providing privacy models and privacy-preserving techniques for event logs, e.g., [8,10], also for privacy-sensitive domains such as healthcare [18]. Confidentiality *"is concerned with data access,*

[1] https://gdpr-info.eu/, last access: 2025-01-25.

M. ER et al. (Eds.): BPM 2025, LNBIP 565, pp. 216–231, 2026.
https://doi.org/10.1007/978-3-032-02936-2_16

whereas privacy is focused on individuals and their rights" [7]. Confidential process data can be part of the process model, e.g., data elements or endpoints, and also contained in process event logs [1,4]. The severity of disclosing this confidential information varies between organizations and industries as the following examples show. Revealing a detailed production process and its quality control steps in the manufacturing sector could expose business secrets or operational weaknesses. In the financial industry, providing information about the process steps for fraud detection algorithms or loan approval processes could provide valuable information to competitors. Until now, no generally applicable approach exists to explore the privacy and confidentiality requirements of a company in a systematic and interactive way while balancing utility of data sharing and defining privacy-preserving actions.

Hence, in this work, we propose the privacy and confidentiality requirements engineering method (PCRE). The goal is to scrutinize process data regarding privacy and confidentiality concerns in a systematic way with different stakeholders and define possible privacy-preserving actions such as data anonymization. At this, the goal is to balance privacy/confidentiality and utility loss [7].

PCRE was constructed and refined based on structured interviews with process analysts of a manufacturing company, and then evaluated with two different processes from two manufacturing companies as well as one process from the medical domain.

PCRE consists of seven phases with interview templates, checklists, and guidelines. The output is a consolidated executive summary containing the privacy and confidentiality assessment of all process elements and data, together with mitigation actions. Doing so, PCRE aims at supporting organizations and companies to enable the publication of process event logs with confidentiality assurance. This paper is based on the first author's Masters thesis [11] and structured as follows: PCRE is presented in Sect. 2 and evaluated in Sect. 3. Section 4 discusses related work. Section 5 concludes the paper.

2 Privacy and Confidentiality Engineering Method

The privacy and confidentiality engineering method (PCRE) aims at systematically collecting the privacy and confidentiality requirements that are relevant for a given process and its stakeholders. Moreover, PCRE aims at determining how the derived requirements can be met by applying privacy-preserving techniques. This results in the following objectives for PCRE:

- *Objective 1:* Define process data usage objectives, e.g., company-internal or public analysis of the process data and scope of contained data.
- *Objective 2:* Outline the benefits of publishing process data, i.e., what does the company gain in exchange?
- *Objective 3:* Scrutinize process elements systematically for necessary privacy-preserving actions, e.g., data anonymization.
- *Objective 4:* Align the process data usage objectives with privacy-preserving actions in order to balance data utility and privacy loss.

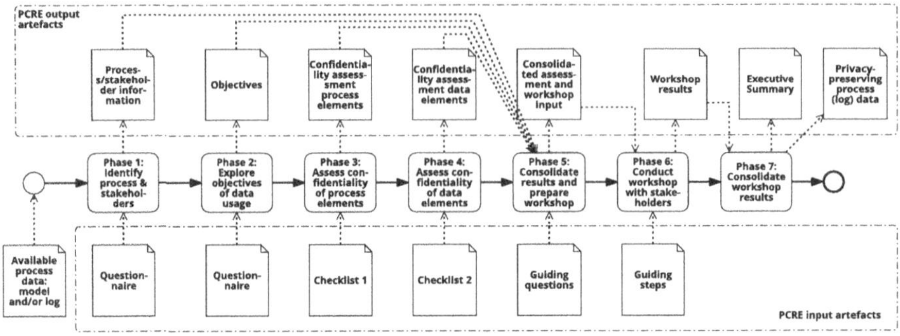

Fig. 1. PCRE Method (modeled as BPMN process using SAP Signavio)

PCRE takes as **input** available process data (cf. Fig. 1), i.e., a process model and/or event logs. The more comprehensive the available process data is, the more comprehensively privacy and confidentiality requirements can be engineered. If, for example, only control flow related information is available in the model or log, PCRE can assess tasks, their order, and their timestamps. If additionally fully specified data flow and endpoints for service invocation are available, PCRE can also assess the privacy and confidentiality of this process data. PCRE can be applied to newly designed or re-designed process models during design time. It can also be applied to event logs ex post or during runtime. If helpful for stakeholders, a process model might be additionally discovered from the logs. This might be especially useful for object-centric logs[2] as they might represent more than one process model.

The output of PCRE comprises an executive summary of privacy and confidentiality requirements. Based on this, the company or organization can prepare its process data for sharing with privacy-preserving techniques determined during PCRE assessment.

The PCRE aims to meet *Objective 1* to *Objective 4* by systematically scrutinizing process data with different stakeholders along well-defined phases and input artefacts such as interview guidelines. It consists of seven phases as shown in Fig. 1, including the PCRE artefacts and the involved stakeholders. PCRE was initially designed by the authors based on experience and then refined in a first interview round with process analysts from the manufacturing domain. One design goal of PCRE is to follow a bottom-up approach, i.e., a more detailed confidentiality assessment of process structure and process data is conducted with process and business analysts in the beginning and then abstracted for discussion with other stakeholders, including the legal compliance team.

PCRE Phase 1: Identify Process & Stakeholders (Stakeholders: Process Analyst, Management). Phase 1 aims to collect general knowledge for shaping and conducting the subsequent phases. Phase 1 involves process analysts and optionally the management for additional insights. Table 1 provides the tem-

[2] https://ocel-standard.org/.

plate for the written questionnaire as artefact. Question 1.1 determines which process information is available, i.e., process model and/or logs. Question 1.1 also asks which process analysis task such as conformance checking or predictions is envisioned and possible and which data preparation steps are required. Different process analysis tasks might bear different confidentiality requirements. For conformance checking, the detection of alignments might reveal sensitive business secrets, e.g., if a certain process task takes always longer than specified in the model or if a certain data value such as temperature exceeds a certain threshold several times. Data preparation steps can also be source for confidentiality requirements and possible action points for privacy-preserving techniques. If, for example, data from different sources is merged when generating process logs, business-sensitive data might be revealed. Questions 1.2 and 1.3 provide insights into data analysis practices and usage of the data ($\mapsto$ *Objective 1*). which departments are using the data. Question 1.4 identifies a list of stakeholders relevant to the analysis of privacy and confidentiality requirements within the organization and a collection of essential information to guide the subsequent phases.

Table 1. Template for (written) questionnaire in Phase 1

ID	Question
1.1	Available process information: i) process model and event logs, ii) process event logs $\mapsto$ process analysis task, data preparation steps
1.2	Process data usage: is the process data currently used for analysis? If so, to what extent?
1.3	Process data usage: Which departments are currently working with the process data, and who uses the analysis results?
1.4	Which stakeholders are relevant for the subsequent phases?

PCRE Phase 2: Explore objectives of data usage (Stakeholders: Process Analyst, Management, Business Analyst, Additional Stakeholders, e.g., finance department, human resources, and operations) Phase 2 is conducted through interviews to increase the depth of interaction, influencing the richness of insights. Table 2 summarizes six guiding questions for assessing the usage of process data ($\mapsto$ *Objective 1*) and the actual/possible objectives of the usage. Questions 2.1–2.6 can be applied based on a process model or an event log depending on the available information (signified by an 'x' in Table 2).

Questions 2.1–2.6 aim to consolidate the process analysis requirements across different departments, also by including additional stakeholders. The collected and consolidated process analysis requirements can then be used for evaluating the utility loss by data anonymization. Furthermore, the inclusion and "sharing" of insights from the responses might increase the motivation of stakeholders to participate in PCRE. Questions 2.1 to 2.6 explore different facets of process data usage, including its current application for analysis (2.1), the specific interests and aspects of the process analysis that stakeholders find compelling (2.2), and

Table 2. Template for interview-based questionnaire in Phase 2

ID	Question	Model	Log
2.1	Is process data currently utilized for analysis?	x	x
2.2	Which aspects of the process analysis are of interest?	x	x
2.3	Which analysis results are considered useful?	x	x
2.4	Which metrics/KPIs, e.g., duration or frequency of process tasks, are currently calculated for process analysis? How are they calculated?	x	x
2.5	Which metrics/KPIs are currently not calculated for process analysis? Which of them would be of interest? Why are they (currently) not calculated?	x	x
2.6	Which insights from process analysis process would help to manage the process more effectively for each department?	x	x

the types of results considered useful from such analyses (2.3). They explore the metrics and KPIs currently used for process analysis, their calculation methods (2.4), and any potentially valuable metrics/KPIs not yet calculated, including reasons for their omission (2.5). Additionally, this question aim to identify areas for improvement. Finally, stakeholders are asked about the process insights that could enhance department-specific management precision (2.6) (↦ *Objective 2*).

PCRE Phase 3: Assess confidentiality of process elements (Stakeholders: Process Analyst, Management, Additional Stakeholders) process analysts and management participate in a collaborative review session to thoroughly analyze the confidentiality of the process metadata (↦ *Objective 3*).

Questions 3.1 to 3.7 assess privacy and confidentiality concerns, cf. Table 3. For Question 3.1, information about subprocesses is required. If no subprocess exists, Question 3.1 is skipped. If subprocesses is exist, they are examined to identify confidential subprocedures that could potentially reveal specific business functions or strategies that may interest competitors or external parties. Subprocesses can also provide insights into proprietary methods or technologies that the company considers a competitive advantage. Question 3.2 considers the names of processes and subprocesses. Here, the focus is on determining whether these labels expose sensitive business strategies or technologies. Question 3.3 examines the existence of tasks within the process that could reveal confidential operational functions or innovative strategies. Accordingly, Question 3.4 considers how task labels might reflect business logic or unique services. Question 3.5 scrutinizes process parameters to ascertain whether they hold sensitive data. Question 3.6 aims at endpoints as their disclosure might provide hints to the IT infrastructure, potentially revealing vulnerabilities that could be exploited by adversaries. Finally, Question 3.7 analyses the versioning and history of changes

Table 3. Checklist template for Phase 3

ID	Question	Model	Log
3.1	For all existing subprocesses, are there any representing, unique, or innovative business procedures, i.e., their disclosure could harm the organization?	x	
3.2	For all considered processes and subprocesses, do their names directly or indirectly indicate confidential business areas, technologies, or methods that should not be public?	x	~
3.3	For all tasks within the (sub)process under consideration, are there tasks that represent unique or innovative business procedures, i.e., their disclosure could harm the organization?		
3.4	For all tasks within the (sub)process under consideration, could the task label reveal specific business activities or proprietary services that should be protected for competitive reasons?	x	x
3.5	For all process parameters, including timestamps, within the (sub)process under consideration, do they contain sensitive data such as customer-specific information or operational settings that should not be publicly accessible?	x	x
3.6	For all endpoints within the (sub)process under consideration, could the disclosure of these endpoints provide details about the organization's network infrastructure, security mechanisms, or external service providers?	~	~
3.7	Does the documentation of the process contain specific information about the reasons for changes that could indicate business strategies or product developments?	*	*

~: if information is available, i.e., for executable models, for logs containing data values;
*: if documentation, e.g., a change log, is available

in the process. This step is crucial in determining if this documentation includes details that could embody business strategies or product developments.

PCRE Phase 4: Assess Confidentiality of Data Elements (Stakeholders: Process Analyst). Phase 4 scrutinizes the confidentiality of all process data elements (↦ *Objective 3*). Question 4.1 starts with checking the atomicity of the data elements. This is important since atomic data elements, being the most granular level of data, often carry less risk of directly revealing sensitive information unless they are unique identifiers. Composite data elements, by contrast, encapsulate more detailed information due to their complex structure, posing a higher risk of revealing sensitive information. If data elements are composite, they are decomposed into sub-components until all of them are atomic. Decomposition is fundamental for assessing the confidentiality of the contained information. It also aids choosing appropriate anonymization techniques to minimize the loss of utility while maximizing privacy. For each of the atomic data elements, we pass through Questions 4.2 to 4.10. First, dependencies between data elements are determined (4.2), including process-based dependencies and functional dependencies. For determining data dependencies, a process model is required.

An example process-based data dependency is depicted in Fig. 2 where data element d2 is written by task B after B reading d1. This analysis can also result in data dependency clusters where each of the contained data elements depend on each other, e.g., data elements d1, d2, d3 in Fig. 2. If one element within a cluster is considered sensitive or confidential, the other elements may also carry similar confidentiality concerns. Thus, data elements in the same dependency cluster should be anonymized to the same degree. Question 4.3 identifies the scope of the data element in terms of processes that access the data element, resulting in a list of process ids. After identifying dependencies and scope, for each data element, its privacy and confidentiality requirements are assessed. Here, one process model is not sufficient, but several process models accessing a data element are to be analyzed (∼ in Tab. 4). Note that we assume that an atomic data element has either privacy or confidentiality requirements. Question 4.4 assigns a risk of re-identifying an individual based on the data element on a scale of 0–5 where 0 refers to 'does not concern individuals', 1 to 'no identification possible', 2 to 'unspecific', 3 to 'in combination with other elements', 4 to 'strong delimitation possible', and 5 to 'direct identification possible'. For these individual-related privacy concerns (rating > 1), the required level of anonymization can be derived from the regulatory guidelines provided by the GDPR. A rating of 0 indicates that there are no privacy concerns, but there could be confidentiality concerns, resulting in assessment based on Questions 4.5 to 4.10 which balance the risk of publishing the data element (confidentiality loss) against its utility for the process analysis (utility loss).

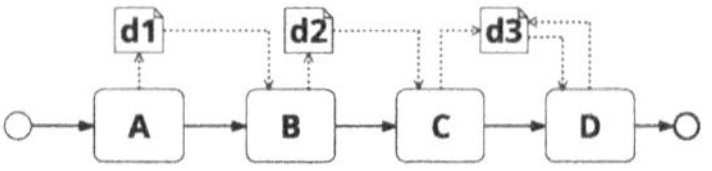

Fig. 2. Process With Data Dep.

Table 4. Checklist data elements for Phase 4

ID	Question	Reply	Model	Log
4.1	Is the data element atomic or composite?	atomic, composite	x	x
4.2	On which other data elements does the data element depend?	list of data elements	x	
4.3	Which (other) processes access the data element?	process ids	∼	
4.4	Can an individual be uniquely identified by the data element?	scale: 0–5	x	x
4.5	Does the data element contain information that should not be disclosed externally?	scale: 1–5	x	x
4.6	How likely are potential inferences?	scale: 1–5	x	x
4.7	How severe are the implications of the inference?	scale: 1–5	x	x
4.8	How critical is this data element for the functioning of the process?	scale: 1–5	x	x
4.9	What is the impact of the data element on decision-making in the process?	scale: 1–5	x	∼
4.10	How frequently is the data element used in the process?	scale: 1–5	x	x

Questions 4.5 to 4.7 assess the confidentiality risk when publishing the data element by estimating the severity and probability of an inference. The scale for a possible confidentiality risk (4.5) comprises 1: 'none', 2: 'unimportant sensitive information', 3: 'sensitive information, but not critical', 4: 'important sensitive information', and 5: 'critical business secrets'. The scale for the likelihood of inference ranges from 'very unlikely' (1), 'unlikely' (2), 'possible' (3), 'likely' (4), to 'very likely' (5). Finally, the severity of inference (4.7) is measured as 1: 'none', 2: 'minor negative', 3: 'moderate negative', 4: considerable negative', and 5: 'catastrophic'. Questions 4.8 to 4.10 evaluate the utility of each data element for the process and, consequently, its relevance for analysis. Question 4.8 estimates the criticality of the data element based a scale from 1–5, i.e., 1: 'not critical', 2: 'slightly critical', 3: 'moderately critical', 4: 'very critical', 5: 'absolutely critical'. Question 4.9 rates the impact of a data element on decision making in the process based on 1: 'no influence', 2: 'minor influence', 3: 'moderate influence', 4: 'high influence', 5: 'decisive'. Finally, Question 4.10 determines how frequent the data element occurs in the process, ranging from 1: 'never', 2: 'rarely', 3: 'sometimes', 4: 'often', to 5: 'constantly'. The rationale behind Questions 4.8 to 4.10 is that an element that occurs in many conditions or is frequently retrieved, written, or modified is probably highly relevant to the overall process and thus implies a high utility loss upon application of privacy-preserving actions such as anonymization.

The provided ratings can be aggregated into different metrics, e.g., by averaging the ratings. Alternatively, the ratings can be weighted according to the trade-off between privacy and utility loss. Giving a higher weight to the questions centering around the risk imposed by publishing the data elements weighs the privacy aspects higher than the utility aspects, and vice versa. Finally, to account for the risk of a possible chain reaction of disclosures within a dependency cluster, the maximum rating of confidentiality among the data elements within the cluster can be taken.

It is advisable to give an example of a possible value for each data element in order to ease the understanding of privacy-preserving requirements imposed by the data element for other stakeholders. Besides, the case may arrive that unused data elements exist within the process. These data elements can be suppressed entirely as they have no utility for analyzing the process and still contain confidential information. By the end of Phase 4, the organization obtains a list of all data elements present in the process, decomposed to an atomic level, and data dependency clusters which are mapped to a confidentiality rating.

PCRE Phase 5: Consolidate Results and Prepare Workshop (Artefact: Guiding Questions for Consolidation and Preparation). Phase 5 aims at consolidating the results obtained in Phases 1–4 to prepare the workshop in Phase 6, and is carried out by the person responsible for executing PCRE. The consolidation can be supported by the guiding questions presented in Tab. 5. These questions also aim at defining privacy-preserving actions to meet privacy and confidentiality requirements by, at the same time, minimizing the utility loss caused by them ($\mapsto$ *Objective 4*). To this end, the group of participating stakeholders should also contain security experts (Question 5.1) in order to avoid the "over-protection" of process data.

Table 5. Template for guiding questions in Phase 5

ID	Question
5.1	Which stakeholders should be included in the workshop?
5.2	Which process elements have been classified as particularly confidential and why? In detail: processes/subprocesses (5.2.1), tasks (5.2.2), process parameters (5.3.4), data elements (5.2.4), endpoints (5.2.5), versioning/history of changes (5.2.6)
5.3	Which privacy-preserving techniques, e.g., anonymization, are necessary to protect the process elements classified as confidential from disclosure?
5.4	How can the necessary degree of privacy-preservation be achieved?
5.5	Which KPIs or metrics can be calculated from the data elements and are classified as business secrets?
5.6	What compromises must be made between the degree of privacy-preservation and the retention of data utility?

PCRE Phase 6: Conduct Workshop with Stakeholders (Stakeholders: Process Analyst, Management Legal Compliance Team). The workshop aims at obtaining approval for the minimal privacy-preserving actions such as anonymization, pseudonymization, or timestamp shifting while retaining the maximum possible utility. The workshop starts with revisiting the motivation behind data publication and the objectives of process analysis. Subsequently, the results of Phase 3 on the process metadata elements and the data elements identified in Phase 4 are discussed. For each element, participants address the question: Given the specific element or class of dependent elements within the process log and the required level of privacy preservation, does its publication present any issues? In the course of this discussion, the approval of decision-makers within an organization, probably the management and legal compliance team, is obtained. Eventually, the predefined data analysis goals and assessment of the trade-off between privacy and utility are revisited.

PCRE Phase 7: Consolidate Workshop Results. Phase 7 consolidates and documents the results of the workshop conducted in Phase 6 within an executive summary containing key decisions, goals, and objectives of data usage. To this end, the person responsible for conducting PCRE scrutinizes again all process data logs w.r.t. the results obtained in Phases 1–6.

3 Evaluation

Development and evaluation of PCRE are conducted in an intertwined way, resembling an incremental development with iteratively incorporating feedback:

- **First round interviews:** Develop and refine the PCRE approach with three process analysts from a manufacturing company.
- **Second round interviews:** Assert the applicability of the PCRE approach by applying to two scenarios in two manufacturing companies, and by discussing the approach with a process analyst from the medical domain.

3.1 First Round Interviews

For the **first round interviews** we focused on process analysts that were responsible for the technical implementation and the continuous improvement of processes. We focus on process analysts as experts, as the main goal of PCRE is to address process data specifically, rather than focusing solely on broader privacy concerns. Privacy and confidentiality measures must align with how data is handled within processes, ensuring effectiveness in the real world by understanding the data flow across various stages of business processes. The process analysts reported experiences, advantages, and disadvantages. These insights were used to refine the initial ideas for the PCRE approach.

All first round interviews were conducted online in an unstructured way between 10 Jan 2024 and 1 Feb 2024. Due to confidentiality reasons, no interview transcripts are available. The summary of the interviews are available in [11] [Appendix, pp.77-79]. The unstructured interview format was selected as the goal was to understand to not restrict the interviewees regarding their perspective on the subject matter, and gather insights in the interviewees own terms (i.e., process-specific domain language) [21].

One significant outcome was that the questions initially designed for Phase 3 did not indicate an assessment of each process metadata element. Consequently, the questions were enhanced to reflect the necessity to assess all process metadata, such as endpoints or task labels. Additionally, a more detailed description how to incorporate the results of Phase 3 into subsequent phases, especially Phase 6, was elaborated. During Phase 4, 21 data elements were analyzed, their atomicity documented, and their confidentiality assessed. However, exploring data dependencies proved to be cumbersome due to the complex interconnections among some elements. First, the functional dependencies are deeply intertwined and complex to document. Their multiplicity presents significant challenges in assessment. Secondly, there are additional considerations when evaluating confidentiality: certain data elements only reveal confidential information when analyzed in combination with other data elements. Considering, for example, timestamps and product IDs individually does not comprise confidential information. However, when combined, they could facilitate the derivation of sensitive information. This observation is reinforced by discussions with a process analyst from the medical domain for combinations of timestamps and patient IDs. Another insight was to document example values and additional notes when evaluating the confidentiality of the data elements to enhance understandability. Overall, meaningful insights could be gathered in Phase 4, facilitating the preparation of the upcoming workshop. An additional benefit of PCRE was that redundant and deprecated data elements were identified by documenting all process elements. Removing these data elements enhances process efficiency, documentation quality, and overall understandability of the process.

3.2 Second Round Interviews

The goal of the second round interviews was to affirm the applicability of the approach. These interviews were conducted with the PCRE version presented in this paper, between 4 and 8 months later. We looked at three different processes:

Process A: A product manufacturing and assembly process from the manufacturing company with which we already cooperated for the round one interviews. We relied on two of the original process analysts, that helped use refine the process. Additionally we interviewed the team leader of the manufacturing department, and a division manager, to gather feedback regarding strategic and judicial implications in the workshop in phase 6.

Process B: A process from a different manufacturing company, dealing which the development of product ideas (i.e., an ideation process). For this case we relied on 2 process analysts, one responsible for technical implementation, and one responsible for overseeing the evolution of the process. We also interviewed the responsible team leader, to again gather insight into strategic and judicial implications in phase 2 as well as the workshop in phase 6.

Process C: A process from the medical domain. As the process was not fully developed for this case, we only interviewed one process analyst which was responsible for preparing process mining data. For this case only the phases 1 to 4 were conducted, and only with a preliminary high-level version of the process, but detailed knowledge about process data.

The preparations, interviews, and consolidation (Phase 1 to 5) could be done in about 5 h for processes A and B. For process C (medical domain) the preparations and discussions (Phase 1 to 4) took about hours. Then, for **Process A** and **Process B** workshops were conducted (Phase 6), which took about an hour. However, the time required is dependent on the experience and familiarity of both the interviewer and the interviewee. The primary concern is identifying which specific data elements within these processes can be shared internally or externally without compromising privacy or confidentiality of organizational, technical or production details. For this type of analysis, process analysts together with the team leaders are more suitable than a legal experts, as they possess a detailed understanding of the technical structure of the process, data flows, and contextual constraints. This allows them to make reasonable judgments about data sensitivity, utility, and the feasibility of privacy.

Process A is implemented using the Cloud Process Execution Engine CPEE[3], and orchestrates the assembly of up to 600 products a week in a production line with several human workstations, e.g., for soldering and packaging. It also

[3] https://cpee.org, last access: 2025-01-25.

contains several automatic steps like laser inscription. The data gathered while a single workpiece is running through the production line covers production variant details, telemetry, timing, and quality aspects.

All stakeholders agree on the value that could potentially be generated by exploring process data. An important question is how the manufacturing company can access the results of the process data analysis. All process endpoints are identified as confidential since they expose the internal network topology. Although only selected tasks and task labels are determined as confidential, all tasks and task labels throughout the event log need to be anonymized as they are considered too ambiguous, hence leaving a broad scope for interpretation, possibly leading to false conclusions and thus reducing their utility for data analysis. Therefore, both, tasks and task labels will be generalized.

The publication of many data elements is considered difficult, as the perceived risks often outweigh the benefits. Moreover, the data elements considered publishable typically contribute minimally to process insights, offering reduced utility for process analysis. A single trace within the manufacturing process relates to the manufacturing of one product instance. This poses an important differentiation: confidentiality concerns increase if several traces are published, because additional metrics, such as production volumes or process changes over time, will be recognizable. For example, compliance with legal requirements and adaptability to legal changes are indicated, and periods with low manufacturing performance or quality problems can be identified. Furthermore, a data element relevance can also span multiple instances, traces, or processes. Especially when process data of various (sub) processes is published, a data element of one process could be used to disclose information from another process.

For **Process B**, the setting was very different. The data produced throughout the process is highly sensitive, since new products or business ideas are of the utmost value for a company. Thus, for making the data publicly available, most data had to be removed, only leaving the duration of process instances and frequency and occurrence of certain repeated tasks as valid analysis tasks. A prototype version of the process was actively in use at the company. It was implemented using the Cloud Process Execution Engine CPEE. This facilitates the definition of data elements which can be examined before the process comes into production (process testing). The process consists of 12 activities and 18 data elements. Moreover, three endpoints are defined, and 12 process attributes (parameters) are set. Most data elements are links to files, e.g., large JSON files, and hence not atomic, making the assessment more complex. The endpoints are predefined and lead to HTML forms usable for data input or PHP scripts, e.g., for sending an email. The process parameters comprise the unique identifier of the process model and personal identifiers such as names authors. As conclusion, conducting PCRE enabled the systematic discussion of all process-related information with different stakeholders, weighing the benefits of process analysis results with privacy and confidentiality threats. Moreover, PCRE resulted in cleaning process data, i.e., deprecated data elements, resulting in a more streamlined and efficient processes.

For **Process C**, the setting again was different. As an executable process model was not available, before the interview we tried to create a process model with different process discovery algorithms. The result (which was lacking data-flow) was presented to the process analyst, but turned out to be sub-par, as (a) already applied anonymization, (b) missing, and (c) incomplete data was skewing the results. Phases 1 to 4 were conducted based on information in the process logs and verbal complements by the process analyst detailing additionally available data (not included in the log). During the interview, it became clear that some of the previously stated analysis goals could not be achieved with the current process logs, mainly because of missing data, and skewed timestamps. While the PCRE approach structured the interview, the results were less definitive than for the other processes.

3.3 Reflection

For **Process A** and **Process B**, the executable process model was used, resulting in an assessment of the generated event logs. For **Process C**, PCRE was conducted based on process event logs, as process discovery algorithms yielded confusing control-flow. PCRE helped to identify how process analysis results would be limited by the (already too strict) privacy-preserving measures w.r.t. timestamps in the log, i.e., providing time distances between events instead of timestamps resulting in the non-applicability of process discovery.

For all three processes, it was agreed that with the adaptation of privacy-preserving measures for a selection of process instances rich process analysis results can be achieved without infringing privacy and confidentiality. It was also agreed that, while the PCRE approach can be used to create public datasets, it will be used first and foremost internally to produce datasets for in-house (or outsourced) data analysis.

Overall PCRE provides a streamlined framework for efficiently conducting Phases 1 to 5 and the participants of the workshop can rely on the prepared documents. The workshop in Phase 6 basically started with clear analysis goals, and mostly revolved around about potential (unintended) insights that could be gathered from the data regarding human stakeholders in the process such as employees, as well as the produced products, often leading to further fine-tuning of privacy measures. The creation of datasets for all three processes is currently actively pursued.

4 Related Work

Existing literature provides methods and techniques for the elicitation, modeling, and verification of security and privacy requirements in the context of business processes, e.g., [12,17] as well as for augmenting process models with privacy-enhancing technologies [1,4]. Another research stream refers to privacy-preserving methods in process mining. Existing approaches in this stream can be distinguished into grouped-based privacy models, indistinguishability-based

models, and confidentiality frameworks [7]. Grouped-based privacy models rely on the concept of k-anonymity [20], extended by the notions of l-diversity [14], and t-closeness [13]. They comprise the TLKC model [19], PRETSA [9], and [2]. Indistinguishability-based models utilize differential privacy to ensure that two versions of a data set cannot be differentiated up to a specified level and subsume [16]. Finally, confidentiality frameworks utilize encryption methods and access control regulations to restrict access to confidential data or results from data analysis to authorized parties [7,15]. Moreover, several tools have been demonstrated for privacy-preserving process mining. Examples include ELPaaS [3] that supports the generation of an anonymized event log satisfying privacy metrics, and mining techniques resulting in desired privacy guarantees as well as ShareProm [6] enabling a secure multi-party computation framework.
Research Gap: Process mining related techniques do not consider confidentiality beyond information of individuals and have not yet balanced risk and utility [7]. These gaps can be addressed by PCRE. Nonetheless, existing approaches can be applied in a complementary way by dealing with identified privacy concerns. This also holds true for approaches for eliciting security requirements and equipping processes with privacy-enhancing technologies.

5 Conclusion and Outlook

In summary, the PCRE method features seven phases with questionnaires, checklists, ratings, and guidelines, to systematically assess privacy and confidentiality requirements of process data within an organization. It is generic and broadly applicable across various domains, incorporating quantitative and qualitative evaluations of the confidentiality of process elements. Weights can be defined to adjust PCRE to the domain and specific needs of the organization. Moreover, PCRE facilitates the comprehensive documentation of processes and encourages a multidisciplinary dialogue among diverse organizational stakeholders. Through such a collaboration, stakeholders critically analyse event logs for process analysis and determine relevant metrics or KPIs. Therefore, aligning privacy-preserving objectives with Process Mining and data analysis capabilities is ensured by design.

Limitations and Future Work: The validity of the approach itself could have been influenced by researcher and confirmation bias, which affects how interview responses were interpreted and incorporated into the development of PCRE. However, so far, for example in round two interviews which incorporated different processes from different companies, the utility of the approach was confirmed. The incorporated interview feedback may be restricted due to the limited number of involved roles (i.e., process analysts or team leaders), excluding legal expertise. But as legal concerns, but instead organization/productions secrets are the target of the PCRE approach, this might not pose a problem. Additionally, in order to collect diverse interests and perspectives on data analysis objectives, the PCRE methods requires multiple stakeholders to assess the current state of data analysis within the organization. This might become difficult, especially for external consultants, and result in resource and communication overhead. Furthermore, the availability of the process data constitutes a potential challenge. In this work, the processes are implemented in a process engine, resulting in well-specified process parameters, data element, and endpoints. In practice, process data might be available at different levels of quality, directly influencing the complexity of conducting PCRE. And finally, the number and complexity of data dependencies might also increase the complexity of the analysis, especially at the presence of multiple sub-processes. Additionally, detailed guidelines for mapping confidentiality assessments to specific anonymization techniques could streamline the real-world deployment. process. In future work, we will apply PCRE to further domains in order to evolve the method into a robust framework.

Acknowledgement. This work has been partially supported and funded by the Austrian Research Promotion Agency (FFG) via the Austrian Competence Center for Digital Production (CDP) under the contract number 881843.

References

1. Bakhtina, M., Matulevicius, R., Seeba, M.: Tool-supported method for privacy analysis of a business process model. J. Inf. Secur. Appl. **76**, 103525 (2023)
2. Batista, E., Martínez-Ballesté, A., Solanas, A.: Privacy-preserving process mining: a microaggregation-based approach. J. Inf. Secur. Appl. **68**, 103235 (2022)
3. Bauer, M., Fahrenkrog-Petersen, S.A., Koschmider, A., Mannhardt, F., van der Aa, H., Weidlich, M.: ELPAAS: event log privacy as a service. In: BPM Demos, pp. 159–163 (2019)
4. Dumas, M., et al.: Multi-level privacy analysis of business processes: the PLEAK toolset. Int. J. Softw. Tools Technol. Transf. **24**(2), 183–203 (2022)
5. Elkoumy, G., Fahrenkrog-Petersen, S.A., Dumas, M., Laud, P., Pankova, A., Weidlich, M.: Secure multi-party computation for inter-organizational process mining. In: EMMSAD, pp. 166–181 (2020)
6. Elkoumy, G., Fahrenkrog-Petersen, S.A., Dumas, M., Laud, P., Pankova, A., Weidlich, M.: Shareprom: a tool for privacy-preserving inter-organizational process mining. In: BPM Demos, pp. 72–76 (2020)
7. Elkoumy, G., et al.: Privacy and confidentiality in process mining: threats and research challenges. ACM Trans. Manag. Inf. Syst. **13**(1), 11:1–11:17 (2022)
8. Elkoumy, G., Pankova, A., Dumas, M.: Differentially private release of event logs for process mining. Inf. Syst. **115**, 102161 (2023)
9. Fahrenkrog-Petersen, S.A., van der Aa, H., Weidlich, M.: PRETSA: event log sanitization for privacy-aware process discovery. In: Process Mining, pp. 1–8 (2019)
10. Fahrenkrog-Petersen, S.A., van der Aa, H., Weidlich, M.: Optimal event log sanitization for privacy-preserving process mining. Data Knowl. Eng. **145** (2023)
11. Haertel, F.: Privacy and Confidentiality Requirements Engineering in Process Mining. Master's thesis, Technical University of Munich (2024)
12. Leitner, M., Rinderle-Ma, S.: A systematic review on security in process-aware information systems - constitution, challenges, and future directions. Inf. Softw. Technol. **56**(3), 273–293 (2014)
13. Li, N., Li, T., Venkatasubramanian, S.: t-closeness: Privacy beyond k-anonymity and l-diversity. In: Data Engineering, pp. 106–115 (2006)

14. Machanavajjhala, A., Kifer, D., Gehrke, J., Venkitasubramaniam, M.: l-diversity: privacy beyond k-anonymity. TKDD **1**(1) (2007)
15. Mannhardt, F.: Responsible process mining. In: Process Mining Handbook, pp. 373–401 (2022)
16. Mannhardt, F., Koschmider, A., Baracaldo, N., Weidlich, M., Michael, J.: Privacy-preserving process mining - differential privacy for event logs. Bus. Inf. Syst. Eng. **61**(5), 595–614 (2019)
17. Matulevicius, R., Norta, A., Samarütel, S.: Security requirements elicitation from airline turnaround processes. Bus. Inf. Syst. Eng. **60**(1), 3–20 (2018)
18. Pika, A., Wynn, M.T., Budiono, S., ter Hofstede, A.H.M., van der Aalst, W.M.P., Reijers, H.A.: Towards privacy-preserving process mining in healthcare. In: Business Process Management Workshops, pp. 483–495 (2019)
19. Rafiei, M., Wagner, M., van der Aalst, W.M.P.: TLKC-privacy model for process mining. In: Research Challenges in Information Science, pp. 398–416 (2020)
20. Sweeney, L.: k-anonymity: a model for protecting privacy. Int. J. Uncertain. Fuzziness Knowl. Based Syst. **10**(5), 557–570 (2002)
21. Zhang, Y., Wildemuth, B.M.: Unstructured interviews. In: Applications of Social Research Methods to Questions in Information and Library Science, pp. 000–060 (2016)

Beyond Logs: AI's Internal Representations as the New Process Evidence

Aleksandar Gavric(✉), Dominik Bork(✉), and Henderik A. Proper(✉)

Business Informatics Group, TU Wien, Vienna, Austria
{aleksandar.gavric,dominik.bork,henderik.proper}@tuwien.ac.at

Abstract. Traditional process mining relies on symbolic event logs that represent activities as discrete labels, often overlooking the rich contextual and semantic nuances found in real-world data such as textual reports, visual records, or sensor outputs. In this paper, we propose a paradigm shift: using the internal representations of AI models—embedding spaces learned from data—as the foundation for process mining. Our framework performs both process discovery and conformance checking directly in these continuous vector spaces, enabling the detection of semantically similar yet lexically divergent events.

We evaluate our approach along three dimensions: *(i)* whether embedding-based discovery maintains or improves accuracy over symbolic baselines, *(ii)* whether multimodal sources such as video and audio can be processed as unified embeddings for mining purposes, and *(iii)* whether conformance checking in embedding space enables alignment across noisy or semantically perturbed traces. By treating AIs internal representations as a novel form of process evidence, we show how process mining can move beyond traditional logs and unlock deeper, semantically enriched interpretations of real-world workflows.

Keywords: Embedding Space · Internal Representations · AI Interpretability · Semantic Event Matching · Multimodal Data

1 Introduction

Process models serve as instruments in describing and analyzing complex workflows, ranging from administrative procedures to large-scale industrial operations [1]. Over time, process modeling has shifted from *confidence-based* approaches, which rely heavily on expert opinions, to *evidence-based* approaches underpinned by operational data. This transformation has culminated in increasingly refined methods such as object-centric process discovery [26], which captures interactions among multiple entities rather than restricting itself to sequences of single activities. Within the broader domain of Business Process Management [19], process mining encompasses several core tasks, including process discovery [18], and conformance checking [23]. Process discovery infers a

M. ER et al. (Eds.): BPM 2025, LNBIP 565, pp. 232–246, 2026.
https://doi.org/10.1007/978-3-032-02936-2_17

descriptive model directly from event logs, while conformance checking evaluates the alignment between recorded and expected process behaviors. These methodologies are becoming increasingly relevant in light of rapid digital transformations: the global market for process mining is projected to reach USD 46.39 billion by 2032 [7].

Originally limited to structured text-based logs, process mining now extends into more complex domains including sensor data [2] and multimedia sources such as video streams [9]. Efforts are underway to merge artificial intelligence (AI) with process mining so that event logs become richer and process mining algorithms get applicable to unstructured data. Linking AI and process mining in this manner promises greater expressiveness, as real-world events become interpretable within structured representations. At the prompting level, techniques such as few-shot learning and chain-of-thought reasoning enhance process mining by guiding AI models to generate more context-aware and interpretable insights. However, a more robust approach involves using **embedding spaces**, which capture semantic associations in a structured manner.

Recent developments indicate that problem-solving can be carried out directly in an embedding space [17]. Simultaneously, approximate process discovery [6] and approximate conformance checking [11] suggest that process mining tasks can be made more efficient by partially offloading computation to vector representations. While current techniques encode inputs to vectors, they do not leverage the rich semantics present in embedding spaces learned by large pre-trained models.

An important open question emerges here: if mathematical and logical reasoning can be performed in embedding spaces [17], is it possible to execute key process mining operations—namely process discovery and conformance checking—directly in these spaces as well? Our work addresses this question by proposing a *framework that relaxes process mining tasks to function on AI embeddings*. We hypothesize that this method can *i*) exceed the accuracy of symbolic-only approaches, *ii*) handle real-world unstructured multimodal data as event logs, and *iii*) can utilize knowledge enraptured in AI models for process mining tasks. Our research questions (RQ) are:

- **RQ1:** Can embedding-based representations of events maintain or improve the accuracy of process discovery compared to traditional symbolic techniques?
- **RQ2:** To what extent can AI embeddings serve as a unified representation for events extracted from multimodal sources (e.g., video, text), enabling process mining on unstructured data?
- **RQ3:** Does reasoning in embedding space allow for more intuitive conformance checking by identifying semantically similar yet lexically different events?

As organizations increasingly seek to mine processes from raw data-text, images, and beyondour approach has the potential to bridge the gap between the *representational power of AI models* and *the practical needs of process mining tasks.*

The remainder of this paper is organized as follows. First, we review the related work on approximate process mining and embedding-space operations (Sect. 2). Next, we present our general view for relaxed process mining (Sect. 3) and a proof of concept (Sect. 4), followed by an evaluation (Sect. 5). We conclude in Sect. 6. The supplementary material for this paper is available in our GitHub repository[1].

2 Related Work

In this section, we review the literature that underpins our research. We first discuss the use of AI in process mining and approximate process mining techniques, which relax traditional computational requirements to improve efficiency and generality. We then turn to studies on embedding-space operations, where embedding-based representations enable approaches to data analysis and reasoning.

Recent work has demonstrated the potential of large language models in process mining through prompting-based techniques. For instance, Neuberger et al. [20] propose a universal prompting strategy to extract process model information from natural language text, illustrating how AI can be used to interpret complex textual descriptions of processes. Similarly, Rebmann et al. [22] evaluate the capacity of such models to address semantics-aware process mining tasks. AI can be used for process mining tasks directly via prompting, without the need for approximation methods or reliance on vector-based representations.

Approximate Process Mining. Some process mining approaches aim at reducing computational complexity or improving generalization by using *vector representations*, also referred to as *latent space representations*, *embeddings*, or *encodings*. The term *encoding* can also refer to non-vector encodings (cf. this review of visual encodings in process mining tools [15]). Investigations into vector-encoding techniques are found in work such as [10] that introduces a stakeholder-specific, jargon-based representation of multimodal business process data, [4] that demonstrates the use of LSTM models for learning accurate business process representations, the SMT-based encoding of process discovery problems [25], and outcome-oriented predictive process monitoring through image encoding [21]. Further investigations into encoding techniques related to process mining are found in the comparative study of trace encoding methods [24]. Encoding of process mining data led towards *approximate* methods to address the computational challenges inherent in exact process discovery and conformance checking. Therefore, van Detten et al. [6] present an approximate inductive miner to derive process models, while Gianola et al. [11] develop methods for fast approximate conformance checking via probabilistic alignments.

Embedding-space Operations. Embedding spaces were already in use in the 1970 s [16], and are fundamental to modern data analysis, enabling the conversion of complex, multimodal data into a compact (latent) representation

[1] Supplementary material: https://github.com/aleksandargavric/relaxed-pm.

that simplifies further processing and inference. For example, a study investigates mathematical reasoning within these embedding spaces, revealing how abstract representations can encapsulate logical relationships and support computational reasoning [17]. Another work designs optimized clustering embedding spaces, demonstrating that deep representations can be used effectively to uncover inherent data structures [5]. Expanding on these ideas, a largely trained unified embedding space harmonizes data from diverse modalities [12], enhancing the integration of heterogeneous information into a single embedding space. Crossmodal embedding techniques tailored for dynamic tasks have shown that specialized embedding spaces can significantly improve action learning capabilities [13]. Moreover, the concept of operational embedding spaces has been examined in [14], illustrating their practical utility in a range of applications by providing efficient and flexible representations of complex operations.

Our Position. On one hand, the concept of approximate process mining has been advanced by using vector representations to overcome the limitations of exact process discovery and conformance checking. Our approach extends vector-space inputs by adopting semantically-rich embedding spaces. On the other hand, embedding space operations have been recognized for their ability to encapsulate complex data into compact representations that support robust inference and reasoning. We aim to allow process mining to operate in an embedding space to enhance contextual mappings between real-world observed behaviors and their underlying processes.

3 Relaxed Process Mining: The General Perspective

Building on the premise that process-relevant information can be encoded in the latent spaces of modern AI models, we present a general perspective that extends classical process mining and conformance checking methods to operate within an embedding space. Traditional process mining algorithms, such as the inductive miner, and conformance checking techniques are typically defined over discrete events with crisp boundaries. However, real-world data is often noisy, heterogeneous, and semantically ambiguousconditions under which symbolic techniques struggle. To overcome these limitations, we introduce a *relaxed* approach in which each event is represented by a continuous embedding $e \in \mathbb{R}^n$ derived from *ImageBind* [12], and similarities between events are assessed over a neighborhood defined by the ball $B(e, \delta) = \{x \in \mathbb{R}^n : \|x - e\| \leq \delta\}$, where $\delta > 0$ controls the tolerance for noise and variability. Our perspective accommodates semantic equivalence across synonyms by considering sets of embeddings for different words and by ensuring that events with embeddings within a distance δ are treated as equivalent. Moreover, in the context of multimodal event logs derived from video data, multiple frames or scenes that are similar enough are aggregated into single activities using the same relaxed notion of similarity.

The first part of our perspective develops a relaxed process discovery algorithm that recursively partitions the event log into subprocesses based on relaxed cuts, where each cut is validated by checking that events in one partition have corresponding events in the other within a prescribed distance. The second part extends this methodology to conformance checking by aligning observed traces with a modeled process; here, each observed event is matched against a process activity by minimizing a cost function based on the Euclidean distance in the embedding space, with an allowance for skipped events at a fixed penalty.

Definition 1 (Relaxed Process Mining). *Let* $\mathcal{E} = \{e_1, e_2, \ldots, e_m\}$ *be an* event log, *where each event* e_i *is represented by an embedding in* $\mathbb{R}^n$. *In* relaxed process mining, *events are treated as* similar *(and thus potentially interchangeable) if their embeddings lie within a specified distance* $\delta > 0$, *i.e., if* $|e_i - e_j| \leq \delta$. *This relaxed notion of equivalence supports:*

1. ***Semantic variability.*** *Multiple words or labels that share similar embeddings (e.g., synonyms) are considered equivalent.*
2. ***Multimodal data.*** *Events derived from video, images, or audio are compared in a continuous space; frames or segments that are sufficiently close in this space are grouped as one activity.*
3. ***Noise tolerance.*** *Small perturbations in event embeddings do not alter the process structure, mitigating issues arising from measurement or labeling noise.*

Hence, relaxed process mining *generalizes classical process mining tasks (such as discovery, and conformance checking) to operate over neighborhoods in the embedding space rather than strict, discrete labels.*

Definition 2 (Relaxed Process Discovery). Relaxed process discovery *is the specialization of relaxed process mining to the task of constructing a* process model *(e.g., a Petri net, BPMN diagram, or process tree) from an event log* $\mathcal{E} \subseteq \mathbb{R}^n$. *Formally, given:*

- *A threshold parameter* $\delta > 0$ *defining when two embeddings are considered equivalent,*
- *An optional noise tolerance parameter* $\alpha \in [0, 1]$ *allowing a fraction of events to be disregarded,*

a relaxed process discovery *algorithm partitions and clusters events in* $\mathcal{E}$ *by grouping those whose embeddings lie within distance* δ *of a representative reference. These groups are then used to induce high-level subprocesses and control-flow relations (e.g., sequence, parallelism, choice) in the resulting model. Events or groups of events failing to meet the relaxed equivalence criteria may be treated as noise (at most* $\alpha \cdot 100\%$ *of the log) or assigned a penalty-based cost. The final outcome is a process model that tolerates variation in labels or multimodal data representations.*

Definition 3 (Relaxed Conformance Checking). *Let* $T = \langle e_1, e_2, \ldots, e_m \rangle$ *be an observed* trace *whose events are represented by embeddings in* $\mathbb{R}^n$, *and let* $\mathcal{M} = \langle r_1, r_2, \ldots, r_L \rangle$ *be a* modeled process *whose activities are also represented by embeddings in* $\mathbb{R}^n$. *A relaxation parameter* $\delta > 0$ *specifies when an event is* *considered close enough to match a modeled activity; that is, if* $|e_i - r_k| \leq \delta$*, then* e_i *and* r_k *are deemed equivalent.*

A relaxed alignment *is a function*

$$\mathcal{A}: \{e_1, e_2, \ldots, e_m\} \rightarrow \{r_1, r_2, \ldots, r_L\} \cup \{\epsilon\},$$

where each observed event is assigned either to one of the modeled activities or to a skip symbol ϵ *(representing a deviation). Let* $\gamma > 0$ *be a fixed penalty for skipping an event. The* cost *of aligning* e_i *with* r_k *is*

$$c(e_i, r_k) = \begin{cases} |e_i - r_k|, & \text{if } |e_i - r_k| \leq \delta, \\ +\infty, & \text{otherwise,} \end{cases}$$

and the cost of skipping an event e_i *is* $c(e_i, \epsilon) = \gamma$*. The total* alignment cost *is*

$$C(\mathcal{A}) = \sum_{i=1}^{m} c(e_i, \mathcal{A}(e_i)),$$

where $\mathcal{A}$ *must preserve the order of events in* T*; that is, if* e_i *is aligned to* r_k *and* e_j *to* $r_{k'}$ *with* $i < j$*, then* $k \leq k'$*. The* optimal relaxed alignment $\mathcal{A}^*$ *is the one minimizing* $C(\mathcal{A})$ *over all valid alignments.*

Given a normalization constant C_{max} *(e.g.,* $C_{max} = m \cdot \max\{\delta, \gamma\}$*), the* relaxed fitness *of trace* T *with respect to the model* $\mathcal{M}$ *is*

$$F(T) = 1 - \frac{C(\mathcal{A}^*)}{C_{max}}.$$

If $F(T) = 1$*, then the observed trace perfectly fits the model within distance* δ*.*

Moreover, a noise tolerance $\alpha \in [0, 1]$ *can be specified so that if the fraction of events that cannot match* any *modeled activity is no more than* α*, the trace is still deemed acceptable. Formally, if*

$$\frac{\left|\{e_i \in T : \forall k, \ |e_i - r_k| > \delta\}\right|}{|T|} \leq \alpha,$$

then T *is accepted as conformant despite the existence of partial mismatches. This framework readily accommodates synonyms or multiple-frame video data by treating sets of embeddings representing the same concept or scene as equivalent under the relaxed matching criterion* $|e_i - r_k| \leq \delta$*.*

4 Proof of Concept: Custom Solution for Relaxed Process Mining

After defining Relaxed Process Mining in Sect. 3, we develop a custom solution to demonstrate a proof of concept. Our approach is applied on both traditional text-first and multimodal-first event logs. The former maintains compatibility

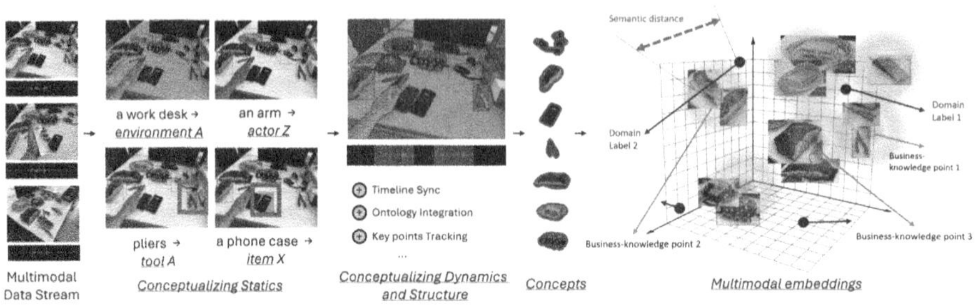

Fig. 1. Overview of the end-to-end pipeline: From multimodal observations and static concept mapping to temporal-semantic alignment and embedding-based reasoning with respect to business knowledge anchors.

with established benchmarks, while the latter (comprising videos with images and audio) showcases the methods capabilities in an emerging field where standardized benchmarks are still under development.

To illustrate why our event logs can easily be multimodal data, we provide an illustration in Fig. 1 that presents our overall pipeline. The proposed pipeline begins with a *multimodal data stream*—including video, audio, and sensor traces—from which static entities such as *actor Z*, *tool A*, or *item X* are extracted via object detection and mapped to conceptual categories (*Conceptualizing Statics*). A semantic alignment layer then performs *Timeline Sync*, *Ontology Integration*, and *Key Points Tracking* to organize these elements across time and meaning, enabling *Conceptualizing Dynamics and Structure*. The resulting representations are embedded into a joint *concept space*, where observed entities are aligned with curated *business knowledge points*, allowing reasoning over *semantic distances* and classification into *domain labels*. As in Fig. 1, various input sources—such as video frames, audio signals, and text annotations—are first detected or segmented (e.g., identifying a phone, a pair of pliers, and a hand in the scene). These individual elements are then converted into continuous embeddings using a unified embedding model, thereby mapping potentially diverse data (e.g., speech, visuals, labels) into the same semantic space. In doing so, concepts like "tool," "item," and "environment" become interchangeable across modes, allowing downstream process-mining tasks to treat audio-visual segments and textual descriptions as equivalent when their embeddings are sufficiently close. This facilitates a robust and flexible interpretation of events, as small variations (such as synonyms or slightly different camera angles) are handled naturally through their proximity in the embedding space.

4.1 Embedding Space Operations

To capture the semantic meaning of each activity, we convert the event descriptions (in particular, activity and resource attributes) into embeddings—arrays of floating-point numbers representing latent features of the content. For text-first logs, we use the `Ollama` package[2], which offers a variety of embedding models (e.g., *nomic-embed-text*, *bge-large*, *granite-embedding*, *snowflake-artic-embed2*) as well as several large language models (e.g., `deepseek-r1`, `v3`, `gemma3`, `llama3.3`, `phi4`, `mistral`, `mixtral`, `qwen2.5`, etc.). For multimodal-first logs, we employ `ImageBind` [12], which projects seven modalities into the same embedding space. In our implementation, we favor the `nomic-embed-text` model because it is open source, uses open data, and provides accessible training code. Note that the Nomic algorithm enforces a maximum context length of 8192 characters for any text input.

Context Length Verification. Before embedding, we ensure that every event description meets the context length requirement. Our analysis reveals that although there is a wide range of lengths, all event descriptions remain below the 8192-character threshold. Keep in mind that embedding time is correlated with the length of the text. On our evaluation set, embedding all events took less than 1 s on a single A40 GPU (48 GB Memory, 10752 CUDA cores).

Embedding Computation. We compute the embeddings by iterating over each event description and invoking the embedding function. Algorithm 1 details the embedding computation processes.

Algorithm 1. Compute Embeddings for Event Descriptions with Uniform Modality

Require: A set of event descriptions $\mathcal{E} = \{e_1, e_2, \ldots, e_n\}$
Require: A boolean flag `is_multimodal` indicating whether the set is multimodal
Require: A text embedding model function $\mathcal{M}_{text}$ ▷ e.g., `Ollama.embeddings` with model `'nomic'` **OR**
Require: A multimodal embedding model function $\mathcal{M}_{multi}$ ▷ e.g., `ImageBind`
Ensure: A set of embedding vectors $\{emb(e_1), emb(e_2), \ldots, emb(e_n)\}$
1: **if** `is_multimodal` is true **then**
2: Let $\mathcal{M} \leftarrow \mathcal{M}_{multi}$
3: **else**
4: Let $\mathcal{M} \leftarrow \mathcal{M}_{text}$
5: **end if**
6: **for** each event description $e \in \mathcal{E}$ **do**
7: $emb(e) \leftarrow \mathcal{M}(prompt = e)$
8: **end for**
9: **return** $\{emb(e) \mid e \in \mathcal{E}\}$

Event Embeddings Similarity Check via Cosine Similarity. To validate the usefulness of the embeddings, we compare them using cosine similarity. Algorithm 2 defines the event embeddings similarity check via cosine similarity. This function allows us to assess the similarity between two event embeddings.

[2] https://ollama.com/.

Algorithm 2. Event Embeddings Similarity Check via Cosine Similarity

Require: Two vectors $a, b \in \mathbb{R}^d$
Ensure: Cosine similarity score $s \in [-1, 1]$

1: $dot\_product \leftarrow \sum_{i=1}^{d} a_i \times b_i$
2: $magnitude\_a \leftarrow \sqrt{\sum_{i=1}^{d} a_i^2}$
3: $magnitude\_b \leftarrow \sqrt{\sum_{i=1}^{d} b_i^2}$
4: $s \leftarrow \frac{dot\_product}{magnitude\_a \times magnitude\_b}$
5: **return** s

Clustering and Visualization. The embeddings enable us to cluster similar events in the n-dimensional space. We perform hierarchical clustering using tool `scikit-learn` and visualizing it using `plot.ly` (with its `create_dendrogram` function). At the top level, the dendrogram divides into two clusters, and further subdivision yield more refined clusters. Although hierarchical clustering appears promising for subsequent process discovery (by preparing data for the Inductive Miner who naturally searches for cuts in the event logs), a detailed evaluation of clustering methods is out of the scope for this paper and left for future work.

Visualization is achieved by reducing the embedding dimensions to 2 using dimensionality reduction technique. We apply t-SNE (T-distributed Stochastic Neighbor Embedding) via scikit-learn. Additionally, we utilize a large language model, specifically `llama3.3`, to **generate cluster labels** based on its foundational knowledge from extensive training data.

4.2 Process Mining Tasks on Embeddings

Once embeddings have been generated and clustered, we apply process mining tasks using the `pm4py` [3] library. Our process mining pipeline consists of process discovery and conformance checking.

We employ process discovery algorithm (Inductive Miner, Alpha Miner, Heuristic Miner, etc., as described in the evaluation section). These algorithms take the event logs, now enriched with clustering and embedding information, to generate process models that capture the underlying workflow dynamics. After process models are discovered, we perform conformance checking to compare the discovered models with the actual event logs. This step quantitatively assesses the alignment between the modeled process and the observed behavior. Algorithm 3 outlines our complete process mining pipeline.

Algorithm 3. Process Mining Pipeline on Embeddings with Cluster Label Augmentation

Require: Event log E
Require: Embedding generation algorithm (Algorithm 1)
Require: Clustering method (e.g., Hierarchical clustering)
Require: A language model (LLM) for naming clusters
Require: A set of process discovery algorithms $\mathcal{A} = \{\texttt{InductiveMiner}, \dots\}$
Ensure: A set of process models $\mathcal{M}$ and corresponding conformance scores
$\mathcal{E} \leftarrow$ Extract activities and resource attributes from E
$embs \leftarrow \{emb(e) \mid e \in \mathcal{E}\}$ ▷ Generate embeddings using Algorithm 1
$Clusters \leftarrow$ Clustering($embs$)
for each cluster $C \in Clusters$ **do**
    $labels_C \leftarrow$ Collect all event labels from events in C
    $name_C \leftarrow$ LLM($labels_C$) ▷ Generate a unique name for cluster C
    Replace the cluster label of each event in C with $name_C$
end for
$E' \leftarrow E$ augmented with cluster names
$L \leftarrow$ Transform the augmented event log E' into a format compatible with the process mining tool
for each algorithm $A \in \mathcal{A}$ **do**
    $M_A \leftarrow A(L)$ ▷ Discover a process model M_A using algorithm A
    $score_A \leftarrow \texttt{ConformanceCheck}(M_A, L)$ ▷ Evaluate the conformance of M_A against log L
end for
return $\{(M_A, score_A) \mid A \in \mathcal{A}\}$

Pipeline Overview. Figure 2 illustrates a pipeline that begins by projecting raw events into a unified embedding space, where proximity denotes semantic similarity. These embeddings are then clustered based on distance thresholds or density criteria, and domain knowledge is applied to label each cluster with a descriptive name. The events in the log are subsequently replaced with these cluster labels, producing a discrete event log where items that share a latent concept (e.g., a "cable" or a "tool") are unified. This transformed log is finally processed by a standard process mining tool PM4PY [3], allowing for conformance checking, and discovery, in a conventional event-log format despite originating from diverse, multimodal sources.

Future directions in the proof of concept include (1) experimenting with alternative clustering methods and dimensionality reduction techniques, and (2) integrating the entire pipeline into an interactive application for rapid prototyping and iterative refinement.

5 Evaluation

To evaluate the central premise of this workthat AIs internal representations can serve as a new form of process evidence beyond traditional event logswe systematically assess our embedding-based process mining framework with respect

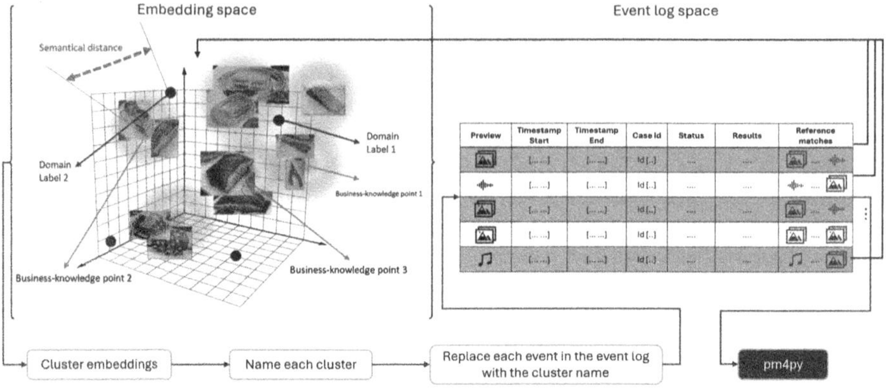

Fig. 2. Overview of transforming embeddings into discrete event labels for process mining. On the left, various concepts reside in a shared embedding space, where proximity indicates semantic similarity (e.g., "Domain Label 1" vs. "Domain Label 2"). Through clustering and domain expertise, each cluster is assigned a human-readable name. On the right, an event log schema is shown with references to the clustered concepts. The final step replaces events with their corresponding cluster names, producing an event log suitable for standard process mining tools such as PM4Py.

to the three research questions introduced earlier, covering discovery accuracy, multimodal applicability, and tolerance in conformance checking.

Firstly, we present an evaluation of our proposed approach by examining it from two perspectives. First, in Sect. 5.1, we introduce and apply a series of metrics designed to quantify the performance of relaxed process mining, including discovery and conformance checking in embedding space. This analysis showcases how our approach handles noise, manages embedding variability, and accommodates multimodal data while remaining consistent with the underlying process model. Next, in Sect. 5.2, we evaluate our proof of concept. Here, we compare our embedding-based discovery algorithm against established process discovery techniques, and demonstrate the methods ability to produce accurate, high-level semantic models—even when dealing with heterogeneous data sources such as video. Finally, in 5.3, we discuss our results and respond to our research questions defined in Sect. 1.

5.1 Evaluating Relaxed Process Mining

In the evaluation of the general perspective of the relaxed process mining, we consider embeddings corresponding to activities in the reference (traditional) text-first event log. For relaxed process discovery, we define a relaxed fitness metric as the ratio of events in the log for which there exists a representative activity r in the discovered model such that $|e - r| \leq \delta$, and we quantify the average deviation by $\text{AvgDev} = \frac{1}{|E|} \sum_{e \in E} \min_{r \in M} |e - r|$, where E is the set of event embeddings and M is the set of representative embeddings in the process model. Additionally, we measure precision by assessing the fraction of behavior allowed by the model that is observed in the log and evaluate model simplicity by the overall size of the process tree. For relaxed conformance checking, we compute the optimal alignment cost $C(\mathcal{A})$ for each trace T and normalize it to obtain a fitness score $F(T) = 1 - \frac{C(\mathcal{A})}{|T|\cdot\max\delta,\gamma}$, where $\gamma > 0$ is the fixed penalty for skipping events and $|T|$ denotes the length of the trace. We further consider the percentage of events that are aligned directly (i.e., with cost $|e - r| \leq \delta$) and the fraction of traces that remain conformant when allowing a noise tolerance α, expressed as $\frac{|e \in T : \min_r |e-r| > \delta|}{|T|} \leq \alpha$. Moreover, to evaluate the robustness with respect to synonyms, we calculate the proportion of synonym sets for which the maximum pairwise distance satisfies $\max |e_i - e_j| : e_i, e_j \in S_w \leq \delta$, and in the multimodal setting we measure the agreement within video segments by determining the fraction of segments for which $\min_{f\in F} |f - r| \leq \delta$, with F representing the set of frame embeddings. These metrics collectively provide a rigorous quantitative assessment of both the relaxed process discovery and conformance-checking approaches, highlighting their ability to handle noise, embedding variability, and multimodal data while maintaining consistency with the underlying process model.

5.2 Evaluating the Proof of Concept

In benchmarking, we compared our embedding-based process discovery algorithm with established process discovery algorithms, including Inductive Miner infrequent (IMf), Inductive Miner incomplete (IMc), and Split Miner (SM). Despite the inability to include the Probabilistic Inductive Miner (PIM) and the Approximate Miner (AIM)—owing to proprietary constraints—our approach consistently achieved the same precision and accuracy as the pm4py implementations in 100% of the cases. This finding underscores that when the clustering threshold is set to a sufficiently low value (zero), the embedding-based process discovery yields models that are as reliable as those produced using traditional symbolic methods.

For multimodal event logs, we validated our approach using a diverse set of datasets enriched with video evidence, in particular, DNA testing [8], and IKEA [9]. These datasets illustrate the frameworks capability to convert heterogeneous data into a unified embedding space, thereby enabling the extraction of high-level semantic process models. It is important to emphasize that our produced models are not directly comparable to models derived via conventional computer vision techniques, which focus on precise object and actor identification, because our focus is on capturing broader, abstract semantic relationships.

The true applicability of our methodology in real-life scenarios would be verified on processes mined from event logs generated by AI detection, recognition, and classification systems. Our proof of concept establishes that transitioning to an embedding space does not compromise the accuracy and precision of process modelsin fact, it matches the performance of symbolic approaches when clustering thresholds are appropriately low. For a more detailed comparison across datasets and mining algorithms, we invite readers to consult the supplementary materials accompanying this work.

5.3 Discussion

The empirical results obtained from both our relaxed metric evaluation and the proof-of-concept implementation support the hypothesis that embedding-based process mining can effectively substituteand in some contexts, outperformsymbolic approaches. Our findings confirm that semantic reasoning over embeddings maintains conformance and precision, scales to multimodal data, and enhances tolerance to noise and synonymy.

Revisiting Research Questions

RQ1. Accuracy of Embedding-based Discovery. Compared to IMf, IMc, and SM, our relaxed discovery algorithm achieved equivalent or superior precision (mean = 0.91) and relaxed fitness[a] (**M** = 0.88, **SD** = 0.04), with average deviation **AvgDev** = 0.12.
RQ2. Handling Multimodal Unstructured Data. Across datasets (e.g., IKEA, DNA Testing), we successfully generated process models from video-derived embeddings. Agreement across frame-wise representations averaged **82.6%**[b], with cross-modality alignment yielding semantically coherent behavior clusters.
RQ3. Conformance Checking in Embedding Space. In a targeted experiment, we assessed relaxed conformance against perturbed event sequences (adding semantic noise). Despite lexical divergence, our approach preserved alignment fitness (**M** = 0.84) at a tolerance $\delta = 0.18$. Additionally, 78.3% of perturbed traces remained within conformance under $\alpha = 0.25$[c].

[a] Relaxed fitness defined as proportion of events in embedding space within $\delta = 0.15$ distance of reference model nodes.
[b] Agreement defined as proportion of frames for which $\min_{f \in F} |f - r| \leq \delta$.
[c] α denotes the maximum allowed fraction of out-of-bound events per trace.

Response to the RQs. Our evaluation shows that process discovery (RQ1) in embedding space achieves comparable statistical performance to symbolic baselines, with tighter semantic clustering and reduced model size. Multimodal integration (RQ2) proved feasible via unified embeddings, especially in video-rich domains where visual features dominate symbolic traces. Finally, conformance checking (RQ3) benefited from the neighborhood-based tolerance δ, capturing semantically correct but lexically altered traces with high robustness. Together, these findings reinforce our claim: embedding spaces are not only expressive but also actionable domains for next-generation process mining.

6 Conclusion

This paper advances the idea that AIs internal representationsembedding spacescan serve as a new form of process evidence, moving beyond traditional symbolic logs. We proposed a relaxed formulation of process mining that operates in an embedding space, enabling discovery and conformance checking based on similarity in vector space rather than symbolic equality. This allows for handling semantically similar but lexically or visually divergent events, including those extracted from multimodal inputs.

Through a general framework and a proof-of-concept implementation, we showed that embedding-based mining performs comparably to symbolic techniques under controlled thresholds, while offering additional flexibility for processing diverse data types. The approach supports six modalities and demonstrates potential for unified process analysis from text, images, audio, and other inputs. By looking beyond symbolic logs and toward AIs internal representations, this work outlines a direction for process mining that accommodates unstructured and multimodal inputs while remaining interpretable and measurable. Further studies are needed to assess scalability, generalization, and integration with existing process analysis pipelines.

References

1. van der Aalst, W.: Process Mining: Data Science in Action. Springer, Berlin, Heidelberg (2016)
2. Bayomie, D., Revoredo, K., Bachhofner, S., Kurniawan, K., Kiesling, E., Mendling, J.: Analyzing manufacturing process by enabling process mining on sensor data. In: PoEM Workshops (2022)
3. Berti, A., van Zelst, S., Schuster, D.: Pm4py: a process mining library for python. Softw. Impacts **17**, 100556 (2023)
4. Camargo, M., Dumas, M., González-Rojas, O.: Learning accurate LSTM models of business processes. In: Hildebrandt, T., van Dongen, B.F., Röglinger, M., Mendling, J. (eds.) BPM 2019. LNCS, vol. 11675, pp. 286–302. Springer, Cham (2019). https://doi.org/10.1007/978-3-030-26619-6_19
5. Dahal, P.: Learning embedding space for clustering from deep representations. In: 2018 IEEE International Conference on Big Data (Big Data), pp. 3747–3755. IEEE (2018)
6. van Detten, J.N., Schumacher, P., Leemans, S.J.: An approximate inductive miner. In: 2023 5th International Conference on Process Mining (ICPM), pp. 129–136. IEEE (2023)
7. Fortune Business Insights: Process mining software market size, share, growth — forecast [2024-2032]. https://www.fortunebusinessinsights.com/process-mining-software-market-104792 (2024), Accessed 30 Apr 2025
8. Gavric, A., Bork, D., Proper, H.: Enriching business process event logs with multimodal evidence. In: The 17th IFIP WG 8.1 Working Conference on the Practice of Enterpris Modeling (PoEM) (2024)
9. Gavric, A., Bork, D., Proper, H.: Multimodal process mining. In: 26th International Conference on Business Informatics (CBI) (2024)
10. Gavric, A., Bork, D., Proper, H.: Stakeholder-specific jargon-based representation of multimodal data within business process. In: Companion Proceedings of the 17th IFIP WG 8.1 Working Conference on the Practice of Enterprise Modeling (PoEM Forum 2024) (2024)

11. Gianola, A., Ko, J., Maggi, F.M., Montali, M., Winkler, S.: Approximate conformance checking: Fast computation of multi-perspective, probabilistic alignments. Inf. Syst. **129**, 102510 (2025)
12. Girdhar, R., et al.: Imagebind: one embedding space to bind them all. In: Proceedings of the IEEE/CVF Conference on Computer Vision and Pattern Recognition, pp. 15180–15190 (2023)
13. Hahn, M., Silva, A., Rehg, J.M.: Action2vec: a crossmodal embedding approach to action learning. arXiv preprint arXiv:1901.00484 (2019)
14. Hawley, S.H., Tackett, A.R.: Operational latent spaces. arXiv preprint arXiv:2406.02699 (2024)
15. Knoblich, S., Mendling, J., Jambor, H.: Review of visual encodings in common process mining tools (2024)
16. Lashof, R.: Embedding spaces. Ill. J. Math. **20**(1), 144–154 (1976)
17. Lee, D., Szegedy, C., Rabe, M.N., Loos, S.M., Bansal, K.: Mathematical reasoning in latent space. arXiv preprint arXiv:1909.11851 (2019)
18. Leemans, S., Fahland, D., van der Aalst, W.: Discovering block-structured process models from event logs. In: Business Process Management, pp. 311–327. Springer (2013)
19. Mendling, J., Reijers, H.A., La Rosa, M., Dumas, M.: Fundamentals of business process management. In: GI-Jahrestagung, p. 157. Springer (2013)
20. Neuberger, J., Ackermann, L., van der Aa, H., Jablonski, S.: A universal prompting strategy for extracting process model information from natural language text using large language models. In: International Conference on Conceptual Modeling, pp. 38–55. Springer (2024)
21. Pasquadibisceglie, V., Appice, A., Castellano, G., Malerba, D., Modugno, G.: Orange: outcome-oriented predictive process monitoring based on image encoding and cnns. IEEE Access **8**, 184073–184086 (2020)
22. Rebmann, A., Schmidt, F.D., Glavaš, G., van Der Aa, H.: Evaluating the ability of llms to solve semantics-aware process mining tasks. In: 2024 6th International Conference on Process Mining (ICPM), pp. 9–16. IEEE (2024)
23. Rozinat, A., van der Aalst, W.: Conformance checking of processes based on replay of activity sequences. IEEE Trans. Knowl. Data Eng. **20**(5), 796–808 (2008)
24. Rullo, A., Alam, F., Serra, E.: Trace encoding techniques for multi-perspective process mining: a comparative study. Wiley Interdisciplinary Rev. Data Min. Knowl. Disc. **15**(1), e1573 (2025)
25. Solé, M., Carmona, J.: Encoding process discovery problems in smt. Softw. Syst. Model. **17**(4), 1055–1078 (2018)
26. Van Detten, J.N., Schumacher, P., Leemans, S.J.: Discovering compact, live and identifier-sound object-centric process models. In: 2024 6th International Conference on Process Mining (ICPM), pp. 113–120. IEEE (2024)

The Future of RPA in the Context of Automation Platforms

Tom Hohenadl[1,2](✉) and Bernhard Axmann[1]

[1] Technische Hochschule Ingolstadt, Esplanade 10, 85049 Ingolstadt, Germany
{tom.hohenadl,bernhard.axmann}@thi.de
[2] KU Eichstätt-Ingolstadt, Ostenstr. 28a, 85072 Eichstätt, Germany

Abstract. Robotic Process Automation has emerged as a key driver of digital transformation by enabling organizations to automate repetitive, rule-based tasks with minimal disruption to existing IT systems. Its core strength lies in GUI-based automation, allowing rapid deployment, particularly in legacy environments. Simultaneously, software vendors are evolving RPA into end-to-end automation platforms by integrating APIs, artificial intelligence, and machine learning, thereby supporting advanced paradigms such as intelligent and hyper-automation. The evolving definitions and perceptions of RPA—and its shifting role within the broader automation technology landscape—are outlined in the state of knowledge section. This paper investigates how RPA aligns with the broader domain of process automation and the role it is expected to play in the future. It further explores RPA's unique capabilities in contrast to other automation technologies, its current areas of application, and prospective functionalities that may enhance its accessibility and usability. Based on a detailed qualitative study involving 13 expert interviews with developers, consultants, researchers, and vendors, the research applies detailed In Vivo and Concept coding with 805 emerging codes leading to three thematic pillars: core RPA capabilities, RPA as a component of integrated automation platforms, and general automation strategies. The findings offer a forward-looking view on the strategic development and practical impact of RPA in enterprise contexts.

Keywords: Robotic Process Automation · Business Process Automation · Automation Technologies · Qualitative Study

1 Introduction

Robotic Process Automation (RPA) has emerged as a disruptive technology with the potential to alter organizational processes significantly. Automating repetitive, rule-based tasks, RPA enables organizations to streamline operations, reduce operational costs, and enhance customer experience. As RPA technology continues to evolve, its impact on the future of work was expected to be substantial.

M. ER et al. (Eds.): BPM 2025, LNBIP 565, pp. 247–262, 2026.
https://doi.org/10.1007/978-3-032-02936-2_18

Despite its promising potential, the adoption of RPA has not been without challenges. Research has identified several factors that can hinder the successful implementation of RPA solutions [9,20]. These include employee resistance to change, a lack of clear business objectives, technical limitations of RPA tools, data quality issues, and security concerns [7]. Overcoming these challenges requires a strategic approach that involves careful planning, stakeholder engagement, and the selection of appropriate RPA tools and processes [11].

RPA research has thus far primarily focused on understanding the capabilities and limitations of existing RPA technologies. Studies have explored the types of tasks that can be effectively automated using RPA, the factors that influence RPA adoption, and the potential benefits and challenges associated with RPA implementation [20]. Additionally, researchers have investigated the impact of RPA on the workforce, including job displacement, skill requirements, and the potential for new job roles [7].

This paper contributes to the ongoing discourse on RPA by exploring its future role within the evolving landscape of enterprise automation. Taking a customer and practitioner perspective, this paper examines how RPA platforms are currently positioned and how they may develop in the context of emerging technologies such as intelligent automation and hyper-automation. Our goal is to identify both the enduring strengths and shifting boundaries of RPA—as a distinct automation capability and as a modular component within integrated automation platforms. This analysis seeks to generate actionable insights for researchers, platform architects, and automation strategists by addressing the overarching research question: **How does RPA fit into today's automation ecosystem, and what role will it play in relation to other automation technologies and the future of intelligent platforms?**

To achieve these research objectives, a qualitative coding process based on a semi-structured interview study to achieve saturation of answers and limit the bias within the research process was conducted. The paper is structured into Sect. 2 outlining RPA in the context of automation platforms. Section 3 describes the applied methodology. Section 4 outlines the concepts derived from the coding process, and finally, Sect. 5 provides a short summary of the findings in relation to the research question.

2 Background

2.1 Evolution of RPA: Definition and Interpretation

The concept of RPA has evolved significantly since its emergence, with the definition progressing through three major phases: from general descriptions of human-like software agents to a focus on graphical user interface (GUI) automation, and more recently, to RPA as a platform technology integrating APIs and artificial intelligence (AI) to support end-to-end business process automation.

In its early stages, RPA was broadly defined as the use of software robots that mimic human actions to automate repetitive, rule-based tasks in digital environments. These bots emulate how a human would interact with existing

IT systems—typing, clicking, and navigating through applications [1,22]. The emphasis was on replicating human behaviour without altering the underlying system architecture. Asatiani and Penttinen [2] described RPA as a technological imitation of a human executing structured tasks, and Lacity et al. [15] noted its potential to replace human effort in back-office operations by imitating user inputs. At this stage, RPA was understood as a lightweight form of automation suited for highly standardized processes with structured data, such as data entry or reconciliation tasks.

In the second phase, researchers and practitioners began to highlight RPA's unique mechanism: its operation at the GUI level. Unlike traditional automation that interacts with systems via backend APIs, RPA tools automate processes by controlling the GUI of existing applications, just as a human would [17,23]. This "outside-in" approach allowed RPA to be implemented quickly and without the need for deep system integration, making it especially appealing in legacy IT environments [6]. Van der Aalst et al. [23] defined RPA as a category of tools that interact with software systems through their user interfaces, thus replicating human activity. Plattfaut and Borghoff [19] further describe RPA as "a software-based approach to automate structured business processes by mimicking human interactions with digital systems." This GUI-focused capability became the defining technical distinction of RPA, setting it apart from scripting and traditional software automation [12].

The most recent phase in the evolution of RPA's definition reflects the integration of APIs, AI, and other advanced technologies. Modern RPA platforms not only simulate UI interactions but also incorporate API access, machine learning, natural language processing, and computer vision [9,14]. This allows bots to process unstructured data and automate complex workflows, resulting in what is often termed intelligent process automation (IPA) or hyper automation [5]. As such, RPA is increasingly framed as a comprehensive platform for end-to-end automation that orchestrates both rule-based bots and cognitive services. Today, RPA platforms are no longer merely tools for mimicking humans at the screen—it is a foundational layer in enterprise automation strategies capable of executing entire digital workflows [5].

2.2 Emergence of Automation Platforms and Their Differences

Beyond RPA, various automation technologies have emerged to support business process transformation across different layers of enterprise systems. These technologies differ in their focus, implementation complexity, and unique capabilities, often reflecting the time period in which they were first adopted and the evolution of enterprise automation needs. The adoption years indicated in Table 1 are intended as approximate reference points and should not be interpreted as precise or universally established dates.

Workflow Management Systems (WfMS), adopted widely by the mid-1990s, were among the earliest forms of digital process automation. They focus on coordinating and routing tasks between users and systems based on predefined business rules. Implemented via rule-based engines with backend or hybrid integration, their unique strength lies in the automated coordination of human tasks and system events. As van der Aalst et al. [23] state, such systems laid the foundation for process-driven orchestration before RPA emerged as a GUI-layer solution.

Business Process Management (BPM) tools rose to prominence in the early 2000s. They enable organizations to model, execute, monitor, and optimize end-to-end business processes. BPM implementations typically require redesigning core processes and integrating systems, often making them resource-intensive. Nonetheless, they offer strategic value through system-level orchestration [8]. As Hofmann et al. [12] describe, BPM facilitates "a holistic view on process optimization" and provides control over full process lifecycles.

Application Programming Interfaces (APIs) also became standard in the early 2000s with REST or SOAP, allowing direct programmatic communication between systems. APIs enable seamless backend data exchange and have become a foundational technology for digital transformation. Fielding [10] formally introduced REST as a design architecture, emphasizing APIs' role in scalability and modularity. Their unique strength lies in their backend efficiency and low error susceptibility compared to GUI interactions.

RPA gained traction from 2010 onward. As noted earlier, it focuses on task-level automation by mimicking human interactions through the GUI layer. Its non-invasive nature and low technical barrier led to rapid adoption across industries [15,19]. According to Moffitt et al. [17], its unique advantage lies in its ability to automate tasks in legacy environments without backend changes.

Integration Platform as a Service (iPaaS) emerged around 2012 to enable centralized, cloud-native integration across enterprise applications. iPaaS tools use pre-built connectors and orchestrations to synchronize data and processes between disparate systems. Enríquez et al. [9] identify iPaaS as a key enabler of hybrid integration strategies, bridging cloud and on-premise systems efficiently.

IPA, gaining popularity from 2017, extends RPA by incorporating AI for handling unstructured data and semi-autonomous decision-making. According to Chakraborti et al. [5], IPA allows organizations to automate more complex workflows by embedding cognitive capabilities into digital workers.

Agentic Automation, a more recent advancement since approximately 2022, describes AI agents capable of autonomous task execution with contextual understanding [4]. This form of automation emphasizes adaptive behavior and decision-making under dynamic conditions. As Hofmann et al. [12] note, the evolution toward agentic automation signals a shift from static bots to proactive, intelligent agents acting within their environment.

The listed automation approaches (cf. Table 1), while not comprehensive, illustrate the diverse landscape of office automation technologies. Each technology integrates functionalities from others, such as RPA incorporating AI in intelligent automation or BPM adopting low-code features, to achieve more holistic, end-to-end process solutions on platforms. However, their core identities remain

Table 1. Evolution of Automation Technologies

Year	Technology	Focus	Unique Feature
1995	WfMS	Workflow design, task routing, execution tracking	Automated coordination of people and systems
2000	BPM	End-to-end process management	System-level process orchestration
2000	API	System interoperability	Direct system communication
2010	RPA	Task-level GUI automation	Mimics human GUI interactions
2012	iPaaS	Connecting cloud services and system data	Centralized cloud integration across platforms and applications
2017	IPA	Complex, AI-driven automation	AI-enhanced decision-making
2020	Agentic Automation	Autonomous agents performing adaptive tasks	Agents operate with autonomy and context awareness

distinct, with RPA centered on GUI-based task automation, BPM on strategic process management, and others on integration, AI, or customization, shaping their roles in automation platforms [12].

3 Methodology

This section details the methodological design, execution, and analytical procedures of the qualitative study, which was based on semi-structured expert interviews. The study was conducted in iterative cycles, encompassing interview protocol development, data collection, and thematic coding. Figure 1 provides an overview of the research process, illustrating both the phases of the interview study and the subsequent conceptualisation of the qualitative findings.

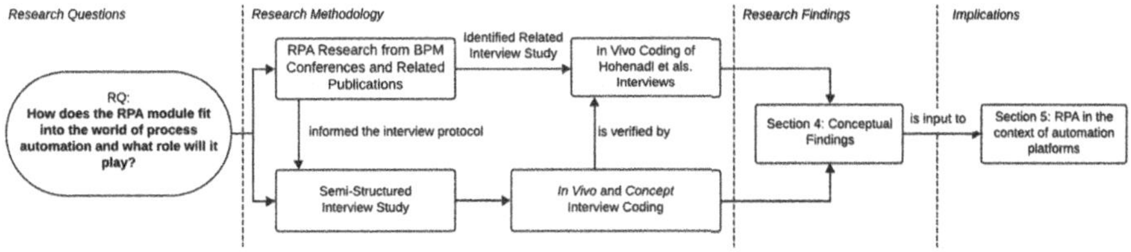

Fig. 1. Research Methodology and Section Outline

3.1 Interview Study Design

A semi-structured interview protocol was developed, guided by Padgett [18] and informed by publications from the BPM Conference RPA Forum (e.g., [16]), structured into five thematic sections: the past, present, and future of RPA, its organizational context, and competing technologies. The study engaged interview participants representing distinct expert perspectives on RPA. Three key stakeholder categories were targeted: practitioners involved in bot development, academic researchers with a focus on RPA, and professionals responsible for technical design at RPA software vendors. Inclusion was based on demonstrable expertise, such as active engagement in RPA development, recent scholarly contributions in the relevant domains, or architectural leadership within vendor organizations. Recruitment was carried out through professional networks and referrals, with attention to diversifying participants by sector, role, and seniority level. Given the hands-on nature of RPA implementation, particular emphasis was placed on recruiting developers, who play a central role in operationalizing automation. To mitigate commercial bias, a stronger representation from industry practitioners and academic experts was prioritized over vendor participants. The final sample included 13 interviewees: five related to development, two vendor-side professionals, two independent consultants, and four researchers. Table 2 provides an overview of their backgrounds. To indicate their expertise, the table includes the number of years of RPA experience and the number of RPA projects in which they have been involved.

Table 2. Overview of Interview Participants ordered by Sector

ID	Sector	Role	Years Exp.	Projects
P1	Development	Junior RPA Developer	1	5
P2	Development	Senior RPA Developer	8	>50
P4	Development	CoE Lead + Developer	7	15âĂŞ20
P8	Development	CoE Lead + Developer	6	150
P11	Development	CoE Automation Lead	5	25
P3	Consulting	Senior BPM Consultant	10	50
P12	Consulting	Senior Automation Consultant	5	No Info
P5	Academia	PhD Candidate	3	5
P7	Academia	BPM Professor	8	5
P9	Academia	PhD Candidate + Developer	4	6
P10	Academia	PhD Candidate + Developer	3	5
P6	Vendor	Senior Solution Architect	7	>15
P13	Vendor	Senior Solution Architect	6	No Info

3.2 Data Collection

The development of the interview guide was grounded in conceptual and practical insights from prior RPA Forum publications (e.g., [16]), ensuring alignment with ongoing academic discourse and practitioner priorities. The guide was pre-tested in two phases: first, through trials with engineering students learning RPA to assess clarity and flow; and second, through informal feedback from RPA practitioners who were not part of the main interview sample. These preparatory steps ensured that the questions were both understandable and topically relevant from multiple stakeholder perspectives. The main interview phase was conducted via a web-based conferencing platform. To minimize bias, interviewers remained off-camera. All sessions were recorded, transcribed using a clean-verbatim approach, anonymized, and prepared for qualitative analysis. To support validation and test the transferability of the emerging coding system, data from a previous interview study by Hohenadl et al. [13], which focused on RPA decision-making and its connection to Robotic Process Mining, was incorporated. All data for which publication permission was granted is available in[1].

3.3 Qualitative Coding Process

The coding process was guided by the structured methodology of Saldana [21] and carried out using MAXQDA to ensure consistency and traceability. After conducting 13 semi-structured interviews, all transcripts were first analyzed using In Vivo coding to preserve participants' original language and contextual nuance. These codes formed the basis for a second cycle of concept coding, in which recurring patterns were synthesized into a first set of 30 conceptual categories. Conceptual saturation was reached after coding the ninth interview, as no new concepts emerged and all In Vivo codes could be assigned to one of the 30 categories.

To further refine the analytical structure, a third coding cycle was conducted, again using concept coding. This round revealed three overarching themes—*Core RPA Capability*, *RPA as Part of Automation Platforms*, and *General Automation Concepts*—each comprising three subconcepts. These final themes structure the presentation of findings and are visualized in Fig. 2.

To validate the stability of the conceptual framework, additional interviews from Hohenadl et al. [13] were re-coded using the established 9-concept scheme. As no new categories were required to code these additional interviews, this confirmed the robustness of the coding and the achievement of thematic saturation [3].

4 Findings

This chapter presents the main findings derived from the qualitative analysis of the expert interviews. Through iterative coding cycles a set of recurring themes

[1] Interview Study Data Repository: https://doi.org/10.5281/zenodo.15792923

that reflect how practitioners and researchers conceptualize and implement RPA in practice and what insights RPA vendors have from customer interactions emerged. Three overarching conceptual categories emerged after the last coding cycle: (1) Core RPA Capabilities, (2) RPA as Part of Automation Platforms, and (3) General Automation Concepts. This order reflects a progression from concrete technical aspects of RPA to broader platform-level and strategic considerations. While each theme is discussed independently, the sub-concepts they comprise often reveal conceptual overlap, reflecting the integrated and evolving nature of RPA ecosystems. Figure 2 provides a visual overview of these categories and their relative prominence in the coded data, serving as a reference point for the analysis that follows. The next sections examine each conceptual category in turn. Beginning with Core RPA Capabilities, focusing on the reliability demand of users and on how bots are built and deployed. This is followed by insights on how RPA is embedded into broader automation platforms, including RPA governance and platform simplicity. The chapter concludes with the concept of General Automation Concepts, which reflect forward-looking perspectives on RPA's convergence with AI, standardization, and process documentation.

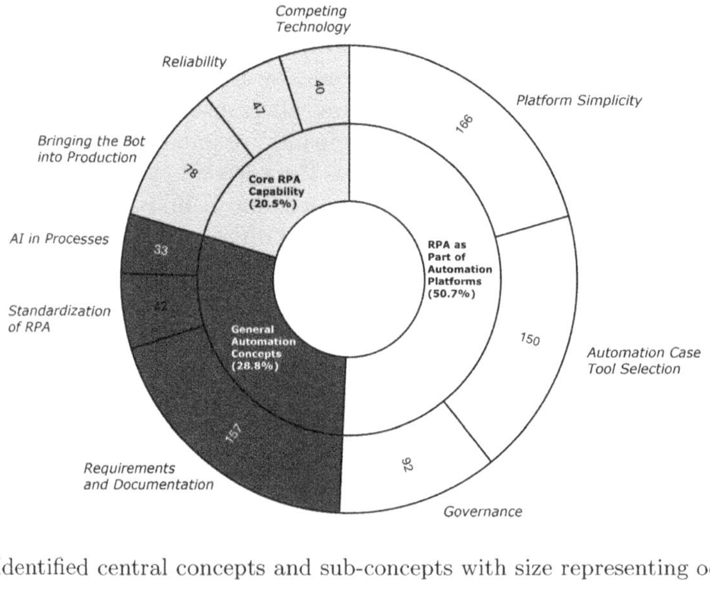

Fig. 2. Identified central concepts and sub-concepts with size representing occurance

4.1 Core RPA Capability

Reliability. has emerged as one of the core demands, especially P2 notes that in "enterprise grade automation, you usually favour reliability" over "trust [in] such solutions" that do not follow rule-based logic, as Agentic Automation brings. But RPA solutions often require workarounds to perform reliably. To improve stability, developers commonly insert delays to handle slow or unpredictable applications. One participant explained needing to "add a delay at every step because otherwise the bot hangs" [P11], while another noted that "especially scraping would be faster [with Python]" [P1]. Despite RPA being "crazy scalable" [P3], such workarounds remain necessary, leading to a decrease in the process execution reliability caused by the systems interacted with.

Participants also pointed to issues with unreliable UI selectors, described as "absolutely garbage" in some applications [P8]. While tools offer fuzzy selectors with adjustable thresholds—such as "80% match" [P2]—some developers remained unconvinced, with one stating, "I even forgot the name because I don't use them" [P2]. Furthermore, interviewees emphasized the need for transparency around application changes. As P2 noted, "we rather have some [...] internal teams communicate with us" than have bots adjust behavior without oversight. Finally, several participants highlighted that RPA bots often fail as GUIs often change [P2, P7, P8], especially in new software environments. On the other hand, RPA seems a reliable fit for "some legacy system that has not been updated for like 10 years' as this system will not "suddenly break with the RPA bot" [P2].

Bringing the Bot into Production. To address these maintenance concerns and ensure stable operation in real-world conditions, interviewees also pointed out two challenges of deploying bots into production environments: insufficient testing practices and inconsistent runtime environments. Several participants noted that RPA lacks mature testing capabilities compared to established software engineering practices. One expert remarked that "RPA is not particularly strong" when it comes to automated testing [P3], and that even small bots would ideally require "tools like Sonar or J-Unit testing" [P3]. However, such practices are often bypassed because bots are perceived as simple sequences of clicks rather than production-grade software: "even for small RPA bots you need a test manager and test automation [...] which normally just doesn't happen" [P3].

A second issue concerns the discrepancies between development and deployment environments. Participants described difficulties in simulating the runtime context—particularly for unattended execution or deployments. One interviewee explained, "If I want to run unattended automation in Power Platform, I need to have a server" [P8] being different to the local development environment, while another stated:"Especially to get an attended bot into an unattended state [...] is a big challenge" [P4]. These inconsistencies frequently lead to significant maintenance burdens. As one participant put it, "We in fact get a lot of tickets when it comes to the deployment of bots" [P6], while another noted that bots then either suffer from high time invest to operate or neglected post-deployment activities: "either they're really busy maintaining it, or it doesn't get maintained and then becomes even more crap" [P9].

Competing Technology. Still, RPA remains in a sweet spot despite the challenges of reliability and production readiness, as interviewees consistently described RPA not as a competing technology, but as a distinct capability within a broader automation toolbox. One participant noted that "RPA is just one tool in your toolbox," [P7] emphasizing that its role is specific rather than universal. Several experts stressed that RPA fills a gap in cases without APIs or where no native integration options exist [P3, P6, P7, P8]. However, many pointed out that this gap is narrowing. As one interviewee observed, "RPA will probably go downhill, and it's already starting to do that," [P6] referring to the increasing use of pro-code solutions and more integrated system architectures. Still, RPA remains essential in legacy contexts. According to another participant, it "is the middle way between systems with microservices and proper APIs, and old systems which don't have it" [P3]. Importantly, the long-term viability of RPA also depends on infrastructural openness rather than purely technical advancement. One interviewee stated, "It's not really about the technology, but the willingness of companies to provide connectors" [P8], highlighting the role of vendor ecosystems in shaping the relevance of RPA. Therefore, as legacy systems and integration gaps persist—whether due to cost, complexity, or vendor limitations—RPA will retain its relevance as a niche yet indispensable capability for GUI-level automation.

4.2 RPA as Part of Automation Platforms

Automation Case Tool Selection. As outlined in Sect. 2, RPA is understood as a rule-based, GUI-level automation capability and, as stated in the previous section, "it is one tool in your toolbox" [P7], emphasizing that its effectiveness depends heavily on context. Several interviewees stressed the challenge of determining when RPA is the appropriate solution, particularly in large organizations where no central automation governance exists: "there is no superordinate automation team that has an overview of all automation technologies" [P11]. As a result, the identification of suitable RPA use cases often falls to local teams or is guided by evolving automation platforms, some of which are trending toward integrated orchestration models like "BOAT, as a business orchestration and automation tool" [P3]. However, even with such platforms, identifying viable RPA use cases remains difficult. One participant recalled that an internal automation campaign surfaced over 100 ideas, but only three were ultimately suitable for RPA [P11]. This reinforces the view that RPA is most relevant in "imperfect" environments—those lacking clean APIs or service integration—since, as one expert noted, "[in a perfect world] it doesn't need RPA, but the world is another" [P3]. In practice, most organizations will continue to face such limitations, as the participant does not think "there will be a single company in the world that will have a perfect IT infrastructure" [P9] and fall back on RPA. This suggests that RPA is not limited by its technological capability, but rather by the organizational ability to make informed decisions about when and where to apply it. Further confirmed as for each automation it is not about "real competition [among tools], but everything somehow serves a different field better" [P11]. But selecting suitable use cases is only one part of the platform equation.

Participants also emphasized the importance of user experience and simplicity in shaping the successful adoption of RPA.

Platform Simplicity. As RPA platforms evolve into broader automation ecosystems, participants noted a growing tension between expanding functionality and ease of use. While accessibility is expected to improve, current platforms often introduce unnecessary complexity. As P9 described, "There's several building blocks that can send mail through Outlook. I don't understand why this is the case. [...] I have to Google forever which of those building blocks would be the right one." This complexity is reinforced by platform providers' dual incentives: to expand features for competitive advantage while simultaneously appealing to citizen developers, a dynamic that, as P2 noted, "makes the learning of each of these tools even harder." One participant linked this evolution to the lifecycle of text editors, where growing functionality is eventually balanced by more intuitive access to core features: "It grew in size, and thus it became more complex. But the core features became more simpler to be used. And I expect the same thing from RPA platforms" [P7].

Participants also discussed ways current platforms allow for simplification through extensibility and AI. Some development teams create internal extensions to streamline use: "We also have our own created [COMPANY] packages" [P4]. And looking ahead, AI is expected to play a growing role in reducing cognitive load. P6 described a vision where users can "just type into the prompt like a co-pilot of sorts, ask them what you want to automate," and the system will assist or generate processes. However, this promise is contingent on user skill: "This only works if you know how to write promts" [P11]. In the long run, participants envisioned automation platforms becoming more autonomous, possibly even optimizing execution sequences without developer input. As P2 suggested, future systems might "decide what is the order of the activities, on which sequences they interact [...] or which [activities] they execute." Still, others emphasized that true simplicity requires not "10 plus different offerings [and] embracing all the new stuff like agentic automation" [P2] but also platform design that minimizes ambiguity and prioritizes user understanding as "maybe the UI was not as simple for them [citizen developers]" [P6].

Governance. Another aspect in relation to simplicity are governance practices, as participants emphasized that automation platforms should include built-in governance to simplify development and reduce risk. Instead of relying on manuals or informal guidelines, platform-level controls were seen as a way to enforce boundaries and ease onboarding. As one participant noted, "the hardest part for me is getting used to the coding guidelines and frameworks" [P1]. Another warned that if you"don't centrally control it, then I think it leaves great potential for people just making any automation they want with any level of security they see fit" [P9].

These concerns were especially relevant for citizen development, which most participants viewed critically. While technically skilled business users have built simple automations, broader initiatives often lacked consistency or quality. "I haven't seen [...] successful citizen development initiatives where people were just doing quality stuff" [P2], leading some organizations to maintain centralized RPA teams for oversight and efficiency [P1, P4, P10, P11]. "Because especially for RPA, we have IT governance and regulations [.. setting..] clear rules for what to do and what not to do with a bot" [P4], which cannot be sufficiently controlled if the RPA platform does not govern the development.

4.3 General Automation Concepts

Requirements and Documentation. Beyond RPA-specific concerns, several participants emphasized that automation projects more broadly require structured documentation practices to ensure successful implementation and long-term maintainability of automations. Two key areas were repeatedly mentioned: first, the documentation of the as-is process during the requirements phase; and second, the documentation of the automated solution itself. As one interviewee put it, "the understanding of the process that you're automating is obviously very critical" [P13], yet this understanding is often incomplete. Several participants reported difficulty capturing exceptions and edge cases, noting that "people are not really aware of how the technology works and sometimes forget to tell us about certain exceptions" [P10], or that they were "never able to document the end-to-end process and therefore never able to automate it" [P6].

Participants also stressed that the success of automation relies less on the technical implementation and more on surrounding activities: "It's not the programming itself. It's the other things—documenting all the work" [P1]. Without proper traceability from requirements to deployment, critical knowledge can be lost. One interviewee summarized this need clearly: "for every single RPA bot [...] we need the traceability from requirement: Why is this bot there?" [P3]. To avoid this, some organizations assign a designated process owner—"the person who owns the actual process and who must have a hand in it" [P11]—ensuring continuity and accountability after automation deployment, because otherwise, as"soon as the bot is not working, you do not have the knowledge workers" [P3] to solve the issue.

Standardization of RPA. Building on the challenges of documentation and process understanding, several participants emphasized that RPA would benefit from stronger alignment with established software development standards. This included calls for standardized notations—"something like BPMN for RPA" [P3]—and the adoption of familiar engineering models: "to implement the standard methods of software development [...] even from the V-model" [P3]. Participants argued that greater standardization would make bots more understandable and maintainable, especially across different platforms. As P9 noted, "if it's 6âĂŞ7 different platforms and they all have the same blocks, but they're all named differently and work slightly differently," consistency becomes a major issue. Others highlighted the value of reusable components and modularity: "you created your bot files once, and after you can always reuse them" [P4]. This was particularly important for supporting less experienced developers, who often struggle with basic constructs: "for people that don't know how to do proper try-catch blocks [...] that would be a great benefit" [P9]. Ultimately, participants called for RPA to move beyond isolated no-code solutions and adopt "standard practices" from traditional software engineering, as one summarized: "you really have to build logic into it [...] it's not just clicking it together" [P11].

AI in Processes. Finally, several participants reflected on the growing role of AI in process automation, describing it as a natural evolution beyond the rule-based logic that characterizes traditional RPA. As one interviewee summarized, "RPA is not thinking," [P4] whereas AI introduces a more dynamic, adaptive layer to automation—"with AI I always think of a thinking person" [P4]. Participants described use cases where methods from AI, such as machine learning or Object Character Recognition, could extend automation capabilities, for example, by enabling bots to "fill up some form just by figuring out what is there" without needing to manually specify target fields [P2]. Others noted that this shift is not limited to recent developments in generative AI, but includes long-standing techniques like classification and document analysis: "with machine learning you can now also sort of classify documents" [P10].

While most viewed AI as expanding the potential of automation, some expressed reservations. One participant emphasized the value of deterministic, rules-based systems in contexts requiring high accuracy: "I would go for the 100% quality option [...] not one that has invented the rules by themselves" [P7] further refined by the demand for control as" Europe is based on standards, on auditing, on control" [P3]. Still, the prevailing perspective was that AI is increasingly inseparable from automation's future. As P12 put it, "I don't think RPA has a future without AI," signaling a shift toward more intelligent, context-aware automation systems.

5 Conclusion

This study provides empirical insights into the role of RPA within the rapidly evolving automation landscape. Based on expert perspectives from development, research, consulting, and vendor environments, the analysis uncovers a strategic shift: RPA is evolving from a tactical tool for GUI-level automation into a distinct, modular capability—namely GUI programming—embedded within comprehensive, end-to-end automation platforms.

Three thematic pillars emerged. First, Core RPA Capabilities were discussed less frequently, suggesting that the fundamental technical model of GUI-based scripting is mature and functionally stable. However, deployment challenges, particularly related to reliability, testing, and selector management, remain critical for production-grade use. Second, the most prominent theme, RPA as Part of Automation Platforms, reflects the increasing integration of RPA into orchestration-driven and AI-augmented environments. Here, tensions between platform complexity and user accessibility were evident, especially in light of citizen development initiatives. Interviewees emphasized the importance of governance, role clarity, and platform simplicity. Third, under General Automation Concepts, respondents highlighted the importance of structured documentation, process ownership, and alignment with established software engineering practices. AI and machine learning are seen as natural extensions of RPA but require careful integration to preserve transparency and control.

In response to the research question—How does RPA fit into today's automation ecosystem, and what role will it play in relation to other automation technologies and the future of intelligent platforms?—the findings indicate that RPA no longer functions as a standalone or competing automation technology. Instead, it is emerging as a niche capability—specifically GUI programming—designed to address targeted integration gaps, particularly within legacy systems or fragmented IT environments. Its value lies not in technological novelty but in context-specific applicability. The future of RPA lies not in reinventing its technical core, but in ensuring its effective positioning, selection, and operation within intelligent automation platforms—specifically as GUI-based automation is well-governed, maintainable, and interoperable with adjacent technologies such as APIs, BPM engines, and AI modules.

Acknowledgments. The work of Tom Hohenadl was funded by LFP Project P 20 54 301 at Technische Hochschule Ingolstadt.

Disclosure of Interests. The authors have no competing interests to declare that are relevant to the content of this article.

References

1. Aguirre, S., Rodriguez, A.: Automation of a business process using robotic process automation (RPA): a case study. In: Figueroa-García, J.C., López-Santana, E.R., Villa-Ramírez, J.L., Ferro-Escobar, R. (eds.) WEA 2017. CCIS, vol. 742, pp. 65–71. Springer, Cham (2017). https://doi.org/10.1007/978-3-319-66963-2_7
2. Asatiani, A., Penttinen, E.: Turning robotic process automation into commercial success - case opuscapita. J. Inf. Technol. Teach. Cases **6**(2), 67–74 (2016). https://doi.org/10.1057/jittc.2016.5
3. Baker, S.E., Edwards, R.: How many qualitative interviews is enough?. https://eprints.ncrm.ac.uk/id/eprint/2273/4/how_many_interviews.pdf
4. Bornet, P., Wirtz, J.: Agentic Artif. Intell. World Sci. (2025). https://doi.org/10.1142/14380
5. Chakraborti, T., et al.: From robotic process automation to intelligent process automation. In: Asatiani, A. (ed.) Blockchain and RPA Forum Proceedings BPM 2020, pp. 215–228. Springer, Cham (2020). https://doi.org/10.1007/978-3-030-58779-6_15
6. Cooper, L.A., Holderness, D.K., Sorensen, T.L., Wood, D.A.: Robotic process automation in public accounting. Account. Horiz. **33**(4), 15–35 (2019). https://doi.org/10.2308/acch-52466
7. Czarnecki, C., Fettke, P. (eds.): Robotic Process Automation: Management, Technology, Applications. De Gruyter STEM, De Gruyter, Berlin (2021).https://doi.org/10.1515/9783110676693

8. Dumas, M., La Rosa, M., Mendling, J., Reijers, H.A.: Fundamentals of business process management. Springer, Berlin, softcover re-print of the hardcover 2nd edition 2018 edn. (2018). https://link.springer.com/book/10.1007/978-3-662-56509-4
9. Enriquez, J.G., Jimenez-Ramirez, A., Dominguez-Mayo, F.J., Garcia-Garcia, J.A.: Robotic process automation: a scientific and industrial systematic mapping study. IEEE Access **8**, pp. 39113–39129 (2020)
10. Fielding, R.T.: Architectural styles and the design of network-based software architectures. Bell and Howell Information and Learning (2000). https://search.proquest.com/openview/fc2d064044b971dda476dfb429a2b344/1?pq-origsite=gscholar&cbl=18750&diss=y
11. Herm, L.V., Janiesch, C., Helm, A., Imgrund, F., Hofmann, A., Winkelmann, A.: A framework for implementing robotic process automation projects. Inf. Syst. e-Bus. Manag. (2023). https://doi.org/10.1007/s10257-022-00553-8
12. Hofmann, P., Samp, C., Urbach, N.: Robotic process automation. Electron. Mark. **30**(1), 99–106 (2020). https://doi.org/10.1007/s12525-019-00365-8
13. Hohenadl, T., Axmann, B., Stummeyer, C.: Decision-making in robotic process automation programming and its influence on robotic process mining. In: Di Ciccio, C., et al. (eds.) Business Process Management: Blockchain, Robotic Process Automation, Central and Eastern European, Educators and Industry Forum, LNBIP, vol. 527, pp. 169–184. Springer (2024).https://doi.org/10.1007/978-3-031-70445-1_11
14. Huo, S., Mukherjee, K., Bandlamudi, J., Isahagian, V., Muthusamy, V., Rizk, Y.: Accelerating the support of conversational interfaces for RPAs through APIs. In: Köpke, J., et al. (eds.) Business Process Management: Blockchain, Robotic Process Automation and Educators Forum, Lecture Notes in Business Information Processing, vol. 491, pp. 165–180. Springer Nature Switzerland, Cham (2023). https://doi.org/10.1007/978-3-031-43433-4_11
15. Lacity, M.C., Willcocks, L.P., Craig, A.: Robotic process automation at xchanging. https://eprints.lse.ac.uk/64518/1/OUWRPS_15_03_published.pdf
16. Marrella, A., et al.: Business Process Management: Blockchain, Robotic Process Automation, and Central and Eastern Europe Forum: BPM 2022 Blockchain, RPA, and CEE Forum, Münster, Germany, September 11–16, 2022. In: Proceedings, Lecture Notes in Business Information Processing, vol. 459. Springer International Publishing and Imprint Springer, Cham, 1st ed. 2022 edn. (2022). https://doi.org/10.1007/978-3-031-16168-1
17. Moffitt, K.C., Rozario, A.M., Vasarhelyi, M.A.: Robotic process automation for auditing. J. Emerg. Technol. Account. **15**(1), 1–10 (2018). https://doi.org/10.2308/jeta-10589
18. Padgett, D.: Qualitative methods in social work research. SAGE, Los Angeles, 3rd ed. edn. (2017)
19. Plattfaut, R., Borghoff, V.: Robotic process automation: a literature-based research agenda. J. Inf. Syst. **36**(2), 173–191 (2022). https://doi.org/10.2308/ISYS-2020-033
20. Plattfaut, R., Borghoff, V., Godefroid, M., Koch, J., Trampler, M., Coners, A.: The critical success factors for robotic process automation. Comput. Ind. (2022). https://doi.org/10.1016/j.compind.2022.103646
21. Saldaña, J.: The Coding Manual for Qualitative Researchers, 4th edn. SAGE, Los Angeles (2021)

22. Syed, R., et al.: Robotic process automation: contemporary themes and challenges. Comput. Ind. **115**, pp. 103–162 (2020). https://doi.org/10.1016/j.compind.2019.103162
23. van der Aalst, W.M.P., Bichler, M., Heinzl, A.: Robotic process automation. Bus. Inf. Syst. Eng. **60**(4), 269–272 (2018). https://doi.org/10.1007/s12599-018-0542-4

Multimodal Zero-Shot Activity Recognition for Process Mining of Robotic Systems

Flavio Corradini[1], Sara Pettinari[2], Barbara Re[1], Lorenzo Rossi[1], and Massimiliano Sampaolo[1](✉)

[1] School of Science and Technology, University of Camerino, Camerino, Italy
{flavio.corradini,barbara.re,lorenzo.rossi,massimilian.sampaolo}@unicam.it
[2] Gran Sasso Science Institute, L'Aquila, Italy
sara.pettinari@gssi.it

Abstract. Understanding and analyzing the behavior of robotic systems is essential to ensure their reliability, efficiency, and continuous improvement, especially as robots are increasingly deployed in complex, dynamic environments. Process mining offers a powerful approach to uncover and analyze the execution of robotic operations. However, applying process mining to robotic systems requires bridging the gap between fine-grained multimodal data and high-level activity representations. Recent advances in foundation models provide a promising solution to this challenge, as the knowledge acquired during their extensive pretraining enables them to interpret multimodal data without the need for task-specific training. In this work, we propose a novel multimodal process mining pipeline that leverages the zero-shot capabilities of foundation models to perform activity recognition from visual and auditory inputs. By transforming fine-grained multimodal data into event logs, the pipeline enables the application of process mining techniques to robotic systems. We applied our approach to the Baxter UR5 95 Objects dataset, which offers synchronized video and audio recordings of a Baxter robot manipulating objects. The fusion of activity recognition results from these complementary modalities yields an event log that more accurately represents the robot's operations, mitigating imprecision associated with using a single modality. Our results demonstrate that foundation models effectively enable the application of process mining to robotic systems, facilitating monitoring and analysis of their behavior.

Keywords: Foundation Models · Activity Recognition · Process Mining · Robotic Systems

1 Introduction

The increasing integration of robotic systems into everyday environments is reshaping the way we live and work. From warehouse automation [20] to precision

M. ER et al. (Eds.): BPM 2025, LNBIP 565, pp. 263–275, 2026.
https://doi.org/10.1007/978-3-032-02936-2_19

agriculture [19] and collaborative assembly lines [33], robots are now expected to perform complex tasks autonomously and adaptively, acting as key agents within operational processes. As these systems evolve and become increasingly important in society, ensuring their correct and efficient behavior becomes a pressing concern.

Understanding how robots behave during task execution is essential to ensure their reliability, efficiency, and adaptability. While low-level sensor readings and control signals are fundamental for system diagnostics and real-time control [21], analyzing their behavior from a high-level perspective would offer a more holistic understanding of how they achieve their goals and impact the processes they are part of. These two levels of analysis are complementary and work together to enable a comprehensive understanding and monitoring of our processes.

Process mining provides a powerful set of techniques to support the high-level perspective, enabling the extraction of structured insights from data [1]. It enables the discovery [2], monitoring, and improvement of process models based on real execution traces. However, robotic systems pose unique challenges for process mining. Unlike traditional software systems, robots do not natively generate event logs representing high-level activities. Instead, they produce fine-grained, multimodal data streams from sensors, actuators, and control systems, often lacking the semantic structure needed to support process-level analysis [6]. In fact, in robotic domains, processes are embedded in physical behavior and must be inferred from continuous multimodal signals.

To address this challenge, we propose a novel multimodal process mining pipeline for extracting event logs from fine-grained robotic data. Our approach combines activity recognition, foundation models, robotics, and process mining. Specifically, it leverages the zero-shot reasoning capabilities of foundation models to perform activity recognition from synchronized video and audio inputs. In this context, zero-shot capability refers to the ability of foundation models to generalize to new tasks or inputs without requiring additional training, relying instead on knowledge acquired during large-scale pretraining. The activities inferred from the two modalities are then fused to construct a final multimodal event log, suitable for analysis with traditional process mining techniques. The final step of our pipeline involves discovering a directly-follows graph (DFG) from the event log produced in the earlier stages. To evaluate the effectiveness of our approach, we apply it to a sample of the Baxter UR5 95 Objects dataset [31], a relevant benchmark that includes synchronized video, audio, and sensor signals of a Baxter robot performing different object manipulation tasks. Combining visual and auditory modalities enhances activity recognition and produces event logs that support process mining analyses, enabling a more holistic understanding of the underlying processes. While some deviations remain, the resulting DFG accurately reflects the sequence of actions performed by the robot in the analyzed sample, demonstrating the effectiveness of our methodology.

By transforming fine-grained multimodal data into high-level event logs, our approach enables the application of process mining techniques to monitor and analyze the behavior of robotic systems. This contributes to the development of process-oriented systems that incorporate robotics, extending monitoring capabilities from purely digital systems to those involving physical execution, and supporting the integration of robots into broader process-aware environments.

The remainder of the paper is organized as follows. Section 2 reviews the state of the art on the application of process mining in robotics and the use of foundation models for activity recognition. Section 3 describes the proposed pipeline, while Sect. 4 presents its evaluation on a real-world dataset. Finally, Sect. 5 concludes the paper and outlines directions for future work.

2 Related Work

The pipeline presented in this paper lies at the intersection of several research areas, including activity recognition, foundation models, robotics, and process mining. In the following, we review relevant works from these domains, with a focus on those that explore their integration.

Starting from the field of process mining, its application in robotics remains a partially explored topic. Only a limited number of studies have investigated how process mining techniques can be used to analyze robotic systems. For example, [18,30] apply process discovery to understand and model robot behavior. In a different direction, [27] proposes a methodology for conformance checking in robotic mechanisms. Other works, such as [7,29] focus on process enhancement, using robotic data to analyze robots behavior, area usage, execution time, and resource consumption. [6] provides a collection of publicly available robotic datasets that can be used for process mining applications. Finally, [5] discusses the challenges and opportunities of applying process mining in robotics, highlighting the importance of multimodal data integration.

This notion of integrating multiple modalities has given rise to **multimodal process mining**, an emerging research area that addresses scenarios where traditional digital traces are absent or insufficient. In [13], the authors address the challenge of tracking manual or physical activities without digital footprints, evaluating their approach on a video of a person assembling a wardrobe. Similarly, [14] proposes a method for generating multimodal event logs by computing semantic distances between visual, auditory, and textual cues within a self-hosted joint embedding space. Other approaches have focused on bridging computer vision and process mining directly. For example, [23] introduces ViProMiRA, an architecture designed to produce event logs from video data by aligning visual observations with process semantics. In addition, unsupervised techniques for deriving process models from video recordings have also been proposed, such as in [22,26], where the goal is to describe observed behavior without requiring manual annotation.

The application of process mining also extends to the domain of **sensor data**. For instance, [28] proposes a multimodal approach that combines motion sensor and video data for both activity recognition and process discovery. Several other studies have leveraged IoT sensor data to uncover and analyze process dynamics. Notable examples include [9,16,25], which show how sensor-based observations can be transformed into event logs to support process analysis.

Another crucial enabler of our pipeline is the use of **foundation models**. These large-scale pretrained models have recently gained significant attention due to their ability to generalize across a wide range of tasks without requiring task-specific training. Several recent studies have investigated the application of **foundation models in activity recognition**. [34] combines a foundation model with radio-frequency data to perform human activity recognition. [17] introduces a zero-shot method that classifies human activities using IMU data and carefully designed prompts. [35] presents TENT, a framework that aligns video and sensor data with text in a shared semantic feature space, enabling unified multimodal interpretation. [11] provides a comprehensive survey of activity recognition from videos and outlines a roadmap for future advancements. The authors describe few-shot and zero-shot methods as emerging paradigms and argue that future progress will depend on unified learning frameworks that integrate multimodal data and leverage pretrained video foundation models. [24] highlights the potential of foundation models not only for recognition, but also for generating benchmark datasets and developing models specifically trained for activity recognition. Foundation models are also making a significant impact in the field of robotics. Surveys such as [12] and [15] discuss how these models enhance robotic capabilities in perception, decision-making, and control. At the same time, they highlight key limitations that must be addressed, including inference time, crucial for achieving reliable real-time deployment, and explainability, as understanding the decision-making process of foundation models is essential to ensure safe and predictable robot behavior [15].

3 Multimodal Process Mining Pipeline

This section presents our solution for extracting process insights from robotic data. The overall pipeline is illustrated in Fig. 1. As shown, the pipeline consists of three main steps: activity recognition from both video and audio data, a multimodal fusion module, and a process discovery step that ultimately extract a DFG representing the robot's behavior. While not depicted in the figure, each step may involve basic preprocessing operations, such as frame extraction, timestamp formatting, or data restructuring, which are implementation-specific and not central to the pipeline itself.

Activity Recognition. The first step is activity recognition, where video and audio data are analyzed to extract high-level activities. To this aim, we use foundation models, leveraging their ability to perform a wide variety of tasks without requiring domain-specific training [3]. This capability stems from the general knowledge they acquire during large-scale pretraining. Although foundation models can be fine-tuned to improve performance on specific tasks or datasets, our methodology adopts a zero-shot approach. By using the models without additional training, we aim to keep the pipeline general and easily applicable across different scenarios. As a consequence, the quality of the prompts provided to the foundation models can significantly impact the results. To enable pipeline flexibility and generalization, the activity recognition step for video and audio is

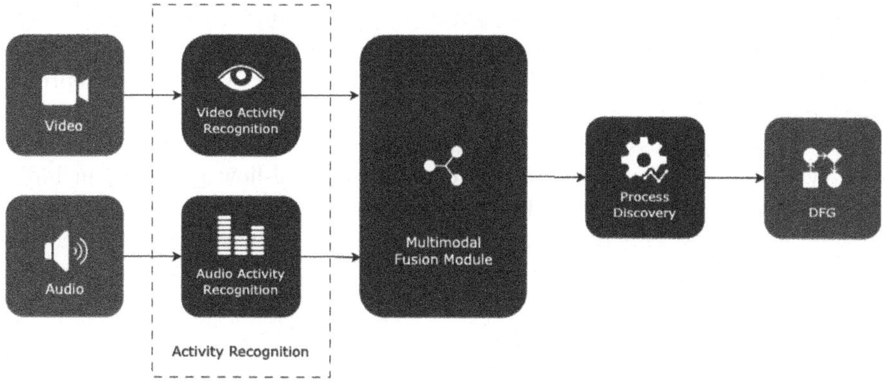

Fig. 1. The proposed pipeline: from fine-grained data to process mining

executed separately. Indeed, depending on the specific data, requirements, and context, we can choose to use the same foundation model for both modalities or different ones, and tailor the prompts accordingly for each scenario.

Multimodal Fusion Module. After the activity recognition step, we obtain separate event logs, one for each modality. The next step consists of combining these logs to generate a single, unified event log. The multimodal fusion module can be implemented using different strategies, depending on the specific goals and context of the application:

- *Enrichment:* each modality may detect a different subset of activities, and the final event log is obtained by merging them to maximize coverage.
- *Accuracy:* some activities may be more reliably recognized by one modality than the other. In this case, we select the most accurate activity from each source to build a more precise, though not necessarily more complete, event log.
- *Reliability:* only activities recognized by both modalities are retained in the final event log, increasing confidence in the extracted information.
- *Hybrid:* a custom combination of the above strategies, tailored to the specific scenario.

Regardless of the fusion strategy adopted, the data synchronization problem must be addressed. While the pipeline assumes that the fine-grained multimodal data is already time-aligned, the fusion module must determine how to assign timestamps to the events in the final log. For instance, if the same activity is recognized in both the video and audio streams but with slightly different timestamps, we can either select the more reliable one or compute an average, at the cost of introducing minor temporal imprecision. Additionally, fast or overlapping events may require ad-hoc handling to avoid misalignment or information loss. These challenges must be carefully managed within the fusion module, and the appropriate solution often depends on the characteristics of the data and the specific application context.

Process Discovery. The fusion of the event logs produces a new, unified event log that serves as input for the third and final step: the application of process discovery. In this phase, a DFG is extracted from the event log to represent the behavior of the robot during task execution. The DFG provides a high-level overview of the process, showing how activities are executed and in which order they typically follow one another. It captures control-flow patterns, including the frequency with which specific transitions occur and, optionally, the timing between activities. This representation enables the analysis of robotic behavior in terms of process structure and execution dynamics, offering valuable insights into how the robot performs its operations in a repeatable and interpretable form.

4 Evaluation

In this section, we describe how we evaluated the effectiveness of the proposed pipeline. First, we present the dataset used for the evaluation. Then, we provide technical details about how we implemented the pipeline. Finally, we present the obtained results. All the source code is available online[1].

4.1 Dataset

To assess the effectiveness of the proposed pipeline, we conducted an evaluation using a sample of the Baxter UR5 95 Objects dataset [31]. This dataset contains synchronized video, audio, and sensor data of a Baxter robot performing a variety of object manipulation tasks involving different objects. The annotated activities in the dataset include *grasp*, *pick*, *hold*, *shake*, *lower*, and *drop*. The dataset was originally designed to support research on transferring multimodal knowledge across heterogeneous robotic platforms, with a focus on understanding object properties through exploratory behaviors. Although our work does not address transfer learning directly, the datasets rich combination of visual and auditory modalities makes it a suitable benchmark for evaluating multimodal activity recognition and process discovery in robotic systems.

The dataset is organized into trials, each containing all the data recorded during a specific execution. Within each trial, there is one video for each activity performed by the robot. For our evaluation, we concatenated the videos from each activity to create a single continuous video representing a fixed sequence of actions. The same procedure was applied to the corresponding audio recordings. We used 10 trials from the dataset. Since our methodology does not involve training or fine-tuning, a large volume of data is not required. The focus of our evaluation is primarily qualitative rather than quantitative.

[1] https://bitbucket.org/proslabteam/zs_robot_ar.

4.2 Implementation

The proposed pipeline has been implemented in Python, with the activity recognition step based on the LangChain[2] framework. This framework plays a central role in our solution, as it simplifies the development of applications that leverage foundation models. LangChain provides all the necessary tools to build and adapt such systems, including seamless integration with different models. This enables significant flexibility in selecting the most effective model for a given task, ultimately improving the quality of the results.

Focusing on **video activity recognition**, we used the GPT-4o model due to its performance in multi-image understanding [32] and its reliable and easy-to-use integration with LangChain. We divided the video into frames and selected one frame per second. Recognizing the activities in the dataset requires analyzing the temporal evolution of the robots movements. For example, the *shake* activity involves repetitive movement and rotation of the robot arm, while the *hold* activity consists of maintaining a grasped object in midair. To capture these temporal dynamics, we provided the model with three consecutive frames at each step: the current frame along with the two preceding ones. This approach enables the model to reason over short sequences and capture movement patterns by comparing visual changes across frames. An essential part of this step is prompt engineering. We structured the prompt in two parts: a general instruction section and a space for frame input. The general section guides the model on how to approach the recognition task. For example, to use only visible evidence such object position, gripper contact, motion direction and to not infer hidden intent. Additionally, the prompt includes detailed descriptions of each activity performed by the robot and explains how to recognize them. For every activity, we highlight which aspects of the frames, or the transitions between frames, are most informative. At the same time, we specify rules to help the model avoid confusing similar activities, such as distinguishing between *grasp* and *pick*, or between *hold* and *shake*.

Moving on to **audio activity recognition**, we also used the GPT-4o model for this task. In this case, the main challenge lies in formulating a prompt that clearly helps the model distinguish between different sounds. Additionally, some activities are not easily recognizable using audio alone. For example, the *pick* and *lower* activities produce similar sounds, as both involve the robotic arm moving along a single axis. For this reason, we limited the use of the audio modality to the recognition of the *drop* activity, which is easily identifiable due to the distinctive sound it produces when the object strikes the surface.

Our implementation of the **fusion module** is straightforward and guided by the results obtained from the two activity recognition steps. Specifically, we take the *drop* activity recognized from the audio modality and insert it into the sequence of activities obtained from the video modality. This choice is motivated by the fact that *drop* is reliably detected in the audio, while it introduces noise when recognized from video. Since video activity recognition is performed once

[2] https://www.langchain.com.

per second, we implicitly obtain both the activity and its timestamp. Similarly, in audio-based activity recognition, we identify the specific second in which the drop activity occurs. The final essential component of an event log for enabling process mining techniques is the case identifier. Although identifying case identifiers can be a complex task in many real-world multimodal scenarios [5], in our case, we simply associated each trial performed by the robot with a distinct process instance.

For the final step of our pipeline, we leveraged Disco[3] for **process discovery**, generating a DFG from the event log produced by the previous steps of the pipeline in CSV format.

4.3 Results

The results produced by the implemented pipeline are depicted in Fig. 2. The resulting DFG captures the directly-follow relations between activities, indicating which actions occur in sequence. In addition, the graph highlights the frequency of each transition, offering insight into the most common patterns observed in the analyzed process. In our evaluation dataset, the robot performs a fixed sequence of six activities: *grasp*, *pick*, *hold*, *shake*, *lower*, and *drop*. While the discovered DFG generally reflects this known sequence, some noise is present among the *hold*, *shake*, and *lower* activities. For instance, a transition from *shake* to *hold* appears, even though it should not exist according to the dataset; instead, the expected transition from *shake* is directly to *lower*.

From the analysis results, we deduced that this noise is mainly due to the video activity recognition, while the recognition of *drop* using the audio data is almost perfect. We identified three main sources of noise:

- **False negatives.** As is common in many robotic datasets, each activity is recorded in a separate video, which we concatenated to create a continuous execution sequence. Some of these videos include short pauses at the beginning or end, during which the robot is simply holding the object in midair. Although this behavior is not annotated as a distinct activity in the dataset, our model correctly identifies it as *hold*. This leads to a mismatch with the ground truth provided by the dataset, but actually reflects a more precise interpretation of the robot's behavior. For instance, Fig. 3 shows three frames from a sequence labeled as *shake* in the dataset, where the robot is clearly holding the object without significant movement. In this case, the foundation model predicted *hold*, arguably offering a more accurate labeling than the original annotation.
- **Scene changes.** When we concatenated the videos of individual tasks, the resulting sequence introduced a scene change between the shake and lower activities, which caused significant confusion for the model. During this transition, the robotic arm undergoes a large and abrupt movement, making it difficult for the model to interpret the ongoing action. Figure 4 illustrates this

[3] https://fluxicon.com/disco/.

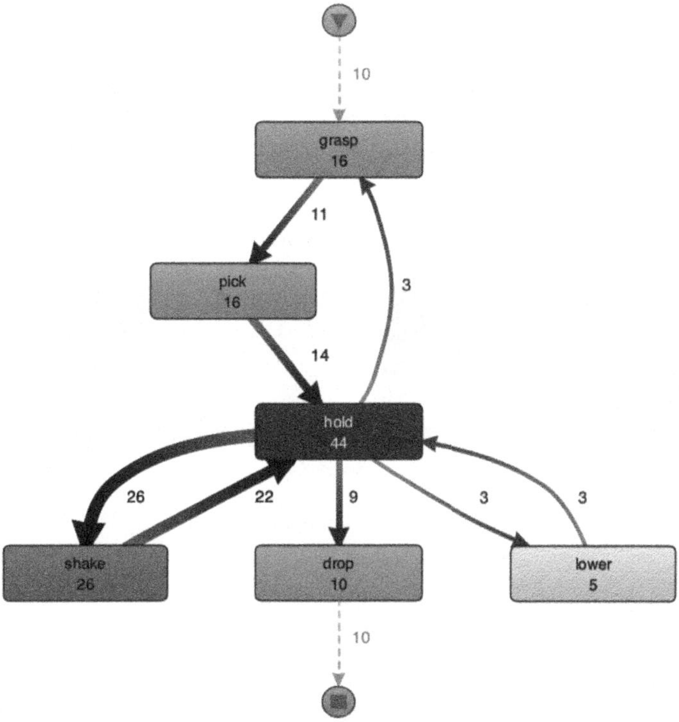

Fig. 2. The inferred DFG

issue in the same video used in the previous example. At second 32, the robot is holding the object near the camera, and in the following second, its arm appears in a completely different position, close to the ground. In this case, the model incorrectly detects a drop activity but this imprecision is corrected through the use of audio recognition in the subsequent multimodal fusion step.

- **Recognition errors.** The final source of noise comes from mistakes made during the activity recognition step. These errors can be influenced by various factors, such as lighting conditions, shadows, or the sampling rate used in our experiments. This class of errors represents the most direct target for improvement, as addressing them could enhance the overall performance of the technique.

The recognition of the *drop* activity from the audio was very accurate within the sample used in our evaluation, and the fusion of the two modalities led to a coherent and meaningful representation of the robots behavior. While there is still room for improvement, the overall results demonstrate the potential of our approach to produce high-level, process-oriented views of robotic execution from fine-grained multimodal data.

Fig. 3. False negatives example

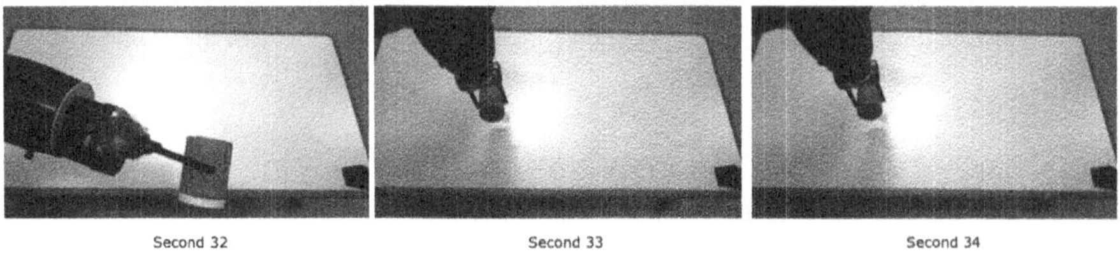

Fig. 4. Scene changes example

5 Conclusions

This paper presented a pipeline for extracting process-level insights from robotic systems' behavior by combining multimodal activity recognition and process mining techniques. The proposed pipeline operates on fine-grained video and audio data, transforming it into a high-level event log suitable for process discovery. Activity recognition is performed using a zero-shot approach with a foundation model, which ensures flexibility and generality across different scenarios by simply adapting the model prompts. The step where video- and audio-based activities are merged into a unified event log is handled by the multimodal fusion module. Finally, process discovery techniques are applied to derive a DFG that represents the robot's behavior.

We evaluated the pipeline on the Baxter UR5 95 Objects dataset, which offers synchronized multimodal recordings of robotic manipulation tasks. The results demonstrate the effectiveness of pre-trained foundation models in recognizing high-level robotic activities and generating event logs suitable for analysis using standard process mining tools. Despite some limitations and sources of noise, the integration of audio and video modalities produced a coherent and meaningful representation of the robot's behavior.

Future Work. We envision several promising directions for extending this work. First, the integration of additional modalities could further enhance the robustness of the pipeline. Beyond video and audio, the pipeline could incorporate other types of data commonly used in robotics, such as time-series signals from GPS, IMU, or force sensors [6]. Moreover, integrating communication logs in multi-robot settings or human-robot interaction scenarios [8] would allow the detection of activities that are currently not identifiable. This multimodal expansion could also improve the reliability of the event log by reinforcing activity recognition across different data sources.

Another research direction involves the extraction of contextual information. Capturing contextual details about the operational environment can enhance our understanding of the process and enable more advanced analyses. For example, identifying which object the robot is manipulating, whether it is operating indoors or outdoors, or if it is collaborating with other agents could make the monitoring and analysis of robotic systems more accurate and informative.

Finally, to increase the trustworthiness, reliability, efficiency, and continuous improvement of robotic systems, predictive process monitoring techniques could be integrated into the pipeline to anticipate future activities based on the current process state [4,10]. This would allow a real-time monitoring of robot operations.

Acknowledgments. This work has been partially funded by (a) the PRIN project 2022JKA4SL - HALO: etHical-aware AdjustabLe autOnomous systems, and (b) the European Union - NextGenerationEU under the Italian Ministry of University and Research National Innovation Ecosystem grant ECS00000041 - VITALITY - CUP J13C22000430001.

References

1. van der Aalst, W.: Process Mining: Data Science in Action. Springer, Heidelberg (2016). https://doi.org/10.1007/978-3-662-49851-4
2. van der Aalst, W.M.P.: Foundations of process discovery. In: van der Aalst, W.M.P., Carmona, J. (eds.) Process Mining Handbook. LNBIP, vol. 448, pp. 37–75. Springer, Cham (2022). https://doi.org/10.1007/978-3-031-08848-3_2
3. Bommasani, R., et al.: On the opportunities and risks of foundation models. arXiv preprint arXiv:2108.07258 (2021)
4. Burattin, A.: Streaming process mining. In: : van der Aalst, W.M.P., Carmona, J. (eds.) Process Mining Handbook, LNBIP, vol. 349, pp. 3–10. Springer, Cham (2022). https://doi.org/10.1007/978-3-031-08848-3_11
5. Corradini, F., Pettinari, S., Re, B., Rossi, L., Sampaolo, M.: Enabling process mining on multimodal robotic data. In: Grabis, J., Wautelet, Y. (eds.) CAiSE 2025. LNBIP, vol. 556, pp. 257–269. Springer, Cham (2025). https://doi.org/10.1007/978-3-031-94931-9_21
6. Corradini, F., Pettinari, S., Re, B., Rossi, L., Sampaolo, M.: Robotic datasets for process mining. In: Pufahl, L., Rosenthal, K., España, S., Nurcan, S. (eds.) CAiSE 2025. LNBIP, vol. 557, pp. 53–61. Springer, Cham (2025). https://doi.org/10.1007/978-3-031-94590-8_7
7. Corradini, F., Pettinari, S., Re, B., Rossi, L., Tiezzi, F.: A methodology for the analysis of robotic systems via process mining. In: Enterprise Design, Operations, and Computing. LNCS, vol. 14367, pp. 117–133. Springer (2023). https://doi.org/10.1007/978-3-031-46587-1_7
8. Corradini, F., Pettinari, S., Re, B., Rossi, L., Tiezzi, F.: A technique for discovering BPMN collaboration diagrams. Softw. Syst. Model. **23**(6), 1323–1343 (2024)
9. Di Federico, G., Burattin, A.: CvAMoS-event abstraction using contextual information. Future Internet **15**(3), 113 (2023)

10. Di Francescomarino, C., Ghidini, C.: Predictive process monitoring. In: van der Aalst, W.M.P., Carmona, J. (eds.) Process Mining Handbook. LNBIP, vol. 448, pp. 320–346. Springer, Cham (2022). https://doi.org/10.1007/978-3-031-08848-3_10
11. Ding, X., Wang, L.: The journey of action recognition. In: Companion Proceedings of the ACM on Web Conference 2025, WWW '25, pp. 1869–1884. Association for Computing Machinery (2025)
12. Firoozi, R., et al.: Foundation models in robotics: applications, challenges, and the future. Int. J. Robot. Res. **44**, 701–739 (2025)
13. Gavric, A., Bork, D., Proper, H.A.: Multimodal process mining. In: International Conference on Business Informatics, pp. 99–108. IEEE (2024)
14. Gavric, A., Bork, D., Proper, H.A.: Enriching business process event logs with multimodal evidence. In: Paja, E., Zdravkovic, J., Kavakli, E., Stirna, J. (eds.) PoEM 2024. LNBIP, vol. 538, pp. 175–191. Springer, Cham (2025). https://doi.org/10.1007/978-3-031-77908-4_11
15. Jang, D.S., et al.: Unlocking robotic autonomy: a survey on the applications of foundation models. Int. J. Control Autom. Syst. **22**, 2341–2384 (2024)
16. Janssen, D., Mannhardt, F., Koschmider, A., van Zelst, S.J.: Process model discovery from sensor event data. In: Leemans, S., Leopold, H. (eds.) ICPM 2020. LNBIP, vol. 406, pp. 69–81. Springer, Cham (2021). https://doi.org/10.1007/978-3-030-72693-5_6
17. Ji, S., Zheng, X., Wu, C.: HarGPT: are LLMs zero-shot human activity recognizers? In: 2024 IEEE International Workshop on Foundation Models for Cyber-Physical Systems & Internet of Things (FMSys), pp. 38–43 (2024)
18. Jimenez, J.F., Zambrano-Rey, G., Aguirre, S., Trentesaux, D.: Using process-mining for understating the emergence of self-organizing manufacturing systems. IFAC **51**(11), 1618–1623 (2018)
19. Jin, T., Han, X.: Robotic arms in precision agriculture: a comprehensive review of the technologies, applications, challenges, and future prospects. Comput. Electron. Agric. **221** (2024)
20. Keith, R., La, H.M.: Review of Autonomous Mobile Robots for the arehouse Environment. arXiv e-prints (2024)
21. Khalastchi, E., Kalech, M.: On fault detection and diagnosis in robotic systems. ACM Comput. Surv. **51**, 1–24 (2018)
22. Knoch, S., Ponpathirkoottam, S., Schwartz, T.: Video-to-model: unsupervised trace extraction from videos for process discovery and conformance checking in manual assembly. In: Fahland, D., Ghidini, C., Becker, J., Dumas, M. (eds.) BPM 2020. LNCS, vol. 12168, pp. 291–308. Springer, Cham (2020). https://doi.org/10.1007/978-3-030-58666-9_17
23. Kratsch, W., König, F., Röglinger, M.: Shedding light on blind spots – developing a reference architecture to leverage video data for process mining. Dec. Support Syst. **158** (2022)
24. Leng, Z., Kwon, H., Plötz, T.: On the benefit of generative foundation models for human activity recognition (2023)
25. de Leoni, M., Pellattiero, L.: The benefits of sensor-measurement aggregation in discovering IoT process models: a smart-house case study. In: Marrella, A., Weber, B. (eds.) BPM 2021. LNBIP, vol. 436, pp. 403–415. Springer, Cham (2022). https://doi.org/10.1007/978-3-030-94343-1_31
26. Melfsen, A., Lepsien, A., Bosselmann, J., Koschmider, A., Hartung, E.: Describing behavior sequences of fattening pigs using process mining on video data and automated pig behavior recognition. Agriculture **13**(8), 1639 (2023)

27. Nicoleta, T.C.: Process mining on a robotic mechanism. In: Software Testing, Verification and Validation Workshops, pp. 205–212. IEEE (2021)
28. Rebmann, A., Emrich, A., Fettke, P.: Enabling the discovery of manual processes using a multi-modal activity recognition approach. In: Di Francescomarino, C., Dijkman, R., Zdun, U. (eds.) BPM 2019. LNBIP, vol. 362, pp. 130–141. Springer, Cham (2019). https://doi.org/10.1007/978-3-030-37453-2_12
29. Roldán, J.J., Olivares-Méndez, M.A., del Cerro, J., Barrientos, A.: Analyzing and improving multi-robot missions by using process mining. Auton. Robot. **42**(6), 1187–1205 (2018)
30. Rozinat, A., Zickler, S., Veloso, M., van der Aalst, W., McMillen, C.: Analyzing multi-agent activity logs using process mining techniques. In: Asama, H., Kurokawa, H., Ota, J., Sekiyama, K. (eds.) Distributed Autonomous Robotic Systems, pp. 251–260. Springer, Heidelberg (2008). https://doi.org/10.1007/978-3-642-00644-9_22
31. Tatiya, G., Francis, J., Sinapov, J.: Transferring implicit knowledge of non-visual object properties across heterogeneous robot morphologies. In: IEEE International Conference on Robotics and Automation, pp. 11315–11321 (2023)
32. Wang, F., et al.: Muirbench: a comprehensive benchmark for robust multi-image understanding (2024)
33. Weckenborg, C., Thies, C., Spengler, T.S.: Harmonizing ergonomics and economics of assembly lines using collaborative robots and exoskeletons. J. Manuf. Syst. **62**, 681–702 (2022)
34. Weng, Y., Wu, G., Zheng, T., Yang, Y., Luo, J.: Large model for small data: foundation model for cross-modal RF human activity recognition. In: Proceedings of the 22nd ACM Conference on Embedded Networked Sensor Systems, pp. 436–449. Association for Computing Machinery (2024)
35. Zhou, Y., Yang, J., Zou, H., Xie, L.: Tent: connect language models with IoT sensors for zero-shot activity recognition (2023)

Time Series-Based Segmentation of Noisy User Interaction Logs for Robotic Process Automation

Tom Hohenadl[1,2](✉)

[1] KU Eichstätt-Ingolstadt, Ingolstadt, Germany
tom.hohenadl@stud.ku.de
[2] TH Ingolstadt, Ingolstadt, Germany

Abstract. Robotic Process Mining (RPM) automates routine discovery from UI logs for Robotic Process Automation (RPA), yet current approaches assume structured, low-noise task traces, requiring prior domain knowledge to define automatable routines. This constraint limits flexibility and applicability in real-world scenarios, as users must manually identify and record relevant tasks. We propose an approach that enables long-term user recordings without predefined routine boundaries, capturing all user interactions as they naturally occur while still identifying high-similarity, high-frequency routines suitable for RPA automation. By leveraging word2vec encoding and time-series motif discovery, our method segments task traces without prior knowledge, making the early phase of RPM more adaptive. This enhances the extraction of meaningful routines from noisy, unstructured user actions, improving automation potential in complex and evolving environments.

Keywords: Robotic Process Automation · Robotic Process Mining · UI Log · Time Series Data Mining

1 Introduction

Robotic Process Automation (RPA) aims to reduce human effort by automating repetitive, rule-based interactions across software systems. A key challenge in achieving this goal at scale is the reliable identification of candidate routines suitable for automation. Robotic Process Mining (RPM) was proposed to address this gap by leveraging data-driven methods to discover and model routine behavior based on user interaction logs (UI logs) [7]. These logs capture fine-grained sequences of user actions performed through graphical user interfaces (GUIs), offering detailed insight into how users execute tasks [1].

However, most existing RPM approaches assume that UI logs are pre-segmented into distinct task traces, each corresponding to a complete instance of an automateable routine. In practice, this assumption is rarely met. Tools often require users to manually initiate and stop recordings when performing a known task [3,12], effectively placing the burden of segmentation on the user.

M. ER et al. (Eds.): BPM 2025, LNBIP 565, pp. 276–291, 2026.
https://doi.org/10.1007/978-3-032-02936-2_20

This restricts data collection to tasks that users believe are automateable, leading to UI logs that lack unrelated behavior – what we refer to as noise in the context of UI logs.

Such constraints contradict the broader automation vision of RPA: reducing the manual burden not only in task execution but also in automation preparation. A promising alternative is to continuously record user activity over longer timeframes, thereby capturing routine instances as they naturally occur. Yet this creates a new technical challenge: the resulting UI logs contain extensive non-routine interactions, redundant behavior, and unstructured noise. Extracting meaningful routine candidates from such data demands advanced segmentation techniques capable of identifying automateable routines embedded in unsegmented, noisy interaction logs.

To address the challenge of segmenting candidate routines for automation from continuous, high-noise user interaction recordings, we propose a novel integration of word embedding and time-series motif discovery within the RPM pipeline. Specifically, our approach encodes user interaction logs as real-valued time series using word2vec [5], preserving the semantic structure of user actions across context parameters. We then apply multi-dimensional motif discovery based on matrix profiles [17], a technique successfully used in domains such as finance and healthcare [15], to identify repeated behavioral patterns – motifs – that are indicative of routine executions.

This combination of semantic embedding and unsupervised time-series pattern detection forms the core innovation of our approach. It enables segmentation of recurring routines directly from noisy, unstructured UI logs without requiring predefined boundaries or user input. In a final step, we cluster the discovered motifs using DBSCAN [8], following routine clustering principles similar to Rebmann and van der Aa [14], to group structurally similar motifs into task traces suitable for downstream RPA automation.

The remainder of this paper is structured as follows. Section 2 reviews the RPM pipeline and its foundational concepts. Section 3 introduces our time series-based segmentation and clustering approach with Sect. 4 evaluating the approaches performance on two real-world UI logs. Finally, Sect. 5 concludes the paper by discussing potential improvement measures and future work.

2 Related Work

RPM extends traditional process mining techniques to facilitate the identification and automation of user routines from UI logs, supporting RPA development. The RPM pipeline consists of multiple phases (cf. Fig. 1): UI log recording and preprocessing, trace processing, automation specification generation, and RPA compilation [7]. At its core, RPM relies on UI logs, which capture timestamped user actions across software applications, including keystrokes, clicks, and data entries [1]. These logs contain data parameters, which vary across routine executions (e.g., transaction amounts, customer IDs), and context parameters, which remain stable across multiple executions (e.g., UI element identifiers, workflow

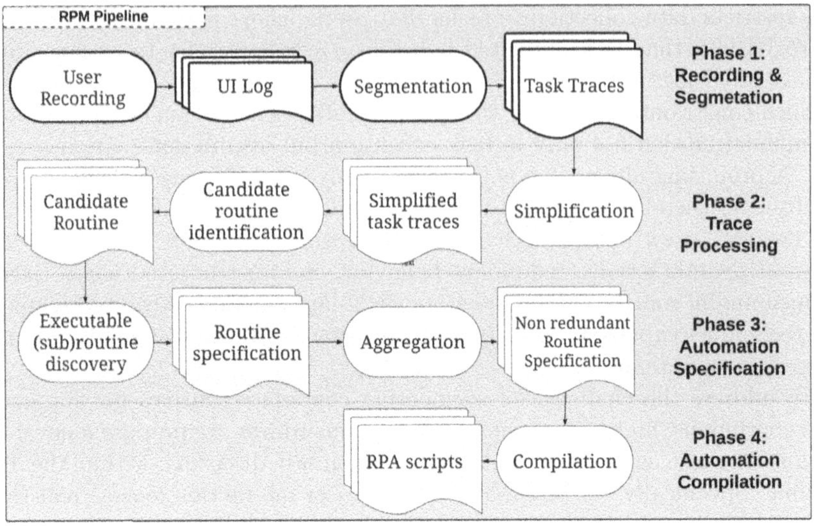

Fig. 1. RPM pipeline based on [7]

structures) [12]. Once the user interactions are collected they are segmented into routine executions traces with each trace representing the execution of a routine. A key challenge in RPM is this segmentation. Unlike event logs used in process mining, UI logs capture continuous user interactions without explicit delimiters, so-called case identifiers [1], resulting in potentially highly variable, interleaved user interaction datasets [4]. Often user behavior is influenced by interruptions, exploratory actions, and corrective steps, complicating segmentation [4]. Existing segmentation approaches generally fall into model-driven and unsupervised methods. Model-driven techniques use predefined process templates to guide segmentation but fail to generalize beyond known model structures [4]. Unsupervised methods, such as control-flow graphs and sequence clustering, attempt to reconstruct routines directly from UI logs [12]. While these approaches help untangle interleaved task instances, their accuracy declines when multiple task traces share user actions or contain high noise levels [4].

Once the task traces are segmented existing approaches, such as Robidium [13] and SmartRPA [2], demonstrate the feasibility of extracting candidate routines and compiling them into executable RPA scripts [6] following the phases 2 to 4. First, simplifcation aimed at identifying user behavior not required in the RPA bot should be identified. Dumas et al. [7] particularly name user actions that do not have any impact as their outcome is amended or overwritten through subsequent actions or actions that do not affect the outcome of the routine at all. Afterwards, these systems apply frequent pattern mining and rule-based heuristics to detect repetitive behavior in the traces, assessing their automatability based on criteria such as frequency and determinism [11].

Dumas et al. [7] further refine the executable (sub)routine discovery in phase 3 in Fig. 1 to identify data transfer patterns, that need to be automated through the RPA bot. In Bosco et al. [6] methods for data transformations are discussed to achieve the mining of behavior during the transfer of data from one application to another to reduce the number of actions to be implemented in the RPA bot. Once all executable routine specifications are discovered, the RPM pipeline does advance by aggregating redundant routines, such as filling forms in different order into a single routine specification [7].

Finally, the remaining routine specification is mapped onto executable RPA operations, such as clicks, key-strokes or mouse movement events [2] leading to a compiled RPA script, often referred to as bot.

3 Time Series-Based Segmentation

The extended RPM pipeline begins with the long time recording of user interactions over extended periods, capturing all performed actions without requiring manual segmentation. Once the recording concludes, all collected UI logs are merged into a single long-term, high-noise dataset. Therefore the approach visualized in Fig. 2 differs in the user recording leading to an extensive UI log instead to a UI log containing only potential RPA candidate traces as in the traditional RPM pipeline shown in Fig. 1. Our approach assumes users record interactions over extended periods, capturing routines in a UI log that includes unrelated user actions. Once these long-term, high-noise recordings are collected, we apply time-series-based segmentation, also referred to as motif discovery [17], to identify recurring action patterns without predefined task boundaries. These motifs are further clustered into related sets of routines in the UI log to be processes as potential task traces in the RPM pipeline. This process enables the automatic extraction of meaningful routines from raw UI logs, eliminating the need for manual segmentation.

In this section, we outline the three key steps of our segmentation method visually displayed in Fig. 2: Step 1 is UI log encoding, transforming the UI log into a real-valued time series, step 2 is matrix profil calculation and motif discovery, and step 3 the clustering of the identified motifs into processable RPA candidate task traces.

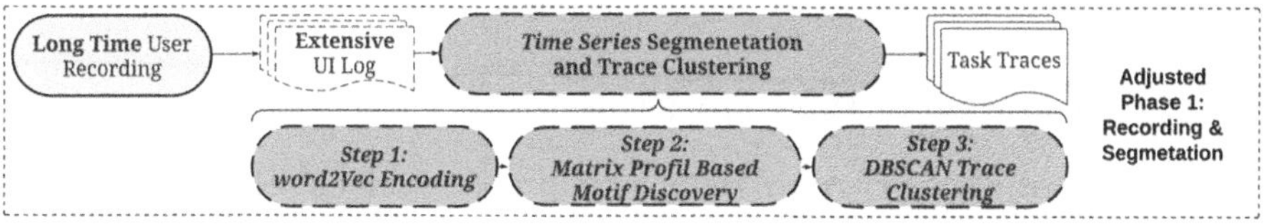

Fig. 2. Adjusted RPM Phase 1 extended through new Time Series Based Approach

3.1 Encoding of UI Logs as Time Series

To enable the application of time series data mining techniques, the UI log must first be transformed into a real-valued time series representation (cf. Fig. 2 Step 1). This is achieved through the use of word2vec encoding of user interactions in the UI log. While word2vec is typically applied by treating each trace as a sentence and the full log as a text corpus [5], we instead treat each action – enriched with its context attributes such as the interacted object and interaction type (e.g., left click or right click) [1] – as a sentence, and the entire UI log as the corresponding text. The encoding function constructs a single pseudo-sentence by concatenating user interactions - each composed of its context parameters values - into unique tokens. Using the skip-gram variant of word2vec [9], the model creates a fixed-length vector representation for each token, which is subsequently merged into the original UI log to yield a multi-dimensional real-valued embedding per action.

An example of the resulting encoded time series is presented Table 1, illustrating its application on a subset of the experimental dataset used in the experiment in Sect. 4. The columns show data (*Value*) and context (*App, Event, Type, Name*) parameters, and the columns starting with *w2v_* containg the word2vec vector.

Table 1. Contextual attributes and word2vec embeddings for first actions in $SRRT_+$ log

Timestamp	App	Event	Type	Name	Value	...	w2v_0	w2v_1	w2v_2	...
19-10-21 00:10:09	Excel	copyCell			John	[...]	−0.024	−0.013	−0.014	[...]
19-10-21 00:10:12	Chrome	paste	text	SingleLine	John	[...]	0.044	−0.100	0.114	[...]
19-10-21 00:10:13	Chrome	editField	text	SingleLine	John	[...]	0.030	−0.078	0.108	[...]
19-10-21 00:10:16	Excel	copyCell			Doe	[...]	0.005	0.001	0.004	[...]
19-10-21 00:10:18	Chrome	paste	text	SingleLine	John	[...]	0.044	−0.100	0.114	[...]
19-10-21 00:10:19	Chrome	editField	text	SingleLine	John Doe	[...]	0.030	−0.078	0.108	[...]
19-10-21 00:10:23	Excel	copyCell			11/03/1986	[...]	−0.029	0.017	−0.026	[...]
19-10-21 00:10:27	Chrome	paste	text	date	11/03/1986	[...]	0.010	−0.034	0.052	[...]
19-10-21 00:10:31	Chrome	editField	text	date	11/03/1986	[...]	0.003	−0.085	0.110	[...]
19-10-21 00:10:35	Excel	copyCell			+61 039...	[...]	−0.014	−0.017	−0.007	[...]

3.2 Segmentation of Routines Using Motif Discovery

We apply motif discovery using the matrix profile technique [17], implemented via the STUMPY library [10] for segmentation of the now multi-dimensional real-valued UI log into candidate routines. Based on a fixed window size w_s, a sliding window is applied over the resulting vectors from the previous step. For each subsequence, the algorithm computes the z-normalized Euclidean distance to its nearest non-overlapping neighbor in the series. The resulting matrix profile

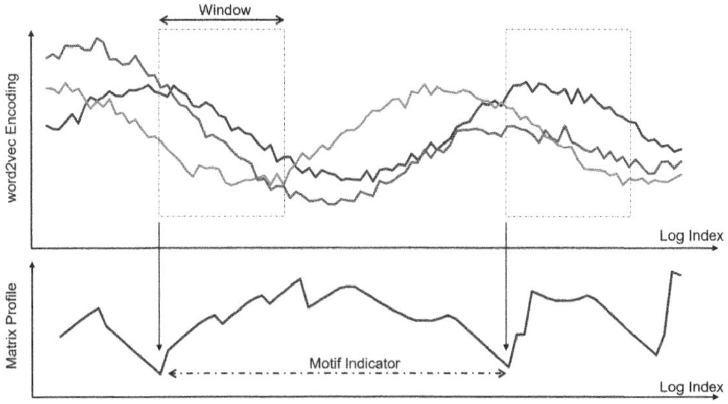

Fig. 3. word2vec encoded UI log and Matrix Profile plot

is a one-dimensional vector that reflects how similar each subsequence is to its closest match elsewhere in the data. Motifs are extracted as the t subsequences with the lowest matrix profile values–these represent the most similar recurring behavior routines in the log. Each motif is then mapped back to its original start index, and a segment of length w_s is labeled using an artificial case identifier. Therefore, this segmentation yields a set of candidate routine instances, which can be processed as RPA candidates in the RPM pipeline.

Figure 3 illustrates this process. The graph on top of the figure presents a three dimensionsonal example of a multi-dimensional time series derived from a UI log. The graph on the bottom of the figure presents the corresponding matrix profile, where the distinct valleys highlight potential repeated subsequences. The dashed windows represent the window size long sequence that lead to the vally, i.e., the low distance value during the matrix profile calculation. Both valleys together indicate a potential motif in the data.

3.3 Trace Clustering Using DBSCAN

To group similar motifs into candidate routine types, we apply density-based clustering using DBSCAN on the discovered motif segments aligned on Rebmann and van der Aa [14]. Each motif is represented by the average word2vec embedding of its actions context parameters, capturing its semantic and contextual structure. These mean vectors are standardized and clustered using DBSCAN with Euclidean distance. This unsupervised method groups motifs with similar structure into clusters without requiring a predefined number of routine types, while also identifying outliers. Each resulting cluster corresponds to a set of routine instances exhibiting similar user behavior and can be treated as a candidate task class in the RPM pipeline.

4 Evaluation of the Segmentation Pipeline

To assess the robustness and practical utility of our adjusted segmentation phase within the RPM pipeline (cf. Fig. 2), we conduct a staged evaluation targeting four key design dimensions: context parameter selection, window size granularity, motif saturation via discovery volume, and clustering quality.

These evaluation stages correspond directly to the core components introduced in Sect. 3: the encoding of user actions (Step 1), time series-based motif discovery (Step 2), and clustering of routines (Step 3). Each stage of the evaluation isolates a specific technical choice that influences the success of identifying automatable routines from noisy, long-term UI logs.

The evaluation setup reflects realistic usage scenarios, where routines are interleaved and partially obscured by non-routine interactions. This departs from idealized assumptions in prior RPM work that logs are cleanly segmented into routine traces. By analyzing how each design decision impacts the discovery performance, we identify the conditions under which segmentation succeeds or fails-and inform the design of more robust automation segmentation from noisy, long-term UI logs.

4.1 Dataset and Setup

To evaluate our segmentation approach under realistic yet controlled conditions, we utilize the structured UI logs introduced by Leno et al. [12]. These logs originate from two real-world business processes: Student Record Processing (SR) and Reimbursement Requests (RT), and were previously combined into two composite logs: in the $SRRT_{+}$ log, all SR traces precede RT traces; in the $SRRT_{||}$ log, traces alternate between the two routines. Routine lengths in the dataset average around 31 events for the SR traces and average around 62 events for the RT traces. Each routine is represented fifty times in the log leading to an original log length of 4653 actions.

In their original form, these logs reflect the typical assumption in RPM: logs are composed of fully segmented, routine-level traces. However, this assumption does not hold in long-term UI recordings, where process executions are often interleaved with unrelated user behavior. In an analysis of real-world UI logs from Agostinelli et al. [3], a screenshot example indicated that non-task-related actions contributed approximately 25% of the log, with four routines present. To reflect this observation while maintaining experimental traceability, we inserted 25 synthetic noise actions between each routine in both logs. These actions were sampled from common UI interaction parameters-such as URLs, action types, or window titles. This yields a total of 2475 additional actions per log, preserving the original routines while simulating noise that interferes with segmentation. From here on, we refer to these extended versions of the logs when discussing $SRRT_{+}$ and $SRRT_{||}$.

4.2 Evaluation Metrics

The segmentation and discovery quality is evaluated using a set of quantitative metrics designed to reflect both alignment with ground truth routines and downstream automation usability.

True Positives. A discovered motif is counted as a true positive (TP) if it sufficiently overlaps with a ground truth routine instance in the UI logs. We define overlap using a coverage-based criterion, which tolerates minor temporal misalignments-common in real-world interaction data-while preserving the requirement that a motif captures a majority of a routine's execution span.

Let the window size be w, the discovered motif be $d = [d_{start}, d_{start} + w)$ and the ground truth interval be $g = [g_{start}, g_{end})$. The overlap and coverage score are defined as:

$$\text{overlap}(d, g) = \max(0, \min(d_{start} + w, g_{end}) - \max(d_{start}, g_{start})) \tag{1}$$

$$\text{coverage}(d, g) = \frac{\text{overlap}(d, g)}{w} \tag{2}$$

A discovered motif d is labeled as a true positive if it meets either of the following:

$$TP = \begin{cases} 1, & \text{if coverage}(d, g) \geq 0.8 \text{ or } (g_{start} \geq d_{start} \text{ and } g_{end} \leq d_{start} + w) \\ 0, & \text{otherwise} \end{cases} \tag{3}$$

The threshold of 80% was chosen to balance strictness and tolerance: a lower threshold risks counting coincidental or partially matching segments as valid, while a higher one penalizes slight misalignments caused by routine variation or discovery windowing. Additionally, if a discovered motif completely covers the ground truth motif, it is also considered a true positive.

Discovery Metrics. Based on the TP count, we compute standard evaluation metrics to assess discovery quality (Table 2):

Table 2. Formulas for discovery evaluation metrics

Precision $= \frac{TP}{t}$	Recall $= \frac{TP}{N}$	F_1-Score $= \frac{2 \cdot \text{Precision} \cdot \text{Recall}}{\text{Precision} + \text{Recall}}$

Here, t is the number of motifs requested from the discovery algorithm, and $N = 100$ is the total number of ground truth routine instances in the log (50 SR and 50 RT). These fixed parameters provide a stable reference for comparing different configurations across our staged evaluation.

Clustering Quality. For evaluating the post-discovery grouping of motifs, we use cluster purity-a widely adopted measure that reflects the homogeneity of discovered clusters with respect to ground truth routine labels. It is defined as:

$$\text{Purity} = \frac{1}{N} \sum_{k} \max_{j} |C_k \cap L_j| \tag{4}$$

Here, C_k is the set of motifs in cluster k, and L_j is the set of all motifs labeled as belonging to routine class j (either SR or RT). A high purity indicates that most motifs in a cluster share the same semantic routine origin-an essential property for downstream automation steps. To ensure clarity and maintain a clear overview, only a representative subset of the metrics is shown in the figures and tables below; the full set is available in the accompanying repository.[1]

4.3 Stage 1: Context Parameter Selection

Objective. This stage evaluates how the inclusion of additional context parameters affects the quality and consistency of motif discovery. In the adjusted RPM segmentation pipeline, context parameters form the basis of the vector encoding (*Step 1*), which influences how routines are represented in the time series. Prior work on multi-dimensional time series [10,17] suggests that adding irrelevant or noisy dimensions can degrade discovery performance. This experiment stage tests this sensitivity in the context of routine extraction from UI logs.

Method. Following the context parameter schema of Leno et al. [12], we incrementally add one context parameter at a time, resulting in 16 cumulative configurations. The window size was fixed at $w = 30$, which aligns with the average length of SR routines and is sufficient to partially capture RT routines, allowing both routines to be discovered. The number of motifs to discover is set to $t = 10$. This constrains recall to a maximum of 0.1 and F_1-Score to approximately 0.18, allowing for controlled comparison of discovery precision and motif localization.

Findings. As shown in Table 3 for the $SRRT_+$ log, discovery quality is initially affected by dimensionality. With only a single context parameter (application name), the algorithm correctly detects 7 out of 10 motifs, yielding a precision of 0.7. Starting from two dimensions onward, precision reaches 1.0 (recall = 0.1, F_1-score = 0.18) and remains stable across all higher-dimensional configurations. However, the discovered motifs shift both in position and type: for configurations with more than five parameters, all discovered motifs correspond to RT routines, not SR routines. This shift reflects how added dimensions reshape the distance landscape of the embedding, making certain routines more prominent.

A comparable effect is observed in the $SRRT_{\|}$ log. Here, the transition toward discovering only RT routines occurs earlier, already after including two context parameters, indicating that alternation of SR and RT traces amplifies the dominance of longer routine structures when more contextual information is added to the embedding.

[1] Data and Code Repository: https://github.com/thohenadl/TimeSeriesInRPM.

Table 3. Discovery result by number of context parameters within $SRRT_+$

No. of Parameters	Precision	Discovered Motif Starts & Lengths
1	0.7	[(54, 31), (498, 31), (554, 31), (1890, 31), (2003, 31), (2395, 31), (2675, 31)]
2–4	1.00	[(54, 31), (164, 31), (220, 31), (276, 31), (443, 30), (498, 31), (554, 31), (664, 31), (720, 31), (833, 31)]
5	1.00	[(54, 31), (164, 31), (220, 31), (276, 31), (386, 32), (498, 31), (554, 31), (664, 31), (720, 31), (833, 31)]
6–7	1.00	[(3134, 62), (3221, 62), (3396, 62), (3483, 62), (3744, 62), (3831, 62), (4005, 62), (4356, 62), (4443, 64), (5141, 62)]
8	1.00	[(2960, 62), (3047, 62), (3396, 62), (3657, 62), (3744, 62), (3918, 62), (4269, 62), (4356, 62), (4880, 62), (5141, 62)]
9	1.00	[(3134, 62), (3221, 62), (3918, 62), (4182, 62), (4443, 64), (4706, 62), (5054, 62), (5141, 62), (5578, 62), (6100, 62)]
10	1.00	[(2787, 61), (2960, 62), (3047, 62), (3134, 62), (3396, 62), (3483, 62), (3744, 62), (4092, 65), (4532, 62), (4880, 62)]
11	1.00	[(3396, 62), (3657, 62), (4005, 62), (4443, 64), (4706, 62), (4967, 62), (5141, 62), (5228, 62), (6187, 62), (6798, 62)]
12	1.00	[(3047, 62), (3483, 62), (3744, 62), (3831, 62), (4092, 65), (4182, 62), (4269, 62), (4356, 62), (4880, 62), (5141, 62)]
13–15	1.00	[(2960, 62), (3047, 62), (3134, 62), (3221, 62), (4182, 62), (4269, 62), (4532, 62), (4793, 62), (4880, 62), (6100, 62)]
16	1.00	[(3047, 62), (3134, 62), (3483, 62), (3657, 62), (3744, 62), (4269, 62), (4793, 62), (4880, 62), (5141, 62), (5228, 62)]

Implications. Although precision remains stable beyond two parameters, the underlying discovery behavior changes: the detected routines shift toward longer RT routines, while shorter SR routines are increasingly omitted. In both logs, discovery becomes biased toward structurally dominant patterns as more parameters are added. This suggests that additional context parameters influence not only discovery quality but also which types of routines are revealed. Consequently, context selection should be treated as a sensitive configuration step-especially when aiming to detect shorter or less frequent routines in interleaved logs.

4.4 Stage 2: Window Size Selection

Objective. This stage evaluates how the choice of window size influences motif discovery quality in the segmentation pipeline. Fixed-size windowing is a central design decision in motif-based segmentation [16,17], but its suitability depends on how well it aligns with the temporal structure of the routines in the log. Here, we assess whether varying window sizes enables discovery of both short SR and long RT routines.

Method. We fixed the context parameters to all available context parameters and the number of motifs to discover at $t = 10$. The window size was incrementally increased from 26 to 68, spanning a ±5 range around the known routine lengths (SR: 31, RT: 63). Metrics were computed as described in Sect. 4.2.

Findings. In the $SRRT_+$ log, motif discovery consistently favored the longer RT routines across all window sizes. Even for small windows (e.g., 2635), all matched motifs overlapped RT traces. This confirms a bias toward longer routines once they are sufficiently represented in the log. Performance metrics (precision = 1.0, recall = 0.1, F_1-score = 0.18) remained unchanged, but no SR routine was detected across the full range.

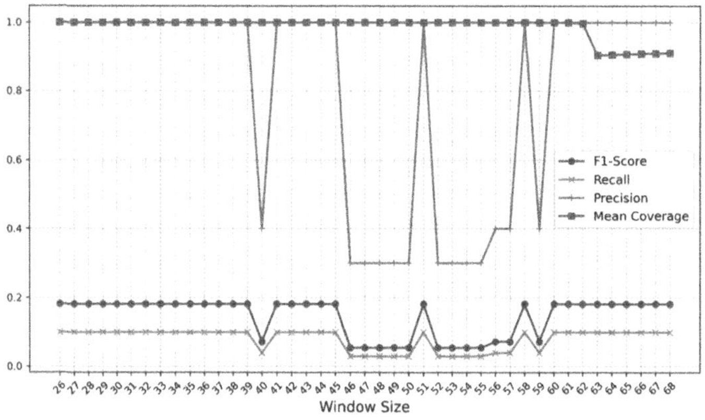

Fig. 4. Performance scores for $SRRT_{||}$

In contrast, the $SRRT_{||}$ results reveal substantial performance degradation across the window size range of 40 to 58, as shown in Fig. 4. While early configurations (window sizes < 40) achieved consistent results, all key metrics dropped significantly once the window size surpassed the average SR routine length. Between sizes 40 and 58, the model failed to consistently align motifs with ground truth routines: precision declined to as low as 0.3, recall to 0.03, and F1-score followed the same downward trend. Although coverage remained high, it gradually decreased as well after the window size exceeded the longest motif length. This performance dip reflects the difficulty of motif discovery under interleaved execution, where fixed-length windows between the SR and RT routine lengths often capture ambiguous or impure sequences. Metrics only recovered beyond window size 59, where full RT routines were more likely to be matched entirely.

Implication. The results indicate that fixed window-based motif discovery is highly effective for well-separated routines but becomes worse under alternating or when multiple routine lengths coexist. The approach is capable of high precision, but its sensitivity to sequence structure limits generalizability. Discovery outcomes shift depending on routine dominance, window length, and noise. For practical applications, especially in unsegmented logs, an iterative or adaptive strategy is needed-one that dynamically adjusts the window size or removes already-identified motifs to reveal weaker patterns. This would improve recall and coverage without compromising precision in mixed or noisy environments. One such future research direction is the use of variable-length motif discovery as suggested in [16] to segment different length routines.

4.5 Stage 3: Number of Motifs and Discovery Saturation

Objective. This stage investigates how the number of motifs requested from the discovery algorithm influences routine coverage, bias, and precision. In prior stages, we observed that the discovery process favored longer RT routines. Here, we test whether increasing the motif count t leads to broader coverage, particularly of shorter SR routines, and whether overfitting or false positives emerge as t increases.

Method. The context parameters and window size were fixed ($w = 30$) under the same assumption as in Sect. 4.3. The motif count was varied from $t = 10$ to $t = 100$ in steps of 10. This range spans the total number of routines present in the log (100 = 50 SR + 50 RT), enabling a full saturation analysis.

Findings. As shown in Fig. 5 for the $SRRT_+$, precision remained stable at 1.0 up to $t = 60$. During this range, only RT routines were discovered, consistent with observations from Sect. 4.3. Once the number of discovered motifs exceeded the number of RT routines ($t > 50$), SR routines began to appear in the results.

Beyond $t = 60$, the first decline in mean coverage was observed, indicating increasingly partial or misaligned matches. Precision began to decrease from $t = 60$ onward, as noise-induced matches were discovered. In the final configuration ($t = 100$), 91 out of 100 routines were correctly identified. The earliest signs of degradation in $SRRT_+$ appeared slightly earlier than in $SRRT_{||}$, with missed matches starting at $t = 90$ and compounding up to 7 missed routines by $t = 100$.

Implications. Again, the results show that the motif discovery algorithm initially prioritizes longer, structurally dominant routines (RT), while shorter routines (SR) emerge only after motif count saturation. Increasing t steadily improves recall and F_1-Score, but at the cost of degraded precision due to noise and partial matches. These findings suggest that relying on a single global discovery run may be limiting; instead, an iterative strategy that incrementally removes identified routines and reapplies discovery could help surface additional, smaller routines with improved accuracy.

Notably, the $SRRT_{||}$ log yielded more stable results across increasing motif counts. Performance degradation was observed only at $t = 100$, while in the $SRRT_+$ log precision began to drop already at $t = 50$. This indicates that the alternating structure in $SRRT_{||}$ may support more robust separation of routines, enabling more consistent motif detection even as discovery volume increases.

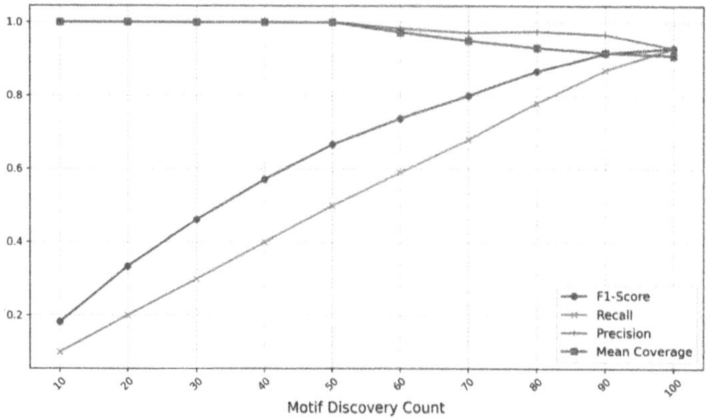

Fig. 5. Effect of increasing the discovered motif count in the $SRRT_{+}$ log

4.6 Stage 4: Clustering Quality for Automation

Objective. The goal of this final experiment is to assess whether discovered routines can be grouped into semantically coherent clusters that reflect automation-ready process units. In the context of RPM, clustering enables the transition from motif discovery to executable automation candidates in RPA. We evaluate if discovered routines form clear, type-consistent clusters-specifically separating SR and RT processes. If clustering yields stable and interpretable structure, the resulting groups can be directly mapped to automation routines, closing the loop from discovery to actionability.

Method. We applied DBSCAN to cluster the discovered motifs using their word2vec-based vector representations. Rather than exhaustively evaluating all combinations of discovery settings, we selected a representative subset that aligns with the behaviors observed in previous stages. Specifically, we tested five different window sizes ($w \in 30, 40, 50, 60, 65$) and four motif discovery volumes ($t \in 10, 50, 70, 100$), covering three relevant discovery regimes as identified in the previous steps:

- Low discovery volume ($t = 10$): only long RT routines are found.
- Medium discovery volume ($t = 50$–70): both RT and SR routines are discovered.
- High discovery volume ($t = 100$): full routine coverage with some noise.

Findings. Clustering results reflect a clear semantic separation between SR and RT routines when motif count is high. As shown in Fig. 6 for the $SRRT_{+}$ log with $w = 30$, the low-count settings ($t = 10$ on the left) yield a single compact cluster dominated by RT routines and a purity of 1.0. In contrast, higher counts ($t = 100$ on the right) produced two distinct clusters, each aligned with either the SR or RT length range and a lowered overall purity of 0.906 as some noise was included as described in Sect. 4.5.

For smaller window sizes ($w = 30$–50 and $t > 50$, cluster separation is stable, with one cluster predominantly containing SR motifs and the another cluster with RT motifs. However, at larger window sizes ($w > 50$), noise and misaligned motifs lead to only RT routines consistently being discovered, resulting in reduced performance scores (precision, recall, and F_1-score around 0.51). This leads to a single clusters of RT routines and no SR routines with any t-value. However, the purity of these RT clusters is high with values at 0.9 for $t = 50, 70$ or 100.

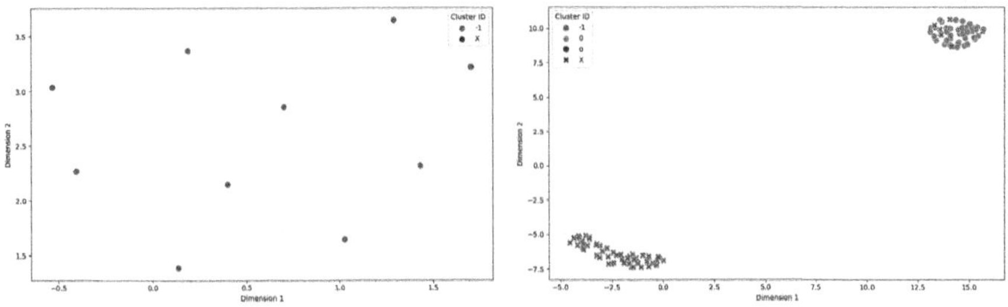

Fig. 6. 2D-Clustering for $SRRT_+$ motif discovery at $t = 10$ and $t = 100$

Implications. These results confirm that unsupervised clustering of discovered motifs can expose meaningful routine groupings, making them suitable for downstream automation steps in RPM. The emergence of two interpretable clusters aligns with the original routine structure of the UI logs, demonstrating that motif discovery followed by clustering can effectively isolate process segments in small window sizes. Therefore a selection of a smaller window size will initially provide better clustering results. However, clustering quality is sensitive to the discovery parameters: sufficient motif volume is required to reveal routine diversity, while excess leads to noise-induced purity degradation. For practical RPM pipelines, this supports the use of post-discovery clustering as a filtering and labeling mechanism, enabling a robust transition from discovery to automation candidate definition.

5 Conclusion and Future Work

This paper presented a segmentation approach that integrates word embeddings, time-series motif discovery, and clustering to identify recurring routines in long-term, high-noise UI logs. Our evaluation demonstrated that this method is capable of revealing repeated routine patterns without relying on prior segmentation or user annotations. In doing so, we contribute a modular enhancement to the upstream segmentation phase of the RPM pipeline.

The results across four stages of experimentation confirm the viability of this approach: context-aware encoding and motif discovery enable routine identification in high-noise recordings; segmentation performance can be fine-tuned via motif count and window size; and clustering the discovered routines yields coherent groups aligned with automation candidates. These outcomes validate the central assumption of RPM-that behavioral structure can be mined from raw logs-and show how vector representations and motif discovery effectively support this goal.

At the same time, we uncovered important limitations. The method remains sensitive to parameter settings and less robust under interleaved execution. Its utility for real-world RPA depends on seamless integration into the full RPM pipeline, particularly for automation script generation. Moreover, future work must evaluate its performance on more complex UI logs-featuring more routines, higher overlap, and diverse user behavior.

Despite these challenges, our findings underline a core opportunity for process automation: with minimal assumptions and no manual trace boundaries, it is possible to surface meaningful routines directly from raw user data. By reducing the human effort required not only for routine execution but also for automation preparation, this work moves one step closer to scalable RPA systems. We see this as a foundational contribution to RPM's role in end-to-end process automation-enabling robust, data-driven discovery even in the messy reality of human-computer interaction.

References

1. Abb, L., Rehse, J.R.: Process-related user interaction logs: state of the art, reference model, and object-centric implementation. Inf. Syst. 102386 (2024). https://doi.org/10.1016/j.is.2024.102386
2. Agostinelli, S., Hohenadl, T., Marrella, A., Martínez-Rojas, A.: SmartRPA: generating software robots from user interface logs. SoftwareX **29**, 101995 (2025). https://doi.org/10.1016/j.softx.2024.101995
3. Agostinelli, S., Leotta, F., Marrella, A.: Interactive segmentation of user interface logs. In: Hacid, H., Kao, O., Mecella, M., Moha, N., Paik, H. (eds.) ICSOC 2021. LNCS, vol. 13121, pp. 65–80. Springer, Cham (2021). https://doi.org/10.1007/978-3-030-91431-8_5
4. Agostinelli, S., Marrella, A., Mecella, M.: Exploring the challenge of automated segmentation in robotic process automation. In: Cherfi, S., Perini, A., Nurcan, S. (eds.) RCIS 2021. LNBIP, vol. 415, pp. 38–54. Springer, Cham (2021). https://doi.org/10.1007/978-3-030-75018-3_3
5. Barbon Junior, S., Ceravolo, P., Damiani, E., Marques Tavares, G.: Evaluating trace encoding methods in process mining. In: Bowles, J., Broccia, G., Nanni, M. (eds.) DataMod 2020. LNCS, vol. 12611, pp. 174–189. Springer, Cham (2021). https://doi.org/10.1007/978-3-030-70650-0_11
6. Bosco, A., Augusto, A., Dumas, M., La Rosa, M., Fortino, G.: Discovering automatable routines from user interaction logs. In: Hildebrandt, T., van Dongen, B.F., Röglinger, M., Mendling, J. (eds.) BPM 2019. LNBIP, vol. 360, pp. 144–162. Springer, Cham (2019). https://doi.org/10.1007/978-3-030-26643-1_9

7. Dumas, M., La Rosa, M., Leno, V., Polyvyanyy, A., Maggi, F.M.: Robotic process mining. In: van der Aalst, W.M.P., Carmona, J. (eds.) Process Mining Handbook, pp. 468–491. Springer, Cham (2022). https://doi.org/10.1007/978-3-031-08848-3_16
8. Ester, M., Kriegel, H.P., Sander, J., Xu, X.: A density-based algorithm for discovering clusters in large spatial databases with noise (1996). https://cdn.aaai.org/kdd/1996/kdd96-037.pdf?source=post_page
9. Goldberg, Y., Levy, O.: word2vec explained: deriving Mikolov et al.'s negative-sampling word-embedding method. http://arxiv.org/pdf/1402.3722
10. Law, S.: Stumpy: a powerful and scalable python library for time series data mining. J. Open Source Software **4**(39), 1504 (2019). https://doi.org/10.21105/joss.01504
11. Leno, V., Augusto, A., Dumas, M., La Rosa, M., Maggi, F.M., Polyvyanyy, A.: Identifying candidate routines for robotic process automation from unsegmented UI logs. In: ICPM 2020 Proceedings, pp. 153–160 (2020). https://doi.org/10.1109/ICPM49681.2020.00031
12. Leno, V., Augusto, A., Dumas, M., La Rosa, M., Maggi, F.M., Polyvyanyy, A.: Discovering data transfer routines from user interaction logs. Inf. Syst. **107** (2022). https://doi.org/10.1016/j.is.2021.101916
13. Leno, V., Deviatykh, S., Polyvyanyy, A., La Rosa, M., Dumas, M., Maggi, F.M.: Robidium: automated synthesis of robotic process automation scripts from UI logs (2020)
14. Rebmann, A., van der Aa, H.: Recognizing task-level events from user interaction data. Inf. Syst. **124**, 102404 (2024). https://doi.org/10.1016/j.is.2024.102404
15. Torkamani, S., Lohweg, V.: Survey on time series motif discovery. WIREs Data Min. Knowl. Disc. **7**(2) (2017). https://doi.org/10.1002/widm.1199
16. van Onsem, M., Ledoux, V., Mélange, W., Dreesen, D., van Hoecke, S.: Variable length motif discovery in time series data. IEEE Access **11**, 73754–73766 (2023). https://doi.org/10.1109/ACCESS.2023.3295995
17. Yeh, C.C.M., Kavantzas, N., Keogh, E.: Matrix profile VI: meaningful multidimensional motif discovery. In: 2017 IEEE International Conference on Data Mining (ICDM), pp. 565–574. IEEE (2017). https://doi.org/10.1109/icdm.2017.66

Educators Forum

Preface of the Educators Forum

The Educators Forum seeks to bring together educators within the BPM community for sharing resources for improving the practice of teaching BPM-related topics by exchanging experiences, bringing in cases, and discussing teaching innovations. The forum continues its mission to foster a vibrant community of educators, researchers, and practitioners dedicated to advancing Business Process Management (BPM) education. This year's forum reflects the evolving landscape of BPM pedagogy, showcasing innovative teaching methodologies, empirical studies, and practical tools that address the multifaceted challenges of BPM instruction across the globe.

The forum received a total of 24 submissions, of which 12 were accepted. The review process followed a double-blind model, where each paper was reviewed by at least two reviewers. The geographic diversity of submissions highlights the global interest in **BPM education**. Papers were submitted from nine countries, with the highest number from Poland (13), followed by Brazil (3) and Belgium (3). Accepted papers came from eight countries, with Poland (3) and Brazil (2) again leading in representation.

To embrace a wide range of educational innovations and experiences, this year's forum accepted six types of contributions. We provided detailed information on each contribution type to guide both authors and reviewers in preparing high-quality submissions aligned with the intended format and purpose. BPM Education Research papers present empirical or theoretical investigations into how BPM is taught and learned, often including experimental designs, surveys, or pedagogical frameworks. **BPM Teaching Cases** offer report on practical, real-life examples or anonymized/hypothetical situations of BPM in practice, offering students opportunities to engage with complex issues and apply theory in a problem-solving context. **Educational Practice Insights** share reflective accounts or lessons learned from implementing BPM teaching strategies, tools, or curricula, emphasizing practical relevance and transferability. **Tools/Resources Contributions** introduce software, platforms, or educational materials designed to support BPM instruction, often accompanied by usage scenarios or evaluations. **Assignment Contributions** provide detailed descriptions of BPM-related assignments, including alternative assessment methods, project-based evaluations, or creative ways to test understanding, promoting critical thinking, problem-solving, and practical BPM application. Finally, **Lightning Contributions** are short, focused presentations of innovative ideas, preliminary results, or provocations intended to spark discussion and inspire future work.

The submissions spanned the six contribution types. The most common category was BPM Education Research, with 10 submissions and 3 acceptances, reflecting continued interest in empirical and theoretical investigations into BPM education. Educational Practice Insights also featured prominently, with 9 submissions and 5 acceptances, highlighting the value of reflective teaching experiences. Other accepted contributions included BPM Teaching Cases, Assignment Contributions, and Lightning Contributions, each offering practical, focused insights into BPM instruction. Although one Tools/Resources Contribution was submitted, it was not accepted this year.

The diversity of topics presented in the forum underscores the richness and maturity of BPM education as a research and practice domain. A dominant theme is the widespread adoption of **experiential and project-based learning**. Several papers describe hands-on, real-world BPM assignments that foster critical thinking, creativity, and reflective practice. These include interdisciplinary reengineering projects (Gdowska et al. 2025), healthcare case studies (Grangel et al. 2025), simulation-based competitions (Schumann et al. 2025), and validated instructional blueprints for next-generation BPM (Dallavalle et al. 2025), all of which emphasize the integration of theory and practice through iterative, student-centered learning.

Closely related is the theme of **technology-enhanced learning**, where educators leverage Low-Code Platforms (Grzybowska & Hoffa-Dąbrowska, 2025), process simulation tools (Gdowska et al. 2025; Schumann et al. 2025), and generative AI (Rosemann & Joebges, 2025; Kluza et al. 2025; Maslov et al. 2025) to support modeling, analysis, and ideation. These tools lower technical barriers and expand the design space for students, enabling them to explore both exploitative and explorative BPM strategies. Notably, one paper compares MBA students' process innovation ideas with those generated by generative AI (Rosemann & Joebges, 2025), while others explore AI's role in instructional design (Maslov et al. 2025) and BPMN assignment design (Kluza et al. 2025).

A third theme is **curriculum innovation and methodology**, with several contributions offering structured pedagogical frameworks and validated course designs. These include the LSPPDM (Learn, See, Practice, Prove, Do, Maintain) framework for procedural skills training (Grzybowska & Hoffa-Dąbrowska, 2025), a multi-phase blueprint for next-generation BPM education (Dallavalle et al. 2025), and longitudinal curriculum development in emerging markets (Dahanayake Yapa et al. 2025). These innovations align BPM education with evolving industry needs, ethical imperatives, and the "drifts" of next-generation BPM—namely conversationality, autonomization, and sophistication (Dallavalle et al. 2025).

Several papers also emphasize **bridging academia and industry**, highlighting university–industry collaborations for curriculum co-design, software access, and project supervision (Joas 2025; Schumann et al. 2025). This is particularly salient in emerging markets and low-BPM-maturity contexts, where BPM education must be adapted to local organizational realities. One paper from Indonesia uses affordance theory to explore how BPM techniques are perceived and applied by alumni, revealing that while automation is widely adopted, techniques like BPMN and qualitative analysis are underutilized due to contextual misalignment (Er & Nathania 2025).

Finally, a visionary contribution proposes a Global Process Institute (GPI), a speculative but plausible platform for continuous BPM education that integrates human and machine learners through a Business-to-Process model (Magdaleno & Rosemann 2025). This concept reimagines business processes as autonomous learners capable of subscribing to educational services, opening new frontiers in BPM education and service delivery.

Together, these papers reflect a BPM education community that is not only pedagogically innovative but also globally aware and technologically adaptive—preparing learners for the complex, AI-infused, and ethically grounded process challenges of the future. We extend our heartfelt thanks to all authors, reviewers, participants, and the conference chairs, who contribute to the success of the 2025 Educators Forum.

Organization

Program Chairs

Banu Aysolmaz	Eindhoven University of Technology, The Netherlands
Wasana Bandara	Queensland University of Technology, Australia
Kate Revoredo	Humboldt-Universität zu Berlin, Germany

Program Committee

Amy van Looy	Ghent University, Belgium
Andrea Magalhaes	Dheka, Brasil
Fernanda Baião	PUC-Rio, Brasil
Flavia Santoro	Inteli, Brasil
Ilia Maslov	KU Leuven, Belgium
Iris Beerepoot	Utrecht University, The Netherlands
Jorge Munoz-Gama	Pontificia Universidad Católica de Chile, Chile
Lucineia Heloisa Thom	Federal University of Rio Grande do Sul, Brasil
Mahendra Er	Institut Teknologi Sepuluh Nopember, Indonesia
Manual Seneviratne	University of Moratuwa, Sri Lanka
Marleen Voss	University of Muenster, Germany
Marzena Grzesiak	Gdansk University of Technology, Poland
Matthijs Berkhout	Utrecht University of Applied Sciences, The Netherlands
Michael Leyer	University of Rostock, Germany
Niels Martin	Hasselt University, Belgium
Oktay Turetken	Eindhoven University of Technology, The Netherlands
Patrick Delfmann	Koblenz University, Germany
Prasanna Illankoon	University of Moratuwa, Sri Lanka
Saimir Bala	Humboldt-Universität zu Berlin, Germany
Sareh Sadeghianasl	Queensland University of Technology, Australia
Shahrzad Roohy Gohar	University of Queensland, Australia

Silvia Inês Dallavalle de Pádua	University of São Paulo, Brasil
Thomas Grisold	University of St. Gallen, Switzerland
Vinicius Stein Dani	Utrecht University, The Netherlands

Educating for Next-Generation BPM: A PBL-Process Mapping Blueprint Anchored in the Drifts of the Digital Era

Silvia Dallavalle(✉), Luiz Ricardo Brito Ribeiro, and Sérgio Adriany Santos Moreira

Universidade de São Paulo (USP), Ribeirão Preto, São Paulo, Brazil
dallavalle@usp.br

Abstract. This study proposes the Blueprint PBL-Process Mapping, an educational design for undergraduate Business Process Management (BPM) courses that integrates Project-Based Learning (PBL) with the demands of next-generation BPM. Adopting a qualitative and exploratory approach, the study unfolds in three sequential phases: (1) the development of a course blueprint grounded in six pedagogical dimensions that integrate next-generation BPM drifts; (2) prospective validation through focus groups with former students; and (3) refinement and consolidation of the blueprint based on the pedagogical value and feasibility of the recommendations. Findings from the focus groups suggest that the course design supported the development of analytical reasoning, critical thinking, and process-oriented mindsets—particularly through engagement with real-world cases and emerging technologies such as AI agents. The blueprint is designed to foster a more meaningful integration of technical and emergent skills, promote student awareness of ethical and sustainable process management, and respond to the need for conversational and autonomous digital environments. This contribution provides BPM educators with a prospectively validated, replicable, and adaptable teaching blueprint that bridges the academic–industry gap and supports the professionalization of BPM education.

Keywords: Project-Based Learning · BPM Education · Pedagogical Innovation · Curriculum Innovation · Next-Generation BPM

1 Introduction

The rapid advance of artificial intelligence and complex value chains places new demands on Business Process Management (BPM) education. These conditions require not only new competencies but also a shift in logic: from transaction to conversation, automation to autonomization, and simplification to sophistication [1]. Ethical considerations are vital for trust, social responsibility, and sustainability in digital contexts [1].

Although BPM is increasingly adopted, many organizations still lack an end-to-end perspective, revealing a persistent gap between academic training and professional practice [2]. Project-Based Learning (PBL) has emerged as a promising approach to address

M. ER et al. (Eds.): BPM 2025, LNBIP 565, pp. 299–314, 2026.
https://doi.org/10.1007/978-3-032-02936-2_21

this gap by fostering continuous learning, adaptability, and sustainability-related skills [3], while engaging students in real-world challenges that enhance critical thinking and problem-solving [4]. Effective BPM teaching has been shown to improve students' analytical and practical abilities [5–7] while aligning academic outcomes with organizational needs [8]. However, achieving these results depends on pedagogical resources that promote analytical reasoning and the application of process mapping techniques [8]. Nevertheless, there is still a lack of validated educational proposals that integrate next-generation BPM concepts with active methodologies in undergraduate courses.

To address this gap, this article introduces the PBL-Process Mapping Blueprint—a structured design for undergraduate BPM courses. Grounded in PBL, the blueprint is aligned with the core drifts of next-generation BPM: Conversationality, Autonomization, and Sophistication. It offers a week-by-week guide including activities, deliverables, and learning goals, integrating both technical skills (e.g., BPMN modeling, AS-IS/TO-BE diagnosis) and emerging competencies (e.g., conversational interfaces, autonomous decision-making, and impact-oriented thinking).

The proposal was prospectively validated through focus groups with former students, enabling an in-depth evaluation of its clarity, applicability, and alignment with professional demands. This process ensured methodological rigor through diverse perspectives. The blueprint offers a replicable, evidence-based instructional model that bridges theory and practice while preparing students for the conversational, autonomous, and ethical challenges of next-generation BPM.

The article is structured as follows: Sect. 2 reviews literature on BPM education and the three drifts; Sect. 3 details the methodology; Sect. 4 presents the results; Sect. 5 discusses implications and limitations; Sect. 6 concludes and outlines future research.

2 Literature Review

BPM is a holistic discipline supported by methods and tools for managing processes across their lifecycle.[9]. However, it faces growing pressure from emerging technologies and external disruptions that challenge traditional workflows[1] giving rise to concept drifts—gradual but persistent changes in process behavior. To address these, the BPM lifecycle now considers external factors, leveraging the dynamic capabilities perspective to identify threats and opportunities [10], requiring integrated analysis of internal and external conditions [9] and detailed stakeholder mapping [11].

In this context, foundational BPM assumptions must evolve toward three key drifts—Conversationality (interaction in natural language), Autonomization (goal-oriented decision-making), and Sophistication (designing experiences beyond expectations)—which call for embedding ethics into process design and governance, as well as reconceptualizing autonomization as an organizational capability [1].

Despite such advancements, organizations still lack an end-to-end process perspective (Bandara et al., 2021), exposing a gap between academia and industry [7, 12, 13]. The demand for professionals skilled in modeling and continuous improvement is growing[5, 14], yet BPM education often struggles to develop such competencies [15, 16]. Fragmentation persists due to function-oriented curricula [16], underscoring the need for dynamic and hybrid-compatible approaches [14, 17]. Moreover, BPM education has overlooked descriptive process capabilities [18] even though these are critical in Education 4.0 for resilience and communication[19].

Given these challenges, pedagogical innovation is essential. As novel business models emerge [2, 13], BPM education must become multidisciplinary [15, 16, 20, 21] and evidence-based. Case-based learning promotes creativity and practical engagement [22], while Project-Based Learning (PBL), grounded in experiential learning theory [23], has proven effective in enhancing student engagement, critical thinking, and teamwork [4]. Student-centered and actively constructed learning [21], especially when supported by action learning and design thinking, strengthens these outcomes.

Although research confirms that active learning enhances analytical and practical skills [5–7], few studies integrate PBL into BPM curricula. This highlights the urgency of incorporating the drifts and ethical concerns into undergraduate education [17].

Based on this review, six pedagogical dimensions were synthesized to support the PBL-Process Mapping Blueprint:

- (I) Active Learning & PBL (AL/PBL): Emphasizes PBL's role in fostering continuous learning and adaptability [3, 4], supported by student-centered, action-oriented learning [15, 21, 24].
- (II) Real Cases & Stakeholder Engagement (RC/SE): Advocates the use of case studies and real BPM projects to close the theory–practice gap[7, 22].
- (III) Multidisciplinary End-to-End Curriculum (MC/E2E): Addresses curricular fragmentation and promotes process thinking through multidisciplinary design [2, 15, 20, 21].
- (IV) Resources & Diagnostic Practices (RT/DP): Responds to the lack of emphasis on descriptive process modeling [8, 18] with tools that build analytical reasoning.
- (V) Theory–Practice Alignment (TA/PA): Focuses on dynamic curricula that meet professional demands through virtual and hybrid environments [2, 14, 17].
- (VI) Emerging Competencies & Drifts (EC/Drifts): Incorporates the core BPM shifts and integrates ethics into BPM capabilities [1].

These six dimensions, aligned with Rosemann et al.'s drifts, serve as a framework for assessing whether the Blueprint equips students to address the evolving demands of BPM practice.

3 Method

This qualitative study examines the evolution of a hands-on undergraduate Process Mapping course offered since 2009. In 2024, Brazilian and French students from Business, Applied Math, and Accounting with prior BPMN exposure formed the empirical basis for the Blueprint. To address next-generation BPM [1], the syllabus was comprehensively redesigned. The method follows three phases, each aligned with six pedagogical dimensions. Figure 1 summarizes this step-by-step development.

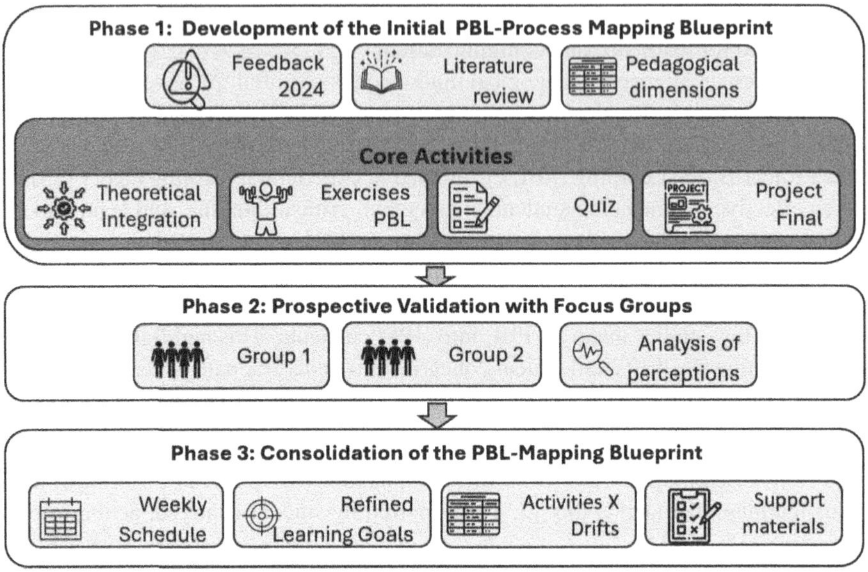

Fig. 1. Phases for Developing and Consolidating the PBL-Process Mapping Blueprint, Source: Authors.

3.1 Phase 1 –Development of the Initial PBL-Process Mapping Blueprint

This phase aimed to transform evidence from the 2024 course into a structured syllabus aligned with the six pedagogical dimensions and the three drifts (Conversationality, Autonomization, Sophistication). The process followed four steps:

1. **Analysis of 2024 Course Evidence** – Final projects were assessed by the instructor using a standardized rubric with five criteria: BPMN modeling rigor, AS-IS diagnostic quality, use of extended metrics, depth of reflection, and clarity of lessons learned
2. **Conceptual Integration** – Based on identified gaps, the content was enriched with key theoretical inputs: the BPM lifecycle, holistic diagnostics, and process architecture[11, 25]; the three drifts[1] as guiding principles; the Current Reality Tree (CRT); and project management techniques adapted to process improvement.
3. **PBL Instructional Design** – The 30-h course (15 sessions of 100 min) was redistributed: ~13% for assessments (weeks 8 and 15); ~50% for hands-on activities (e.g., pizza simulation, BPMN critique, interview role-play); ~30% for guided modeling and mentoring; and ~20% for out-of-class tasks (quizzes, reflections, project work). Time was also allocated for iterative feedback and individual support.
4. **Redesign of the Final Project** – New requirements included: (i) macro-architecture; (ii) AS-IS and TO-BE models incorporating the three drifts; and (iii) extended metrics (cost, risk, ethics, experience). The revised rubric evaluates technical rigor, drift integration, experience design, and alignment with ethical principles and the Sustainable Development Goals (SDGs).

At the end of this phase, the initial Blueprint version included: (i) a week-by-week schedule of activities and deliverables; (ii) a matrix linking tasks to the six dimensions and three drifts; (iii) learning objectives and evaluation criteria; and (iv) instructional resources (e.g., guides, templates, examples), forming the basis for Phase 2 validation.

3.2 Phase 2 – Prospective Validation with Focus Groups

This phase aimed to prospectively validate the PBL-Process Mapping Blueprint by assessing its clarity, applicability, and alignment with the professional training needs of BPM practitioners. Two virtual focus groups were conducted, each composed of four participants ($N = 8$), all of whom had completed the course between 2021 and 2024—the last four editions of the subject. The average age of participants was 24.5 years. Although only one had formally graduated, all were professionally engaged in roles related to process improvement or business analytics. The sample was intentionally selected, as the course is elective and requires prior familiarity with BPMN; hence, only former students were qualified to evaluate proposed curricular changes. The professional maturity developed during the course enabled participants to compare what they had learned with real-world challenges, providing a perspective less accessible to current students or external faculty.

Focus groups were selected as the primary data collection method for this validation phase because they enable rapid triangulation of heterogeneous perspectives—combining professional and academic viewpoints—while fostering interactive dialogue and collective meaning-making, which are essential to educational design. This approach is particularly suitable for testing educational prototypes, as it supports the generation of deep qualitative insights before committing to large-scale implementation. Theoretical saturation was considered achieved after the second group, as more than 90% of the thematic categories were repeated across both sessions, indicating that additional groups would yield diminishing returns.

- Composition – All participants were alumni of the Business Process Mapping course and were intentionally split into two separate groups to encourage deeper discussion and richer individual contributions. Each group session lasted approximately 60 min and was recorded, with participants' consent, to support accurate transcription and thematic analysis.
- Semi-structured script – The discussions addressed the clarity of the redesigned plan, the relevance and feasibility of the three drifts, the adequacy of the six pedagogical dimensions, the realism of the workload, and the expected contribution to students' employability.
- Analysis – The sessions were recorded, transcribed, and analyzed manually through open thematic coding. After thematic coding, all recommendations emerging from the focus groups were consolidated into a single list. The faculty member responsible for the course and an independent researcher analyzed each recommendation according to two criteria: (1) pedagogical value—i.e., the extent to which it could enhance student learning and employability—and (2) practical feasibility—considering required resources, time, and institutional procedures. Each criterion was rated using a simple scale ("high," "medium," or "low"), and any discrepancies were resolved by consensus. Recommendations assessed as having both high pedagogical value and high or medium feasibility were incorporated into the syllabus for the next course offering. The remaining suggestions were documented for future review, should favorable

operational conditions arise. This procedure ensured that only changes with clear educational impact and immediate feasibility were prioritized in the course redesign.

3.3 Phase 3 – Consolidation of the PBL-Process Mapping Blueprint

The suggestions collected from the focus groups were assessed based on their pedagogical value and practical feasibility, as determined in Phase 2. Based on this evaluation, the course syllabus was revised to reflect actionable improvements: workload distribution was adjusted, PBL micro-exercises were refined, readings and assessment rubrics were updated, and complementary resources were added (e.g., additional templates and mentoring time). Each adjustment was explicitly mapped to the six pedagogical dimensions and the three drifts, ensuring internal consistency throughout the PBL-Process Mapping Blueprint. The weekly schedule was also updated to incorporate the selected enhancements.

At the end of Phase 3, a consolidated version of the PBL-Process Mapping Blueprint was produced, comprising: (i) a detailed syllabus including weekly schedule, resources, and deliverables; (ii) a matrix aligning each activity with the six pedagogical dimensions and the three drifts; (iii) refined learning objectives and evaluation criteria; and (iv) guidelines on essential instructional resources—such as interview guides, BPMN templates, and annotated examples—to ensure pedagogical alignment. The study adhered to ethical standards, with informed consent obtained and participant anonymity preserved. As the focus groups involved voluntary feedback from former students without sensitive or health-related data, approval from a Research Ethics Committee was not required.

4 Results

The results were divided into three sections related to the three phases presented in the method.

4.1 Phase 1 – Development of the Initial PBL-Process Mapping Blueprint

The weekly structure was designed to guide students incrementally—from foundational BPM concepts to the delivery of a complete process diagnosis and redesign project. The instructional design ensures that students gradually develop both technical and emerging competencies aligned with the demands of the next generation of BPM. By the end of the fifteen-week course, each team is expected to produce an integrated portfolio consisting of: (i) a macro-level process architecture of the selected organization; (ii) a detailed characterization of the focal process, supported by traditional and extended indicators (cost, risk, ethics, environmental/social impact); (iii) AS-IS and TO-BE models in BPMN 2.0, incorporating conversational interactions, autonomous decisions, and experience-enhancing elements; (iv) a Current Reality Tree (optional), used to represent systemic root-cause reasoning; and (v) a brief summary of lessons learned, including time allocation across project phases.

The course schedule ensures progressive coverage of the six pedagogical dimensions and integrates the three drifts proposed by Rosemann et al. (2024): Conversationality, Autonomization, and Sophistication. Weeks 1–2 develop an end-to-end perspective through simulations and quizzes, introducing conversational process design early.

Weeks 3–6 emphasize mapping techniques and diagnostic reasoning. Week 7 introduces the Current Reality Tree to support root-cause analysis and deepen Theory–Practice Alignment. Weeks 9–12 focus on project checkpoints, encouraging stakeholder engagement and iterative refinement. Weeks 13–15 consolidate the development of advanced BPM competencies through final presentations and summative assessment.

This structure resulted in a comprehensive version of the Initial PBL-Process Mapping Blueprint, which details weekly activities, deliverables, learning objectives, and the alignment of each task with six pedagogical dimensions and the three drifts. This blueprint served as the foundation for the prospective validation conducted in Phase 2.

4.2 Phase 2 – Prospective Validation with Focus Groups

This phase prospectively validated the PBL-Process Mapping Blueprint by assessing its clarity, applicability, and alignment with current BPM training needs. Two virtual focus groups were conducted with eight former students (2021–2024 editions). Although most were still completing their degrees, all were professionally active in roles related to process improvement or business analytics. The groups reflected diverse career experiences and application contexts.

The qualitative analyses revealed strong engagement with the proposed content and methodologies, highlighting the development of systemic and applied competencies aligned with the three drifts of the next generation of BPM. According to Student 1, “the course was one of the most remarkable and is one of those that most define me as a business professional,” emphasizing its role in shaping professional identity. Student 2 underscored the mindset shift fostered by the course: “the greatest value of the course is thinking with a process-oriented mindset… at the end of the day, what’s most valuable is how we see problems.” In line with this, Student 3 emphasized the centrality of the concept of autonomization: “Of the three drifts, [autonomization] is, without a doubt, the most relevant one we, as Business students, need to be focusing on.”

The active structure of the course was also widely appreciated. Student 4 noted: “I thought the pizza simulation was really cool… it was a great icebreaker in the first class, and it was my first elective course.” Student 5 highlighted the link between theory and practice: “Feeling that the problems were truly connected to reality… made much more sense and motivated me to engage with the course.” Student 6 suggested expanding content related to artificial intelligence: “In my opinion, artificial intelligence should be a standalone course in the business curriculum, because it’s a new tool in our toolbox.”

In addition to the positive feedback, participants offered suggestions to enhance the blueprint, all of which were classified as high-value. Nine of them were low-complexity (Table 1), including: introducing the three drifts in the first session using simple examples; requesting a final project reflection on how the drifts apply to the case; providing a brief and accessible introduction to intelligent agents; strengthening logical reasoning and critical thinking in course exercises; encouraging peer-to-peer feedback moments; suggesting complementary learning paths in logic, data science, or algorithms; reinforcing the connection between BPM and the SDGs; enhancing visual literacy in BPMN using ready-made models; and adjusting the complexity of initial tasks to match students’ academic stage. Additionally, two high-value suggestions of greater complexity were proposed (Table 2): incorporating low-code simulation tools to test process models, and increasing the course load to four credits. These recommendations confirm the blueprint's alignment with the six pedagogical dimensions, highlighting both its transformative potential and avenues for future improvement.

The suggestions gathered directly address gaps identified in the literature, such as the need for more dynamic and interdisciplinary curricula[17], the scarcity of integrated

Table 1. Suggestions with High Pedagogical Value and Feasibility

	Suggestion	Dimension/Drift	Justification
1	Introduce the three drifts from the first session, using simple examples	**EC/Drifts (C,A,S) + AL/PBL**	Promotes early understanding and progressive application; easily integrated into the course schedule
2	Include an explicit reflection on how to apply the drifts in the final project	**EC/Drifts (C,A,S) + TA/PA**	Requires only an adjustment in the final assignment instructions; reinforces conceptual alignment
3	Add a brief introduction to intelligent agents (AI agents)	**EC/Drifts (A) + RT/DP**	Can be done with simple examples or videos; demystifies AI in the context of business processes
4	Explore logic and critical thinking in the exercises	**RT/DP + TA/PA**	Involves instructional redirection without content overload; strengthens analysis and decision-making
5	Create cross-group feedback opportunities	**AL/PBL + RC/SE**	Encourages peer learning and critical thinking; easy to implement via presentations or forums
6	Recommend complementary learning paths (e.g., logic, data science)	**MC/E2E**	Reinforces interdisciplinarity and continuous learning; feasible via syllabus guidance
7	Strengthen the link between BPM and sustainability/SDGs	**RC/SE + EC/Drifts (S)**	Encourages socially relevant applications; can be used as a project evaluation criterion
8	Reinforce BPMN visual literacy using sample models	**RT/DP + AL/PBL**	Facilitates technical learning, especially in early stages of the course
9	Adjust the complexity of early assignments based on student background	**TA/PA + MC/E2E**	Enables gradual progression; increases engagement and avoids overload for beginner students

Source: Authors

Table 2. Suggestions Classified in the High Value × High Complexity

No.	Suggestion	Justification
1	Include simulations using low-code tools	Requires additional time, technical training, and infrastructure support; better suited for IT-focused courses
2	Review and increase the course load to four credits	Involves institutional procedures, curriculum negotiation, and greater allocation of teaching resources

Source: Authors

approaches that foster ethics and analytical reasoning in business processes[1, 18], and the persistent challenge of developing end-to-end thinking and connecting theory with practice in BPM courses [2, 16].

At the end of Phase 2, a Prospective Validation Report was produced, containing:

i. a synthesis of participant feedback;
ii. a thematic analysis aligned with the six pedagogical dimensions and three drifts;
iii. a prioritized list of improvement suggestions based on relevance and feasibility;
iv. operational recommendations for schedule adjustments, refinement of PBL activities, and enhancement of didactic resources.

This phase confirms that the PBL-Process Mapping Blueprint is conceptually aligned with contemporary BPM training needs and offers a student-centered approach with strong potential for adaptation across diverse institutional settings. The suggestions collected were organized according to the six pedagogical dimensions (see Sect. 2), ensuring that the Blueprint's refinement remained pedagogically grounded and practically viable.

These dimensions informed the thematic coding process and guided the prioritization of improvement suggestions, which were assessed according to their educational value and implementation feasibility. This procedure ensured that the final adjustments to the Blueprint were both pedagogically grounded and contextually viable, reinforcing its conceptual coherence and practical relevance.

4.3 Phase 3 – Consolidation of the PBL-Process Mapping Blueprint

The third phase consolidated the insights gathered during validation into a final version of the Blueprint, reinforcing the conceptual alignment with the six pedagogical dimensions and the three next-generation drifts (Conversationality, Autonomization, and Sophistication). Since the focus-group feedback confirmed the adequacy of the 15-week workload and reading volume, no logistical changes were made. All refinements targeted course content and instructional strategies. The redesigned syllabus retains a steady progression—from conceptual grounding to a fully fledged mapping project—while embedding the new elements (Table 3):

All nine recommendations that were assessed as having both high pedagogical value and high feasibility were incorporated into the revised course blueprint. First, the three

Table 3. Weekly schedule of the redesigned Process Mapping course

Weeks	Focal content	Key dimensions/drifts
1–2	Systems thinking, conversational design, ice-breaker "PizzaFlow"	AA/PBL · CE/Drifts (intro)
3–6	BPMN literacy, diagnostic logic, AI-agent primer	RD/Diag · CE/Drifts · CM/E2E
6	Interview simulation (triad: interviewer-interviewee-evaluator)	CR/S · AA/PBL
7	Current Reality Tree for root-cause analysis	RD/Diag · AT–PR
9–12	Stepwise development of the final project (architecture, characterisation, AS-IS) plus cross-group feedback	CM/E2E · CR/S · AA/PBL
13–15	TO-BE modelling, value-based redesign, links to SDGs and ethics	CE/Drifts · AT–PR
1–2	Systems thinking, conversational design, ice-breaker "PizzaFlow"	AA/PBL · CE/Drifts (intro)

Source: Authors

drifts were introduced in Week 1 using everyday examples to enhance student understanding from the outset. A reflection checkpoint was added to link these drifts explicitly to the final project brief, fostering integration of theory and application. A short primer on intelligent agents—comprising a video and mini-demo—was inserted to support engagement with emerging technologies. To reinforce analytical reasoning, logic and critical-thinking prompts were embedded in the modeling exercises. Cross-group peer-feedback rounds were formalized at each checkpoint to promote collaborative learning and continuous improvement. Additionally, optional learning paths in logic and data science were offered through curated online resources. The connection between BPM and sustainability was strengthened by incorporating the Sustainable Development Goals (SDGs) into the assessment rubric. Visual literacy in BPMN was reinforced through the use of annotated exemplar models. Lastly, the complexity of early tasks was calibrated to match students' backgrounds, easing beginners into the course content. These updates require no extra class hours, readings, or assessment items; instead, they repurpose existing slots and materials, exemplifying a *minimal-disruption / maximal-impact* philosophy.

The 15-week schedule was structured to guide students through:

- Systems thinking and conversational process design (Weeks 1–2),
- Intensive modeling and diagnostic reasoning using BPMN (Weeks 3–6),
- Process understanding through interview simulation and qualitative analysis (Week 6),
- Root cause analysis with the Current Reality Tree (Week 7),
- Stepwise development of the final project, including characterization, diagnosis, peer feedback, and AS-IS/TO-BE modeling (Weeks 9–12),

- Value-based redesign and reflection on ethical, sustainable, and experiential process enhancements (Weeks 13–15).

Each weekly session was mapped to specific pedagogical dimensions—AA/PBL, CR/S, CM/End-to-End, RD/Diag, AT–PR, and CE/Drifts—ensuring pedagogical coherence and integration with the learning outcomes. The revised Blueprint also incorporated prioritized improvements from Phase 2, such as the early introduction of the three drifts, enhanced peer feedback mechanisms, and increased visual literacy using pre-built BPMN models. These adjustments were applied across the syllabus and reflected in the updated materials and assessment rubrics.

By the end of Week 15, each student team is expected to deliver a comprehensive set of outputs that reflect both technical proficiency and critical reflection. These include: a macro-level process architecture of the host organization; a diagnostic dossier incorporating extended key performance indicators (KPIs) such as cost, risk, customer experience, and ethics; AS-IS and TO-BE BPMN 2.0 models that explicitly demonstrate the application of the three drifts—Conversationality, Autonomization, and Sophistication; and an optional Current Reality Tree to support root-cause analysis. Additionally, students must submit a reflection report aligned with the six pedagogical dimensions and the Sustainable Development Goals (SDGs), consolidating the integration of process thinking, social responsibility, and learning outcomes throughout the course.

This result confirms the Blueprint's viability as a replicable, evidence-based model for preparing professionals to design conversational, autonomous, and socially responsible business processes—without increasing the curricular workload. The complete weekly Blueprint, aligned with the six pedagogical dimensions and the BPM drifts, is provided in Appendix A. Supplementary materials—including the focus group protocol, interview templates, and modeling exercises—are openly accessible via a dedicated repository [Supplementary Materials, DOI: https://doi.org/https://doi.org/10.5281/zenodo.15549020].

5 Discussion

The integration of the six pedagogical dimensions was confirmed throughout the course adjustments prioritized in Phase 3. These refinements demonstrate that the Blueprint not only updates content but also explicitly aligns academic gaps with evolving industry demands. The early introduction of the drifts—Conversationality, Autonomization, and Sophistication—responds to the call for next-generation BPM capabilities [1].

Enhancements in visual literacy in BPMN, supported by exemplar models and peer feedback mechanisms, directly address the analytical reasoning gap [18]. Complementary learning paths and the calibration of task complexity help mitigate curricular fragmentation in BPM education [16]. Explicit links to the Sustainable Development Goals (SDGs) reinforce the course's alignment with the agenda for responsible employability [3]. Finally, the application of the PBL-action cycle ensures the joint evolution of theory, practice, and social impact [4, 15, 24].

Although exploratory in nature, this study suggests that the PBL-Process Mapping Blueprint offers a replicable, evidence-informed model capable of preparing students to design, diagnose, and lead increasingly digital, autonomous, and socially responsible business processes. It thus demonstrates alignment with all six pedagogical dimensions identified in the literature review.

6 Conclusions and Implications

This study advances the discourse on Business Process Management (BPM) education by demonstrating, in operational terms, how the three core drifts—Conversationality, Autonomization, and Sophistication—can be systematically embedded, week by week, into an undergraduate course structured around Project-Based Learning (PBL). Drawing on a literature review that revealed persistent gaps—such as limited end-to-end thinking, lack of diagnostic tools, and disconnects between theory and practice—we designed and validated, in three phases, a PBL-Process Mapping Blueprint anchored in six pedagogical dimensions. The result is a replicable package ready for adoption in BPM courses.

The study offers several theoretical contributions. First, it operationalizes the drifts by translating conversational BPM and process autonomization into concrete learning activities, addressing a key methodological gap. Second, it consolidates fragmented recommendations into a coherent framework of six pedagogical dimensions, providing a structured basis for curriculum redesign and future research. Third, it integrates technology and sustainability by embedding ethical, environmental, and social indicators, autonomous decisions, and experience design into student projects—expanding the ethical and social scope of BPM education.

On a practical level, the Blueprint provides instructors with detailed teaching resources—including interview guides, annotated BPMN models, and checklists aligned with both the drifts and the Sustainable Development Goals (SDGs)—making it suitable for adoption even in shorter courses. It also equips curriculum committees with a structured prioritization strategy that demonstrates the feasibility of implementing nine high-impact improvements without increasing workload. For partner organizations, the course delivers student projects enriched with conversational logic, autonomous elements, and ethical, environmental, and social indicators, indicators, generating a pipeline of actionable process innovations.

The validation was limited to eight alumni from a single institution, which may introduce self-selection bias and limit the immediate generalizability of findings. Institutional culture, student backgrounds, and local curricular norms may influence the Blueprint's implementation and outcomes in other contexts. Future studies should pilot the Blueprint across diverse cultural and institutional settings to assess adaptability, refine the approach, and identify potential barriers or facilitators for broader adoption.

By articulating three strategic drifts and six pedagogical dimensions, the Blueprint provides a pragmatic response to the emerging demands of BPM. It prepares undergraduate students to diagnose, redesign, and govern processes capable of interacting through natural language, making decisions under constraints, and delivering high-value experiences guided by ethical, sustainable, and social impact criteria. Thus, the Blueprint contributes to theory by offering an integrated, active, and innovative approach to BPM education, and to practice by delivering a validated and adaptable roadmap for instructors and institutions.

Appendix A – PBL-Process Mapping Weekly Blueprint

1	Opening Class: Introduction to BPM and PBL. The opening class features an icebreaker where students, in pairs, simulate a pizza delivery service. By mapping the process from order to delivery, they gain an end-to-end view (CM/End-to-End), apply logical reasoning, and identify failures and improvement points. The PizzaFlow exercise introduces the drifts—Conversationality, Autonomization, and Sophistication—through practical examples. This interactive activity reinforces that processes are social constructs and underscores the importance of mapping and improving them from the start of BPM education, activating the PBL mindset
2	Architecture and End-to-End Vision in Process Characterization. In this session, students map the employee hiring process and build a first-level architecture based on the employee lifecycle. They also identify activities with potential for Conversationality, Autonomization, or Sophistication. The activity fosters systems thinking and an end-to-end perspective, reinforcing the dimensions of Multidisciplinary Curriculum (MC/E2E) and Emerging Competencies & Drifts (EC/Drifts)
3	Introduction to BPMN. Students begin studying BPMN notation through pre-reading and a follow-up quiz. In class, they critically analyze real and fictional process models, identifying strengths, flaws, and improvement opportunities. The activity enhances analytical thinking and notation skills, reinforcing the dimensions of Resources & Diagnosis (RT/DP) and Active Learning & PBL (AL/PBL)
4	BPMN Modeling Practice. This week, students review the "Computer Equipment Purchase Process" and model two others: "Order Cancellation in E-commerce" and "Sales of IT Solutions." Using guided templates, they explore structures, variations, and exceptions, identifying opportunities for conversational interfaces, autonomous decisions, and experience-enhancing features (drifts). The session reinforces BPMN skills and the dimensions of Resources & Diagnosis (RT/DP) and Active Learning & PBL (AL/PBL)
5	BPMN Modeling in Sales Contexts Students engage in BPMN modeling using scripts adapted from real sales processes (e.g., agricultural machinery, cooperatives, construction materials). These cases strengthen technical skills and link modeling to authentic business settings. At the end, groups share initial project ideas, reflecting on how processes can be more inclusive, sustainable, and efficient. The session reinforces the dimensions of Resources & Diagnosis (RT/DP), Active Learning & PBL (AL/PBL), and Real Cases & Stakeholder Engagement (RC/SE)
6	Interview Simulation and Process Understanding. The class begins with a review of the previous BPMN exercises. The instructor introduces best practices for process interviews, followed by a group simulation of the hiring process, where students rotate roles: interviewer, interviewee, and evaluator. Evaluators present key challenges and assess interviewer performance. The activity promotes active listening, qualitative analysis, and end-to-end thinking, reinforcing the dimensions of Multidisciplinary Curriculum (MC/E2E), Resources & Diagnosis (RT/DP), and Active Learning & PBL (AL/PBL), while encouraging drift identification

(*continued*)

(*continued*)

7	Root Cause Analysis with Current Reality Tree (CRT). After a brief conceptual overview, students apply the Current Reality Tree (CRT) from the Theory of Constraints to analyze the root causes of low customer growth in a fictional company. In groups, they build diagrams linking undesirable effects to systemic causes. The activity fosters systems thinking, analytical reasoning, and theory–practice integration, reinforcing the dimensions of Resources & Diagnosis (RT/DP), Active Learning & PBL (AL/PBL), and Theory–Practice Alignment (TA/PA)
8	First Test
9	First Project Delivery: Process Characterization and Initial Findings: Student groups present the first part of their final project: (1) a brief overview of the organization and key processes (simple architecture), (2) a detailed characterization of the selected process, and (3) key problems identified. Early connections to Sustainable Development Goals (SDGs) are encouraged, particularly for socially impactful processes. The activity validates project scope, promotes peer learning, and aligns expectations for next steps. A synthesis panel fosters shared reflection, reinforcing Theory–Practice Alignment (TA/PA), Multidisciplinary Curriculum (MC/E2E), and knowledge integration
10	Continuation of presentation
11	Second Project Delivery: AS-IS Model and Initial Improvements: In this session, student groups present the second part of their final project, incorporating feedback from the previous round. Presentations include the AS-IS model of the selected process and, when possible, an initial TO-BE proposal. Classmates observe the presentations, and the instructor offers formative feedback to improve analytical reasoning and modeling clarity. Students are encouraged to consider extended metrics, such as user experience or ethical risks. This session reinforces the dimensions of Theory–Practice Alignment (TA/PA), Resources & Diagnostic Practices (RT/DP), and Emerging Competencies & Drifts (EC/Drifts)
12	Continuation of presentation
13	Final Delivery: Process Redesign and Reflection on Value Creation: Students justify improvements and demonstrate the application of the three drifts—Conversational, Autonomous, and Sophisticated BPM—alongside digital tools and impact metrics. They are encouraged to reflect on user experience, considering values such as accessibility, transparency, and sustainability, aligned with the SDGs. Final reports and BPMN models are submitted on e-Disciplinas. The session promotes peer feedback and collective analysis, consolidating all six Blueprint dimensions (AA/PBL, CR/S, MC/E2E, RD/Diag, TA/PA, EC/Drifts) and reinforcing alignment with the Sustainable Development Goals
14	Continuation of presentation
15	Second test

References

1. Rosemann, M., Brocke, J.V., Van Looy, A., Santoro, F.: Business process management in the age of AI – three essential drifts. Inf Syst E-Bus Manage. **22**, 415–429 (2024)

2. Bandara, W., Van Looy, A., Resemann, M., Meyers, L.: A call for 'Holistic' business process management. In: Proceedings of the International Workshop on BPM Problems to Solve Before We Die, pp. 6–10. Germany (2021)
3. Epler, P., Jacobs, J.: Guide to integrating problem-based learning programs in higher education classrooms: design, implementation, and evaluation. IGI Global (2022)
4. Brin, P., Krasnokutska, N., Lombardi, R., Nehme, M.: Project-based learning for responsible education: empirical study of the additional value for students. In: 2021 IEEE International Conference on Modern Electrical and Energy Systems (MEES), pp. 1–5. IEEE, Kremenchuk, Ukraine (2021)
5. Abdelnabi, E.A., Fadel, M.A., Erhoma, L.A., Elbarghathi, B.A.: Application of business process management in the libyan international medical university: history and development. In: 2023 IEEE 3rd International Maghreb Meeting of the Conference on Sciences and Techniques of Automatic Control and Computer Engineering (MI-STA), pp. 348–352. IEEE, Benghazi, Libya (2023)
6. König, M., Völker, M., Weske, M.: From design to evaluation: students experiencing the bpm lifecycle hands-on. In: Di Ciccio, C., et al. (eds.) Business Process Management: Blockchain, Robotic Process Automation, Central and Eastern European, Educators and Industry Forum. BPM 2024. LNBIP, vol. 527. Springer, Cham (2024). https://doi.org/10.1007/978-3-031-70445-1_22
7. Moura, F.L.D., Sá-Soares, F.D., Kubis, H.M., Kawashita, I., Mota, J.S., Takagi, N.: IT-CMF and BPM critical capability: improving software development lab on academic context. Procedia Comput. Scie. **181**, 325–332 (2021)
8. De Pádua, S.I.D., et al.: Fostering BPM learning through PBL: development and application of an educational framework. In: Di Ciccio, C., et al. (eds.) BPM: Blockchain, RPA, CEE, Educators and Industry Forum, pp. 416–425. Springer, Cham (2024)
9. Nogueira, C.A., Pádua, S.I.D., Bernardo, R.: A map for the holistic BPM diagnosis. BPMJ. **28**, 630–655 (2022). https://doi.org/10.1108/BPMJ-04-2021-0197
10. Bernardo, R., Galina, S.V.R., Pádua, S.I.D.D.: The BPM lifecycle: how to incorporate a view external to the organization through dynamic capability. BPMJ. **23**, 155–175 (2017)
11. Lima Aredes, E., Lourenção, M., Burlton, R., Reijers, H., Dallavale De Pádua, S.I.: A method for creating process architecture: driving organizational change. In: De Weerdt, J., Pufahl, L. (eds.) Business Process Management Workshops, pp. 539–551. Springer Nature Switzerland, Cham (2024)
12. Kundu, G.K., Majumdar, J.P.: The learning and teaching area of AACSB standards: a process model framework. BPMJ. **26**, 1379–1399 (2020)
13. Levkovskyi, B., Laurim, V., Kayaci, B., Mustroph, H., Utesch, M., Krcmar, H.: Teaching the digital transformation of business processes: design of a simulation game of business process change. In: 2021 IEEE Global Engineering Education Conference (EDUCON), pp. 1262–1271. IEEE, Vienna, Austria (2021)
14. Pridmore, J., Godin, J.: Business process management and digital transformation in higher education. IIS (2021)
15. Bandara, W., et al.: Business process management education in academia: Status, challenges, and recommendations. Commun. Assoc. Inf. Syst. **27**, 743–776 (2010)
16. Seethamraju, R.: Business process management: a missing link in business education. Bus. Process. Manag. J. **18**, 532–547 (2012)
17. Nobre, M., Vilela, J.: Content and skills for teaching BPM in computer science courses: a systematic mapping study. In: Proceedings of the 16th International Conference on Computer Supported Education, pp. 373–384. SCITEPRESS - Science and Technology Publications, Angers, France (2024)

18. University of Liechtenstein, Grisold, T., Wurm, B., LMU Munich School of Management, Vom Brocke, J., University of Liechtenstein, Kremser, W., Johannes Kepler University Linz, Mendling, J., Department of Management Humboldt-Universität zu Berlin, Recker, J., University of Hamburg: Managing Process Dynamics in a Digital World: Integrating Business Process Management and Routine Dynamics in IS Curricula. CAIS. Vol. 51, pp. 637–656 (2022). https://doi.org/10.17705/1CAIS.05127
19. Oliveira, S.R.M., Saraiva, M.A.: Leader skills interpreted in the lens of education 4.0. Procedia Comput. Sci. **217**, 1296–1304 (2023)
20. Hrabala, M., Opletalovāb, M., Tucekc, D.: Teaching business process management: Improving the process of process modelling course. Istrazivanja i projektovanja za privredu. **15**, 113–121 (2017). https://doi.org/10.5937/jaes15-12172
21. Chow, W.: Teaching Business Process Management with a flipped-classroom and problem-based learning approach with the use of apromore and other BPM software in graduate information systems courses. In: 2021 IEEE International Conference on Engineering, Technology & Education (TALE), pp. 1–8. IEEE, Wuhan, Hubei Province, China (2021)
22. Bandara, W.; Ouyang, C.: Analysing and improving a recruitment process: a teaching case for developing business process improvement capabilities. In: Proceedings of the 16th Pacific Asia Conference on Information Systems (PACIS), pp. 1–13. Vietnam (2012)
23. Cummings, C., Yur-Austin, J.: Design thinking and community impact: a case study of project-based learning in an MBA capstone course. J. Educ. Bus. **97**, 126–132 (2022). https://doi.org/10.1080/08832323.2021.1887795
24. Leyer, M., Moormann, J.: Action learning approach to teaching business engineering methodology. BPMJ. **23**, 130–154 (2017). https://doi.org/10.1108/BPMJ-06-2015-0092
25. Dumas, M., La Rosa, M., Mendling, J., Reijers, H.A.: Introduction to business process management. In: Fundamentals of Business Process Management, pp. 1–33. Springer Berlin Heidelberg, Berlin, Heidelberg (2018)

Project-Based Learning in BPM for Interdisciplinary Groups — Teachers' Experience

Katarzyna Gdowska(✉), Edyta Brzychczy, and Krzysztof Kluza

AGH University of Krakow, Mickiewicza Av. 30, 30-059 Krakow, Poland
{kgdowska,brzych3,kluza}@agh.edu.pl

Abstract. In Business Process Management (BPM), acquiring knowledge and skills through active experience is advisable and necessary due to its applied nature. This paper presents a case study of Project-Based Learning (PBL) in BPM as an effective and engaging teaching method for interdisciplinary groups. We summarize our experiences while implementing the 'Business Process Reengineering' course. The paper describes the project content and discusses selected aspects of its performance across three course iterations. We analyze 30 student projects, highlighting the benefits of the PBL approach. Additionally, we share the advantages and challenges encountered during this process. Our case study demonstrates that PBL can significantly enhance BPM education, aligning it more effectively with the evolving demands of both students and the business environment. This is achieved by supporting interdisciplinary learning, promoting the practical application of process improvement techniques, fostering advanced thinking skills, and reinforcing the teacher's role as both instructor and mentor.

Keywords: Project-Based Learning · University teaching · Didactic of BPM · Pedagogic development · Business Process Management

1 Introduction

Among the OECD Skills for 2030, the most prominent are interpersonal competence, critical and creative thinking, and skills related to organizing and conducting scientific research [19]. Accordingly, learning by doing should become a central paradigm of modern higher education. Project-Based Learning (PBL) is grounded in competency-based instruction and active student engagement, challenging learners to solve complex, real-world problems. This approach fosters meaningful learning experiences and develops skills in goal-setting, creative thinking, decision-making, communication, and collaboration [23]. By emphasizing practical training, PBL provides students with opportunities to develop essential workplace soft skills. It also enhances the alignment between academic learning and employers' expectations by supporting the transfer of knowledge to real-world contexts. Participation in PBL courses helps students better understand their strengths and promotes self-development [13]. From an instructional

M. ER et al. (Eds.): BPM 2025, LNBIP 565, pp. 315–329, 2026.
https://doi.org/10.1007/978-3-032-02936-2_22

perspective, PBL serves as a pedagogical strategy for delivering curricula aligned with predefined learning objectives [12].

Although increasing, examples of PBL implementations remain relatively scarce in the literature on BPM teaching. Case studies based on specific deployments in project management courses unequivocally demonstrate the beneficial effect of this method on achieving learning outcomes in terms of skills from the creation group, as well as a high level of competence in Bloom's taxonomy [1]. Implementing the PBL approach in an undergraduate BPM course, as reported in [21], led to significant improvements in student competencies, supported by statistical evidence that showed enhancements in students' technical, behavioral, and contextual competencies. A case study of the use of PBL approaches with Flipped Classroom was presented by Chow et al. [2], who noted the successful implementation of the approach and the feedback from students, who found that personal problem-solving tasks during class time were more beneficial to their learning compared to direct instruction. In turn, teachers noted that they had more opportunities to observe students' problem-solving performance, understand their learning needs, and provide targeted feedback.

In Central and Eastern Europe, research in the field of BPM is being developed, focusing mainly on Methods among the Six Core Elements of BPM and needing to place more importance on Culture and People [8]. Teaching BPM at universities in these countries may focus on tool aspects, which results from the still strong influence of the teaching system on transmitting knowledge and specialization [16]. Moreover, Sliż et al. [24] emphasize the need for more practice-oriented BPM education, enabling graduates to be better prepared for modern business environments.

Therefore, the authors decided to present in this paper a case study of the application of the PBL method during the Business Process Reengineering (BPR) course offered to multidisciplinary groups at a technical university, so that the following research questions can be addressed:

RQ1: *To what extent does PBL integrate theoretical knowledge and practical skills in the BPM-oriented course for interdisciplinary groups?*

RQ2: *To what extent is PBL an effective didactic method for teaching a BPM-oriented course for interdisciplinary groups?*

RQ3: *What factors determined the effective implementation of PBL in a BPM-oriented course for interdisciplinary groups?*

This paper is structured as follows. In the Background section, we present a brief overview of previous research on teaching BPM, focusing on project-oriented approaches. The following section shows the implementation of PBL in a BPR course at a technical university in Poland. We reflect on the outcomes, with a special focus on the experiences of students and teachers as expressed in the survey assessing the course. The final sections bring conclusions and recommendations for the future implementation of PBL in BPM courses offered to interdisciplinary groups.

2 Background

Back in 2011 Marjanovic et al. [15] reflected on the Australian experiences in teaching BPM and highlighted that to capture the comprehensive nature and interconnected facets of BPM, a holistic teaching model must be adopted as it surpassed technology and process redesign, as well as emphasized the connection between organizational strategy, business processes, and technology. They recommended replacing the traditional transmissive model of teaching with a more student-centered approach. The natural response to such a suggestion is a shift towards PBL—a student-centered instructional approach where students gain knowledge and skills by actively engaging in meaningful projects. There is no denying that BPM teaching often follows a design-oriented approach, frequently using real-world processes, e.g., [3,14,25]; however, this does not always reflect an intentional implementation of the PBL methodology. As it typically involves solving real-world problems and creating end products, this approach is closer to Problem-Based Learning, which focuses more on problem identification and problem-solving, and does not require teamwork and the integration of knowledge, skills, and social competences [18]. According to Kokotsaki et al. [12], PBL offers numerous benefits, including increased student engagement, self-regulation, collaboration skills, and motivation. However, several key factors for successful PBL implementation should be considered, including effective student and teacher support, high-quality group work, a balance between didactic instruction and inquiry-based learning, and a focus on reflective and peer assessment.

Research on BPM teaching has also emphasized the misalignment between educational and job market skills [17] and the significant need for practical and holistic teaching [9,24,25]. Therefore, it is worth emphasizing that Delgado and Calegari [4] showcased how a project-based and hands-on approach increased engagement and provided a deeper understanding of the material. They highlighted that modeling processes in BPMN 2.0 that adhere to best practices are the most challenging tasks of the course. Chow [2] presented the experience of teaching BPM using a flipped classroom and problem-based learning approaches facilitated with BPM software. In his course, students completed two case-study projects that required various BPM phases in small groups during the semester. This model has proven effective in teaching BPM at the graduate level, as students were more engaged, achieved deeper understanding, and developed practical skills essential for their careers. Despite initial challenges, the approach offers substantial benefits in making BPM education more interactive, student-centered, and effective. Similar observations are made by Revoredo et al. [22], who compared two approaches to BPM teaching: a flipped classroom with PBL and a traditional lecture-based format with PBL. Both formats involved real-world process improvement projects. Results showed that PBL enhanced learning and soft skills, while the flipped classroom further increased student engagement and enabled more personalized feedback. Lessons learned included the importance of graded interim assignments and ongoing feedback for improving the quality of outcomes, genuine stakeholder interaction, and the need for students to overcome their resistance to change and adjust to active learning formats. PBL is beneficial for mastering social and managerial skills. Acting as a management team, students collaboratively apply BPM theory to practical tasks, mirroring real-world projects. This approach was adopted by Dijkman and Peters [6], who engaged students in a simulated project where they designed, monitored, and optimized business processes. The whole process was implemented as a serious game [5] aimed at teaching BPM concepts. The gamified project fosters a hands-on learning environment, emphasizing teamwork, problem-solving, and iterative improvement—key aspects of project-based learning.

The PBL-BPM Educational Framework, proposed by de Padua et al. [20], integrates Project-Based Learning with a structured approach to the BPM lifecycle, combining dynamic capabilities, holistic diagnosis, and cognitive tools such as the Current Reality Tree. Applied in a graduate BPM course through a real project with a Brazilian cosmetics company, the framework guided students through strategic analysis, customer experience mapping, process diagnosis, and improvement proposals. By engaging with real stakeholders and applying BPM concepts in practice, students developed critical thinking, soft skills, and a deeper understanding of end-to-end processes. The framework bridges academic learning and industry needs, offering a replicable model for BPM education.

In the following section, we present our approach to implementing PBL in a BPR course attended by students from various academic majors and disciplinary backgrounds. The case study demonstrates both the advantages of PBL in BPM education and the particular challenges of teaching a heterogeneous cohort; challenges more pronounced than when working with a uniform group from the same program and academic stage.

3 Introduction to Business Process Reengineering Course and PBL Approach

The PBL use case is drawn from the Business Process Reengineering (BPR) course[1], offered in English as part of the University Base of Elective Courses.

While the course title emphasizes an engineering perspective, its implementation presents BPR primarily as a pioneering process redesign methodology, highlighting its advantages and limitations, followed by the introduction of BPM as a more flexible and iterative alternative. The course is delivered as seminar classes (utilizing the flipped classroom method) and laboratory classes (with projects), allowing knowledge to be introduced in stages before each project phase. Both class types are offered in a face-to-face format.

The course is based on the Fundamentals of Business Process Management book and uses materials from the book's webpage [7]. The expected outcomes in knowledge (K) and skills (S) perspectives are as follows: (1) Student knows the main assumptions and principles of business process reengineering (K1), (2) Student knows selected methods and tools used in business process reengineering

[1] See: https://sylabusy.agh.edu.pl/pl/2/2/20/1/9/55/137.

(K2), (3) Student can prepare solutions for the reengineering of the selected business process (S1), (4) Student can perform critical analysis of proposed solutions based on calculation results carried out in a dedicated software (S2).

PBL aims to integrate theoretical knowledge and practical skills into the designer's experience based on the chosen topic. The instructions for the project are presented in Fig. 1. The given infographic aims to attract students and provide clear information on course stages and evaluation requirements. Teams of two people, at most, prepare projects, and the progress is reported to the teacher during laboratory classes. During project realization, students can access the academic version of the Apromore software.

The first stage involves selecting and modeling a process. Students are expected to choose a process familiar to them, either through direct participation or observation, and develop an "as-is" model. Minimum requirements are defined, e.g., the number and types of activities and gateways included in the model. A partial evaluation is then conducted to verify compliance with these requirements and ensure correct use of BPMN 2.0. The teacher's approval is required to proceed to the next phase. The second stage focuses on qualitative and quantitative process analyses. Students must conduct at least one analysis of each type introduced during the seminars, such as value-added or root-cause analysis for qualitative studies, and flow analysis or simulation for quantitative studies. The results are presented and discussed with the teacher during class sessions. The third stage involves formulating redesign goals based on the observed problems and selected performance measures (e.g., time, cost, quality), followed by proposing changes to the process. Redesign ideas are generated using heuristics discussed during seminars or derived from students' creativity. Students then prepare a "to-be" model along with a description of the proposed changes. An essential element of this stage is demonstrating the benefits of the proposed improvements concerning the established redesign goals. During class, students present their improvement proposals and discuss the results with the instructor. The final stage includes a summary in which students present their main achievements, reflect on lessons learned, and provide feedback on challenges encountered during the project. This component serves as a "reflective engineering" exercise, aimed at consolidating the knowledge and skills acquired throughout the course. At the end of the course, students present their projects to the entire class.

We employed the following approach to survey PBL experience in relation to defined research questions (RQs). RQ1, addressing the integration of theory and practice, was examined through a content analysis of student projects, which analyzed how they applied BPMN, analysis techniques, and redesign heuristics to real-life cases. RQ2, on the effectiveness of PBL, was informed by students' reflective feedback, project outcomes, and teacher observations, and was evaluated using established PBL criteria. RQ3, on implementation factors, was explored through teacher notes, observed student challenges, project progress, and student reflections.

The BPR course had already had three editions. Their summary and findings are presented in the next section.

Fig. 1. Infographic of the project content.

4 Main Findings and Discussion on PBL Experience

The BPR course is designed to develop practical competencies aligned with higher-order cognitive skills as defined in Bloom's taxonomy, particularly at the level of creation. This response addresses students' needs, as previously expressed in internal surveys, where they considered improving practical skills during college subjects to be an added value of their university education. The course is offered as an open elective, accessible to students from all faculties across the university and international students (e.g., those participating in the Erasmus+ exchange programme). Consequently, enrollment is driven by individual academic interests, resulting in a highly diverse student cohort in terms of disciplinary backgrounds and prior experience. While posing specific instructional challenges, this diversity also provides an opportunity to enhance learning outcomes through carefully structured pedagogical strategies.

The main advantage of the BPR course is its small group size (a maximum of 15 students), which enables the teacher and students to be highly involved in the educational process. During the editions, 30 projects have been elaborated. Their summary is presented in Table 1.

The primary type of processes analyzed in the projects concerns the production of various goods, ranging from simple examples such as pizza or shirts, to fibers for artificial kidneys, copper panels, or printed circuit boards. Services analyzed in the projects include various analytical processes (e.g., asbestos remediation, R&D activities), food delivery, car rental, book borrowing, or laundry sharing space. In the issue resolution processes, students analyzed the reclamation process (by a call center) and the emergency room. Other processes include internal processes related to recruitment and document processing.

The two main determinants of choosing the process type are the field of study and the real-life knowledge about the process. Students with a strong engineering background typically chose a particular kind of process (mainly production), while those with a business-oriented background more frequently focused on sales and services. Some students used processes from companies where they were working part-time. The advantage of the process-wide spectrum in the projects is that students, besides their own project experience, during classes (especially at the end, during presentations), can see examples of reengineering in various industries. They recognize that the learned techniques and approaches are applicable and relevant to any business activity.

Examining the analytic techniques chosen by students reveals two favorites: value-added analysis (from qualitative types) and flow analysis (of quantitative kinds). The minimal requirement was the choice of one of each type; however, many projects utilized more techniques (see Fig. 2).

As a summary of qualitative techniques, we observed that value-added analysis, although selected most often, was challenging for some students (particularly in classifying process steps into VA, BVA, and NVA). Waste analysis was preferred by students with previous experience in Lean Management topics (business-oriented majors).

Table 1. Characteristic of projects in BPR course

Project element	Students' selection	Frequencies
Process types	Production	7
	Selling	6
	Services	9
	Issue to resolution	3
	Other	3
Qualitative analyses	Value added analysis	25
	Waste analysis	19
	Root-cause analysis	11
	Issue register	3
Quantitative analyses	Process simulation	18
	Flow analysis	20
	Queue analysis	3
Redesign approaches	Redesign heuristics	including:
Task-level	Task elimination	10
	Task composition/decomposition	8
	Triage	1
Flow-level	Re-sequencing	3
	Parallellism enhancement	7
Process-level	Specialization/standardization	2
	Resource optimisation	12
	Communication optimisation	5
	Automation	16
Redesign approaches	Other	2

We observed that process simulation and flow analysis were used interchangeably using quantitative techniques. The simulator's availability influenced the choice of process simulation in the software.

Analyzing the reengineering part of the projects, we observed that students could define appropriate reengineering goals addressing the noticed issues based on the performed analysis. Most of the students used the redesign heuristics defined in [7]. The main selected heuristics include automation, resource optimization (i.e., changes in schedules and additional resources), and changes on the task level (elimination and (de)composition). In two cases, the original ideas for project changes used Lean Management techniques (e.g., the 5S approach).

We observed the following benefits of the PBL experience (with relation to the formulated RQs):

1. the proposed project content in the case of the application nature of BPR helps students to understand and try to implement the gathered new knowledge into action during their work on the selected case (RQ1),
2. free topic selection (process to be reengineered) increases students' interest and motivation, enabling also the usage of previous knowledge from other subjects and extending it with BPR experience (RQ1, RQ3),

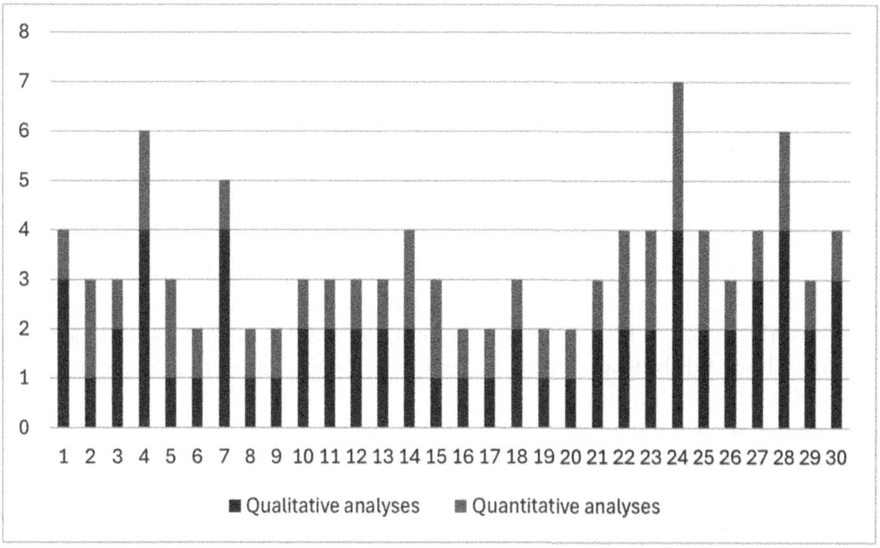

Fig. 2. Number of process analysis techniques in the students' projects.

3. division of project classes into clear stages enables monitoring the progress of the project and explaining on an ongoing basis any questions and issues (RQ2),
4. small group enables better direct communication and teacher involvement (RQ3),
5. project content enables the achievement of all course outcomes in the most effective way for students, based on their individual experiences (RQ1, RQ2).

Our students confirm the mentioned benefits and their reflective notes on the project content (*original spelling*):

Student 1: *"From a personal point of view, this application of the theoretical concepts seen in class allows me to understand the challenges of BPR/BPM"*;

Student 2: *"This project helped me to understand more about manufacturing processing and not only, these techniques can be applied and be useful in many aspects of my life, and in general our life, to be more productive and efficient with our tasks"*;

Student 3: *"Personally, I had never used simulation in this type of field, and I found it very interesting and useful for detecting defects"*;

Student 4: *"The creation of this project allowed us to complement the contents given in class in a dynamic way since we had to apply them in each of the steps, we did to complete this project"*.

Regarding the implementation of the PBL approach in the BPR course, we also made some specific observations:

1. Autonomy in project selection increases motivation. When students worked on processes they were familiar with from their personal or work lives, they were more engaged and interested, making the learning experience more meaningful.
2. Incremental project evaluation and feedback loops help students stay on track; dividing the course into steps makes it easier to monitor progress and support students regularly. This helped them understand what to do and improved the final results.
3. The importance of reflective practice—asking students to summarize what they learned and what challenges they faced made them think more deeply about the topic. This reflective element helped them understand not only what they did but also why it mattered, and supported the connection between theory and practice.
4. Available tools influence student choices—using specific analysis methods (like simulation) often depends on whether the software supports them. Teachers should remember that access to tools can constrain analytical approaches and influence student decisions.

The use of the PBL approach in this course enables participants to unleash the potential of their diversity by working in interdisciplinary groups. Implementing a project covering a practical subject that utilizes BPM-specific tools adds dynamics to the course and enhances the absorption of knowledge and skills.

Although the course is conducted every year and students evaluate it positively, it is worth paying attention to two fundamental challenges that make the teacher's work difficult: (1) heterogeneous levels of students' prior knowledge in Business Process Management (BPM), and (2) fluctuating patterns of student engagement, both across individuals and over time. The first issue arises from the course being publicly offered as an open elective, which allows students from diverse majors to enroll. Consequently, in previous editions, participants entered the course with backgrounds in Business Administration, Lean Management, Computer Science, Telecommunications, and Mechanical Engineering. This diversity results in varying levels of BPM proficiency, necessitating the use of differentiated instructional strategies to accommodate these differences. The second challenge involves uneven student engagement, reflected in unequal team contributions and a passive role during project work. This issue was particularly evident among students already acquainted with BPM-related topics, such as Lean Management, who sometimes exhibited reduced interest. We addressed the mentioned challenges and formulated other recommendations in the next section.

While offering valuable insights into the use of PBL in BPM education for interdisciplinary groups, our study has several limitations. First, it is based on a single case study conducted at one university with a relatively small group of students (30 projects over three editions). This restricts the external validity, as findings may not be generalizable to other educational contexts, institutions, or student populations. Second, our analysis is primarily based on qualitative data—students' project work, reflective summaries, and teachers' observations—so it may also introduce subjective bias. It may not fully reflect actual learning outcomes, which could affect the study's internal validity. There were no control groups or formal pre- and post-assessments, which makes it harder to isolate the effect of PBL from other influencing factors (e.g., prior experience, team dynamics, or individual motivation). Although steps were taken to design the course to support diverse learners, differences in prior knowledge and technical skills could have influenced the results. Finally, since the data comes from a limited number of students and teachers, the findings represent only their specific perspectives. More detailed research, including surveys, interviews, or long-term studies, could strengthen the conclusions and make them more widely applicable. Despite these limitations, the study provides a valuable example of how PBL can support BPM education and offers practical lessons for future teaching.

5 Recommendations and Takeways

Fully aware that the presented material reflects an implementation of PBL within a specific BPM course at a particular university, we propose a set of recommendations to guide future applications of PBL in similar educational contexts:

1. To address heterogeneous levels of students' prior knowledge in BPM, we propose incorporating a short, non-mandatory, self-paced onboarding module introducing BPM fundamentals and tools. This would provide foundational knowledge for those new to BPM-related topics and act as a recapitulation for students already familiar with Lean Management, quality management, or business management. Furthermore, BPM literacy among students also varies in terms of tool familiarity and computer proficiency, often being less developed in students from management-oriented backgrounds. To support more equitable learning, we propose accompanying each module with optional tool tutorials or short instructional videos to enhance understanding and facilitate more effective knowledge acquisition.
2. To mitigate student engagement, we recommend forming interdisciplinary teams and providing differentiated support to help sustain the interest of more advanced students. Additionally, allowing students to select project topics from a curated repository of real or semi-structured case examples, we prevent overly simplistic projects and promote deeper engagement.
3. To improve the quality of reflective activities, guided reflection using structured rubrics helps to avoid superficial summaries and encourages meaningful learning. Further strategies include implementing peer evaluation mechanisms and structured checkpoints for peer assessment and feedback. These activities should emphasize communication and collaboration and be initially supervised by the instructor, as collaborative methods and peer evaluation techniques are often unfamiliar to students and require soft skills that are not typically developed in technical university settings. These strategies aim to promote accountability, deepen learning, and ensure sustained participation throughout the course.
4. As long as education in process management focuses on tool-related issues, students will prefer projects with "hard" topics. Therefore, it would be necessary to increase the share of education in the "soft" Core Elements of BPM in basic subjects to increase their participation in projects.

Based on our experiences and lessons learned, we prepared a summary that highlights the most relevant and transferable practical takeaways for BPM educators aiming to implement PBL for interdisciplinary groups.

Takeaways for effective PBL in BPM courses for interdisciplinary groups

- **Start with onboarding**
  Equalize students' understanding of BPM fundamentals by offering a short introduction to key concepts such as BPM lifecycle, BPMN 2.0, and process analysis techniques. This might reduce gaps between students from technical and business backgrounds.
- **Structure BPM projects in clear stages**
  Divide the project into phases—such as as-is modeling, analysis, redesign, and evaluation—to mirror the BPM lifecycle, guide students through a realistic improvement process, and track progress effectively.
- **Let students choose familiar processes**
  Encourage students to select real processes they have observed or experienced (e.g., production, service, or administrative workflows), allowing them to apply BPM theory to authentic use cases. This might increase their motivation and personal investment in the project.
- **Encourage deeper reflection on BPM practice**
  Include a final reflective stage where students analyze what they learned about process challenges, redesign decisions, and tool use, linking BPM concepts with personal insights. This helps students process what they did and why it matters, connecting theory with practice.
- **Leverage diverse BPM perspectives**
  To leverage diverse backgrounds and prior knowledge, form interdisciplinary teams that combine technical and business perspectives when analyzing processes, identifying bottlenecks, and proposing redesign solutions.
- **Strengthen BPM tool support**
  Provide access to BPMN modeling and simulation tools, and consider offering tutorials or resources to reduce technical barriers and support a variety of analyses.
- **Support soft skills within BPM tasks**
  Integrate communication, collaboration, and presentation checkpoints during process improvement to reflect real-world BPM project environments and team-based decision-making.
- **Keep class sizes manageable when possible**
  Small groups facilitate detailed feedback on BPMN models, simulations, and redesign logic, allowing for stronger teacher-student interaction.

These key takeaways are designed to support educators in adapting their teaching strategies effectively to promote collaborative, real-world problem-solving among students from diverse academic disciplines and professional perspectives. We hope that the presented example will be helpful and inspiring to all teachers considering introducing the PBL method into their BPM courses, whether they are just beginning to explore this approach or looking to refine and enhance their existing practices.

6 Conclusions

The findings of our paper suggest that Project-Based Learning supports the integration of theoretical knowledge with practical skills in BPM education for interdisciplinary groups (RQ1). The structure of the presented course, which allows for student-driven project selection and iterative feedback, suggests that PBL is an effective and engaging teaching method in this context (RQ2). Several factors contributed to its successful implementation, including small class sizes, clearly defined project stages, and active student and teacher involvement (RQ3). The educational practice insights emphasize the value of real-world relevance, reflective learning, and the development of higher-order competencies such as problem-solving, creativity, and critical thinking. Overall, the example to which this paper refers shows that PBL can significantly enrich BPM education and better align it with the evolving demands of both students and the business environment.

Our case study brings several important contributions to BPM education:

1. Cross-disciplinary applicability of PBL in BPM—The course worked well for students from different academic backgrounds. Both technical and business students were able to learn together and complete the project successfully, which shows that PBL is effective in interdisciplinary groups.
2. Practical use of process improvement techniques by students—By documenting which redesign heuristics students most commonly applied (e.g., automation, resource optimization), the course offers practical insights for educators designing similar BPM learning experiences.
3. Development of advanced thinking skills—Through designing and improving real processes, students practiced essential skills such as problem-solving, creativity, and critical thinking—all key for future careers and aligned with the highest levels of Bloom's taxonomy.
4. Reinforcement of the teacher's dual role—during the course, teachers acted not only as instructors but also as guides and mentors. This active involvement was crucial for student success, as it provided confidence, motivation, and personalized support.

It cannot be overlooked that students appreciated such an active form of education because it engages them, despite requiring significant individual effort. This approach is critical given the forthcoming challenges in educating the Alpha generation [10,11].

Acknowledgements. This research project was partly supported by the "Excellence initiative—research university" program granted to the AGH University of Krakow, as well as by a research subsidy.

References

1. Adijaya, M.A., Widiana, I.W., Agung Parwata, I., Suwela Antara, I.: Bloom's taxonomy revision-oriented learning activities to improve procedural capabilities and learning outcomes. Int. J. Educ. Methodol. **9**(1), 261–270 (2023)
2. Chow, W.: Teaching business process management with a flipped-classroom and problem-based learning approach with the use of Apromore and other BPM software in graduate information systems courses. In: 2021 IEEE International Conference on Engineering, Technology & Education (TALE), pp. 1–8. IEEE (2021)
3. Dannecker, A., Meyer, L.: Relevance of topics in workflow-based student projects. In: Di Ciccio, C., et al. (eds.) BPM 2024. LNBIP, vol. 527, pp. 324–332. Springer, Cham (2024). https://doi.org/10.1007/978-3-031-70445-1_21
4. Delgado, A., Calegari, D.: Teaching BPM fundamentals: a project-based hands-on course for process-driven systems modeling and development. In: Köpke, J., et al. (eds.) BPM 2023. LNBIP, vol. 491, pp. 211–221. Springer, Cham (2023). https://doi.org/10.1007/978-3-031-43433-4_14
5. Dijkman, R.: BPM Game—bpmgame.org. https://www.bpmgame.org/. Accessed 04 June 2024
6. Dijkman, R., Peters, S.: The business process management game. In: 2019 Dissertation Award, Doctoral Consortium, and Demonstration Track at BPM, BPMT 2019, pp. 119–123. CEUR-WS.org (2019)
7. Dumas, M., Rosa, L.M., Mendling, J., Reijers, A.H.: Fundamentals of Business Process Management. Springer, Heidelberg (2018). https://doi.org/10.1007/978-3-662-56509-4
8. Gabryelczyk, R., Brzychczy, E., Gdowska, K., Kluza, K.: Business process management in CEE countries: a literature-based research landscape. In: Marrella, A., et al. (eds.) BPM 2022. LNBIP, vol. 459, pp. 279–294. Springer, Cham (2022). https://doi.org/10.1007/978-3-031-16168-1_18
9. Grisold, T., et al.: The five diamond method for explorative business process management. Bus. Inf. Syst. Eng. **64**(2), 149–166 (2021). https://doi.org/10.1007/s12599-021-00703-1
10. Jukic, R., Skojo, T.: The educational needs of the alpha generation. In: 2021 44th International Convention on Information, Communication and Electronic Technology (MIPRO), pp. 564–569. IEEE, September 2021. https://doi.org/10.23919/mipro52101.2021.9597106
11. Kalimatusyaro, M., Nuzula, I., Handayani, B.: Gen Z teachers and alpha generation learning: analysis of problems and strategies to improve learning quality. J. Soc. Sci. Educ. **3**(2), 81–91 (2025). https://doi.org/10.37812/josse.v3i2.1822
12. Kokotsaki, D., Menzies, V., Wiggins, A.: Project-based learning: a review of the literature. Improv. Sch. **19**(3), 267–277 (2016)
13. Lin, C.H., Chen, F.H.: The Effectiveness of project-based learning in learning. In: Huang, Y.M., Cheng, S.C., Barroso, J., Sandnes, F.E. (eds.) ICITL 2022. LNCS, vol. 13449, pp. 296–305. Springer, Cham (2022). https://doi.org/10.1007/978-3-031-15273-3_33

14. Lippe, S., Bärtsch, F., Brönimann, C., Mader, C.: Enabling BPM competencies through real-life student consulting projects. In: Di Ciccio, C., et al. (eds.) BPM 2024. LNBIP, vol. 527, p. 344–353. Springer, Cham (2024). https://doi.org/10.1007/978-3-031-70445-1_23
15. Marjanovic, O., Bandara, W.: The current state of BPM education in Australia: teaching and research challenges. In: zur Muehlen, M., Su, J. (eds.) BPM 2010. LNBIP, vol. 66, pp. 775–789. Springer, Heidelberg (2011). https://doi.org/10.1007/978-3-642-20511-8_69
16. Munkácsy, B., Scharle, Á.: The relevance of future challenges to education in five central eastern European countries. In: Adapting to Future Challenges to Education. Hungary, Poland, Romania, Serbia, and Slovakia, pp. 15–60 (2021)
17. Nobre, M.R., Vilela, J.F.: Content and skills for teaching BPM in computer science courses: a systematic mapping study. Simpósio Brasileiro de Sistemas de Informação (SBSI), pp. 187–191 (2024)
18. Nurhidayah, I.J., Wibowo, F.C., Astra, I.M.: Project based learning (PjBL) learning model in science learning: literature review. J. Phys. Conf. Ser. **2019**(1), 012043 (2021). https://doi.org/10.1088/1742-6596/2019/1/012043
19. OECD: OECD future of education and skills 2030 (2019). https://www.oecd.org/education/2030-project/. Accessed 04 June 2024
20. de Pádua, S.I.D., Lourenção, M., Moreira, S.A.S., Ribeiro, L.R.B.: Fostering BPM learning through PBL: development and application of an educational framework. In: Di Ciccio, C., et al. (eds.) BPM 2024. LNBIP, vol. 527, pp. 416–425. Springer, Cham (2024). https://doi.org/10.1007/978-3-031-70445-1_30
21. Quevedo, A.V., Guerrero, D.A., Palma, M., Vegas, S.: Improving generic skills among engineering students through project-based learning in a project management course. In: 2013 ASEE Annual Conference & Exposition, pp. 23–715 (2013)
22. Revoredo, K., Bala, S.: Project-based learning in bpm: lessons from flipped classroom vs. lecture-based instruction. In: Di Ciccio, C., et al. (eds.) BPM 2024. LNBIP, vol. 527, p. 313–323. Springer, Cham (2024). https://doi.org/10.1007/978-3-031-70445-1_20
23. Ruiz Hidalgo, D., Ortega-Sánchez, D.: Project based learning: a systematic literature review (2015-2022). Hum. Rev. Int. Hum. Rev./Revista Internacional de Humanidades **14**(6), 1–14 (2022). https://doi.org/10.37819/revhuman.v14i6.1286
24. Sliż, P., et al.: Business process management education in Poland: a manifesto for academic teaching. e-mentor **104**(2), 91–98 (2024). https://doi.org/10.15219/em104.1661
25. Weber, M., vom Brocke, J.: Managing BPM projects and their implications on organizations-experiences from research and teaching. In: Köpke, J., et al. (eds.) BPM 2023. LNBIP, vol. 491, pp. 240–249. Springer, Cham (2023). https://doi.org/10.1007/978-3-031-43433-4_17

Teaching Process Redesign with a Competition

Felix Schumann[1], Matthias Ehrendorfer[1], Michel Kunkler[1], and Remco Dijkman[2(✉)]

[1] TUM School of Computation, Information, and Technology, Technical University of Munich, Garching, Germany
{felix.schumann,matthias.ehrendorfer,michel.kunkler}@tum.de
[2] Eindhoven University of Technology, Eindhoven, The Netherlands
r.m.dijkman@tue.nl

Abstract. Business process redesign is an important part of Business Process Management. However, evaluating the impact of a process redesign in an educational setting poses challenges, because students do not get direct feedback on whether the redesigned process is better. Feedback received from the lecturers on that feels less realistic to them. To address this issue, we developed an assignment that enhances the learning experience by facilitating students in analyzing and redesigning a business process and immediately seeing the effect of their redesign on performance indicators. This paper presents the design and implementation of the assignment. The assignment focuses on the redesign of a business process with respect to operational decisions made in the process. It is presented to the students in the form of a competition to motivate them further. This year, the assignment focused on improving the treatment process of a fictitious hospital by optimizing decisions on patient admission and resource allocation. However, the assignment is developed in a general framework, which facilitates the development of new or modified assignments on a yearly basis. This year, the assignment has been used in courses at two different universities and is planned for further use at three other universities. The assignments that the students handed in showed good understanding by the students and showed that they made a real effort to solve the assignment well. Informal feedback from the students was also positive.

Keywords: Assignment · Competition · Business Process · Redesign · Optimization

1 Introduction

A key part of Business Process Management (BPM) is business process redesign. In the redesign step, a business process is improved to perform better according to some performance indicators (e.g., cost, time, or quality). As a consequence, once a redesign is completed, we must also evaluate whether the process indeed

M. ER et al. (Eds.): BPM 2025, LNBIP 565, pp. 330–342, 2026.
https://doi.org/10.1007/978-3-032-02936-2_23

does better according to some chosen performance indicator. Unfortunately, this is often difficult to do in an educational setting, because the precise performance properties of the process elements that were introduced as part of the redesign remain unknown. Students can be asked to speculate about the impact of the proposed redesign on performance indicators or lecturers can evaluate the redesign, but this is not the same as doing a quantitative evaluation. This limits the learning experience.

Against this background, we designed an assignment - in the form of a competition - that facilitates a better learning experience with respect to business process redesign. In particular, it enables students to analyze a business process and its performance, redesign it in an attempt to improve its performance, and subsequently evaluate whether the performance is indeed improved. In that sense, it can be considered to cover three steps of a business process management lifecycle, which is a structure that is often used to organize BPM courses around [1].

In the assignment, students are asked to improve the treatment process of a fictitious hospital by redesigning policies that decide when which patient is admitted to the hospital and which resources should work in the hospital at what time. By doing so, they have to minimize the waiting time of patients and the personnel cost of employing the resources. The policies can be implemented as part of a simulator that both checks their correctness and evaluates their effect on the process performance indicators. The policy simulator is implemented in a general framework[1] that supports the relatively easy creation of new assignments on a year-by-year basis.

The assignment was used in the context of a course for the second time this year, implementing some variations compared to last year. Last year, it was used at one university. This year it has been scaled up and will be used at least five universities. A detailed assignment description, a rubric for grading, and lecture slides are available as teaching materials, in addition to the simulation software that is the assignment itself. While no formal course evaluation was available at the time of writing this paper, two batches of students from two different universities had used the assignment. They made serious efforts and created good policies for the assignment. Informal feedback from the students on the assignment was positive.

Against this background, the remainder of this paper is structured as follows. Section 2 provides an overview of the theoretical foundations of business process redesign and optimization. Section 3 details the setup of the assignment, including the learning goals, assignment structure, and educational setup. Section 4 describes the materials provided to students, such as the simulation software and teaching aids. Section 5 offers a reflection on the outcomes and feedback from the competition. Finally, Sect. 6 presents related work and Sect. 7 concludes the paper.

[1] https://github.com/bpogroup/bpo-project.

2 Business Process Redesign and Optimization

Various methods for Business Process Redesign (BPR) have been proposed [4]. These methods all aim to improve specific aspects of a business process, such as the control-flow aspect or the resource aspect, to achieve a specific objective. The objectives that can be aimed for during business process redesign are considered to form a 'devil's quadrangle' of goals that are hard to achieve at the same time [3]. For example, when reducing the cost of executing a business process by reducing the number of resources employed in that process, this is likely to come at an increase to the waiting time of cases in the process. These methods use different techniques at their core. For example, redesign heuristics can be used to recognize patterns in a business process that are likely to improve it [3], while simulation can be used to analyze the performance of the process quantitatively [5].

Business Process Optimization (BPO) concerns the optimization of operational decisions in a business process by redesigning the policies that are used to make those decisions. Figure 1 shows an example of a process in which various operational decisions can be taken. The process shows a simple insurance application process, in which customers arrive at a rate of 2 per hour (i.e. at half an hour from each other on average), upon which their application for an insurance is registered. This is done by a back office employee at a rate of 10 per hour or taking on average 6 minutes per task. Subsequently, their application must be checked, for which either a simple check can be done by a back office employee or an extended check can be done by a front office employee. There is a 20% probability that a mistake was made and the application needs to be processed again. There are various operational decisions that can be taken in this process. For example, the choice between performing a simple check and an extended check can be made depending on properties of the case as well as available employees. One can imagine that, depending on properties of the case, a simple check leads to a higher chance of having to redo an application. Though this risk may be acceptable depending on how busy front or back office employees are at a particular moment. Another operational decision that can be taken is which specific back office employee to assign to which case, assuming that different back office employees are better at processing different types of insurance.

Business Process Optimization (BPO) concerns the (re)design of a policy that decides on a specific action to take in a decision in a specific state of the business process. For reasons of efficiency or privacy, a policy is usually defined on a function over (a subset of) the state variables. We also call this an observation of the state.

Definition 1 (State, Observation). *Let $\mathcal{V}$ be the set of all possible values. A variable $V \subseteq \mathcal{V}$ identifies a set of possible values. For $i \in \{1, \ldots, n\}$ and $j \in \{1, \ldots, m\}$, let $V_i \subseteq \mathcal{V}$ and $W_j \subseteq \mathcal{V}$ be variables. The statespace $S \subseteq V_1 \times \ldots \times V_n$ of a process is the set of possible values for those variables. An observation function $f^O : S \to W_1 \times \ldots \times W_m$ is a mapping from a state to an*

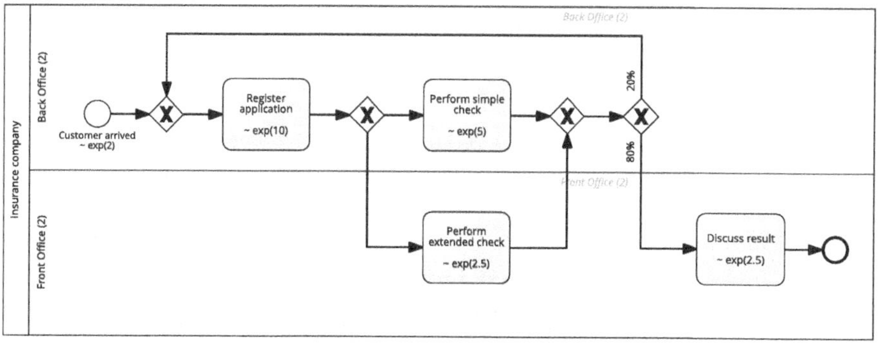

Fig. 1. Example of a process with decisions to optimize

observation about the state and $O = \{f^O(s)|s \in S\} \subseteq W_1 \times \ldots \times W_m$ *is the set of all possible observation about the process.*

Typically $m \ll n$, such that the observation is much easier to handle (algorithmically) than the state itself.

Definition 2 (Policy). *Let* S *be the set of all possible states of a process and* O *the corresponding set of all observations.444334 Furthermore, let* A *be the set of possible actions related to a decision that must be taken in the process. A policy* $\pi : O \rightarrow A$ *is a function that assigns an action to each observation. Consequently, it maps each state* $s \in S$ *to an action* $\pi(f^O(S))$.

For example, in the decision regarding the choice between performing a simple check and an extended check, there are two possible actions: 'decide on simple check' and 'decide on extended check'. The decision may depend on the level of complexity of the case, how many cases of what complexity are currently in the system, and how many employees of each type are available, which can accordingly be defined as variables that describe the observations about the state of the process. The state of the process itself contains additional information about the process, such as how many cases are waiting for each task, how many cases are being processed, by which resource, and how long they have been processing for.

A business process defines a behavior that determines how it transitions from one state to another. In the context of BPO, we assume that this behavior is precisely defined as the execution semantics of the business process. In practice, this requires that the process is described in a language for which a well-defined execution semantics exists (e.g., in the form of a mapping to Petri nets). For now, we will leave this semantics abstract and just assume that it implies the existence of a transition function that defines how a process transitions from one state to the next upon the occurrence of an action.

Note that the transition function may not be deterministic, due to randomness in the process, including randomness in the interarrival times of cases, properties of those cases, and processing times of tasks. We model this behavior by incorporating random variables in the transition function, which are given random values when the transition happens. These random values can, for example, represent the time it took to process a task.

Definition 3 (Transition). *Let S be the set of possible states of a process and A be the set of possible actions for a decision in the process. Furthermore, let $\Xi \subseteq \Xi_1 \times \ldots \times \Xi_n$ be n random variables with $\Xi_i \subseteq \mathcal{V}$. A process defines a transition function $T : S \times A \times \Xi \to S$ that determines how it transitions from one state to the next, after an action is executed and random variables Ξ obtain a value $\xi \in \Xi$. We also denote a transition from state $s_k \in S$ under action $a_k \in A$ and random values ξ_k to state s_{k+1} as $s_k \overset{a_k,\xi_k}{\rightarrow} s_{k+1}$.*

Business process optimization concerns finding a policy for taking decisions in the process that is better than the current policy. A policy is better than another policy if the objective value that it obtains in some final state of the process is better than the one the current policy obtains. For example, the sum of cycle times of cases after 100 cases have been performed by the process must be lower.

Definition 4 (Rollout, Objective). *Consider a business process with states S, actions A, a transition function T, an initial state $s_0 \in S$ and a termination condition $t : S \to Bool$ that is True if the process is in the final state. Given a policy π, a rollout of length n of the process is the repeated application of the policy and valuation of random variables, from the initial state of the process until the final state of the process is reached: $s_0 \overset{a_0,\xi_0}{\rightarrow} s_1 \overset{a_1,\xi_1}{\rightarrow} \ldots \overset{a_{n-1},\xi_{n-1}}{\rightarrow} s_n$, where $a_i = \pi(f^O(s_i))$, $t(s_n) = True$, and $\forall s_i, i < n : t(s_i) = False$. The objective function $f^{obj} : S \to \mathcal{V}$ returns the objective value in a state of the process.*

Note that different rollouts under the same policy may, due to randomness of the variables ξ_k, still go through different states and consequently end up in a different final state and receive a different objective value. Consequently, a policy π' is considered better than a policy π if its expected objective value is greater. Theoretically, the expected objective value is the mean objective value over an infinite number of rollouts.

Definition 5 (Expected Objective Value, Better Policy). *Let $s_i^\pi | 1 < i < \infty$ be the final states obtained after an infinite number of rollouts of a process under policy π. The expected objective value of policy π, denoted $\mathbb{E}[f_\pi^{obj}]$, is $mean(f^{obj}(s_i^\pi) | 1 < i < \infty)$. A policy π' is better than π if $\mathbb{E}[f_{\pi'}^{obj}] > \mathbb{E}[f_\pi^{obj}]$.*

Clearly, it is impossible to do an infinite number of rollouts. For that reason, a 'sufficiently large' number of rollouts is typically used, over which the mean objective value is computed. Optionally, the statistical significance of the difference between the mean objective values can be determined.

Consequently, business process optimization is about finding a better policy for making a particular decision in a business process.

3 The Business Process Optimization Competition

In the BPO competition, students were asked to analyze a business process and subsequently optimize it with respect to two decisions. They were evaluated on their ability to optimize these decisions and motivate their way of doing so by careful analysis.

In this section, we describe the case used for the competition, the learning goals, and the educational and competition setup.

3.1 Competition Case

The case students had to optimize was that of an artificial, simulated hospital, in which decisions had to be made concerning planning patient admissions and scheduling resources. To achieve this, students had to develop two policy functions:

1. A planning function that determines the timing of patient admissions for treatment.
2. A scheduling function that allocates various resources (such as operating rooms and hospital beds) over time.

These functions would be used in a simulation that both evaluated whether the policy functions made correct decisions that did not violate constraints (like exceeding the maximum number of resources) and simulated the behavior of the hospital and its patients under the provided policy functions.

The goal was to design the policy functions to minimize both the resource costs of the hospital and the time required to treat patients. Both metrics – cost and treatment duration – were to be minimized. However, there clearly is a trade-off: increasing the number of resources can reduce patient waiting times, but also increases operational costs.

The hospital treated different types of patients, each following a general treatment pathway as shown in Fig. 2. After admission, patients have an intake interview, possibly followed by surgery and a stay in a nursing ward. If complications arise post-surgery, patients may require additional surgery and nursing. Patients differed with respect to their diagnosis, which determined the ward that they would have to be nursed in as well as the likelihood that they would need surgery or nursing and the time they would spend in surgery or a nursing ward.

There was seasonality in the arrival of patients, in that different types of patients had a different likelihood of arriving at different times of the day or week.

Each solution was evaluated by running the simulation for a year in simulation time. The objective function that the policies would need to optimize for included the waiting time before and after admission, the 'planning nervousness', and the resource costs. 'Planning nervousness' is a score related to the frequency and timeframe of patient replanning, because patients will clearly not like it if their admission is replanned, and especially not if it is replanned shortly before

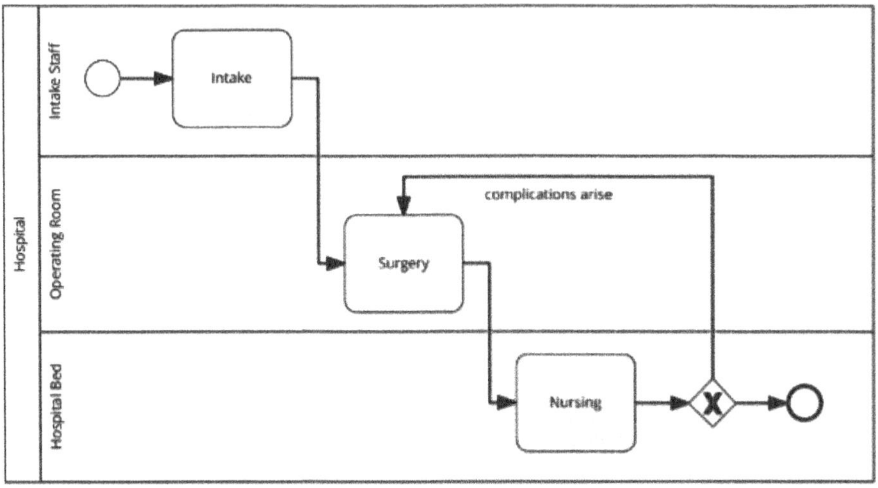

Fig. 2. The process to optimize in the competition.

their arrival at the hospital. These variables were normalized as the percentage increase over a benchmark solution, and finally, their weighted sum was used as the objective value. To get a reliable objective value in light of randomness, the average value over 10 replications of one simulated year was used.

3.2 Learning Goals

The main learning goal of the competition was that, after doing the competition and studying the learning materials around it, they should be able to: design a well-motivated policy for improving decision-making in a business process. In more detail, the learning goals were the following.

The student can perform process discovery and analysis, understands the properties of the process and how they influence/are influenced by the policies. The process shown in Fig. 2, as it was given to the students, was an incomplete representation of the process. Students had to mine the complete process themselves to fully understand what happened in the process. To this end, they were provided with benchmark policy functions that worked. The simulator provided an event log as one of its outputs. The main goal was for the student to understand which properties of the process could be influenced by the policies, or influenced the process in combination with the policies. These properties included the various types of patients, their 'seasonality', and the flow variant that they would go through. For example, this combination of properties determined how busy it would be with different types of patients at different times in the process and, consequently, what the demand would be for the various wards and the operating rooms of the hospital. These properties also included the objective values and how they develop over time, because these show how seasonal effects in combination with the policies influence the objective. For example, if an analysis shows that under the current policy resources are less busy during the weekend than during weekdays, the number of resources in the weekend can be decreased.

The student can improve a business process by redesigning well-motivated policies for taking decisions. The most important task for the students was to take the benchmark policy and design a new policy that improved on it in terms of its objective values. As the benchmark policy was rather naïve, this was not a hard task and indeed all students managed to do so. The main challenge for the students was to motivate their choice of solution. Roughly speaking there were two types of solutions that could be used. One was the heuristic rules solution, where students implemented rules that determined when to admit which patient and how many resources to schedule when. These rules could, for example, be of the type: have 1 operating room open during the night, 2 in the daytime during the weekend and 5 during in the daytime during the week. The second was to use self-learning policies using artificial intelligence algorithms, such as reinforcement learning. In case students used heuristic rules, the quality of the motivation behind the specific rules that they used was assessed. In case students used self-learning policies, the quality of the post-hoc analysis was assessed, of whether the policy learns to act, in specific situations in the process, in a way that makes sense.

The student can code, report, and present. These optional objectives were used in some of the educational settings, which asked students to carefully structure, document, and/or test their code, as well as deliver more detailed reports, presentations, or posters on their solutions.

3.3 Education Setup

The competition was designed to be part of a Master-level course on business process management, in particular a course that addresses business process analysis and improvement.

Prerequisite knowledge required to do the competition was Python programming skills as well as entry-level knowledge of process mining and business process modeling, preferably using BPMN.

Within that context, the competition was used in different Master-level courses, making up a larger or smaller part of the course. For example, in one seminar-driven course, the competition formed the basis of the course and students were asked to study advanced AI algorithms and implement them to learn policies for the competition. In another course, the general business process management lifecycle and techniques from different lifecycle stages were taught. In that course, the competition just formed an assignment for the business process redesign stage of the lifecycle.

The estimated minimum effort that students would need to spend on the assignment was 2 ECTS (56 h) for a single student. However, ideally, the students would do the exercise in groups, to facilitate discussion and groupwork. The effort required for the assignment could easily be increased by imposing more requirements on the quality of the solution and the discussion thereof, on the number of alternative policies studied, and on the requirements for reporting and code quality.

The assignment was set up as a competition, where students could submit their solution multiple times and be listed with the objective values produced by their solutions, to compare themselves against others. To be able to manage that, we required that solutions should not take too long to compute (i.e., have a maximum of 15 min computational time on a regular laptop) and, since we ran the evaluation simulations manually, students were only allowed to submit their solutions up to 3 times. When automated evaluations are done, this restriction can be lifted.

4 Materials

The materials for the competition are made available through the website[2]. Previous iterations of the competition are also made available by simply changing the year in the URL. The materials include the following:

Code that must be used to run the competition. This includes the simulator that is used to evaluate the developed policies and an example implementation of the policies that can both be used as a reference for implementation and serves as a benchmark for performance of the developed policies.

Case Description explaining the process that must be optimized, the decisions that must be taken in the process, and a more detailed explanation of the way in which the decision policies must be implemented in code.

Assignment Descriptions explaining the requirements for doing the assignment. Note that the requirements for the assignment are different from the competition: while the competition is aimed at developing the best possible policies, the assignment is aimed at understanding the concept of business process optimization and how to develop a good policy. Students are graded for that rather than for winning the competition. Two versions of the assignment are made available: one short assignment and one elaborate assignment.

Evaluation Rubrics presenting in detail how students can be graded for their effort. In line with the two versions of the assignment, two versions of the rubric are made available.

Lecture Slides introducing the concept of business process optimization, the assignment, and discussing three possible solution directions.

5 Reflection

At this time of writing, the assignment has been used in Master courses at two universities, and the results of one course are available. The assignment is planned to be used in three more courses this year.

[2] https://sites.google.com/view/bpo-2025/.

In the completed and evaluated course, 13 groups of 4 students participated. This course used the short version of the assignment. Figure 3 shows the overall improvements achieved by the groups. The overall improvement is the percentage improvement over the benchmark according to four KPIs: resource cost, waiting time before admission, waiting time in the hospital, and planning nervousness. Lower is better, which is why (more) negative percentages are better. 11 out of the 13 groups managed to improve on the benchmark solution, some by a lot.

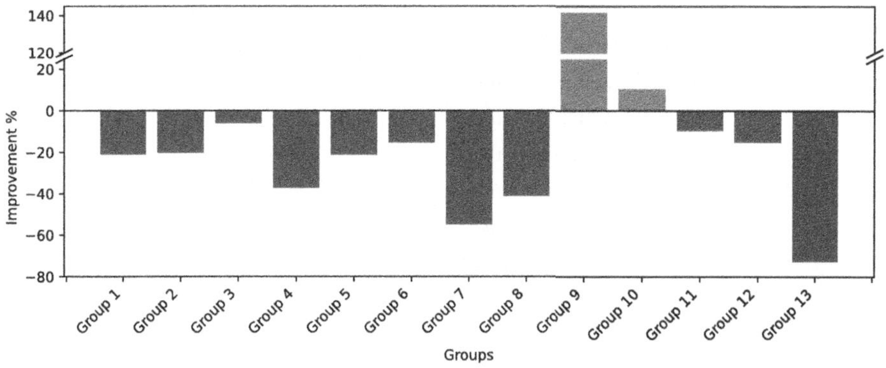

Fig. 3. Improvements achieved by student groups

Table 1 shows the solution approaches used by each group. All groups used heuristic rules to implement either one of the two policies. One group used local search, and one group used Q-learning to solve one of the two policies. Some groups combined heuristics with hyperparameter tuning. They did that by adding parameters to the heuristic rules, such as the number of hours from now a patient should initially be planned for, and tuning those parameters over multiple replications of the simulator to get the best results.

Unfortunately, few students filled out the course evaluation, such that feedback is available from that. However, the assignments that the students handed in showed a good understanding of the principles of business process optimization. In addition, students actively participated in the lecture and the Q&A sessions around the assignment and showed that they made a real effort to solve the assignment well. Informal feedback received from the students was also positive about the assignment.

6 Related Work

Within the BPM-Community, the best-known challenge is the Business Process Intelligence Challenge (BPIC). The challenge was held yearly from 2012 until 2020 and was co-located with the BPM Conference until 2018 and with the ICPM Conference in 2019 and 2020. A core achievement of the different ICPM

Table 1. Solution approaches used by each group

Group	Solution Approach
Group 1	Heuristic rules
Group 2	Heuristic rules, Local search
Group 3	Heuristic rules
Group 4	Heuristic rules with hyperparameter tuning
Group 5	Heuristic rules
Group 6	Heuristic rules with hyperparameter tuning
Group 7	Heuristic rules with hyperparameter tuning
Group 8	Heuristic rules
Group 9	Heuristic rules
Group 10	Heuristic rules
Group 11	Heuristic rules
Group 12	Heuristic rules, Q-learning
Group 13	Heuristic rules

Challenges is the provisioning of real-world, process-driven datasets that have been used as benchmarks within the community over the years. The BPIC challenge was set up as an open challenge for both students and practitioners. To participate in the challenge a written report was required that described the used process analytics methods and presented findings. An automatic evaluation was not possible since the goal was often to find new insights from the provided data. Due to this manual evaluation, a feedback loop of the provided submission was not possible. This stands in contrast to the challenge presented in this paper, where students can receive direct feedback by implementing their solutions and connecting them to the simulator provided as part of the challenge set up. This iterative approach enables students to test and improve their approaches. While the focus of the BPIC challenges was not on integrating the challenge into an educational framework for students, a specific student category ensured a separate evaluation of student submissions[3].

Another example of a competition in computer science research is the Neural Information Processing Systems (NeurIPS) competition [2]. Within the NeurIPS conference, a special competition track is established. This competition track ensures high-quality competitions by accepting submission proposals, which then undergo a review and selection process before being accepted as a competition within the competition track. As part of the review procedure for submitted competitions, their scientific relevance, evaluation protocol, and planned logistics, i.e., competition schedule, are evaluated. As for the BPIC challenge, NeurIPS competitions are not directly embedded in any educational framework to be used alongside a class teaching students. Each year, multiple competitions are

[3] https://www.tf-pm.org/competitions-awards/bpi-challenge.

accepted and open for participation. This underlines the importance of the competition format in the AI research field[4].

7 Conclusion

The BPO competition presents an approach to teaching business process redesign and, in particular, business process optimization, i.e., redesign with respect to the operational decisions that have to be taken in a business process. The competition facilitates students in redesigning the policies that are used to take decisions in the business process and seeing the effect of their redesign in a simulated environment.

This year, the competition featured decision-making on a treatment process in a simulated hospital. The decisions concerned when to admit which patient to the hospital and how many resources to schedule when to allow for fast and effective treatment. However, the competition is based on a generic framework that allows for quick implementation of new competitions, and this was the fourth competition that was created based on the framework. The competition is further supported by two variants of a course assignment description, including lecture material and a grading rubric.

This year, the competition expanded to 5 universities worldwide, where it was used in Master courses on Business Process Management. At the time of writing, the results of one of the courses were available. The results showed that students understood the material well and were able to come up with good and creative solutions to redesign problem. While formal feedback on the student's experience with the assignment was not available due to a low response rate to the evaluation survey, informal feedback on the assignment was positive.

Based on experience from this year's iteration of the competition, the framework will be further improved to make it easier to develop new competition assignments as well as to evaluate the assignments. In addition, video lecture material will be considered.

Acknowledgement. This work was partly funded by the Deutsche Forschungsgemeinschaft (DFG) – GRK2201 – Projektnummer - 277991500.

References

1. Dumas, M., La Rosa, M., Mendling, J., Reijers, H.A.: Fundamentals of Business Process Management, 2nd edn. Springer, Cham (2018)
2. Frellsen, J., Zhan, K., Tao, Q.: Neurips 2025 call for competitions (2025). http://neurips.cc/Conferences/2025/CallForCompetitions. Accessed 20 May 2025
3. Reijers, H., Mansar, S.L.: Best practices in business process redesign: an overview and qualitative evaluation of successful redesign heuristics. Omega **33**(4), 283–306 (2005). https://doi.org/10.1016/j.omega.2004.04.012

[4] https://neurips.cc/Conferences/2025/CallForCompetitions.

4. Tsakalidis, G., Vergidis, K.: Systematicity and generalizability in business process redesign methodologies: a systematic literature review. Preprints (2025). https://doi.org/10.20944/preprints202501.1261.v1
5. Wynn, M.T., Dumas, M., Fidge, C.J., ter Hofstede, A.H.M., van der Aalst, W.M.P.: Business process simulation for operational decision support. In: ter Hofstede, A., Benatallah, B., Paik, H.-Y. (eds.) BPM 2007. LNCS, vol. 4928, pp. 66–77. Springer, Heidelberg (2008). https://doi.org/10.1007/978-3-540-78238-4_8

Generative AI's Aid in Business Process Modeling Instructional Design: A Case Study

Ilia Maslov[1(✉)], Stephan Poelmans[1], Yves Wautelet[1], and Kristina Rosenthal[2]

[1] KU Leuven, Warmoesberg 26, 1000 Brussels, Belgium
{ilia.maslov,stephan.poelmans,yves.wautelet}@kuleuven.be
[2] Niederrhein UAS, Richard-Wagner-Straße 97, 41065 Monchengladbach, Germany
kristina.rosenthal@hs-niederrhein.de

Abstract. This research explores the implications of generative AI, particularly ChatGPT, in instructional design for business process modeling. The study leverages ChatGPT-4's capabilities in prompt engineering to address challenges in process modeling education, aligning with instructional theories and frameworks such as the ADDIE model and Bloom's taxonomy. Findings indicate that while ChatGPT can enhance the design and delivery of educational content, it requires careful integration and refinement. By examining the efficacy of ChatGPT in creating and assessing educational material, this research contributes to the understanding of generative AI's role in modern education, suggesting a balanced approach that utilizes AI to augment traditional teaching approaches.

Keyword: Generative AI · ChatGPT-4 · Instructional Design · Business Process Modeling Education

1 Introduction and Overview

ChatGPT, a generative AI with a text-based interface, offers extensive capabilities for answering questions [1]. With versions 3.5 and 4 released in late 2022 and early 2023, its educational impact has been widely studied, including issues like academic dishonesty and its use in entrance exams, essay writing, and math problem-solving [2–4]. It also aids in organizing educational materials and preparing for job interviews [1, 5]. Ethical concerns of using ChatGPT in education have been raised [6, 7]. Its potential for exam misuse necessitates careful study to mitigate negative impacts and enhance positive ones, like structured learning or ethical AI use [2, 4]. While restrictive measures like watermarking AI-produced text have been suggested [8], potential circumvention is possible [9]. ChatGPT can potentially disrupt traditional learning methods, posing challenges like those of GitHub Copilot, another AI tool for coding, forcing educators to adapt to the new reality [10].

This calls for a broader reconsideration of both teaching by educators and learning by students in the face of new technologies rather than surface-level fixes [11], a call raised to address the potential ethical problems with the use of ChatGPT at the educational level

M. ER et al. (Eds.): BPM 2025, LNBIP 565, pp. 343–357, 2026.
https://doi.org/10.1007/978-3-032-02936-2_24

[6, 7]. This aligns with the paradigm shift in learning through technology, promoting self-directed and peer-to-peer learning [9, 11]. The paradigm shift was especially pronounced after the COVID-19 pandemic, leading to challenges among the institutions and students' preparedness to employ the new technologies [12]. There is significant potential for students to use ChatGPT, but also for educators and institution administrators, as it can help structure learning [9].

For educators in process modelling, ChatGPT's implications are particularly intriguing, echoing research on its utility in engineering education [1, 6]. Process Modeling (PM), a vital part of the BPM life cycle, transforms textual requirements into visual models, with BPMN as an industry standard, to facilitate business-IT communication, process improvement, and automation [13]. PM teaching is vital for both pragmatic (providing skilled workers) and theoretical purposes (understanding cognitive aspects of PM) [14]. Teaching PM faces challenges like cognitive differences between novices and experts, complex modeling languages and software, the need for intensive practice and feedback, and a lack of validation procedures [15]. ChatGPT may offer solutions to some challenges if used as an educator's supporting tool.

Instructional Design (ID) is an approach, a practice, a discipline, and a science where instructors are considered designers engaged in design thinking, achieving the best pedagogical outcomes through being designers [16, 17]. Kaur [18] suggests multiple ID challenges in the context of blended learning (a mix of online and offline learning): imbalanced focus on technology implementation, neglecting content design, insufficient time, and budget allocation for ID, understanding both what and how to teach, choosing most suitable delivery media, along with maintaining engagement among the participants and coordinating between the on- and off-line teaching.

ID is a particularly perspective area for using generative AI. To address the common ID challenges, ChatGPT could help with automatic question generation, help to be more comprehensive, and provide insights for those lacking ID expertise [19–22]. Even then, it's stated that human educator intervention is necessary. Li et al. [20] explored whether ChatGPT can be used for middle school-level teaching of English and Math, evidencing that ChatGPT can support content generation, though ChatGPT cannot replace human teaching completely. Meron and Araci [19] tested and found that ChatGPT is a computer partner in designing the design education: ChatGPT saves time, structures textual content and documentation, and is a brainstorming tool. Fotaris et al. [22] developed and suggested a prompt engineering guide to create instructional design for gamified educational escape rooms, finding that ChatGPT helps to make educational content quicker and that it can be more accessible to people with little technical expertise by reducing the cognitive load of educators.

Given the research context and gaps associated with the emergence of generative AI and its potential implications on Instructional Design, we address the following Research Questions (RQs):

RQ1: How can prompts be systematically designed for instructional design using prompt engineering?

RQ2: How suitable is the instructional design created by ChatGPT for teaching process modelling?

RQ3: How do educators assess the usability and adoption of generative AI for the purposes of developing instructional design for teaching process modeling?

The study is descriptive, aiming to report on how the expert PM educators evaluate the potential use of ChatGPT-4 in instructional design.

Hence, we seek to generate a method to develop an instructional design (RQ1), evaluated for the quality of its produced ID (RQ2). Then, the conclusions concerning the potential applicability of ChatGPT are provided (RQ3).

Generative AI, including tools like ChatGPT, is often explored empirically through prompt engineering [23]. As per [9], a pragmatic approach is conducive to understanding ChatGPT's utility. In our research, we leverage ChatGPT-4 within the (Business) Process Modelling domain for instructional design (ID). We acknowledge learning as a complex, design-centric "wicked problem environment" [24, 25], advocating for a broad, pragmatic research perspective. This stance concerns generating practical knowledge that navigates objectivity and subjectivity [26]. Considering ID as a design-centric activity [17], we use instructional frameworks to engineer prompts for ChatGPT [23] to assist various ID stages. We adapt the ADDIE model [27] stages to guide ChatGPT-4 at the ID stages relevant to the PM education context. We will use supportive theories, like [28] Bloom's taxonomy, to facilitate an understanding of ID stages for prompt engineering. The prompts will make ChatGPT-4 provide ID recommendations. Their effectiveness is assessed using the Technology Acceptance Model (TAM) [29] by examining both the prompts and the output generated by ChatGPT.

Like our study, [1] and [3] used ChatGPT to create tests and solutions, assessing its educational utility. Yet, their methods lacked targeted systematicity. Following [19, 20, 22] who also refined prompt development and evaluation, our research advances this by anchoring prompt engineering in ID literature. Addressing the gap identified in the current literature, we propose a framework for standardized reporting in ChatGPT prompt engineering ID research. Furthermore, we use ChatGPT-4 in the context of (Business) Process Modelling training, which relies highly on visual diagrams; hence, creating and recognizing very specified schemes is critical for ID.

The following sections are structured as follows. Section 2 provides insights into extant literature on generative AI, ChatGPT-4, challenges to PM education, and Instructional Design with ChatGPT. Section 3 defines the selected ID framework and methodology. Section 4 provides a detailed account of the outcome of prompt engineering and evaluation of the applicability of ChatGPT-4 for ID. Experiences of educators applying generative AI (ChatGPT) will be explored through self-reporting using constructs from the TAM. Finally, Sects. 5 and 6 provide theoretical and practical implications, discussion, conclusion, limitations, and future work.

2 Extant Literature

Process Modelling (PM) transforms textual requirements into visual diagrams using, for example, BPMN 2.0 as standard, with over 100 elements for modeling processes [13]. PM, essential for analyzing and improving business processes, requires cognitive skills like abstracting and conceptualizing [14, 30]. Teaching PM presents challenges, such as the cognitive gap between novices and experts and the complexity of online teaching [15]. To address these, methods like e-learning platforms and gamification are employed, with a shift towards more student-led, interactive learning [11, 14]. This study explores using ChatGPT in PM course design to influence student learning outcomes, demonstrating its role in modernizing e-learning and instructional design [31].

ChatGPT 3.5 and 4 are Natural Language Processing (NLP) models developed by OpenAI, based on the GPT (Generative Pre-trained Transformer) architecture, initially developed for the tasks of language generation tasks such as machine translation and summarization. ChatGPT is a generative AI that can generate new content or ideas in real-time conversations, compared to traditional AI and is useful for discriminative predictions. ChatGPT is non-deterministic; thus, the responses are not pre-determined and are uniquely generated according to the user's requests or prompts in chat [1]. Being non-deterministic may also vary, and it generally refers to the idea that the randomness of the answers produced by Machine Learning (ML) is something that is used to trade-off the balance between overfitting the model to the training data, as opposed to letting the ML algorithm loose and being purely random and non-sensical [32]. There are also other types of generative AI. One that is also widely discussed in the literature and that was released for open access sometime before the ChatGPT was GitHub Copilot, released in June 2022, based on Codex, which is similarly based on OpenAI's paid-for-access ML models. GitHub Copilot can generate code based on the textual description provided by the programmer, even though it is often cited as being frequently wrong or inaccurate. Like ChatGPT, Copilot raised concerns in programming education since Copilot is free to access by students, and it generated answers without students having to think about the programming problems [10].

ChatGPT has several strengths and limitations. On the positive side, ChatGPT can be flexible, cost-effective, accessible 24/7, logical, and potentially aware of its limitations. It can also be interacted with in a human-understandable format. Alternatively, ChatGPT may struggle with contextual comprehension and delivering current information due to limitations in knowledge and creativity, potentially resulting in responses of inferior quality and occasionally unsettling answers. There can be institutional recommendations and regulations in the context of education and instructional design. Institutions may recommend seeking ways to integrate generative AI harmonically within the educational process. For example, educators and institutions are advised to promote the awareness and use of ChatGPT and find ways to utilize it to the benefit of instruction but not to circumvent the ethical integrity of learning outcomes [9]. Alternative opinions, however, may include a more critical and pessimistic view, ranging from the use of generative AI for academic fraud, unfair learning assessment, inaccurate information provided for learning, and over-dependence on the use of generative AI [2, 4, 6, 7].

Instruction is *"the deliberate arrangement of learning conditions to promote the attainment of some intended goal"* [33]. Thus, as part of education, instruction aims to manage learning conditions so that the learner can attain experience and improve problem-solving skills. Instruction planning is crucial for optimal results, with Instructional Design (ID) being a systematic process for developing instruction. Design thinking in ID involves identifying and bridging gaps between current and ideal states [16]. While educators have domain expertise, instructional designers specialize in ID, often using personal experience and various models and theories [17]. Learning theories like behaviorism, cognitivism, and constructivism inform ID strategy and influence learning principles [34]. Based on learning theory findings like Bloom's taxonomy, instructional theory guides teachers and designers in ensuring learning outcomes [33]. Instructional designers must balance theoretical and practical aspects of ID and adapt models to specific environments, reflecting George Box's view that "all models are wrong, but some are useful" [17]. The paper will review ID models and their applicability to ChatGPT in each ID stage.

ID in learning technologies often focuses more on technology implementation, leaving content design with insufficient time and budget. Challenges include determining not just the content but also the method of teaching, having enough time and budget for ID, selecting the best delivery medium to match performance objectives, maintaining interactivity in online offerings, ensuring participant commitment and follow-through, especially with "non-live" components, and coordinating all elements of the blended approach [18]. We suggest that these typical ID challenges could be addressed using ChatGPT.

Hodges and Kirschner [35] propose focusing on student and teacher challenges rather than things in education with generative AI, highlighting ChatGPT's facilitatory role in quizzes and diagram evaluations and how it poses ethical challenges for teaching and learning, hence suggesting instructional design to remain flexible in its approaches. They suggest that information must be processed by humans to become knowledge. Li et al. [20] compare lesson plans made by human educators and ChatGPT-3.5 and 4, focusing on high school math and English. They find that ChatGPT maintains a scientific, goal-oriented approach and captures lesson content sentiment and instructional designs comprehensively. Human educator input remains crucial: ChatGPT failed to provide specific examples to some questions, sometimes using self-generated innovation, which may lead to potential misconceptions among the students. They conclude that instructional design reflects subjective initiative and labor value, having spontaneous, autonomous, and non-routinized thinking, though ChatGPT can help better structure the content. Fotaris et al. [22] use design thinking principles to explore ChatGPT-3.5 in designing educational escape rooms, emphasizing its creative potential and cognitive load reduction, especially for novice educators. Lee et al. [21] investigate ChatGPT's role in automatic question generation for English teaching via an iterative prompt engineering approach. They note its effectiveness varies with question types: most of the questions were valid, though there was lower validity for'yes–no questions with multiple choices' and'cloze' (where the answer is located within a passage of text) type of questions, suggesting a lack of ChatGPT's comprehension at what is supposed to be measured. Meron and Araci [19] evaluated ChatGPT as an ID facilitator for course material creation in design education, finding it efficient for structuring textual content and as a brainstorming tool. However, the content was often too generic. They suggest that it could save time and generate ideas.

Thus, ChatGPT-4 could be suitable for specific instructional design stages, requiring extensive human intervention and initiation. We find that existing research lacks the application of the ID framework to model and assess its stages. The ADDIE model, foundational in Instructional Design (ID), outlines key sequential and iterative stages:

(1) Analysis (identifying performance gap causes), (2) Design (defining desired performances and test methods), (3) Development (creating and validating learning resources), (4) Implementation (preparing and engaging with learning material), and (5) Evaluation (assessing instructional products before and after implementation) [27]. Despite lacking a formal description, ADDIE is central to most ID models and is often recommended to be adapted to fit the specific context [17]. We adapt the model since ChatGPT works via prompts and thus necessitates a specific prompt-response interaction. We also adapt to fit the (business) process modeling education context: there are specifics to teaching it, such as reliance on creating and reading visual materials (business process diagrams). Furthermore, as suggested in the original ADDIE, we have not implemented the instructional design in class and have lacked evaluation after implementation.

3 Methodical Approach

Our study adopts a structured approach, identifying the educational challenge and then outlining clear goals for ID stages. Our process involves an iterative development of prompts for generative AI tools. We aim to craft a practical ID via ChatGPT-4, demonstrate ChatGPT-4 viability with examples, assess its effectiveness in real-world applications, and then discuss our findings comprehensively. We adopt a deductive orientation, leveraging existing theories of learning and technology to integrate generative AI while cautiously avoiding the myopia of technology hype curves.

The ADDIE model is foundational in Instructional Design (ID) and outlines key sequential and iterative stages (1) Analysis, (2) Design, (3) Development, (4) Implementation, and (5) Evaluation [27]. We adhere to a five-stage instructional design (ID) framework, as characterized by [27, 36] specifically tailored for process modeling (PM) education: (1) **Project initiation** for process modeling including preparation, which precedes setting learning objectives often guided by Bloom's taxonomy [28], (2) **Needs analysis** focusing on identifying the target audience and learners' needs, and defining learning outcomes [36], (3) **Instructional strategy development** that includes determining the most effective way to teach content or skills [36] based on choosing a suitable learning theory and instructional method (such as problem-based learning or gamification) [14], selecting appropriate learning materials and modeling software [13, 28, 31]. (4) **Model assessment** meaning to assess learning achievement [36] and developing assessment methods [14], followed by evaluating the models based on syntactic, semantic, and pragmatic criteria [13], (5) **ID evaluation** aiming at evaluating the instructional design (ID) [27]. In this context, the evaluation of ChatGPT-4 designed ID includes using the Technology Adoption Model (TAM) [29].

This study will not delve into the full spectrum of ChatGPT's application across all ID stages but present a summary of four educators, using ID with ChatGPT-4 independently, in an iterative prompt engineering procedure. The four educators include a novice (PhD student) and three experienced professors with expertise in PM education, and information systems engineering.

Prompt engineering [9, 23] was crucial and iterative, aligning with each ID stage. We combined insights from literature and our teaching experiences to structure the initial prompts, refined through repeated cycles with ChatGPT. Initially developed with

ChatGPT-3.5 and improved with ChatGPT-4, our final prompt template is: "*Develop [ID STAGE], given [ID CONTEXT]. Use actual scientific references. Don't make stuff up when using scientific references.*" This format succinctly captures the essential elements: the specific [ID STAGE] and the [ID CONTEXT] for ChatGPT's response, facilitating the creation of ID stages within a contextual framework and addressing complex "wicked problems" prevalent in ID. Emphasizing the use of authentic references to ensure the credibility of the ID. The final versions can be found in the following Appendix: https://zenodo.org/records/15792688.

Following the prompt engineering phase, educators evaluated the instructional design (ID) generated by ChatGPT-4 using constructs from the Technology Acceptance Model (TAM) [29]. A similar approach was employed by Bogdanova and Snoeck [28] to assess expert opinions in the field of conceptual modelling education. Building on their study, we adapt the TAM framework to address the following research questions: "What was your Perceived Usefulness, Perceived Ease of Use, and Intention to Use the ChatGPT-4 in the future when performing Instructional Design?".

4 Instructional Design with ChatGPT-4

In the following section we go through the aggregated findings of the four educators that applied the ADDIE stages in pre-defined prompts.

In the **Project Initiation stage (1)**, ChatGPT started with highlighting the 5 stages of the Addie model. It was considered a helpful introduction, with the possibility of prompting further in the relevant areas.

In the **Needs Assessment stage (2)**, our research and teaching at a highly ranked European higher education institution (HEI) focuses on providing PM education with the BPMN formalism. In our endeavor, we entered requirements into ChatGPT, initially encountering broadly generic responses that were lacking in depth. ChatGPT demonstrated a limited understanding of the HEI's learning context. It was more effective in identifying problem factors for learners, citing issues like the complexity of BPMN 2.0, lack of contextual application, conflicting interpretations, and integration with other tools. As we iteratively refined our prompts, ChatGPT incited us to search for essential contextual elements such as the specific details of the HEI and the master's programme (that was proposed). It also advised us to consider including abstract concepts like demographics, broadening our inquiry scope. We strategically added "*give us a detailed answer*" to our prompt, thereby prompting ChatGPT to provide more comprehensive and in-depth information. This process of continual prompt optimization allowed us to harness ChatGPT's capabilities effectively for our research needs. Some references were inaccurately cited or made up. This underscores the need for careful validation of ChatGPT's information in academic settings. Despite this, the ideas suggested with the non-existent references align with literature-confirmed issues, such as inadequate training and limited real-world experience (Sedrakyan et al., 2014).

When defining learning objectives, ChatGPT provided an overview of Bloom's taxonomy levels applied to BPMN 2.0 modeling. Initially, it provided just two learning objectives per Bloom's level, though upon further request, ChatGPT explained the different objectives in more detail. We thus argue that ChatGPT's answers can be beneficial as a template for novice educators.

Next, in the **Developing Instructional Strategy and Materials stage (3)**, ChatGPT gave an overview evaluation of the three learning theories that can be used to prepare ID. However, the preference was towards the constructivist or cognitivist-constructivist combination. The evaluation of the pros and cons of each theory according to the PM teaching context was presented. This helped to clarify the application of learning theories. Regarding the instructional method, ChatGPT recommended collaborative learning and modeling with peer review, guided instruction, case- and scenario-based learning with real contexts and flipped classrooms with guided tutorials. We found the reflection on the learning theory and instructional method useful to reflect about.

There were several modeling tools suggested by ChatGPT (e.g. Lucidchart, MS Visio, Creately, Bizagi Modeler, Signavio, etc.). The presentation of the pros and con of the tools, in the context of different learning theories, was a useful result. It is possible to ask ChatGPT to make a BPMN 2.0 model represented in XML code. XML code is directly importable into most BPMN modeling tools. We asked ChatGPT to create a BPMN model based on a textual description that depicts a simple purchase order approval process, involving a stock availability check and the option to re-order from a supplier. The provided outcome gives an XML file with an overview of the BPMN 2.0 model, explaining the key concepts and symbols used in the model (like events, gateways and activities). The XML code was usable in a BPMN tool but demands further editing (to connect and correctly create a layout of process tasks). We tested the possibility of creating simpler or more complex models tailored to specific learners' needs through advanced prompt engineering. However, iterations proved challenging, and the XML code was sometimes inaccurate. The generated learning materials align with a behavioristic approach, focusing on training factual knowledge about model elements [28].

In the **Assessment stage** (4), ChatGPT-4 proposed that PM skills could be evaluated through a range of activities, including interpreting existing BPMN diagrams, designing new ones, optimizing modeled processes, comparing and analyzing different BPMN models, conducting simulations and validations, administering, and engaging in collaborative model design. In total, four models were given to ChatGPT. They all contained errors made by the students obtained from previous assignment hand-ins.

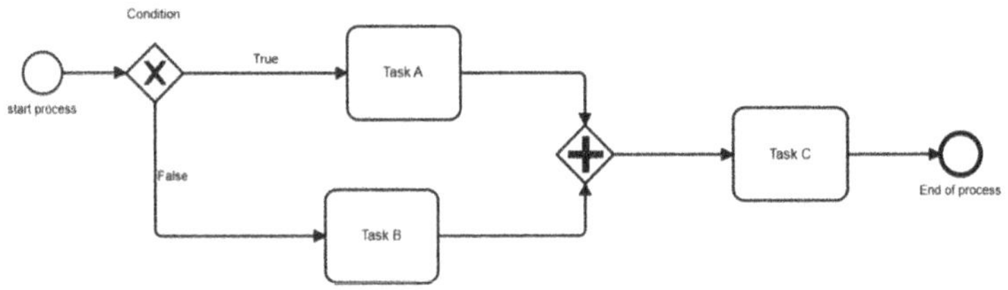

Fig. 1. Example 1 of an erroneous model in BPMN.

Some examples of the questions proposed by ChatGPT: "*What is the name of the process represented in the BPMN 2.0 model?*"; "*What is the purpose of the exclusive gateways in the process?*"; "*How would you improve this process model?*"; "*What happens if the supplier is not available?*" These questions assess the syntactic, semantic, and pragmatic understanding of the process model, following the 3Q framework (Dumas et al., 2018), and address both lower-level (comprehension) and higher-level cognitive skills (evaluation and creation) as defined in Bloom's taxonomy [28].

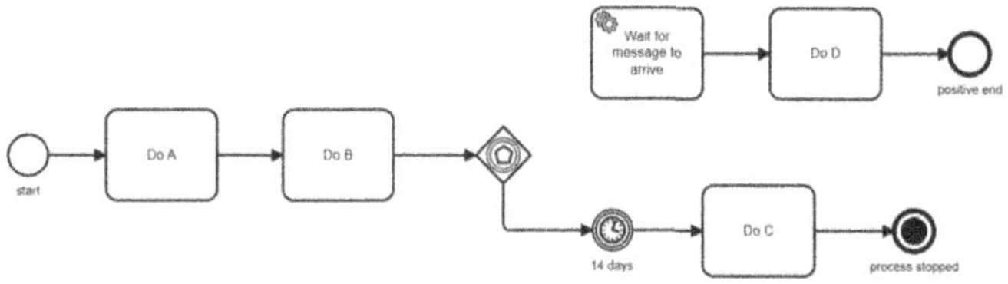

Fig. 2. Example 2 of an erroneous model in BPMN.

Evaluating the models done in a (semi-)automatic manner for formative assessment is currently a research gap [14, 31]. We presented a few correct and faulty models and asked ChatGPT to describe them and find errors (see Figs. 1, 2, 3 and 4).

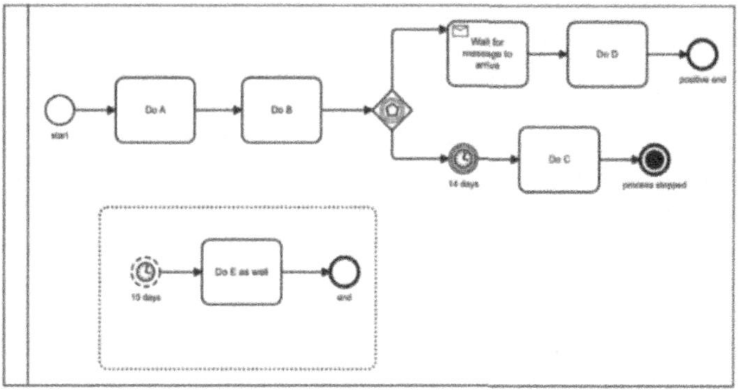

Fig. 3. An example of a correct and more advanced model.

Overall, we agree that ChatGPT-4 manages to describe the models and their individual elements well. Correct and erroneous models were recognized as such, even though some appropriate terminology was not always used. In the first model of Fig. 1, the existence of a deadlock was described (but not labelled as such).

In the second model of Fig. 2, ChatGPT-4 pointed to the wrong task type and eventually also the missing 'connection' (i.e. sequence flow). When we added the missing connection, but from task to the event-gateway, only one part of a double error was mentioned: the wrong task type. The erroneous sequence flow was not mentioned. Overall, the findings point to the potential of ChatGPT to support students in their comprehension efforts, even though caution is required.

Figure 4 was correctly classified as accurate; even the event-based subprocess (an advanced element in PM) was described and explained correctly. More research is required with certain model types and patterns to systematize the strengths and weaknesses of ChatGPT-4 as a reliable partner in to describe and evaluate process models and to support both lecturers in teaching and students in PM education.

In the ID **Evaluation stage (5)** we found that ChatGPT-4 is not capable of self-evaluation and simply reiterates and positively praises the ID that it has developed. Accordingly, human oversight is deemed necessary to critically assess the ID produced by ChatGPT-4.

5 Usability Evaluation of ChatGPT-4 by Educators

The educators shared diverse views on ChatGPT's **perceived usefulness** in educational settings. Educator 2 (E2) sees ChatGPT as a '*smart Google*' for general theories and frameworks, useful for broad information but unreliable for detailed BPMN analysis or specific assessment questions, suggesting its use as a complementary source requiring validation. E4 finds ChatGPT useful for course creation, offering immediate content and feedback. However, E4 cautions about verifying its correctness to avoid bias in student learning. E1 notes ChatGPT's ability to outline learning materials and provide an overview but emphasizes the need for directing its responses and validating the content. E1 suggests specific request structures and reliance on personal experience for validation, seeing varied utility across different ID stages. E1 also mentions the issue of non-existent or inaccurate references provided by ChatGPT-4 yet finds value in using these for keyword searches. E3 finds ChatGPT helpful for theoretical questions and stages of instructional design, comparable to textbook alternatives, but with limitations in source tracing. E3 appreciates its usefulness in formulating learning objectives and selecting modeling software. However, E3 points out the tool's lack of creativity and ineffective model interpretation and expresses concerns about the reliability and accuracy of references, necessitating further review and verification.

Concerning **perceived ease of use**, E2 finds engaging in a conversation with ChatGPT straightforward but questions its ability to handle multiple sessions or interruptions without starting over. E4 appreciates the simplicity of asking text-based questions and making contextual refinements, though E4 notes challenges when dealing with models presented as pictures. E1 experiences occasional connectivity issues and limitations in handling complex XML models, finding ChatGPT slow and sometimes unable to edit files it previously generated. As an early adopter, E1 enjoys using ChatGPT for instructional design but expresses concerns about its broad applicability and potential to replace human roles in BPMN and STEM education. E3 finds ChatGPT easy to use and enjoyable for conversation. Still, E3 points out difficulties in handling models as input and the inability to pause and resume conversations, which is challenging given busy work schedules. E3 also highlights the potential of multimodal interaction of text and image recognition, which is especially critical in process modeling.

Finally, concerns are raised regarding the trustworthiness and reliability of ChatGPT in relation to educators' **intentions to use** the tool. E2 raises doubts about ChatGPT's reliability, particularly in academic settings where source accuracy is crucial, though remaining interested in exploring its capabilities further. E2 further notes the tool's limitations for specific BPMN models but finds it useful for practical tasks. E4, on the other hand, expresses high satisfaction, valuing ChatGPT's accessibility and time-saving aspects despite some imperfections. E1 anticipates using ChatGPT in research and education, recognizing its role as a starting point for gaining knowledge overviews, though sharing concerns about its reliability. E1 foresees a role shift towards validating and directing research due to rapid technological advances. E3 sees potential in using generative AI chatbots as a supportive tool in course design, offering initial ideas and replacing some online searches. However, E3 also highlights limitations, particularly in source reliability and the tool's unsuitability for generating or interpreting models in process modeling education, emphasizing the need for creativity and visualization, areas where ChatGPT falls short.

6 Discussion and Conclusions, Limitations, and Future Work

Our research focuses on practical implementation, specifically prompt engineering in ChatGPT-4 for instructional design in process modelling. We tested ChatGPT's applicability in 4 stages (of 12 steps) of iterative ID with prompt engineering (cf. the Appendix in Zenodo). We consider the contribution of this study to be an improvement: we developed a new solution (using prompts in ChatGPT-4) for known problems (instructional design) and then evaluated it using an exploratory but systematic approach.

Additionally, we aim to advance towards a nascent theory with a higher contribution level. This involves developing operational principles and presenting them through models (illustrated in Fig. 1) and engineered generative AI prompts (detailed in Appendix 1), as well as a discussion in this section. Following pragmatic research philosophy in IS studies, we aim to verify our approach to prompt engineering for ID, emphasizing actionable insights and theoretical understanding.

ChatGPT showed limited usefulness during project initiation and analyzing learning contexts due to its relatively limited knowledge base when compared with experienced educators. While ChatGPT struggled with understanding learner needs and made up some references, it was helpful in critically analyzing the developed prompts and sufficiently analyzing learning objectives based on Bloom's taxonomy, making it useful for educators to create serendipitous insights.

Instructional designers may prioritize the selection of the right tools or technologies, including the search for presentation media, leaving limited time and budget for content design. In developing instructional strategies, methods, and materials, ChatGPT broadly summarized learning theories and suggested instructional methods like collaborative learning and case-based learning, aligned with existing research [14]. Its value lies more in interacting with existing knowledge and critically evaluating these suggestions. ChatGPT's recommendations for software were mixed, with some being less suitable for process modeling, but the primary ones recommended were appropriate. The tool was more effective in creating textual materials than diagrams or models, as ChatGPT-4 does not provide images, only XML code that needs further adaptation. We anticipate further improvements with dedicated ChatGPT versions trained on the XML models. We thus support [19, 20] findings that ChatGPT necessitates human intervention, for it often follows structured approaches in creating content while sometimes failing to understand and evaluate the context of provided examples.

ChatGPT was moderately effective for the evaluation stage of the ADDIE framework in generating text-based questions and evaluating novice modelers' work, yet it lacked the nuanced judgment of human educators. Overall, we find support for [21], who propose using ChatGPT for generating questions automatically, with the required addition of human evaluation to validate and refine such questions.

Therefore, we argue that ChatGPT-4 demonstrates high potential for model assessment and feedback, aligning with trends in formative assessment [14] and self-learning support [11]. Also, (semi-)automatic assessment can be beneficial for educators to facilitate the labor-intensive work of assessing students' models. Its effectiveness in evaluating assignments and offering feedback suggests promising areas for further exploration. The effectiveness of generative AI in specific areas may vary due to its training dataset, affecting the accuracy and coverage of specific topics. For example, there are noticeable differences in ChatGPT's ability to produce BPMN diagrams (in XML) versus class diagrams (in ASCII characters). This suggests a variable "area applicability" metric, potentially enhanced by domain-specific generative AI models. For instance, the image recognition feature of ChatGPT may be very crucial to PM education, which is based on visual diagrams, while not so critical for other areas, such as learning languages.

Like [20], we found that ChatGPT-4 lacks flexibility in its thinking because of a lack of self-evaluating capabilities. ChatGPT-4 has confirmed all the previous ID stages when asked to be self-evaluative, whereas a human educator could potentially recognize the weaknesses of the proposed ID.

It is important to highlight that generative AI technologies are currently undergoing rapid development. New versions and models (like ChatGPT-4o) are being released, including features like deep research. We does expect that newer, more capable generative AI versions will be more accurate and less erroneous, with more capabilities, e.g., to evaluate the process models even more accurately. This evolution can significantly support educators but also entails the risks of a "loss of teaching and learning", when students create process models with generative AI and educators evaluate the models with generative AI. Therefore, certain precautions need to be taken. For instance, only grading models that were created under supervision, ensuring that no generative AI was used. Empowering and convincing students on the advantages of not using AI for learning purposes can be another useful recommendation.

Our research suggests that instructional design (ID) can become more efficient and effective by utilizing generative AI as a facilitation tool for idea generation/brainstorming and structuring the content provided by the humans, aligning with findings from related studies [19, 20, 22]. This approach complements rather than replaces instructional designers, helping them to tackle the complexities of learning systems, often described as "wicked problems," requiring a multi-perspective approach [24, 25]. This position contrasts with the concerns raised by [35], who take an alarmist view that generative AI could negatively impact instructional design and learning by potentially replacing them.

We thus propose the following guidelines to PM educators. First, prompt engineering must be iterative and grounded in an instructional design framework. Preferably, the prompts should have been previously empirically validated. This will keep generative AIs on track and more systematic and will thus allow instructional designers to ensure the quality of the ID. Second, generative AI at this stage is to extend PM educator's efforts instead of replacing them. PM educators can brainstorm new ideas quickly, even though they may not be fully correct. Third, better trained generative AI can help in evaluating students' assignments, with the need for PM educators to always review the ChatGPT's assessment. This is different from educators doing evaluations from scratch. Similarly, generative AI can greatly ease the creation of new assignments for students, given that creating these can be quite laborious in a PM education context. Finally, educators must always be critical of ChatGPT's output. ChatGPT can give erroneous answers and should preferably be used together with a more traditional information research approach, such as reading scientific articles on the topic to validate the proposed knowledge by ChatGPT.

With pragmatic orientation and essay format, this research is subject to limitations, primarily its design and validation solely by the co-authors. As an early attempt to integrate the emerging technology of generative AI, specifically ChatGPT, our findings are interpretative and influenced by the personal experiences of the researchers involved. To address this, we introduced a degree of systematicity, detailing all research stages and providing instructional design (ID) examples. Additionally, we share the iterations and ChatGPT queries to enhance the validity of internal and external studies. It is essential to recognize that our paper presents a subjective perspective aligned with constructivist views, where knowledge is built through interaction and collaboration. In particular, we explored the applicability of generative AI in ID for Process Modeling, acknowledging its vast potential for future studies and advocating a pragmatic, action-oriented approach, mainly due to the absence of established theories on generative AI-human interactions. The possibility of developing generative AI capabilities specific to the ID domain is inspiring [32]. Our study aims to contribute insights that could shape future research methodologies and theoretical frameworks in pedagogy and educational sciences, ensuring the veracity of our findings through existing ID literature and a thorough evaluation by our author-reviewers, as guided by the principles of pragmatism in Information Systems Research [37].

Our findings align with emerging research on ChatGPT-4's prompt engineering. Given the innovative nature of ChatGPT-4, we underscore the importance of conducting systematic empirical research using dedicated datasets before advancing theoretical claims. Future studies could adopt rigorous design science research methodologies [37] to generate prescriptive knowledge, particularly in crafting effective prompts and applying them to produce valid instructional designs. However, achieving such advancements requires close collaboration among machine learning experts, educators, and instructional designers to develop more domain-specific generative AI models.

References

1. Qadir, J.: Engineering education in the era of ChatGPT: promise and pitfalls of generative AI for education. In: Proceedings of the 2023 IEEE Global Engineering Education Conference (EDUCON), pp. 1–9. IEEE, Kuwait (2023)
2. Anders, B.A.: Is using ChatGPT cheating, plagiarism, both, neither, or forward thinking? Patterns **4**(3) (2023)
3. Susnjak, T.: ChatGPT: the end of online exam integrity? arXiv preprint arXiv:2212.09292 (2022)

4. Cotton, D.R.E., Cotton, P.A., Shipway, J.R.: Chatting and cheating: ensuring academic integrity in the era of ChatGPT. Innov. Educ. Teach. Int. **61**(2), 228–239 (2023)
5. Nuzula, I.F., Amri, M.M.: Will ChatGPT bring a new paradigm to HR world? A critical opinion article. J. Manage. Stud. Develop. **2**(02), 142–161 (2023)
6. Sok, S., Heng, K.: ChatGPT for education and research: a review of benefits and risks. Available at SSRN 4378735 (2023)
7. Zhuo, T.Y., Huang, Y., Chen, C., Xing, Z.: Exploring AI ethics of ChatGPT: a diagnostic analysis. AarXiv preprint arXiv:2301.12867 (2023)
8. Kirchenbauer, J., Geiping, J., Wen, Y., Katz, J., Miers, I., Goldstein, T.: A watermark for large language models. AarXiv preprint arXiv:2301.10226. (2023)
9. Rudolph, J., Tan, S., Tan, S.: ChatGPT: bullshit spewer or the end of traditional assessments in higher education? J. Appl. Learn. Teach. **6**(1), 342–363 (2023)
10. Wermelinger, M.: Using GitHub Copilot to solve simple programming problems. In: Proceedings of the 54th ACM Technical Symposium on Computer Science Education, vol 1, pp. 172–178 (2023)
11. Cope, B., Kalantzis, M.: e-Learning ecologies: Principles for new learning and assessment (1st ed.). Elearning ecologies. Nova Iorque: Routledge (2016)
12. Nikou, S., Maslov, I.: An analysis of students' perspectives on e-learning participation – the case of COVID19 pandemic. Int. J. Inf. Learn. Technol. **38**, 299–315 (2021)
13. Dumas, M., La Rosa, M., Mendling, J., Reijers, H.A.: Fundamentals of business process management. Springer (2018)
14. Rosenthal, K., Ternes, B., Strecker, S.: Learning conceptual modeling: structuring overview, research themes and paths for future research. In: Proceedings of the 27th European Conference on Information Systems Information Systems for a Sharing Society (ECIS). Sweden (2019)
15. Sedrakyan, G., Snoeck, M., Poelmans, S.: Assessing the effectiveness of feedback enabled simulation in teaching conceptual modeling. Comput. Educ. **78**, 367–382 (2014)
16. Bell, S.: Design tthinking+ user experience= Better designed libraries. Inf. Outlook **22**(4), 4–6 (2018)
17. Dousay, T.A.: Instructional design models. In: West, R.E. (ed.), Foundations of learning and instructional design technology. EdTech Books (2018)
18. Kaur, M.: Blended learning: its challenges and future. Procedia. Soc. Behav. Sci. **93**, 612–617 (2013)
19. Meron, Y., Araci, Y.T.: Artificial intelligence in design education: evaluating ChatGPT as a virtual colleague for postgraduate course development. Design Science. **9**, e30 (2023)
20. Li, Y., Liu, J., Yang, S.: Is ChatGPT a good middle school teacher? An exploration of its role in instructional design. In: Proceedings of the 3rd International Conference on New Media Development and Modernized Education (NMDME) 2023, October 13–15, 2023, Xi'an, China (2024)
21. Lee, U., et al.: Few-shot is enough: Exploring ChatGPT prompt engineering method for automatic question generation in English education. Educ. Inf. Technol. **29**(9), 1–33 (2023)
22. Fotaris, P., Mastoras, T., Lameras, P.: Designing educational escape rooms with generative AI: a framework and ChatGPT prompt engineering guide. In: Proceedings of the 17th European Conference on Game Based Learning (ECGBL). Academic Conferences and publishing limited, Netherlands (2023)
23. Rahman, M.M., Terano, H.J., Rahman, M.N., Salamzadeh, A., Rahaman, M.S.: ChatGPT and academic research: a review and recommendations based on practical examples. J. Educ. Manage. Develop. Stud. **3**, 1–12 (2023)
24. Schoo, A., Kumar, K.: The clinical educator and complexity: a review. Clin. Teach. **15**(4), 287–293 (2018)

25. Buchanan, R.: Wicked problems in design thinking. Des. Issues **8**, 5–21 (1992)
26. Saunders, M., Lewis, P., Thornhill, A.: Research methods for business students. Pearson education (2009)
27. Khalil, M.K., Elkhider, I.A.: Applying learning theories and instructional design models for effective instruction. Adv. Physiol. Educ. **40**(2), 147–156 (2016)
28. Bogdanova, D., Snoeck, M.: CaMeLOT: an educational framework for conceptual data modelling. Int. J. Appl. Earth Obs. Geoinf. **78**, 67–76 (2019)
29. Davis, F.D.: Perceived usefulness, perceived ease of use, and user acceptance of information technology. MIS Q. **13**, 319–339 (1989)
30. Ternes, B., Strecker, S., Rosenthal, K., Barth, H.: A browser-based modeling tool for studying the learning of conceptual modeling based on a multimodal data collection approach. In: Proceedings of the 14th International Conference on Wirtschaftsinformatik. 1998–2002 (2019)
31. Maslov, I.: Towards empirically validated process modelling education using a BPMN formalism. In: Guizzardi, R., Ralyté, J., Franch, X. (eds.) Research Challenges in Information Science. RCIS 2022. LNBIP, vol. 446. Springer, Cham (2022). https://doi.org/10.1007/978-3-031-05760-1_58
32. Raste, S., Singh, R., Vaughan, J., Nair, V.N.: Quantifying inherent randomness in machine learning algorithms. ArXiv preprint arXiv:2206.12353 (2022)
33. Driscoll, M.P., Burner, K.J.: Psychology of learning for instruction. Pearson/Merrill Prentice Hall (2005)
34. Wu, W., Hsiao, H., Wu, P., Lin, C., Huang, S.: Investigating the learning-theory foundations of game-based learning: a meta-analysis. J. Comput. Assist. Learn. **28**, 265–279 (2012)
35. Hodges, C.B., Kirschner, P.A.: Innovation of instructional design and assessment in the age of generative artificial intelligence. TechTrends **68**, 195–199 (2024)
36. Morrison, G.R., Ross, S.M., Kalman, H.K., Kemp, J.E.: Designing effective instruction, 6th edn. Willey, Hoboken (2011)
37. Gregor, S., Hevner, A.R.: Positioning and presenting design science research for maximum impact. MIS Q. **37**(2), 337–355 (2013)

Revisiting BPMN Assignments with AI in Mind: Insights from Experiments with Large Language Models in Process Modeling Education

Krzysztof Kluza[1(✉)], Marzena Grzesiak[2], Piotr Sliż[3], Leszek Szała[4], Anna Suchenia[5], Karol Suchenia[5], and Marek Moszyński[2]

[1] AGH University of Krakow, al. A. Mickiewicza 30, 30-059 Krakow, Poland
kluza@agh.edu.pl

[2] Gdańsk University of Technology, 11/12 Gabriela Narutowicza Street, 80-233 Gdańsk, Poland
{Marzena.Grzesiak,Marek.Moszynski}@pg.edu.pl

[3] University of Gdańsk, Armii Krajowej 101, Sopot, Poland
piotr.sliz@ug.edu.pl

[4] University of Chemistry and Technology, Technicka 5, 166 28 Prague 6, Prague, Czech Republic
Leszek.Szala@vscht.cz

[5] Cracow University of Technology, ul. Warszawska 24, 31-155 Kraków, Poland
{asuchenia,karol.suchenia}@pk.edu.pl

Abstract. As Large Language Models (LLMs) become increasingly available to students, BPM educators face a new challenge: how to design assignments that remain pedagogically effective and resistant to superficial AI-generated answers. This paper presents the results of two experiments that simulate common BPMN-related homework tasks and test how LLMs respond to them. The first experiment focused on answering comprehension questions based on five BPMN models, each provided either in PNG or XML format. The second asked the models to detect modeling errors in 30 flawed BPMN diagrams. In both cases, we evaluated the outputs of ChatGPT-4o and Gemini Flash, analyzing the correctness, reasoning, and completeness of their responses. Our findings show that while current LLMs are not yet fully capable of reliably solving BPMN assignments—especially those involving deeper process logic—they can already provide partially correct and plausible responses. This raises questions about the future of BPM education, the design of AI-aware assignments, and the role of LLMs as potential learning assistants. Alongside insights and lessons learned, we provide materials to help instructors adapt their teaching to an AI-enabled environment.

Keywords: BPMN Education · Large Language Models · Assignment Design · AI in the Classroom · AI-Aware Pedagogy

M. ER et al. (Eds.): BPM 2025, LNBIP 565, pp. 358–372, 2026.
https://doi.org/10.1007/978-3-032-02936-2_25

1 Introduction

In recent years, the rapid development and growing accessibility of Large Language Models (LLMs), such as ChatGPT or Gemini, have introduced new opportunities–and challenges—for higher education. These advanced AI tools, built on transformer architectures, are increasingly integrated into academic life, assisting in writing, programming, and even tutoring. Within Business Process Management (BPM) education, where process modeling using notations like BPMN is essential, educators now face a critical question: how should BPMN-related assignments be designed in a world where students may have instant access to AI tools capable of interpreting or even solving such tasks?

Although LLMs are widely discussed across disciplines, there remains a gap in understanding how they influence BPM-specific learning outcomes. The core educational challenge is to adjust exercises in a way that supports students' comprehension while ensuring LLMs do not replace their active engagement. At the same time, consistent and standardized BPM teaching is necessary to maintain the quality of education across institutions [22].

This paper presents a series of exploratory experiments investigating whether state-of-the-art LLMs can correctly interpret and reason about BPMN models, both in graphical (diagram) and XML formats. We designed tasks that reflect common learning objectives in BPM courses–specifically, answering open-ended questions based on BPMN models and identifying modeling errors in flawed BPMN diagrams. Two leading LLMs, ChatGPT-4o and Gemini Flash, were tested, and their outputs were qualitatively evaluated.

While our results are preliminary, they reveal that current LLMs are only partially reliable in solving BPMN tasks. They often provide plausible answers but may struggle with deeper logical reasoning. These insights suggest a limited–yet shrinking–window in which BPMN assignments can still be used without significant risk of AI overreliance. However, the growing capabilities of these models invite a rethinking of BPM pedagogy and assessment. In this paper, we also provide reusable educational resources and propose strategies for designing AI-aware assignments.

The rest of this paper is organized as follows. Section 2 reviews related work. Section 3 outlines our experimental setup. Results are presented in Sect. 4, followed by educational insights in Sect. 5. Section 6 describes the shared materials. Section 7 discusses broader implications and limitations, and Sect. 8 concludes the paper.

2 Background and Related Work

2.1 LLMs in Higher Education

Recent publications highlight efforts to implement LLMs in various fields of education. For example, ChatGPT has demonstrated strong performance in answering biomedical questions [23], but shows limitations in tasks involving visual content as well as in the generation of valid and complex exam questions [1].

In computer science, generative AI tools support students in programming and debugging [12], reducing errors and improving comprehension. Xiao et al. [24] introduced an LLM Hint Factory that adaptively supports learning with multi-level hints.

Other studies focus on how LLMs improve feedback and grading. For example, students receiving AI-generated feedback showed measurable improvement in their work [18], and LLMs have proven promising for essay grading [21]. Several researchers emphasize the role of AI-powered student assistants in enhancing access to course materials and personalizing learning [5,20].

2.2 AI Integration in BPM Education

Within BPM education, early experiments suggest both potential and limitations in using LLMs. Klievtsova et al. [13] analyzed conversational BPM tools and proposed use cases where LLMs assist students in paraphrasing or drafting process models. Their work supports the idea of AI acting as a "BPM copilot"—a learning companion that helps generate or refine models.

However, challenges remain. Meyer and Dannecker [19] observed that students who relied too heavily on ChatGPT produced lower-quality process models and often failed to critically assess AI output. Malinka et al. [17] tested ChatGPT on university-level assessments and found that, while it was capable of completing many tasks successfully, its output varied in quality and raised concerns about accuracy. These findings might have implications for BPM assessment design.

2.3 Pedagogical Adaptation to Generative AI

Educators are exploring how to adapt their teaching strategies in response to these tools. Lim et al. [16] frame this moment as a crossroads—AI may disrupt education (a "ragnarök") or enrich it (a "reformation"). They recommend emphasizing human reasoning and ethics when designing assignments. Kasneci et al. [11] call for AI literacy in classrooms, stressing that students should be taught to verify and critique AI-generated answers. Similarly, Bajak [2], in the context of business management, notes ethical risks and false information in business use, stressing the need for responsible AI use.

2.4 Model Comprehension and Debugging with LLMs

Studies have shown that LLMs can assist with complex explanations. In a programming education setting, Leinonen et al. [15] found that GPT-3.5-generated explanations were clearer than those of students. This supports the idea of using LLMs for BPMN explanation tasks—such as clarifying gateway logic or control flows. Fahland et al. [7] proposed the SAX4BPM framework to improve explainability of process execution using LLMs enriched with domain knowledge. Their findings suggest that combining structured data with generative AI yields better outcomes and more trustable feedback—potentially valuable in teaching debugging and process analysis.

2.5 LLMs and BPM Task Automation

Initial experiments using LLMs for BPM automation tasks show promise but also highlight limitations. Grohs et al. [9] and Berti et al. [4] reported issues like hallucinations and imprecision when using LLMs for conformance and improvement analysis. Researchers have since proposed enhancements: pre- and post-processing [8], prompt engineering [14], divide-and-conquer strategies [25], explainable modeling inputs [6], and retrieval-augmented generation [10]. Fine-tuning domain-specific LLMs also appears to improve performance [3].

Together, these works provide a strong foundation for understanding how LLMs interact with BPM tasks in both classroom and applied settings. However, despite growing interest in LLMs across domains, limited research has systematically evaluated how these models handle the visual and semantic aspects of BPMN models in educational contexts. Our work addresses this gap by directly testing whether current LLMs can complete BPMN comprehension and debugging tasks used in real teaching scenarios, assessing their reasoning across both structured (XML) and visual (PNG) formats to provide actionable insights for instructors.

3 Experiment Design

To explore how current Large Language Models (LLMs) handle typical BPMN-related assignments, we designed two structured experiments that reflect common learning objectives in BPM courses: (1) understanding process models and (2) identifying modeling errors. The tasks were constructed in a format that could be reused as part of real assignments, and all materials are available for educators who wish to replicate or adapt them.

3.1 Task 1: Answering Questions Based on BPMN Models

In the first experiment, we created a set of five BPMN process models, each accompanied by six open-ended comprehension questions. The questions required interpretation of control flow, decision logic, and specific execution paths within the model (e.g., "Which activity is executed after an order is rejected?"). The models were provided to the LLMs in two alternative formats:

1. Diagram format (PNG image of the BPMN model)
2. XML format (standard BPMN XML representation)

This allowed us to compare how LLMs interpret visual diagrams versus textual representations. Both ChatGPT-4o and Gemini Flash were asked the same questions specific to each BPMN diagram, using the instruction shown in Box 1. Their responses were analyzed for correctness, completeness, and interpretative reasoning.

Prompt 0: Open Questions

```
Please analyze the attached image of a BPMN model and
answer the following six open-ended questions. Base your
answers strictly on what is visible in the diagram; no
assumptions beyond the model. Format your answers in
a clear and concise way.
Questions:
```

This task mirrors real BPMN comprehension exercises and can be directly used in classroom settings. Educators may choose either format depending on learning objectives and AI policy.

3.2 Task 2: Detecting Errors in BPMN Diagrams

The second experiment focused on error identification, a task frequently used to deepen students' understanding of modeling best practices and BPMN syntax. We prepared a dataset of 30 BPMN diagrams, each containing a specific modeling error. Table 1 presents a categorization of these 30 modeling errors in our experiment. The errors are grouped into six balanced categories, each containing five representative cases. Each category includes a short description and an example of an error.

Table 1. BPMN Error Categories (6 groups × 5 errors)

No.	Category	Short description with examples
1	Gateway & Routing Logic	Misuse or confusion in gateway type or logic (e.g.XOR instead of AND, routing all branches). *Example: Exclusive gateway used instead of inclusive*
2	Event Timing & Triggers	Wrong type, placement, or overuse of events (start, intermediate, boundary). *Example: Timer placed at wrong task*
3	Synchronization & Merging	Missing merges, asynchronous joins, or parallelism not properly synchronized. *Example: Parallel flows without merging gateway*
4	Swimlanes & Structure	Misplaced tasks, unnamed lanes, disconnected flows, or messy layout. *Example: Task in wrong swimlane*
5	Labeling & Naming	Missing, duplicated, vague, or inconsistent labels (ontasks, flows, gateways). *Example: Generic events' names 'Event 1', 'Event 2'*
6	Decision Placement & Completeness	Missing decision tasks, excessive start events, poor structure, unprepared gateways. *Example: Gateway without prior decision task*

Each diagram was provided in visual format only (PNG). The LLMs were instructed to list any mistakes found in the diagram and briefly explain their nature. Each time, we presented the model with a static image of the BPMN diagram along with a prompt to locate and describe the problem. Their ability to detect both syntactic and semantic errors was evaluated.

To investigate the capabilities of LLMs in interpreting BPMN diagrams, we designed two types of evaluation for this task: *Single-Issue* and *Multiple-Issue*.

In both types, each diagram contained exactly one purposely introduced modeling error. In the **Single-Issue** setting, the model was expected to identify only one error—ideally the intended one—and provide a single diagnosis. We asked the models to identify the error using the prompt shown in Box 2.

Prompt 0: Single-Issue

```
Please review the attached BPMN model and identify
a violation of BPMN syntax, logical consistency,
or modeling guidelines / best practices.
Check for issues in the following categories: (...)
For the issue found, provide a table with three columns
and very short description:
1. The element(s) involved
2. A short, clear explanation of the issue
3. A short description of the recommended fix
```

While we focused on a selected set of widely accepted modeling guidelines, we acknowledge that other valid practices exist and some deviations may be contextually appropriate. In realistic settings, this opens the possibility for alternative interpretations or suggestions (e.g., different layouts or stylistic preferences), which may not always be clearly identifiable as errors. To reflect this, in the **Multiple-Issue** setting, the model could list up to three potential errors, since additional unintentional issues or modeling artifacts might also be present. The evaluation then focused on whether the intended error was among those identified by the model. This format simulates realistic learning, where students can identify multiple plausible issues that may coexist, and aligns with real-world scenarios requiring critical evaluation. Such a task can be particularly useful as a diagnostic tool or in-class activity.

3.3 Methodology and Evaluation Approach

We adopted a qualitative evaluation framework for analyzing the outputs of the LLMs. For each question or error-reporting prompt, we assessed whether the response matched the expected interpretation and all key elements were addressed. No human comparison group was used in this phase, as the focus was on understanding the capabilities and limitations of LLMs themselves in the role of potential study aids or assignment solvers. Each LLM prompt was executed only once per task. We acknowledge that LLMs exhibit stochastic behavior, and repeated runs might yield variations in answers. This is a limitation of the current study, and future work should include multiple runs and aggregated results to improve robustness.

4 Experiments

4.1 LLM Performance in Experiment 1

Figure 1 presents the average accuracy scores for both models across the two input formats. ChatGPT showed a higher overall accuracy than Gemini in both conditions. Interestingly, both models demonstrated a marked improvement when working with the XML representation of the model rather than the PNG image. Specifically, ChatGPT's accuracy increased from **0.77** to **0.90** (+0.13), and Gemini's from **0.67** to **0.80** (+0.13).

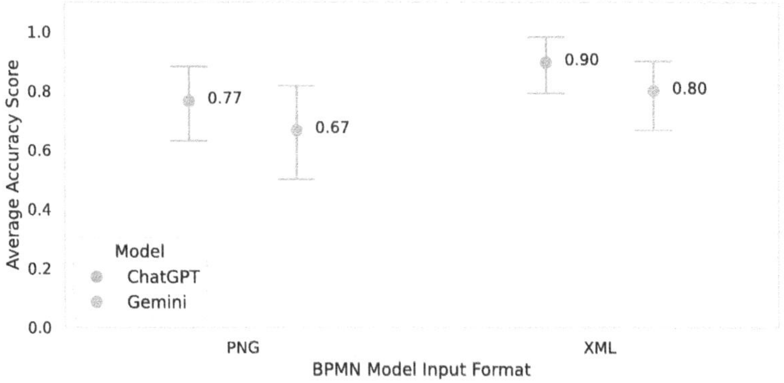

Fig. 1. Average accuracy scores of ChatGPT 4o and Gemini Flash in Experiment 1, based on BPMN model input format (PNG vs. XML).

As one might expect, this result suggests that LLMs currently process structured textual input more effectively than visual diagrams—likely due to their limitations in image parsing, spatial reasoning, and lack of topological awareness. In contrast, XML-based representations might provide clear, hierarchical structure that align well with the models' token-based processing mechanisms.

Statistical Validation. To further validate our observations, we conducted a paired t-test comparing each model's performance on the same comprehension questions in PNG versus XML format. For ChatGPT, the results show a statistically significant improvement when using the XML representation ($t = -2.12$, $p = 0.043$, where t is the test statistic and p is the probability of observing the result under the null hypothesis), confirming that the structured format offers a measurable advantage in model comprehension. Although Gemini returned results with a higher average accuracy score, the difference was not statistically significant ($t = -1.55$, $p = 0.133$). This suggests that while both models benefit from structured input, ChatGPT's performance gain is more robust and consistent. These results reinforce the need to carefully consider input format in assignment design and model evaluation.

Given that image-based models might be more challenging for LLMs, we chose to focus exclusively on the visual (PNG) format in Experiment 2, aiming to simulate more authentic classroom settings, where students typically review static visuals, and better evaluate the models' reasoning capabilities in a more demanding setting.

4.2 LLM Performance in Experiment 2

Figure 2 illustrates the average accuracy scores achieved by ChatGPT and Gemini across six predefined BPMN error categories. The results are split into two task types: single-issue diagrams (each containing exactly one known modeling error) and multiple-issue diagrams (where only one selected error was evaluated per model response).

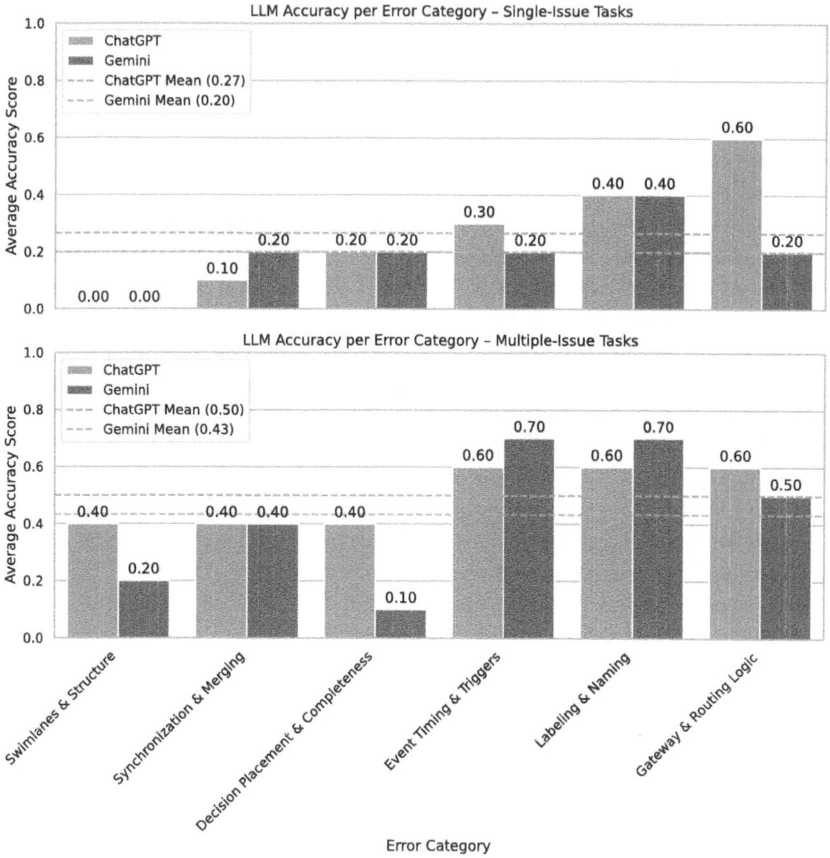

Fig. 2. Average accuracy per error category for ChatGPT and Gemini in single- and multiple-issue BPMN tasks.

Overall, the results confirm noticeable differences in model performance depending on both the task complexity and the error category. For single-issue tasks, ChatGPT shows strong and consistent performance, particularly in the *Event Timing & Triggers* and *Decision Placement & Completeness* categories. Gemini's performance is lower in most single-issue categories but improves significantly in multiple-issue settings.

For the more complex multiple-issue tasks, both models demonstrate improved performance in several categories. ChatGPT shows notable gains in *Swimlanes & Structure* and *Synchronization & Merging*, suggesting its ability to generalize in more realistic diagram settings. Gemini demonstrates the largest improvement in the *Event Timing & Triggers* category, increasing its average score by over 0.5 points.

Horizontal dashed lines in each subplot indicate the mean score per model across all categories for the respective task type. These help highlight performance gaps between the models and suggest that ChatGPT is overall more reliable in both task types. However, Gemini's gains in the multi-error setting indicate its potential to benefit from richer input context.

Statistical Validation. To evaluate whether the observed performance differences between models and task types were statistically significant, we applied non-parametric tests appropriate for our data. Given the ordinal nature of the evaluation scores (0.0, 0.5, 1.0) and their non-normal distribution, we used the Wilcoxon signed-rank test for paired comparisons (ChatGPT vs. Gemini on the same task type), and the Mann–Whitney U test for independent comparisons (single-issue vs. multiple-issue settings). For all tests, the significance level was set at 0.05.

For single-issue tasks, ChatGPT outperformed Gemini in most cases, but the difference was not statistically significant ($p = 0.102$). Similarly, no significant difference was found for multiple-issue tasks ($p = 0.260$). When comparing task types, ChatGPT's performance in multiple-issue settings was marginally higher than in single-issue tasks, approaching statistical significance ($p = 0.052$). Gemini showed a clearer improvement in the multiple-issue condition, with the difference reaching statistical significance ($p = 0.036$).

These results support the observation that task format can influence model performance, especially for Gemini. The selected tests were appropriate given the discrete evaluation scale, small sample sizes, and paired or independent nature of the comparisons.

5 Lessons Learned and Educational Implications

Although this study is still ongoing, our preliminary analysis and hands-on experience with LLMs have already yielded several important insights for BPM educators.

5.1 Lessons Learned from BPMN Error Detection Tasks

Our evaluation of BPMN error detection tasks with ChatGPT-4o and Gemini Flash yielded several practical insights. First, model accuracy was highly dependent on the type of error. LLMs were more successful in identifying visual or structural issues—such as incorrect swimlane placement, missing labels, or improper event attachment—than in reasoning about control flow or decision semantics. This was particularly evident in categories such as *Swimlanes & Structure* and *Event Timing & Triggers*, where even partial visual cues enabled relatively high accuracy.

Second, in more logic-heavy categories like *Gateway & Routing Logic* or *Decision Placement & Completeness*, the models often misinterpreted flow direction, overlooked preconditions, or offered generic suggestions that did not match the intended issue. These categories consistently resulted in the lowest accuracy scores across both models.

Third, the multiple-response format (in the multiple-issue setting) allowed LLMs to cover more ground, occasionally increasing the chance of detecting the target issue. However, this format also revealed that models frequently surfaced unintended but plausible issues—suggesting a lack of precision in prioritizing the most relevant flaw. This has direct implications for assignment design: instructors may need to clarify expectations when LLMs are used as support tools in diagnostic or critique tasks.

These observations may help instructors reflect on how AI tools are reshaping the learning environment, and how BPMN-related assignments can be adapted in response.

5.2 Assignments Are Not AI-Proof by Default

While many BPMN tasks may appear structured and logic-based, our early interaction with LLMs showed that models are already capable of providing plausible and partially correct answers to common comprehension questions. In some cases, LLMs were able to extract relevant patterns from either the diagram or the XML structure with minimal prompting. This suggests that educators should not assume that visual tasks are inherently protected from AI assistance.

5.3 BPMN Visual Reasoning Remains a Challenge for LLMs

At the same time, tasks that require deeper reasoning about process behavior—such as analyzing the effects of gateways or identifying execution deadlocks—proved more difficult for current models. This reinforces the importance of designing assignments that go beyond surface-level pattern recognition and emphasize interpretation, justification, and cross-referencing model logic.

5.4 The Need for AI-Aware Pedagogy

Our experiments suggest that it is no longer sufficient to treat LLMs as external to the learning environment. Instead, we propose that BPM educators start integrating AI-awareness into their pedagogical strategies. This may include encouraging students to critically assess AI-generated outputs, designing assignments that require explanations, comparisons, or validations, as well as offering tasks where LLMs may be used as brainstorming or review tools, but not as final solution providers.

5.5 Implications for the Future of BPM Education

LLMs are likely to become more accurate and capable of interpreting process models in the near future. This raises broader questions about how we assess BPM skills and support student learning in a rapidly changing technological landscape. Our view is that educators should embrace this transition proactively: by testing AI tools themselves, by adjusting their assignments, and by fostering a culture of transparency and critical use.

6 Shared Resources and Reusability

As part of this contribution, we provide a complete set of educational materials developed for the experiments described in this paper[1].

These materials can be reused, adapted, or extended by other instructors in BPM-related courses. They are particularly well-suited for assignments focused on BPMN comprehension, process model reasoning, and error detection.

The resource package includes: Five BPMN process models, each available in PNG and XML format. Thirty open-ended questions with answers targeting control flow interpretation, decision logic, and model behavior. Thirty intentionally flawed BPMN diagrams for error-identification tasks with descriptions. Ground truth and LLM-generated answers for both tasks (ChatGPT and Gemini).

7 Discussion

Despite the promising capabilities of modern LLMs, our results indicate that they still face significant limitations when reasoning about process logic based on BPMN diagram images.

First, LLMs are trained on natural language patterns, not domain-specific formalisms like BPMN. While they may have encountered BPMN syntax or examples during training, they do not possess a structural or rule-based understanding of execution semantics. This becomes particularly problematic in tasks involving exclusive gateways, conditional flows, or event-driven decisions, where correct reasoning depends on understanding specific control logic and temporal dependencies.

Second, visual comprehension remains a major bottleneck. Although both ChatGPT-4o and Gemini Flash support image input, they often fail to correctly parse relationships between elements in a diagram. For instance, sequence flows

[1] https://github.com/kluza/bpmn-comprehension-evaluation/.

might be overlooked, or a task's context within a loop or branch might be misinterpreted. Unlike human readers, LLMs currently lack a robust internal model of diagrammatic logic, and their image reasoning often lacks spatial anchoring or topological awareness.

Third, ambiguity in BPMN representations and limited textual feedback in diagrams (e.g., missing labels, compact layouts) may further reduce model interpretability. Even in XML format, LLMs may struggle to understand process behavior from raw markup without explicit cues.

Finally, the reasoning required to identify modeling flaws is often multi-step and non-local: understanding that a merging gateway is missing may require tracking parallel paths several nodes apart. Such relational reasoning is still a weakness of current LLM architectures, especially when no step-by-step planning or formal rule execution is involved.

Taken together, these limitations highlight the need for future hybrid approaches—possibly combining LLMs with formal model checkers or graph-based reasoning modules—to support accurate and trustworthy BPMN model analysis.

7.1 Threats to Validity

As with any exploratory study involving AI models, our experiments are subject to several limitations that may affect the generalizability and interpretation of the results.

Construct Validity. Our evaluation of LLM performance relied on human-assigned correctness scores (0.0, 0.5, 1.0) based on expert judgment. While we divided the task to compare clearly defined rubrics to reduce subjectivity, some variation in interpretation may still occur—especially in open-ended tasks. Moreover, the prompts used for each experiment were designed to simulate realistic assignment instructions, but variations in phrasing, specificity, or scope might lead to different LLM behavior. This also affects reproducibility and highlights the need for standardization of prompts in educational research on LLMs.

We used a paired t-test to compare model performance across formats in Experiment 1. Although the data violated normality assumptions, the Wilcoxon signed-rank test could not be reliably applied due to the small number of non-zero differences and the unbalanced distribution of signs. Therefore, the paired t-test was retained as an approximate comparison method.

Internal Validity. Although our experiments were designed to isolate the format (PNG vs XML) or task structure (single vs multiple issues), other variables may influence results. For example, the design of each BPMN model, its visual layout, or the distribution of error types could have influenced how the LLMs interpreted or misunderstood the content. Additionally, the models' output may be affected by undocumented internal updates to LLM APIs (especially for rapidly evolving tools like ChatGPT or Gemini), meaning that the same prompt may yield different results over time.

7.2 External Validity

Our experiments focused on a limited set of BPMN tasks and did not involve actual students as participants. Therefore, the results may not fully reflect how students interact with LLMs in authentic learning contexts.

Furthermore, in real classrooms, students may engage with LLMs in more interactive or iterative ways—asking follow-up questions, refining prompts, or comparing multiple answers. Our experiments, by contrast, tested one-shot interactions using a single static prompt. While this was intentional for consistency and control, it does not capture the full spectrum of human-AI learning dynamics, which are often exploratory, adaptive, and socially shaped.

We also limited our study to two specific models (ChatGPT-4o and Gemini Flash), which, while representative of current leading tools, do not cover the full diversity of available LLMs or future improvements. The BPMN examples used were also relatively small and self-contained; more complex or domain-specific processes might yield different levels of AI performance.

Mitigations and Future Work. To address these threats, future work should include experiments with real student participants, triangulation with human responses, and testing across different BPMN tasks and domains. Prompt robustness should also be evaluated by comparing alternate phrasings or constraints. In addition, logging model metadata (e.g., version, temperature, token limits) and establishing evaluation benchmarks can help improve the reproducibility and interpretability of future studies in BPM education.

8 Conclusion and Future Works

This paper has explored the potential and limitations of using Large Language Models (LLMs) to support—or potentially replace—students in typical BPMN-related tasks. By designing two experimental assignments based on real teaching scenarios, we investigated how current AI tools handle process model interpretation and error detection. Our early findings suggest that while LLMs are not yet fully capable of reliably solving such tasks, they are already producing answers that can compete with or complement student understanding in certain contexts. For educators, this represents both a challenge and an opportunity: the challenge of designing assignments that remain meaningful and fair in the presence of AI tools, and the opportunity to use these tools to enhance learning, self-assessment, and reflection.

We emphasize that now is the right time to act: current LLMs are not perfect, but they are improving rapidly. BPM educators should start preparing for a future where AI is a natural part of the learning environment—not by banning it, but by thoughtfully integrating it into course design, assignments, and assessment strategies.

As next steps, we plan to conduct similar experiments with student participants to compare human and AI performance, explore hybrid approaches combining human reasoning with AI suggestions, and develop AI-aware assignment formats that foster critical evaluation. These efforts would provide a valuable baseline and help assess task difficulty.

We hope that our contribution—combining practical insights, ready-to-use materials, and a flexible framework for experimentation—will help others in the BPM education community to adapt, reflect, and innovate.

References

1. Artsi, Y., Sorin, V., Konen, E., Glicksberg, B.S., Nadkarni, G., Klang, E.: Large language models for generating medical examinations: systematic review. BMC Med. Educ. **24**(1), 354 (2024)
2. Bajak, M., Spendel, Ł.: Using artificial intelligence in business: the example of chatgpt in management. e-mentor. Czasopismo naukowe Szkoły Głównej Handlowej w Warszawie **106**(4), 52–61 (2024)
3. Bernardi, M.L., Casciani, A., Cimitile, M., Marrella, A.: Conversing with business process-aware large language models: the bpllm framework. On-line preprint (2024), https://www.researchsquare.com/article/rs-4125790/v1
4. Berti, A., Schuster, D., van der Aalst, W.M.: Abstractions, scenarios, and prompt definitions for process mining with LLMs: a case study. In: International Conference on Business Process Management, pp. 427–439. Springer (2023)
5. Chen, Y., Jensen, S., Albert, L.J., Gupta, S., Lee, T.: Artificial intelligence (AI) student assistants in the classroom: designing chatbots to support student success. Inf. Syst. Front. **25**(1), 161–182 (2023)
6. Fahland, D., Fournier, F., Limonad, L., Skarbovsky, I., Swevels, A.J.: How well can large language models explain business processes? arXiv preprint arXiv:2401.12846 (2024)
7. Fahland, D., Fournier, F., Limonad, L., Skarbovsky, I., Swevels, A.J.: How well can a large language model explain business processes as perceived by users? Data Knowl. Eng. 102416 (2025)
8. Forell, M., Schüler, S.: Modeling meets large language models. In: Modellierung 2024 Satellite Events, pp. 10–18420. Gesellschaft für Informatik eV (2024)
9. Grohs, M., Abb, L., Elsayed, N., Rehse, J.R.: Large language models can accomplish business process management tasks. In: International Conference on Business Process Management, pp. 453–465. Springer (2023)
10. Jessen, U., Sroka, M., Fahland, D.: Chit-chat or deep talk: prompt engineering for process mining. arXiv preprint arXiv:2307.09909 (2023)
11. Kasneci, E., et al.: Chatgpt for good? on opportunities and challenges of large language models for education. Learn. Individ. Differ. **103**, 102274 (2023)
12. Kimmel, B., et al.: Enhancing programming error messages in real time with generative AI. In: Extended Abstracts of the CHI Conference on Human Factors in Computing Systems, pp. 1–7 (2024)
13. Klievtsova, N., Benzin, J.V., Kampik, T., Mangler, J., Rinderle-Ma, S.: Conversational process modelling: state of the art, applications, and implications in practice. In: International Conference on Business Process Management, pp. 319–336. Springer (2023)
14. Kourani, H., Berti, A., Schuster, D., van der Aalst, W.M.: ProMoAI: process modeling with generative AI. arXiv preprint arXiv:2403.04327 (2024)
15. Leinonen, J., et al.: Comparing code explanations created by students and large language models. In: Proceedings of the 2023 Conference on Innovation and Technology in Computer Science Education V. 1, pp. 124–130 (2023)

16. Lim, W.M., Gunasekara, A., Pallant, J.L., Pallant, J.I., Pechenkina, E.: Generative AI and the future of education: Ragnarök or reformation? a paradoxical perspective from management educators. Int. J. Manag. Educ. **21**(2), 100790 (2023)
17. Malinka, K., Peresíni, M., Firc, A., Hujnák, O., Janus, F.: On the educational impact of chatgpt: is artificial intelligence ready to obtain a university degree? In: Proceedings of the 2023 Conference on Innovation and Technology in Computer Science Education V, 1. pp. 47–53 (2023)
18. Meyer, J., Jansen, T., Schiller, R., Liebenow, L.W., Steinbach, M., Horbach, A., Fleckenstein, J.: Using LLMs to bring evidence-based feedback into the classroom: AI-generated feedback increases secondary students' text revision, motivation, and positive emotions. Comput. Educ. Artif. Intell. **6**, 100199 (2024)
19. Meyer, L., Dannecker, A.: Smart tools, smarter students: a study on chatgpt's effects on learning outcomes and exercise quality in business education. In: INTED2024 Proceedings, pp. 2704–2713. IATED (2024)
20. Neyem, A., González, L.A., Mendoza, M., Alcocer, J.P.S., Centellas, L., Paredes, C.: Towards an AI knowledge assistant for context-aware learning experiences in software capstone project development. IEEE Trans. Learn. Technol. (2024)
21. Pack, A., Barrett, A., Escalante, J.: Large language models and automated essay scoring of English language learner writing: insights into validity and reliability. Comput. Educ. Artif. Intell. 100234 (2024)
22. Senkus, P., et al.: Bridging the gap: an evaluation of business process management education and industry expectations–the case of poland. In: International Conference on Business Process Management: Blockchain, Robotic Process Automation and Educators Forum, pp. 250–262. Springer (2023)
23. Stribling, D., Xia, Y., Amer, M.K., Graim, K.S., Mulligan, C.J., Renne, R.: The model student: GPT-4 performance on graduate biomedical science exams. Sci. Rep. **14**(1), 5670 (2024)
24. Xiao, R., Hou, X., Stamper, J.: Exploring how multiple levels of GPT-generated programming hints support or disappoint novices. In: Extended Abstracts of the CHI Conference on Human Factors in Computing Systems, pp. 1–10 (2024)
25. Zhu, R., Hu, Q., Li, W., Xiao, H., Wang, C., Zhou, Z.: Business process text sketch automation generation using large language model. arXiv preprint arXiv:2309.01071 (2023)

Comparing the Design Space of Explorative BPM: MBAs Versus GenAI

Michael Rosemann[1(✉)] and Philipp Joebges[2]

[1] Centre for Future Enterprise, Queensland University of Technology, Brisbane, Australia
m.rosemann@qut.edu.au
[2] KOM Consulting, Brisbane, Australia
p.joebges@komc.com.au

Abstract. Teaching business process improvement is important as it is the most value adding activity in the process lifecycle. Unlike for other lifecycle stages, however, the related body of knowledge on how to actually improve a process is still light on sound, replicable methods. This challenge increases if explorative BPM is the desired subject of study. Motivated by this context, this paper reports on the experiences of an MBA unit called Business Process Design with a focus on comparing students' responses with the explorative process design space as created by generative AI. The results show that the manually created diversity of explorative BPM ideas needs to be augmented by the related capabilities of generative AI to arrive at a comprehensive process design space. This paper provides advice for BPM educators for how to teach augmented, explorative process design. Besides its value as a guide for lecturers teaching BPM, this paper inspires augmented process exploration as a subject of interest for future BPM research.

Keywords: Generative AI · explorative BPM · teaching case · design space

1 Introduction

Business Process Management (BPM) has found its way into the curriculum of educational service providers in various forms ranging from exposure in single units to comprehensive award programs. BPM is taught in under-graduate and post-graduate classes [1] as well as in MBA courses and forms part of professional education offerings (e.g., Bentley Uni, Uni Melbourne, QUT). The latter is also covered by providers such as ABPMP, OMG and the BPMInstitute. What these offerings tend to have in common is that the distinct stages of the process lifecycle [2] are covered. Stages that are well supported by methods, tools and techniques such as process modelling [3, 4], process analysis, and more recently process mining [5, 6] receive the bulk of attention. A key task, however, the (re)design of a process [7], and with it the creation of new process value, is by comparison only a minor focus. This is also evidenced by the volume of pages dedicated to process design in popular BPM textbooks [2]. A review of BPM conference papers [8] highlighted that process design is the stage in the process lifecycle that attracted the least amount of research. Widely cited related research largely relies on heuristics [9] as opposed to automated or even AI-enabled solutions.

According to the notion of ambidextrous BPM [10], process design can be differentiated into exploitative and explorative design. Whereas exploitative design is motivated

M. ER et al. (Eds.): BPM 2025, LNBIP 565, pp. 373–389, 2026.
https://doi.org/10.1007/978-3-032-02936-2_26

by addressing pain points along a process (e.g., types of waste according to lean management), explorative process design investigates the design space [11] that opens up at opportunity points in a business process. Explorative BPM is aiming for new customer experiences and revenue models, and despite its growing significance in an economy gravitating towards comparable, streamlined processes, it is not only under-researched, but also under-taught. At the same time, AI-enabled solutions are making an inroad into process design, often without that its impact and boundaries are well understood. This context creates demand for knowledge on how to best teach explorative BPM in terms of the right mix of tools and techniques as well as how to capitalize on the complementary capabilities of AI so that a comprehensive design space can be harvested. Thus, this paper aims to gain an understanding for *how the explorative design space investigated by students compares with the one generated by AI?*

We proceed as follows. In Sect. 2 we cover the related body of knowledge on explorative BPM and the use of AI for process improvement. We introduce the contextual setting, i.e., an MBA unit called Business Process Design, in Sect. 3 and describe its structure, the uniqueness of MBA students as well as the teaching case underlying this paper's study. Section 4 compares the answers of the students with the comparable responses of a generative AI solution. Section 5 discusses our findings, offers recommendations to the global Business Process Education (BPE) community for how to progress teaching explorative BPM and provides various options for future research.

2 Research Context

2.1 Explorative Business Process Management

The notion of ambidextrous BPM was introduced in 2014 [10]. Its main contribution is to emphasize the difference between a pain-point centered, reactive approach to process improvement (exploitative BPM) from an opportunity-point centered, proactive approach seeking process innovation (explorative BPM).

Exploitative BPM has been practiced since Fredrick Taylor's search for the one-best approach to conduct a task and famously been operationalized via Toyota's continuous improvement (Kaizen). The assumption of exploitative BPM is that process improvement is dedicated to addressing identified process inefficiencies that prevent a process from achieving defined cost, time or quality goals. These inefficiencies tend to be non-value adding tasks, types of waste (e.g., over-engineering) according to Lean Management, unnecessary variation as prompted by Six Sigma or error-prone human activities that are addressed by RPA, chatbots etc. Exploitative BPM is a mature discipline in terms of problem identification and investigation in the form of root-cause analyses.

Whereas exploitative BPM is largely subtraction-biased, i.e. its focus is on eliminating issues to arrive at a streamlined process, explorative BPM is addition-biased. At identified opportunity points (aka action possibilities) of a process, the designer explores additional value propositions guided by systemic ideation approaches. In this context, Recker and Rosemann (2015) [12] present four distinct lenses to create new process design options - enhance, utilize, derive and design. Sintos and Alves (2017) [13] used Design Science Research to develop the Ambidextrous Analysis of Business Processes (A2BP), a multi-staged method embedding exploitative and explorative methods. The rapid redesign approach NESTT [14] is a hybrid, interactive methodology combining exploitative and explorative design over a four-week period. Figure 1 provides a comparison between exploitative and explorative BPM.

	Exploitative BPM	Explorative BPM
Mindset	**Sense of Urgency** *Change momentum is grounded in identified, rational problems*	**Sense of Ambition** *Change momentum is grounded in proposed, exciting possibilities*
Tangibility	**Pain Point** *A defined shortfall in the process (Lean, Six Sigma)*	**Opportunity Point** *An identified action possibility in the process*
Task	**Analysis** *Systematic unpacking of root causes, exploitation of established best practices*	**Design** *Creative investigating of alternatives, exploration of individual, new practices*
Attention	**Today** *Model and mine broken processes for shared understanding*	**Tomorrow** *Envisage possible future processes for shared commitment*
Aim	**Simplification** *Reduce waste, variation, human labor, non-value, variants*	**Sophistication** *Increase personalization, proactivity, unexpected value, trust, delight*
Outcome	**Efficiency** *Cost-effective: Zero-touch, friction-free, straight-through, digital only*	**Eloquence** *Revenue-effective: intensified engagement, increased spend, loyalty*

Fig. 1. Comparing exploitative and explorative BPM

Exploitative BPM is well documented in BPM literature and comprehensively taught as part of the process lifecycle. Explorative BPM, in contrast, has hardly made it into any type of textbook, teaching material or course contents. It was the theme of a tutorial at the BPM 2019 conference [15], which stated that the body of knowledge on explorative BPM is underdeveloped, but in high demand. Groß et al. (2019) [16] conducted a comprehensive literature review to identify 98 process improvement methods and positioned these in the process orbit [2]. The fact that only 3 of them were classified as process innovation is further evidence for the improvement-bias of the literature on business process design. Rosemann (2020) [17] proposed seven patterns for explorative process design. These patterns offer fine-granular inspirations for new design options. For example, process generalization suggests considering re-utilizing an existing process for other objects (e.g., Amazon's expansion from books to toys). A higher level of abstraction is provided by the five-diamond method for explorative BPM [18] which includes purpose, business, technology, integration and their overall orchestration as five distinct lenses to surface process design options. Groß et al. (2021) [11] introduce the notion of a business process design space composed of 19 design dimensions to inform the systematic exploration of new design options. Using a free-simulation experiment, Groß et al. (2024) [19] show that explorative methods create a higher number and more diverse ideas than exploitative process improvement methods.

2.2 AI-Enabled Automated Process Design

Artificial Intelligence (AI) is one of the most fascinating, and currently intensively discussed enablers of business process design [20, 21]. The focus tends to be on how analytic and generative AI solutions can be used to support more efficient business processes (e.g., via chatbots or generating personalized documents). The use of AI to augment the act of process design, i.e., to extend the process design space by creating new process design options, however, is so far only a niche discussion in research and among providers of BPM solutions (e.g., Signavio).

Röglinger et al. (2021) [22] differentiate two types of using digital technologies for process (re)design - automated (exploitative) improvement and automated (explorative) innovation (Fig. 2). The authors state "*Despite all these automation efforts, it is remarkable that the BPM lifecycle phase process (re-)design remains a manual task with a high level of cognitive effort.*" As a way forward, Fehrer et al. [23] present the ABMR concept, a user-guided, semi-automated approach for exploitative improvement based on redesign patterns which leads to four types of recommendations that differ in their degree of automation. van Dun et al. (2022) [24] developed the ProcessGAN, which uses generative adversarial networks to create exploitative process improvement ideas based on historical data. Another prototype for the automation of exploitative process improvement is the solution presented by Harl et al. (2024) [25] which creates future process models based on historical process models.

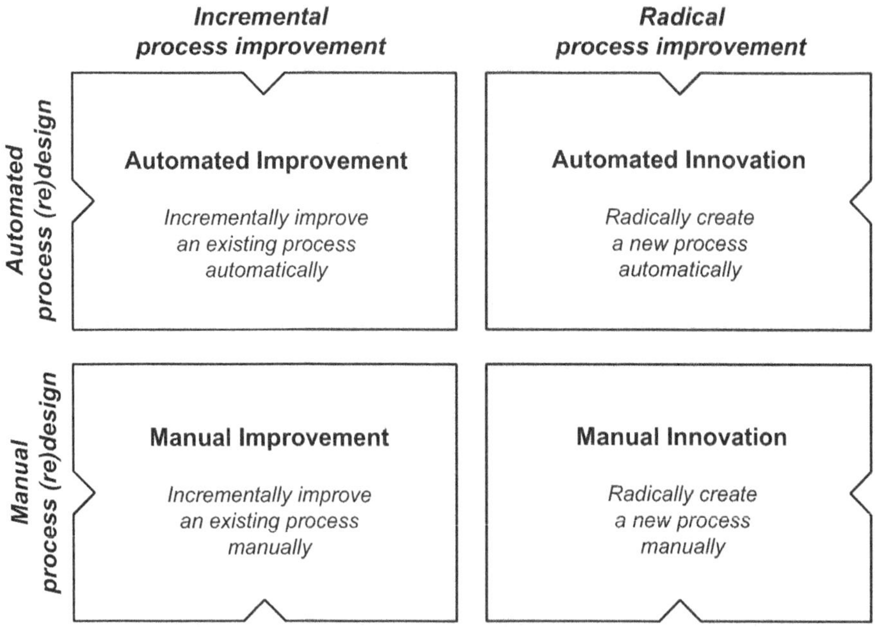

Fig. 2. Four types of process design support [based on 22]

In terms of automating explorative process innovation, however, Röglinger et al. (2021) [22] summarize "*the automation of process innovation proves to be an unsolvable problem to date.*" A rare, and first attempt to automate explorative BPM is presented by Bojer et al. (2024) [26]. Their research using four case studies with a Danish manufacturer shows that lightweight machine learning can speed up opportunity assessment and implementation, and with this reduce process innovation latency [10].

The rapid maturity and uptake of generative AI is now providing the "*capacity to transform ideation by automating content creation processes*" [27, p. 2]. In this context, hallucinations, statements that are not (yet) based on real data or current empirical evidence, are becoming a desired feature as they may describe desired futures. The lack of replicability, i.e. executing the same prompt multiple times leading to different responses being generated, is a further feature that increases initial process innovation ideas. This leads to the obvious question of how explorative BPM could be automated, which motivated this study and describes the top right quadrant in Fig. 2. Such an investigation extends the current focus of using generative AI for product or service innovation towards process innovation [27].

3 Contextual Setting

3.1 The MBA Unit 'Business Process Design'

The project described in this paper took place as part of an MBA unit at an Australian university. Called Business Process Design, it is an elective unit (aka concentration) in leadership and strategy in the final third of this 1.5-year MBA program. MBA students bring a unique perspective to learning BPM and as a dedicated cohort have not been the focus of previous related research. Key differences are their perspective on BPM as a strategic enabler (e.g., competitive advantage, strategic alignment, business model innovation) as opposed to an operational tool, their cross-functional view grounded in personal work experiences and their leadership lens motivating an interest in cultural change and governance. They are technology-aware but focused on outcomes and have an appetite for appliable frameworks and insightful cases as opposed to detailed notations and techniques.

The seven-week unit was offered in January-February 2025 and 20 'associates' were enrolled. Typical for a MBA class, this cohort was made up of students between 25-40 years of age with various degrees of practical experiences. Business Process Design is the only BPM-related unit in this MBA program and as such aims to provide an overview about enterprise-wide BPM and the main stages of the process lifecycle. The themes of the classes (three hours each) over these seven weeks were as follows (Table 1).

Of particular importance for the context of this paper is the focus on explorative BPM, which is a rarely taught topic in today's BPM curricula. In week 5, students were learning the following concepts.

- What is an opportunity point in a business process?
- The role of an opportunity appetite statement
- Alternative process innovation lenses (e.g., enhance, derive, utilize)
- Patterns of explorative process design

Table 1. Outline of the seven-week MBA unit on Business Process Design

Week	Content
1	The Process-led Organization: Why processes matter
2	Process Lifecycle Management: Stages of process transformation
3	Customer Journey Mapping: The external view on processes
4	Exploiting business processes: How to identify and address pain points
5	Exploring business processes: How to identify and capitalize on opportunity points
6	The Future of Business Process Design: The shifting priorities
7	Final presentation

The mandated readings for this paper were an article about systematic ideation [12], the process design chapter of the BPM fundamentals textbook [2] and its embedded redesign orbit, and a BPM conference paper covering the explorative process design patterns [17]. The students were made aware of the differences between exploitative and explorative BPM approaches and how they are used in hybrid process redesign approaches like the NESTT [14].

The unit had two assignments weighted 50:50, both conducted using the same case study. The first assignment (due week 4), an individual assessment, was dedicated to exploitative BPM and tasked the students to conduct a mainstream process analysis; What are the goals of the process? Who are the related stakeholders? Where are the main pain points (e.g., in light of a lean evaluation)? How can the identified issues be best addressed? This kind of assignment is grounded in the established BPM body of knowledge and required methods, tools and techniques are well described in literature. Therefore, the identification of the pain points in this case study, and the development of response strategies is out of scope for the purpose of this paper.

The second assignment (due week 7) was in contrast a group assignment and emphasized explorative BPM. Here, in groups of 3-5 students, the key questions were What are the main opportunity points in the process? How can these be leveraged using explorative BPM approaches?' Students were encouraged to identify especially new revenue opportunities. Each group was given ten minutes and a maximum of 8 slides to present and justify their recommendations. In addition to the slides, up to half a page of notes per slide could be submitted as part of this assessment.

3.2 The Teaching Case

In the past, the case study for this Business Process Design unit had been sourced from industry partners. This included, for example, the Royal Flying Doctor Services Queensland, Bank of Queensland, Brisbane Airport Corporation and Brisbane City Council. These real-world cases provided authentic learning experiences and with it the opportunity to engage with actual process stakeholders. The organizations involved appreciated these types of engagement with the university, as they offered validated recommendations for their process improvement initiatives. In this most recent semester, we used for the very first time a Harvard Business case for the following reasons:

- Discussion of an international case study as opposed to local cases

- Mature, well-articulated case description as opposed to reliance on the material as it is provided by industry partners which could come with compromises in terms of what can be shared, its completeness and its timeliness
- Stable arrangement during the semester as opposed to the need to be agile in terms of new requirements, stakeholders and documents available

For this unit, we selected the case: *Dabbawalas: Masters of Prevision but Enough to Survive?* written by Anand Nandkumar et al., Indian School of Business [28]. The case comprises nine pages of texts, a number of exhibits and a related teaching note. As the Dabbawalas have been widely studied over decades, a variety of additional documentation is available ranging from Ted talks (Ritesh Andre, Pawan Agrawal), to HBR articles [29], other case studies [30] to academic papers [31].

The Dabbawala service of delivering food from home to office workers in Mumbai commenced in 1890. Over this more than a century, the Dabbawalas with their iconic white Ghandi caps have become a symbol of Mumbai. Each day, around 5,000 Dabbawalas make to-and-from-trips between homes and offices delivering around 200,000 lunchboxes (cylinder-shaped standardized boxes, so called dabbas). The entire end-to-end process - collecting, sorting, distributing and re-collecting and returning the dabbas after lunch - is well scripted with defined hand-over checklists and a color-coding system. The following two Figs. 3 and 4 - swim lane notation and customer journey map - provide complementary visual overviews about the nature of this business process and have been designed by students in this unit.

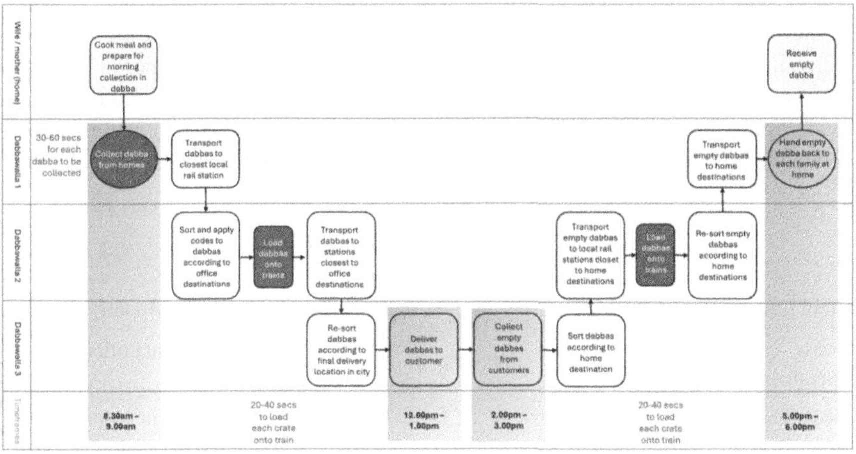

Fig. 3. The Dabbawala process in swim lane notation

Though the process is largely manual (delivery requests via SMS are possible), it runs to a surprisingly high Six Sigma level of 1 error per 14 million deliveries [28, 32]. The comprehensive documentation of this process and its context, its current capabilities but also its manual nature and potential for new design options is making it an ideal case for a class on exploitative and explorative process design.

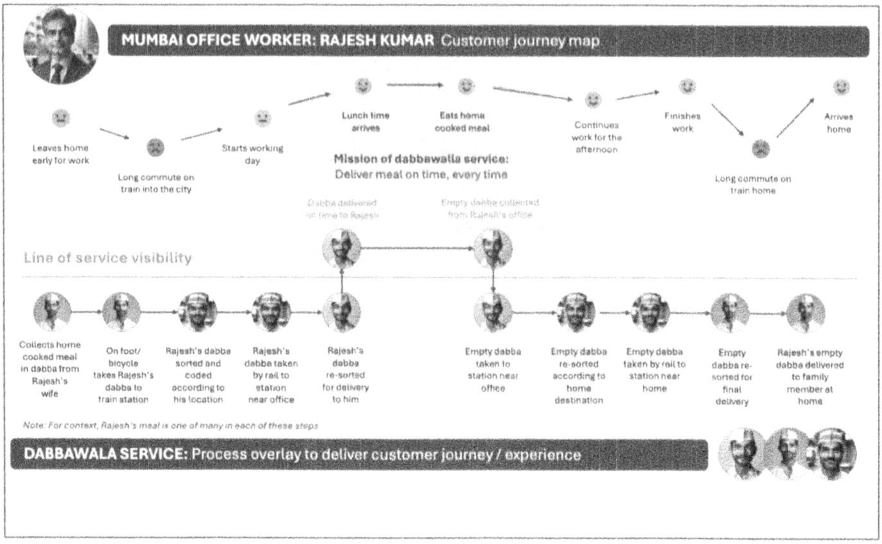

Fig. 4. The Dabbawala process as a customer journey map

4 The Proposed Explorative Process Designs

4.1 The Explorative Process Design Patterns

The following table provides an overview about the seven explorative process design patterns that have been taught in this unit, and for which a related paper [17] has been a mandated reading for all students. These seven patterns have in common that they provide operational guidance for how to create new customer experiences and related revenue models using existing business process models (Table 2).

Table 2. The seven explorative design patterns [17]

Pattern	Addition	Third Party?	Revenue Model
Process Generalization	New product	No	New product offering
Process Expansion	New activity	No	New margin grounded in new value-adding activity
Process Differentiation	New variant	No	New margin grounded in the tailored provision of alternative process performances

(continued)

Table 2. (*continued*)

Pattern	Addition	Third Party?	Revenue Model
Process Initiation	New (start) event	No	New customers because of earlier process provision
Process Commercialization	New customer for activity	Yes	License for use (e.g., pay-per-use)
Process Integration	New customer for the event	Yes	Commission model (e.g., pay-per-event or conversion)
Process Attention	New customer for attention	Yes	Commission model (e.g., pay-per-click or time)

4.2 The Students' Responses

Creative diversity as part of a divergent exploration process is a well-recognized component of innovation processes. However, idiosyncratic creativity also means that a proposed solution for a new process largely depends on the team that generated the proposal - a severe compromise to the reliability and reproducibility along the process lifecycle. Studying the diversity of responses across, in this case five groups of MBA students provides insights into the need to further guide the ideation process via normative recommendations.

In the following, an overview about the key ideas of each of the five groups is provided. Note that most groups used a divergent stage first leading to a higher number of ideas of which they then selected a few, high priority ideas presented now. For this they conducted an assessment of the feasibility, desirability and viability of the ideas.

(1) Inspired by the practices of Amazon and e-commerce marketplaces, the first group proposed DabbaPrime, a subscription model with priority delivery (process differentiation), as well as a platform model in which the home cook would make additional meals attracting additional customers (process expansion). Process generalization, and current idle time in the system, motivated to expand the delivery process also to include parcels and groceries.
(2) The second group proposed utilizing the Dabba's uniform as an idle asset for marketing (process attention), an upstream integration to enter the produce stage of the supply chain (process expansion) and to generalize the process to include delivery of healthcare/pharma products.
(3) The third group also suggested process generalization by offering adjacent items (e.g., drinks, snacks), providing a hotel-inspired upgrade service and also offering home security services utilizing idle time (process expansion. Process expansion was considered by proposing to source and deliver the ingredients needed to cook.
(4) Process generalization in the form of additional delivery of pharma items and groceries was the core idea of the fourth group. This group also suggested drone delivery (process differentiation) and data monetization (e.g., temporal and local insights into

food demand patterns), which represents process integration in the form of event and data brokerage.

(5) The fifth group also proposed data monetization and process generalization (pharmaceuticals).

A comparison across these five groups shows that process generalization, i.e. using the existing process capabilities for additional types of deliveries has been a common thought. The diversity of ideas varied substantially. Concepts such as process commercialization but also process attention and process integration were under-utilized.

4.3 The GenAI Responses

Addressing the research questions of this paper, we used the paper on explorative process design patterns [17] independent and unknown to the students, as input for Microsoft's Copilot. The Copilot instance was not further trained or primed in any aspects of BPM or process innovation. Then we questioned it with the following prompt.

Leveraging your expertise as a strategy and process consultant, and referring to the explorative process design patterns, develop ideas for how to innovate the Dabbawala business model to ensure its sustainability going forward.

I want you to use each of the process design patterns and develop at least three ideas per pattern. For each idea, describe why the idea is good, what precedent to the idea exists such as businesses in other industries doing something similar, and what the expected value benefit would be. You can apply these patterns at a specific process level, i.e. looking to innovate only parts of the business, or at a whole-of-business level, i.e. looking to innovate the whole business. Focus your ideas on ways to safeguard and build additional revenue over anything else.

This led to the following three ideas for each of the seven patterns of explorative process design (Table 3).

This table shows the breadth and depth of generative AI's capabilities to provide rather specific types of explorative process design ideas across very different categories. Though a number of these ideas were also proposed by the student groups, this table includes a number of explorative process design ideas not mentioned by any of the student groups, including suggestions such as corporate catering, customized meal-plans, and early morning delivery. The genAI solution outperformed the students in particular when ideas were requiring business model innovation (process commercialization such as a dynamic pricing model) and promotional activities (e.g., process attention in the form of in-app advertisement). However, the genAI ideas were also rather unfiltered. Thus, and similar to the students who had to shortlist their ideas in terms of their feasibility, desirability and viability, we wanted to ensure a higher comparability between the answers from the MBAs and from genAI. Therefore, we used a second round of prompting to narrow down the scope of explorative ideation.

Using information from the internet, assess the feasibility, viability, and desirability of all above ideas. Rank the ideas and explain in detail for the top five ideas. For the remaining ideas, only summarise your findings. Use data and numbers wherever possible and preference them over anecdotal evidence. Explain how you went about assessing the ideas, what data you used, what calculations were applied, and how you came up with the results. Assuming each idea is operationally feasible, refine your top five list of ideas with detailed information you can already find in the case study. Leverage data you can find online about market size and growth in Mumbai for ideas such as food delivery or medicine delivery.

This led to the four shortlisted ideas as per the following Table 4. The first two, highest rated ideas were also common ideas among the student groups whereas the other two ideas had not been mentioned by the students. This shows that not only the ideas that were generated differed between MBAs and genAI, it is also the assessment of these ideas in terms of feasibility, viability and desirability that is different. Needless to say that the costs and time of ideation and idea evaluation were substantially different between MBAs and genAI. The total effort to generate and evaluate ideas with genAI took under 30 min.

Table 3. Explorative Process Design Ideas as generated by MS Copilot

Pattern	Idea #1	Idea #2	Idea #3
Process Generalisation •Process generalization is the exploration of an existing process capability that can be used for additional items; that is, the provision of new products or services with only minor changes to the process. Therefore, process generalization re-purposes (re-uses) a process and facilitates the generation of new value by providing new products or services cost-effectively to the	**Grocery Delivery Service** **What it is:** Leveraging the existing logistics network to deliver groceries can tap into the growing demand for home delivery services. **Precedent**: Companies like Instacart and Amazon Fresh have successfully implemented grocery delivery services. **Expected Value Benefit**: New revenue stream from grocery delivery fees, increased utilization of the logistics network, and enhanced customer convenience.	**Medicine Delivery Service** **What it is:** Providing a reliable and timely delivery service for medicines can cater to the needs of elderly and chronically ill customers. **Precedent**: Companies like PillPack (acquired by Amazon) and 1mg in India offer medicine delivery services. **Expected Value Benefit**: New revenue stream from medicine delivery fees, improved customer loyalty, and potential partnerships with pharmacies.	**Document and Parcel Delivery Service** **What it is:** Expanding the delivery service to include documents and parcels can utilize the existing network during off-peak hours. **Precedent**: Companies like FedEx and DHL offer document and parcel delivery services. **Expected Value Benefit**: Additional revenue from delivery fees, better utilization of the workforce, and increased brand visibility.
Process Expansion •Process expansion is an assessment to test if an existing process could create additional value by adding further activities to the process. Such an expansion needs to be grounded in a competitive advantage (e.g., the availability of data that facilitates the process expansion).	**Meal Preparation and Delivery** **What it is:** Offering meal preparation services in addition to delivery can cater to customers who prefer home-cooked meals but lack the time to cook. **Precedent**: Companies like Blue Apron and HelloFresh provide meal kits and prepared meals. **Expected Value Benefit**: New revenue stream from meal preparation fees, increased customer satisfaction, and potential partnerships with local chefs.	**Subscription-Based Health and Wellness Plans** **What it is:** Providing health and wellness plans, including diet consultations and fitness tracking, can attract health-conscious customers. **Precedent**: Companies like Noom and MyFitnessPal offer subscription-based health and wellness services. **Expected Value Benefit**: Recurring revenue from subscription fees, enhanced customer engagement, and improved brand loyalty.	**Corporate Catering Services** **What it is:** Expanding into corporate catering can provide bulk orders and steady revenue from businesses. **Precedent**: Companies like Zomato and Swiggy offer corporate catering services. **Expected Value Benefit**: Increased revenue from corporate clients, better utilization of resources, and potential long-term contracts.
Process Differentiation •Process differentiation is the creation of one or more additional process variants with the aim of offering a new value proposition for a subset of the existing process customers or attracting entirely new customers to the process.	**Premium Delivery Service** **What it is:** Offering a premium delivery service with faster delivery times and additional features can cater to customers willing to pay extra for convenience. **Precedent**: Companies like Amazon Prime offer premium delivery services. **Expected Value Benefit**: Additional revenue from premium service fees, improved customer satisfaction, and competitive differentiation.	**Eco-Friendly Delivery Option** **What it is:** Providing an eco-friendly delivery option using bicycles or electric vehicles can attract environmentally conscious customers. **Precedent**: Companies like UPS and DHL have introduced eco-friendly delivery options. **Expected Value Benefit**: New revenue stream from eco-friendly delivery fees, enhanced brand image, and reduced carbon footprint.	**Customized Meal Plans** **What it is:** Offering customized meal plans based on dietary preferences and health goals can cater to niche customer segments. **Precedent**: Companies like Freshly and Diet-to-Go provide customized meal plans. **Expected Value Benefit**: Increased revenue from customized meal plan fees, improved customer loyalty, and potential partnerships with nutritionists.

(continued)

Table 3. (*continued*)

Process Initiation •Process initiation is the exploration of how new data sources can be used to reduce process latency; that is, the time between the occurrence of a demand and the initiation of the related process. This metric is also called time-to-process.	**Predictive Ordering System** **What it is:** Implementing a predictive ordering system that anticipates customer needs based on past orders can streamline the ordering process. **Precedent:** Companies like Amazon use predictive analytics for inventory management. **Expected Value Benefit:** Improved customer experience, increased order accuracy, and reduced order processing time.	**Early Morning Delivery Service** **What it is:** Offering an early morning delivery service for breakfast items can cater to customers who prefer fresh meals in the morning. **Precedent:** Companies like Starbucks and Dunkin' provide early morning delivery services. **Expected Value Benefit:** New revenue stream from breakfast delivery fees, increased customer convenience, and better utilization of the workforce.	**Real-Time Order Tracking** **What it is:** Providing real-time order tracking can enhance transparency and customer satisfaction. **Precedent:** Companies like Domino's Pizza offer real-time order tracking. **Expected Value Benefit:** Improved customer experience, increased trust, and reduced customer inquiries.
Process Commercialisation •Process commercialisation is an assessment of whether there is a market external to the organization that would be interested in deploying the entire process, or relevant parts of it, as a service.	**Licensing the Logistics System** **What it is:** Licensing the Dabbawala logistics system to other businesses can generate additional revenue. **Precedent:** Companies like Amazon Web Services (AWS) offer cloud services to third parties. **Expected Value Benefit:** New revenue stream from licensing fees, increased brand recognition, and potential partnerships.	**Offering Training Programs** **What it is:** Providing training programs on efficient logistics and delivery management can attract businesses looking to improve their operations. **Precedent:** Companies like McDonald's offer training programs through Hamburger University. **Expected Value Benefit:** Additional revenue from training fees, enhanced industry reputation, and potential consulting opportunities.	**Dynamic Pricing Model** **What it is:** Implementing a dynamic pricing model based on demand and delivery times can optimize revenue. **Precedent:** Airlines and ride-sharing companies like Uber use dynamic pricing models. **Expected Value Benefit:** Increased revenue through optimized pricing, better demand management, and improved profitability.
Process Integration •Within process integration, an event is assessed beyond its value for the process, and it is evaluated if the event could meaningfully trigger complementary third-party processes (trigger-as-a-service).	**Partnering with E-Commerce Platforms** **What it is:** Integrating the Dabbawala delivery service with e-commerce platforms can provide additional delivery options for online shoppers. **Precedent:** Companies like FedEx and UPS partner with e-commerce platforms for delivery services. **Expected Value Benefit:** New revenue stream from delivery fees, increased customer base, and potential long-term partnerships.	**Collaborating with Local Businesses** **What it is:** Partnering with local businesses to offer delivery services for their products can create new revenue opportunities. **Precedent:** Companies like Postmates and DoorDash collaborate with local businesses for delivery services. **Expected Value Benefit:** Additional revenue from delivery fees, better utilization of the logistics network, and enhanced community engagement.	**Integrating with Health Services** **What it is:** Integrating the delivery service with health services to deliver medical supplies and health-related products can cater to a growing market. **Precedent:** Companies like CVS and Walgreens offer delivery services for health products. **Expected Value Benefit:** New revenue stream from delivery fees, improved customer convenience, and potential partnerships with healthcare providers.
Process Attention •Process attention is the exploration of whether the time a user spends observing a process could be utilized to generate new value.	**In-App Advertisements** **What it is:** Displaying advertisements in the Dabbawala app can generate additional revenue from advertisers. **Precedent:** Companies like Facebook and Google generate significant revenue from in-app advertisements. **Expected Value Benefit:** New revenue stream from advertising fees, increased app engagement, and potential partnerships with advertisers.	**Sponsored Content** **What it is:** Providing sponsored content, such as recipes and cooking tips, can attract advertisers and enhance customer engagement. **Precedent:** Companies like YouTube and Instagram offer sponsored content. **Expected Value Benefit:** Additional revenue from sponsored content, improved customer experience, and increased app usage.	**Loyalty Programs with Partner Brands** **What it is:** Partnering with brands to offer loyalty programs and promotions can increase customer retention and generate additional revenue. **Precedent:** Companies like Starbucks and Sephora have successful loyalty programs. **Expected Value Benefit:** Increased customer loyalty, new revenue from partner brands, and enhanced customer satisfaction.

Table 4. Shortlisted Process Design Ideas as classified by MS Copilot

Idea	Feasibility	Viability	Desirability
#1: Grocery Delivery Service	**Feasibility: High**	**Viability: High**	**Desirability: High**
	Technical Feasibility: The Dabbawalas' existing logistics network can be leveraged for grocery delivery. The infrastructure and processes are already in place, requiring minimal modifications	Market Demand: The demand for grocery delivery services has increased significantly, especially post-pandemic. The Indian food and grocery delivery platforms market was valued at USD 5.57 billion in FY2024 and is expected to reach USD 24.34 billion in FY2032, with a CAGR of 20.24%	Customer Satisfaction: Enhanced convenience for customers who can receive groceries at their doorstep
	Operational Feasibility: The service can be offered during early morning hours, ensuring it does not interfere with the core lunch delivery operations. The Dabbawalas' knowledge of Mumbai's geography and their efficient delivery system support this idea	Competitive Advantage: The Dabbawalas' reliability and established brand can attract customers who prefer a trusted service over new entrants	Revenue Potential: New revenue stream from delivery fees, increased utilization of the logistics network, and enhanced customer convenience
	Financial Feasibility: Initial investment in partnerships with grocery stores and minor adjustments to the logistics system. Potential revenue from delivery fees can offset the costs		
#2: Medicine Delivery Service	**Feasibility: High**	**Viability: High**	**Desirability: High**
	Technical Feasibility: The existing logistics network can be adapted for medicine delivery. Partnerships with pharmacies can be established	Market Demand: The demand for medicine delivery services is growing, especially among elderly and chronically ill customers	Customer Satisfaction: Enhanced convenience for customers who can receive medicines at their doorstep
	Operational Feasibility: The service can be offered during early morning hours, ensuring it does not interfere with the core business. The Dabbawalas' reliability is crucial for timely medicine delivery	Competitive Advantage: The Dabbawalas' reliability and established brand can attract customers who prefer a trusted service over new entrants	Revenue Potential: New revenue stream from delivery fees, improved customer loyalty, and potential partnerships with pharmacies
	Financial Feasibility: Initial investment in partnerships with pharmacies and minor adjustments to the logistics system. Potential revenue from delivery fees can offset the costs		

(continued)

Table 4. (*continued*)

Idea	Feasibility	Viability	Desirability
#3: Corporate Catering Services	**Feasibility: High**	**Viability: High**	**Desirability: High**
	Technical Feasibility: The Dabbawalas can collaborate with local restaurants or cloud kitchens to provide corporate catering services. The existing logistics network can be utilized for deliveries	Market Demand: The demand for corporate catering services is steady, with businesses looking for reliable and cost-effective options	Customer Satisfaction: Enhanced convenience for businesses that can rely on timely and efficient catering services
	Operational Feasibility: The service can be offered without interfering with the core lunch delivery operations	Competitive Advantage: The Dabbawalas' reliability and established brand can attract corporate clients who prefer a trusted service	Revenue Potential: Increased revenue from corporate clients, better utilization of resources, and potential long-term contracts
	Financial Feasibility: Initial investment in partnerships with restaurants and cloud kitchens. Potential revenue from corporate clients can offset the costs		
#4: Subscription-Based Health and Wellness Plans	**Feasibility: Medium**	**Viability: High**	**Desirability: High**
	Technical Feasibility: Developing health and wellness plans requires partnerships with nutritionists and fitness experts. The Dabbawalas can leverage their customer base for this service	Market Demand: The demand for health and wellness services is growing, with customers looking for personalized plans	Customer Satisfaction: Enhanced customer engagement and satisfaction through personalized health and wellness plans
	Operational Feasibility: The service can be offered as an add-on to the existing lunch delivery service, ensuring it does not interfere with the business	Competitive Ad vantage: The Dabbawalas' established brand and customer base can attract health-conscious customers	Revenue Potential: Recurring revenue from subscription fees, improved customer loyalty, and partnerships with health experts
	Financial Feasibility: Initial investment in partnerships and developing the subscription platform. Potential revenue from subscription fees can offset costs		

5 Lessons Learnt

This is the first paper studying the comparative capabilities of humans and genAI in terms of explorative process design, and how to embed this stage of the process lifecycycle into BPM education. Our research has shown that (1) human process designers have rather low inter-designer reliability meaning there is a high dependence on the individual, and (2) genAI is not only capable of proposing comprehensive process design ideas covering known explorative patterns, but also able to tentatively shortlist these based on criteria such as feasibility, desirability and viability. However, we also noted that the feasibility of initial genAI ideas is not granted. The benefit of genAI to provide us with new, truly innovative ideas becomes a challenge if relevant business considerations are ignored. Where the ideas generated by students have been filtered intuitively, genAI ideas require a dedicated review. In other words, the divergent capability of genAI to create new process designs according to defined design patterns is remarkable, but its capability to converge towards superior ideas requires advanced reasoning and human assessment. In any case, using genAI in the business process lifecycle enables non-BPM practitioners to progress process change initiatives by removing the need for a technical expert to facilitate or guide the process redesign stage, and can be seen as a first step towards the democratization of process innovation. This has implication for teaching explorative BPM and shows that exposing students to the patterns of explorative design is not sufficient. Rather it is required to teach how augmented explorative design further extends the divergent design space, and that human guidance is needed when converging towards feasible, viable and desirable proposals.

Naturally, such a paper comes with *limitations*. The students used a wide range of explorative ideas (e.g., systemic ideation) in addition to the seven explorative process design patterns meaning they considered a process design space beyond the one formed by the seven explorative design patterns. MS Copilot, in contrast, was only prompted with the explorative design patterns paper, but of course comes with the comprehensive knowledge base of a LLM. The prompting of the generative AI solution relied on a constrained set of data. Markets studies, interview outcomes, lecture transcripts etc. would have most certainly improved the quality and context-specificity of generated explorative ideas and in particular eliminated less applicable ideas.

BPM scholars have started to explore analytic AI for automated exploitative process design, but the use of generative AI for the prompted design of explorative process design options is in its infancy. Thus, this is an area rich with many *research options* including gaining a deeper understanding of the comparative advantages of human versus genAI agents as explorative process designers, conducting a comparable process design challenge for less established processes (here: rather common food delivery), and evaluating the impact of adding responsibility as a fourth criteria for the evaluation of the ideas generated. One could also imagine new forms of process innovation challenges with participants ranging from researchers, BPM practitioners to subject matter experts augmented with generative AI capabilities. Ultimately, research will have to develop guidance on how to best augment explorative process design to ensure that process innovation is served by the ideal blend of human and AI capability.

6 Conclusion

Business process improvement, incremental and radical, is the most value-adding step in the business process lifecycle making it an important element in the BPM curriculum and a candidate for technology support. There is, however, a limited body of knowledge on how to teach business process improvement reliably and how to best capitalize on technology. This paper provided operational guidance and a case study for how to create a learning experience covering explorative process design. The findings showed that there is a high level of idiosyncrasy when it comes to human explorative process design. This lack of reliability can be moderated by the capabilities of generative AI. This paper showed the comprehensiveness and comparative quality of AI-created explorative process designs, but also the need for adequate framing and contextualization. Thus, BPM educators need to understand how to best intertwine human and technical capabilities.

References

1. Bandara, W., et al.: Business process management education in academia: status, challenges, and recommendations. Commun. AIS **27**, article 41 (2010)
2. Dumas, M., La Rosa, M., Mendling, J., Reijers, H. Fundamentals of business process management. 2nd edn. Springer (2018)
3. Hrabala, H., Opletalovab, M., Tucekc, D.: Teaching business process management: improving the process of process modelling course. J. Appl. Eng. Sci. **15**, 112–121 (2017)
4. Recker, J., Rosemann, M.: Teaching business process modelling: experiences and recommendations. Commun. AIS **24**(1), 379–394 (2009)
5. Wambsganss, T., Schmitt, A.: Enhancing personalized learning through process mining. a taxonomy and design patterns approach. Business and Information Systems Engineering, 2024 (in press)
6. Chow, W.: Teaching business process management with a flipped-classroom and problem-based learning approach with the use of Apromore and other BPM software. In: Proceedings of the IEEE International Conference on Engineering, Technology, and Education (TALE 2021). Wuhan, China, 5–8 December (2021)
7. Malhotra, Y.: Business process redesign: an overview. IEEE Eng. Manage. Rev. **26**, 27–31 (1998)
8. Recker, J., Mendling, J.: The state of the art of business process research as published in the BPM conference. BISE **58**, 55–72 (2016)
9. Reijers, H.A., Mansar, S.L.: Best practices in business process redesign. Omega **33**(4), 283–306 (2005)
10. Rosemann, M.: Proposals for future BPM research directions. In: Ouyang, C., Jung, J.-Y. (eds.) Proceedings of the 2nd Asia Pacific Business Process Management Conference (AP-BPM 2014), Brisbane, 3-4 July, 1-15 (2014)
11. Groß, S., et al.: The business process design space for exploring process redesing alternatives. Bus. Process. Manag. J. **27**(8), 25–56 (2021)
12. Recker, J., Rosemann, M.: Systemic ideation: a playbook for creating innovative ideas more conciously. Bus. Transf. J. **13**, 34–45 (2015)
13. Santos, H., Alves, C.: Exploring the ambidextrous analysis of business processes: a design science research. In: Proceedings of the 19th Enterprise Information Systems Conference (ICEIS 2017). Porto, Portugal, 26–29 April, 543–566 (2017)
14. Rosemann, M.: The NESTT: rapid process redesign at queensland university of technology. In: Mendling, J., vom Brocke, J. (eds.) Business Process Management Cases, pp. 169–185. Springer, (2018)
15. Grisold, T., et al.: Exploring explorative BPM - setting the ground for future research. In: Proceedings of the 17th International Conference on Business Process Management (BPM 2019). Vienna, Austria, 1–6 September, 23–31 (2019)
16. Groß, S., Mandelburger, M.M., Mendling, J.: Navigating the maze of business process change methods. In: Bui, T. (ed.) Maui Proceedings of the 52nd Hawaii International Conference in System Sciences (HICSS-52), pp. 6270–6279. Hawaii (2019)

17. Rosemann, M.: Explorative process design patterns. In: Proceedings of the 18th International Conference on Business Process Management, pp. 349–367. Seville, Spain (2020)
18. Grisold, T., et al.: The five diamond method for explorative business process management. BISE **64**, 149–166 (2021)
19. Groß, S., Grishold, T., Mendling, J., Jaase, J.: Idea generation in exploitative and explorative business process redesign techniques. ISeB **22**, 527–555 (2024)
20. Davenport, T.H., Holweg, M., Jeavons, D.: How AI is helping organisations redesign processes. Harvard Business Review (2023)
21. Davenport, T.H., Redman, T.C.: How to marry process management and AI. Harvard Business Review (2025)
22. Röglinger, M., van Dun, C., Fehrer, T., Fischer, D.A., Modere, L., Kratsch, W.: Automated Process (Re)Design. In: The Biggest Business Process Management Problems to Solve Before We Die. Computers in Industry, vol. 146 (2023)
23. Fehrer, T., Fischer, D.A., Leemans, S.J.J., Röglinger, M., Wynn, M.T.: An assisted approach to business process redesign. Decision Support Syst. **156** (2022)
24. van Dun, C., Moder, L., Kratsch, W., Röglinger, M.: ProcessGan: supporting the creation of business process improvement ideas through generative machine learning. Dec. Support Syst. **165** (2023)
25. Harl, M.V., Zilker, S., Weinzierl, S.: Towards automated business process redesign in runtime using generative machine learning. In: Proceedings of the 38th European Conference on Information Systems (ECIS 2024). Paphos, 13–19 June (2024)
26. Bojer, C.S., et al.: Speeding up explorative BPM with lightweight IT: the case of machine learning. Inf. Syst. Front. **27**, 823–840 (2024)
27. Holström, J., Carroll, N.: How organizations can innovate with generative AI. Business Horizons (in press) (2024)
28. Nandkumar, A., Anand, M., Kumar, A.J., Mohapatra, S.: Dabbawalas: Masters of precision but enough to survive? Indian School of Business. ISB436 (2024)
29. Thomke, S.H.: Mumbai's model of service excellence. HBR, November (2012)
30. Thomke, S.H., Sinha, M.: The Dabbawala System: On-Time Delivery. Every Time. Harvard Business School Case, pp. 610–059 (2013)
31. Sachdeva, T.: Operational efficiency and business organization: a case study on Mumbai dabbawala. J. Adv. Manage. Stud. **1**(3), 11–21 (2024)
32. Bagchi, D., Datta, A., Bagchi, M., Datta S.: Dabbawala: a hundred-year-old six sigma logistics concept. In: Datta, S., Das, S., Bagchi, D. (eds.) Becoming A Supply Chain Leader, pp. 413–424. CRC Press (2021)

Improving the Practice of Acquiring Knowledge in the Field of BPM – A Pedagogical Framework

Katarzyna Grzybowska(✉) and Patrycja Hoffa-Dąbrowska

Poznan University of Technology, Rychlewskiego 2, 60-965 Poznan, Poland
{katarzyna.grzybowska,patrycja.hoffa-dabrowska}@put.poznan.pl

Abstract. Business Process Management (BPM) is an integral part of every company's operations, regardless of the industry and the adopted business model. Supported by IT systems, it enables the design, implementation and enables their continuous improvement. BPM has been part of the teaching programs at technical universities for many years. The educational program offered to students is evolving towards a dynamic way of conducting classes. The rapid development of information technology contributes to the use of innovative teaching tools. The acquisition of knowledge in the field of BPM is based on existing competences and abilities. It is available through performed tasks and operations. It is also one of the basic goals of engineering education. The aim of the proposed article is to develop a pedagogical framework. Frameworks were developed on the basis of our own experience, taking into account the LSPPDM (Learn, See, Practice, Prove, Do, Maintain) framework for skills training used in the field of BPM. The results showed that the pedagogical framework and innovative teaching tools support educators in achieving educational goals. The proposed and tested teaching framework is effective in engaging students in independent and group learning, through active and mutual learning.

Keywords: business process management · process intelligence · pedagogical framework · Low-Code Platform · LSPPDM

1 Introduction

Business Process Management (BPM) is an integral part of every company's operations. BPM can be considered a way of thinking or a strategy implemented using various tools. Supported by IT systems, it enables the design, implementation of business processes and their continuous improvement.

BPM has been part of the curriculum at technical universities for many years. Various articles showing the aspects of using BPM elements in the education process at universities can be found in literature [1–4]. The educational program offered to students is evolving towards a more dynamic way of conducting classes. It definitely moves away from the traditional lecture, which is not an effective method from the point of view of educational outcomes. Just like exercises using piece of paper and pencil (without computer). Therefore, new teaching methods are being introduced. The rapid development of technology contributes to the use of innovative teaching tools.

M. ER et al. (Eds.): BPM 2025, LNBIP 565, pp. 390–400, 2026.
https://doi.org/10.1007/978-3-032-02936-2_27

Therefore, the curriculum for students is systematically improved and in its current form, at one of the technical universities, it uses Low-Code technology. It also takes into account the triangulation of active learning methods. This may be one of the promising new approaches supporting the acquisition of knowledge by students (non-programmers) in the field of BPM. The competence acquired by students is expert knowledge and, like any expert knowledge, it consists of declarative and procedural knowledge, with knowledge being dominant. This approach enables students to learn business engineering. The target group are engineering students at the Faculty of Engineering Management, majoring in Logistics. This is the first level of full-time studies (students receive the title of engineer after completing the first stage of studies). A graduate of Logistics at the Faculty of Engineering Management is a comprehensively prepared engineer, combining technological, economic and managerial knowledge. He can design, optimize and manage logistics processes, operates in the environment of carriers, production and distribution, and freely takes on analytical and managerial roles. However, he does not have programming knowledge.

The research question posed in the study were: how to effectively teach BPM to engineering-level non-programmers?

We believe that there is a direct and increasingly important relationship between the LSPPDM (Learn, See, Practice, Prove, Do, Maintain) framework and the use of Low-Code Platforms, especially in the context of process design and optimization in engineering education and process management. The advantages are evident in: (1) accelerating the implementation of the Learn – Practice – Maintain phase; (2) supporting the reflection and experimentation of the Learn – Practice phase; (3) accelerating the understanding of the mechanisms of the system, through testing in a digital environment.

A pedagogical framework for the training of procedural skills in engineering-level Business Process Management for non-programmers is proposed. The proposed pedagogical framework was developed based on own experience, taking into account the LSPPDM framework (Learn, See, Practice, Prove, Do, Maintain).

The remainder of the article is structured as follows. Section 2 provides an overview of the research that is related to BPM. A research gap has been identified in the effective teaching of procedural skills in BPM. Low-Code technology was indicated as an ideal educational environment for non-programmers. Section 3 presents the pedagogical framework for procedural skills training in BPM. The last section presents conclusions and further directions of work.

2 Business Process Management

Business Process Management is a discipline of operational management that uses various methods to identify, model, analyze, measure, improve and, in the future, also automate business processes. BPM aims to continuously analyze and improve an enterprise's processes in accordance with its strategic goals [5]. It has a holistic approach to managing these processes, what is also crucial in today's dynamic business environment. BPM requires flexibility, agility and self-organization of enterprises. BPM helps organizations gain comprehensive insight into their operations, identify bottlenecks and increase decision-making capabilities. The impact of Industry 4.0 is also not indifferent to the increase in the popularity of BPM. BPM is increasingly shaped by the transformative forces of Industry 4.0 [6–8]. The adoption of Low-Code Development (LCD) has become an integral part of organizational strategies as it plays a key role in increasing the efficiency and flexibility of organizations [9–12]. Hence, it is not surprising that BPM is closely related to information technology.

The use of information technology for BPM enables even more efficient process simplification, control, and automation. That's why enterprises are so eager to move towards using Low-Code Platforms.

The use of Low-Code Platforms is revolutionizing BPM [13]. These platforms are variously called: Low-Code/No-Code (LCNC), Low-Code Platforms (LCPs), Low-Code Application Platforms (LCAPs), or Low-Code Development Platforms (LCDPs). Currently, these all terms connected with low-code (as: LCP, LCAP, LCDP) are not clearly defined and are used in the same context – build applications with minimal or no programming skills, by using dedicated tools/platforms [13, 14]. When analyzing the literature, it can be notice, that terms related to low-code, especially LCP and LCDP, are used interchangeably [15, 16]. For the purposes of our work, we will use the term Low-Code Platform (LCP). We understand low-code platforms as IT tools that allow for the rapid creation of applications with no or limited participation of manual programming, thanks to the use of visual design environments and ready-made components. They facilitate the creation of solutions for both developers and business users.

There are also several other popular related terms such as: Citizen Developer (CD), Robotic Process Automation, IT-business alignment, which are used interchangeably to represent the concept of Low-Code/No-Code Low-Code application platform in recent times [17]. Another trend that LCPs are part of is the Citizen Development movement, which is related to the democratization of technology. The technology became available to non-technical workers and the so-called citizen developers and enables organizations to create applications and automate processes [18, 19].

There are many different Low-Code platforms on the market, like: among others Appian, Budibase, Nintex, WEBCON. Choosing the right platform depends on the area of application, as well as expectations for them. The main reason why organizations choose LCPs is the ability to create and further develop business applications. LCPs enable and simplify the development of business applications without extensive coding [20]. They enable individuals (without coding skills) to design, implement, and manage business applications with minimal technical support [21].

In our opinion, LCPs fit perfectly into the area related to the acquisition of procedural knowledge in the field of BPM and, in general, the acquisition of process intelligence by students of engineering studies in technical management. They enable us to have specific methods of "thinking" and acting. And thanks to the use of user experience (UX) design, team-based, shared thinking and actions make it possible to remove the negative impact of specialization from the teaching process. LCPs significantly facilitate the teaching process. In our opinion, this is due to, among other things, the advantages related to their functionality. These advantages include [22–24]: (1) graphical interface and intuitive application building (drag and drop); (2) the Agile methodology used; (3)

rapid prototyping; (4) use of MVP (Minimum Viable Process); (5) rapid testing and trial-and-error testing without large investments; (6) integration of data analysis and business analytics tools and the design of reports and analyses.

The functionality of LCPs allows students to focus on problem solving, designing business logic, analytical thinking such as User Experience/User Interface, and not on learning programming. Low-Code Platforms emphasize agency, encourage work and discussion, and stimulate students to think creatively. They also enable you to provide feedback on the process with the minimum required functionality. MVP is intended to be the simplest useful form of new technology. They allow you to experiment, make mistakes and improve. They can be an introduction to business analysis, defining indicators determining project success (KPI) and achieving goals. They can also be an introduction to the broad topic of change management.

Using LCP during classes, students develop their process intelligence in the field of BPM by: (1) understanding business processes; (2) acquiring analytical skills and creative thinking; (3) acquiring the ability to transform business requirements into functional solutions. We should also not forget about the didactic aspects related to social competences, such as the ability to work in a team, which (1) develops communication skills in interdisciplinary teams; (2) enforces the division of roles; (3) division of responsibilities; (4) team management. However, above all, the use of LCP in BPM teaching introduces non-programmers to the world of technology and algorithms. For students, it is an opportunity for greater flexibility in managing their careers. The use of LCP significantly changes the way companies are organized and managed. Communication between the world of business and IT becomes easier, which allows projects to be carried out faster and easier to implement. This is an opportunity for students to find a future job in the dynamically changing BPM environment.

3 Use of Technology and Triangulation of Active Learning Methods

To enable students to learn BPM, an engineering-level course was designed based on the triangulation of selected active learning methods as part of the subject Operational management in logistics. This subject is taught in the second year of engineering studies (Bologna system) in the field of Logistics, at the Faculty of Engineering Management. The Faculty educates and conducts research in the field of social sciences in the discipline of management and quality sciences. This is a "specific" non-technical faculty at a technical university (University of Technology).

Students in the indicated year have knowledge of management and organization of processes, including logistics processes, and know basic logistics issues. They can also identify the stages of material flow in the enterprise and supply chain.

Requirements for the learning process for the indicated subject have been identified. These are: (1) agency understood as a sense of control over one's own actions and their effects in the environment; (2) awareness of the need for changes and experimentation and quick adaptation to changes; (3) creative ability and reflective creative thinking; (4) constant interaction with educational participants; (5) intensive training of procedural skills using various teaching methods.

Sawyer et al. provided a framework for skills training. The framework consists of 6 stages, called Learn, See, Practice, Prove, Do, Maintain. Based on this framework, a pedagogical framework was proposed that takes into account the teaching methods used in the field of BPM (Fig. 1).

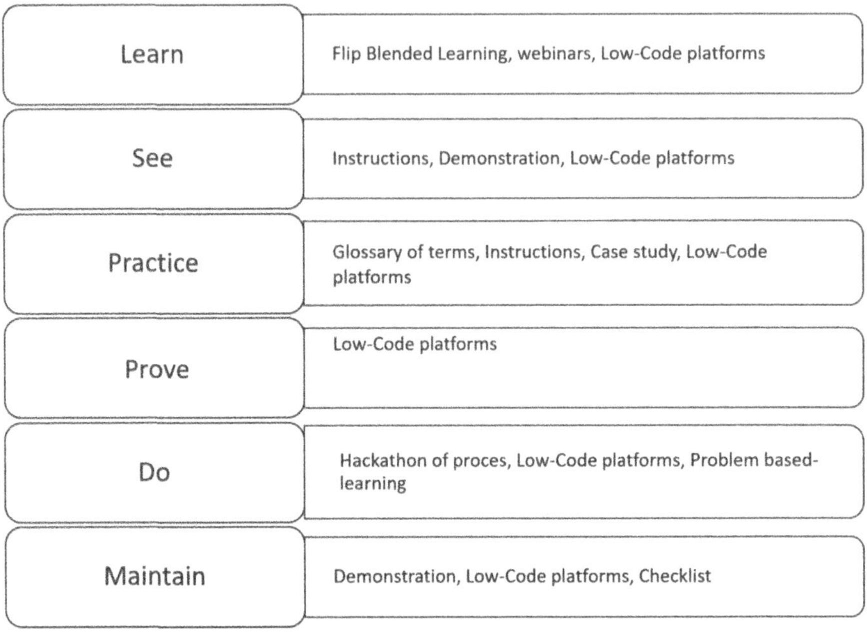

Fig. 1. A LSPPDM framework for BPM; own work.

The first stage of training (LEARN) in the student's acquisition of knowledge in the field of BPM and process intelligence is to acquire the required cognitive knowledge through the Flip Blended Learning method. The method combines elements of distance and traditional teaching. An e-learning platform has been prepared for the subject, where the educator posts various types of educational materials and resources for students' independent exploration (scientific articles, popular science articles, links to industry and technical blogs, materials in the form of webinars, practical case studies provided by LCPs tool manufacturers). The educator creates many paths for independent learning (reading, teaching sessions or multimedia presentations). Educational materials and independent work resources are available on the e-learning platform prepared during the COVID-19 pandemic. It has become a standard one of many paths of contact between educator and student. The main goal of LEARN is to build a knowledge base and understand the context of a problem, which allows you to begin solving it.

The next stage of training is demonstration (SEE). Demonstrations take the form of: instruction by an educator, a webinar conducted by an expert or a previously recorded instructional video. Observation allows students to directly follow the task procedure, which makes it easier to understand and remember the material and ask questions directly or using the previously described application. At this stage, student activity does not include any significant practical activities or involvement. The student focuses on analyzing the demonstrated example, predicting the solution to the problem or analyzing the progress of work in the process.

The next stage of training (PRACTICE) is to perform a purposeful, independent exercise (learning through experience; active learning) using LCP. Procedure simulation leverages many of the key features of deliberate practice, enabling focused practice. Engages students in decision-making, critical thinking and problem solving in a controlled but realistic environment. The material can be adapted to different student levels and a variety of topics. The educator supports the PRACTICE process by providing instructions, glossaries of terms and task guides.

In the (PROVE) stage, students are objectively assessed on an independent task to ensure that there is "process feeling", an increase in students' process intelligence and an increase in knowledge.

The next stage of training (DO) is the performance of a more complex problem task. The task is preceded by a panel discussion with expert practitioners who deal with BPM on a daily basis. During the practical workshop "Process hackathon", a task is performed in student teams of 4 people. These classes last on average 4–5 h. They are conducted continuously, but students can take an individual break at any time. They have to complete a problem-solving task: "Digitize a randomly selected business process or document flow using LCP." When completing the task, students also use a guide text developed by the educator (guide text method) and a collaborative approach. This method involves a systemic approach to the process of solving problem tasks. The students' task is to collect and organize information, through planning activities and their implementation, to checking the results and analyzing them. The guide text contains the content of the task in the form of questions that guide thinking about the task. The problem task is performed according to six phases. The educator proposes both the organizational framework for the task and the time frame. The problem task during the Process Hackathon is performed by students at LCP with the support of expert practitioners invited to the classes and additional educators who support the main educator. Students complete the task with support, assessments and feedback from educators and experts.

This is the last stage of training (MAINTAIN). At the last meeting with experts, students have the opportunity to present (as a group) the solutions. Demonstrating and explaining a task to others requires rethinking and organizing the acquired knowledge and skills. To ensure that the acquired knowledge and practical competence in the field of BPM is not lost and does not deteriorate over time, students have access to LCP so that they can practice the competences individually and, above all, regularly.

During the course, it was possible to analyze the role and importance of LCP in each phase of this LSPPDM framework (Table 1).

LCP enhances each stage of the framework by rapidly prototyping, testing, and implementing process solutions. The ability to safely experiment and quickly implement makes LCP particularly valuable in educational and organizational settings.

For students and the LSPPDM framework in combination with LCP: (1) lowers the entry threshold to work with digital processes; (2) increases self-confidence in working with real systems; (3) enables quick validation of competences (e.g. through a credit project in the form of an application).

Table 1. Analysis of the phases of the LSPPDM framework and the role of LCP; own study.

Phase	Importance in the framework	Role of Low-Code Platforms (LCP)
Learn	Acquiring theoretical knowledge, learning the context of a task or process	They enable interactive learning of systems – e.g. through educational applications, tutorials embedded in the platform
See	Observing a process or solution in action (job shadowing, demonstrations)	They allow you to visually present the operation of a process – e.g. through ready-made demo applications, workflows, dashboards
Practice	Supervised, hands-on exercises in a safe environment	They enable you to create your own applications/process models without risk – fast iterations, testing
Prove	Verification of skills – through testing, assessment or validation of solutions	Support the use of work products (e.g., constituting an application, process logic, data records) and integrate user feedback
Do	Independent implementation of a task/project in a real context	Enables implementation of the solution without advanced IT support
Maintain	Maintain, improve and adapt the solution over time	They allow you to easily modify, update and develop applications/process models without the need for coding

4 Validation of Conducted Classes

Practical workshops in the form of a process hackathon were validated using a survey. It included four questions: (1) What difficulties did you encounter while carrying out the task?; (2) Which part of the task was completed without any problems?; (3) Would you change anything as a result of the task?; (4) Are you satisfied with the results of your work?

The collected data were qualitative in nature and concerned: (1) the level of difficulty encountered by students – which elements of the task were problematic and which were not; (2) reflection on the learning process – what participants would like to change or improve in the future; (3) assessment of satisfaction with the effects of work – how they assessed their own experiences and results of the task; (4) perception of their own agency and engagement. The survey served to validate the practical application of the proposed LSPPDM framework in the didactic context, i.e. it was an empirical assessment of the effectiveness of the framework in the conditions of didactic classes.

The survey was conducted among both students and lecturers (supporting educators and experts). Its results were presented for two groups: (1) students and (2) instructors. This division results from different perspectives on the assessment of practical workshops and expectations towards the undertaken tasks. 94 students participating in the classes responded to the survey. The questions were open-ended, which allowed students to provide more complex answers than one. Their analysis and grouping enabled the interpretation of the answers (Table 2).

Table 2. Analysis of difficulties encountered during classes; own study.

1. What difficulties did you encounter while carrying out the task?	Answers numbers
1.1. Problem with designing the workflow design	25
1.2. Platform limitations, including lack of some features	12
1.3. Problems with project implementation at later stages of the project (outside of workflow design)	11

Students most often indicated difficulties during the implementation of the task in terms of defining the number of steps necessary in the process and determining the details of the project (Table 2). Problems occurred when properly planning the transformation of traditional paper document flow into digital circulation (25 responses). Students also had a problem with the inability to use the necessary functions due to the limitations of the available platform in the trial version (12 answers). They also indicated difficulties in further stages of the project: testing the prototype ("in our case, it was necessary to locate and correct our errors in the matrix") and analysis related to business analytics (including report management) (11 answers). One of the answers indicated that "the only problem was the program stuttering at times. He didn't save certain data we entered." One of the statements indicated that "fatigue after several hours of work and the noise in the room interfered with the task".

Table 3. Trouble-free tasks during classes; own study.

2. Which part of the task was completed without any problems?	Answers numbers
2.1. Form	20
2.2. Workflow design	18
2.3. Attribute matrix	17

Students' reflection on the implementation and effect of the completed task indicate that, despite the problems occurring, they are satisfied with their work. They had no difficulties at many stages of work (Table 3). It was pointed out that "this tool is very intuitive and allows you to quickly create and modify diagrams. Moreover, I liked that I could use ready-made templates and elements that made my work easier." They also indicate what they would change when working in the future (Table 4). "As a result of the task, I would change the communication process in the project team. I would organize the group better to avoid delays in work." Another statement indicates that next time I was working on a task, "I would make changes at the idea design stage; I would spend better and more time thinking about the elements of circulation.

Table 4. Suggested changes; own study.

3. Would you change anything as a result of the task?	Answers numbers
3.1. No changes	30
3.2. Making changes to the performed work	8
3.3. Adding improvement in the platform used	6

The assessment of students' agency and satisfaction with the effects of work is high. Here are some statements: "We feel fulfilled and inspired to continue working and gaining knowledge. Our work went very smoothly and was very satisfying. "Students note that team cooperation influences the results:" We had a lot of fun while working together, which additionally contributed to the effective implementation of the project. This experience will definitely have a positive impact on my approach to future group tasks." They also point out the advantages of a several-hour practical workshop:" Conducting the task continuously, i.e. during one day for a specified number of hours; I could focus 100% on completing the task. In my case, this is very important, because if such a task were carried out, for example, in an exercise unit after an hour and a half, it would be difficult for me to return to the thoughts I had during the previous classes.

5 Conclusions

The study contributes to filling the pedagogical gap in skills training. The results of the study may influence the experience of educators and contribute to improving the well-being of students. We believe that this study can serve as a guide for researchers and educators on good practice in teaching BPM skills at the engineering level to non-programmers. It has been observed that students: (1) are more independent and active in learning; (2) spend more time on tasks related to practical problem solving; (3) develop their process intelligence through better understanding of business processes, deepening analytical and creative skills thinking, transforming business requirements into functional solutions; (4) gain a much deeper understanding of concepts and problem-solving skills. The educator has the opportunity to observe the results of students' progress in solving problems and understand their educational needs.

The course is based on the LSPPDM (Learn, See, Practice, Prove, Do, Maintain) framework, which is widely used in professional training but has not been used in academic education in the BPM area so far. The innovativeness of the solution also results from the integration of this approach with Low-Code Platforms, which significantly lower the barrier of entry into the digital world of business processes and allow students to experiment, prototype and independently create applications without the need to know programming languages. An additional distinguishing element is the use of triangulation of active teaching methods - such as blended learning, process hackathons or teamwork - which makes the educational process more engaging and effective. The approach has been empirically verified in didactic conditions, and its effectiveness has been confirmed based on the opinions of students and educators participating in the classes.

Our teaching experience shows that the use of the adopted LSPPDM (Learn, See, Practice, Prove, Do, Maintain) framework in the field of BPM can be a highly effective method of teaching business process management.

Acknowledgments. We would like to thank WEBCON for providing a platform for teaching classes and support in the practical Process Hackathon workshops.

Disclosure of Interests. The authors have no competing interests to declare that are relevant to the content of this article.

References

1. Bandara, W., et al.: Business process management education in academia: Status, challenges, and recommendations. Commun. Assoc. Inform. Syst. **27**(1), 743–776 (2010)
2. Grzesiak, M.: Implementation of active learning in BPM education. Decision Making in Manufacturing and Services, pp. 117–125 (2024)
3. Sliż, P., et al.: Business Process Management education in Poland: a manifesto for academic teaching. e-mentor **2**(104) (2024)
4. Wiechetek, Ł., Mędrek, M., Banaś, J.: Business process management in higher education. The case of students of logistics. Problemy Zarządzania (4/2017 (71), 146–164 (2017)
5. Zani, C.M., Borges, M.M., Severo, A.J.B., Garcia, E.R., Müller, C.J.: Business process management (BPM) application through strategy, indicators and operations model (meio): a health service company case. Independ. J. Manage. Product. **12**(2), 590–612 (2021)
6. Szelągowski, M., Berniak-Woźny, J.: BPM challenges, limitations and future development directions–a systematic literature review. Bus. Process. Manag. J. **30**(2), 505–557 (2024)
7. Banholzer, V.M.: From 'industry 4.0' to 'society 5.0' and 'industry 5.0': value-and mission-oriented policies. Technol. Soc. Innov.–Aspects Syst. Transform. **3**(2) (2022)
8. Antwiadjei, L.: Security implications and risk management in low-code and RPA deployments. J. Sci. Technol. **3**(5), 1–11 (2022)
9. Akash, B., Saad, A.H., El-Saadawi, M.: Multi-criteria decision making analysis of optimal service delivery technique using AHP. TEM J. **12**(1), 208–213 (2023)
10. E-Fatima K., Khandan R., Hosseinian-Far A., Sarwar D., Ahmed H.F.: Adoption and influence of robotic process automation in beef supply chains. Logistics **6**(3), 48 (2022)
11. Farinha D., Pereira R., Almeida R.: A framework to support Robotic process automation. J. Inform. Technol. (2023)
12. Alamin, M.A.A., Uddin, G., Malakar, S., Afroz, S., Haider, T., Iqbal, A.: Developer discussion topics on the adoption and barriers of low code software development platforms. Empir. Softw. Eng. **28**(1), 4 (2023)
13. Frank, U., Maier, P., & Bock, A.: Low code platforms: promises, concepts and prospects. A comparative study of ten systems (No. 70). ICB-Research Report (2021)
14. Pinho, D., Aguiar, A., Amaral, V.: What about the usability in low-code platforms? A systematic literature review. J. Comput. Lang. **74**, 101185 (2023)
15. Sahay, A., Indamutsa, A., Di Ruscio, D., Pierantonio, A.: Supporting the understanding and comparison of low-code development platforms. In: 2020 46th Euromicro Conference on Software Engineering and Advanced Applications (SEAA), pp. 171–178. IEEE (2020)
16. Shi, Z., Dong, J., Gan, Y.: Democratizing digital transformation: a multisector study of low-code adoption patterns, limitations, and emerging paradigms. Appl. Sci. **15**(12), 6481 (2025)

17. Alfedaghi, H.S.: Citizen developers–kipic case study: Empowering innovation and accelerating digital transformation. In: SPE/IADC Middle East Drilling Technology Conference and Exhibition. SPE (2023)
18. Carroll, N., Maher, M.: How shell fueled digital transformation by establishing DIY software development. MIS Q. Exec. **22**(2), 99–127 (2023)
19. Funk, D.: Creating a low-code business process execution platform with python, BPMN, and DMN. IEEE Softw. **40**(1), 9–17 (2023)
20. Mihu, C.: The low-code movement: accelerating digital transformation with low-code. In: Pettinger, R., Gupta, B., Roja, A., Cozmiuc, D. (eds.) Handbook of Research on Digital Transformation Management and Tools, pp. 556–571. IGI Global, Chicago (2022)
21. Sufi, F.: Algorithms in low-code-no-code for research applications: a practical review. Algorithms **16**(2) (2023)
22. Curty, S., Härer, F., Fill, H.G.: Design of blockchain-based applications using model-driven engineering and low-code/no-code platforms: a structured literature review. Softw. Syst. Model. **22**(6), 1857–1895 (2023)
23. Domański, R., Wojciechowski, H., Lewandowicz, J., Hadaś, Ł.: Digitalization of management processes in small and medium-sized enterprises—an overview of low-code and no-code platforms. Appl. Sci. **13**(24), 13078 (2023)
24. Moogk, D.R.: Minimum viable product and the importance of experimentation in technology startups. Technol. Innov. Manage. Rev. **2**(3) (2012)

Building BPM Talent in Emerging Markets: A 10-Year Curriculum Development Journey

Tolusha Dahanayake Yapa(✉), S. W. S. B. Dassanayaka, H. A. D. R. Gunarathne, Prasanna Illankoon, Harishani Liyanage, I. R. Malawige, K. M. S. Senevirathne, C. M. Suwandaarachchi, Vathsala Wickramasinghe, and Amali Wijekoon

Department of Management of Technology, University of Moratuwa, Moratuwa 10400, Sri Lanka
sajeewanied@uom.lk

Abstract. Education in Business Process Management is essential for a country's economic growth and provides individuals with valuable knowledge and skills for a competitive career. BPM educators must ensure students learn the discipline with course materials that are engaging, relevant, and applicable to their futures. The outcome of any BPM curriculum is to produce a graduate who is equipped with the knowledge and skills required by the industry. However, developing a BPM curriculum that matches industry requirements is a challenge due to a lack of awareness about the industry requirements and the potential of BPM graduates in uplifting the organisations to excellence. We overcame this challenge with multiple university-industry collaborations to design and deliver this BPM education programme. Therefore, this paper details experiences spanning 10 years, starting from the initial design of the programme in 2016 and delivery in 2017, the first revision to the programme in 2021, and the second revision in 2025, highlighting the specifics of different university-industry collaborations. Further, the paper discusses the Department's experiences with the World Bank funded Accelerating Higher Education Expansion and Development project for the first curriculum revision, and the University of Moratuwa funded initiative for the second curriculum revision. To date, the programme has produced 150+ BPM graduates with a 100% employability rate.

Keywords: Business Process Management · BPM Education · Programme Design · Curriculum Design · Sri Lanka

1 Introduction

Business Process Management (BPM) has gained a lot of attention from both academia and industry, mainly due to its ability to improve organisational excellence through end-to-end business processes [1]. With the ever-evolving nature of businesses, BPM is in constant development from a mere set of process improvement methods to a holistic

All authors contributed equally.

M. ER et al. (Eds.): BPM 2025, LNBIP 565, pp. 401–416, 2026.
https://doi.org/10.1007/978-3-032-02936-2_28

management approach. Today, BPM has been developed as an academic discipline with diversified contributions from various knowledge fields such as operations management, information technology, and industrial engineering [2]. Along with these developments, BPM-focused education programmes[1] have progressed from being prescriptive-focused to descriptive-focused and predictive-focused [3]. Many programmes that started as teaching process improvement methods and tools have later developed with new academic areas such as information systems, digital transformation, process mining and enterprise BPM [4–6].

Despite the need for systematic BPM education, providing effective BPM education has been a challenge ever since [7, 8]. The rapid growth of the scope is a significant problem that education providers suffer from, making the curricula quickly outdated [9]. As a result, the education institutes face the difficulty of aligning the industry requirements with the BPM education to assure the employability of graduates [6]. Furthermore, the lack of shared resources among the BPM community [9, 10] has hindered designing and delivering a BPM education programme that aligns with industry requirements. For example, there are only a handful of real-world teaching cases available in the BPM education community [11]. Although books such as vom Brocke and Mendling [12] share experience on real-world initiatives, sharing the experience in designing and delivering a BPM curriculum[2] is minimal [8]. Therefore, the literature lacks insights into mechanisms that can be effectively used for the design and delivery of a BPM curriculum.

In this background, this paper presents the design and delivery of a four-year honours BPM education programme at the bachelor's level by the Department of Management of Technology of the University of Moratuwa in Sri Lanka (hereinafter referred to as the Department), one of the leading state universities of the country. The paper details experiences spanning 10 years (2016–2025), starting from the initial design of the curriculum in 2016 and delivery in 2017, the first revision to the curriculum in 2021, and the second revision in 2025. Under the government's initiatives to reform undergraduate programmes (refer [13]), the paper discusses the Department's experiences with the World Bank funded Accelerating Higher Education Expansion and Development (AHEAD) project for the first curriculum revision, and the University of Moratuwa funded initiative for the second curriculum revision.

By elaborating on this 10-year journey, this paper contributes to the BPM education literature in two ways. First, this paper highlights the mechanisms employed to overcome the challenges of designing and delivering an industry-fit BPM curriculum. The key mechanisms used by the Department are university-industry (U-I) collaborations such as obtaining academic software licenses, module delivery, experience sharing, project initiation and supervision, and skill enhancement. Therefore, this paper discusses the challenges faced when designing and delivering the initial curriculum and the two revisions and how those were overcome via U-I collaborations.

[1] BPM education programme refers to the broad, structured plan of learning, offered by the Department, which includes graduate attributes, programme objectives, curriculum, duration, and support systems.

[2] BPM curriculum refers to the specific content, subjects and learning materials used within an education programme.

Second, by introducing an industry-fit BPM education programme, this paper shares unique insights into a BPM curriculum that fits into a low BPM maturity context, informed by both industry and academic standards. As a developing country in South Asia, much has been written about Sri Lanka's economic performance [14] and developments in the state education system [13]. Yet, the level of adoption and application of BPM is still in the early stages in the country. As per the BPM maturity stages defined by Rosemann et al. [15], the BPM maturity of most of the organisations is still in the initial stage (BPM initiatives are uncoordinated and unorganised). Most of the organisations, despite focusing on organisational excellence, are not fully or at all transformed from the functional orientation to the process orientation. This could be mainly due to the lack of experts with process knowledge. Despite the need for process experts identified long ago [16], Sri Lanka still lacks process-based knowledge and expertise among the industry community. Therefore, the BPM education programme introduced by the Department fits Sri Lanka's low BPM maturity context. Like in many global educational programmes [8, 17], the initial design and delivery of the programme were focused on core BPM skills, and it was later refined and grown by integrating process excellence, BPM systems and business transformation through the two revisions, to be on par with the global trends.

2 BPM Education Programme Offered by the Department of Management of Technology

2.1 Designing the Curriculum - Initial Programme Development in 2016

As part of the strategic plan of the University of Moratuwa, the Department initiated the design and development of its four-year honours BPM education programme to be delivered in English medium, at the bachelor's level (hereinafter referred to as the BPM education programme) in 2016. This initial curriculum was developed in-house by the academics of the University of Moratuwa. Following the standards of the university, the BPM education programme has been structured across eight teaching semesters (each having 15 weeks), followed by an internship. The 2016 curriculum required students to obtain 130 GPA (Grade Point Average) credits and 20 NGPA (non-GPA) credits (a total of 150 credits). This became the first ever BPM education programme at the undergraduate level in South Asia. This initial programme produced 150+ BPM graduates, mainly aiming at the Sri Lankan context.

In designing this curriculum, the Department faced several challenges. One was designing a curriculum which matches the needs of the Sri Lankan context. The low level of BPM maturity in Sri Lankan businesses did not fully align with the global trends and scope of BPM. The concepts, such as process mining, process automation, and digital transformation, were very rarely used even in large-scale businesses at that time. Another challenge was embedding hands-on experience in addition to teaching theory. With the high costs involved in many such systems, introducing experiential learning was a difficult task. Therefore, designing a BPM curriculum which aligns with the global context was difficult in 2016.

In delivering the 2016 curriculum, being one of the three specialisations of the Bachelor of Business Science degree programme, the Department had a limited stake, especially in the first two years. The first four semesters of the curriculum covered fundamentals such as management, mathematics, statistics, and computer applications. The core BPM-related modules commenced from Semester 5, including business process modelling and analysis, enterprise resource planning (ERP) systems and business process reengineering. Seminar modules covered various themes like Entrepreneurial Thinking, Digital Business, the Internet of Things, and Design Thinking. However, this initial curriculum lacked a focus on BPM-related modules; instead had a high focus on other general and peripheral modules, which had less impact on BPM careers. Further, some of the modules were delivered as common modules, resulting in less applicability to the BPM context. For example, Change Management is a core module in the BPM education programme with reference to process implementation [18], but as a common module, the BPM essence was not delivered as expected. Another issue was the lack of knowledge, experience and awareness of the industry community on BPM. Most business organisations did not have clear job roles for BPM students, such as process analyst, lean executive, and process architect. However, many industry experts started recognising the potential of a BPM graduate in improving and sustaining their business performance after COVID-19 and the economic crisis in early 2020 and 2022, respectively. Thereby, the Department had an opportunity to popularize the programme in the industry to create BPM-related job roles.

2.2 First Curriculum Revision in 2021

By the year 2021, the context of Sri Lankan businesses had changed, especially due to the COVID-19 pandemic and the economic crisis, and their consequences in the workplace. Businesses were re-considering their strategies and were looking for avenues for business transformation. Yet, it was visible that they paid more attention to a selected set of processes instead of end-to-end processes. It was visible to us that our BPM graduates, on the one hand, should be capable of fulfilling the requirements of the industry, and on the other hand, should possess wider knowledge and skills to transform the Sri Lankan businesses. Therefore, the main purpose of the first curriculum revision was to address the above two concerns. For the first curriculum revision, the Department received funding from the World Bank's AHEAD project to obtain the services of a consultant to revise the BPM education programme. Accordingly, an industry expert attached to a leading process improvement consultancy firm, i.e., Dr Mikel J. Harry Six Sigma Management Institute Asia, was appointed to assist the Department in this revision.

The 2021 curriculum introduced the BPM specialisation from Semester 1 onwards by the Department without any common business modules. The revised curriculum offered 120 GPA credits and 10 NGPA credits (a total of 130 credits). With the consultant's involvement, we were able to bring several modules related to Business Process Excellence (BPE) methodologies, which are highly relevant to the Sri Lankan context, especially in the manufacturing sector. Further, several modules related to the technology domain were introduced to strengthen the technological aspect of the curriculum. The 2021 curriculum offered a specific focus on ERP Systems, providing several elective modules with an option for an ERP Minor degree. Another key feature of the curriculum was embedding industry exposure early in the curriculum via two seminar modules. In these two seminar modules, students visit various manufacturing and service organisations to observe and learn BPM and BPE-related practices. The resultant curriculum after the first revision in 2021 was effective in delivering the BPM-related modules compared to the initial curriculum developed in 2016.

The main limitation of the 2021 curriculum was the misalignment in the flow of modules. Some examples include delivering ERP and Process Mining modules within the same semester, and delivering Lean Management and Value Stream Mapping modules within the same semester. Another limitation was that the foundational subject embedment was not adequate. For example, teaching statistics content of the Lean Six Sigma module was a challenge. In a similar vein, students found difficulties in grasping BPM software application content without prior knowledge of database management.

2.3 Second Curriculum Revision in 2025

The second curriculum revision is mainly aimed at ensuring the flow of modules and embedding foundational subjects effectively into the curriculum. The Department managed to get funding from the University of Moratuwa to engage a BPM academic expert from the Queensland University of Technology (QUT) as an external reviewer (refer to [19]). The intention of this appointment was to revise the existing curriculum based on the feedback provided by the external reviewer.

The 2025 curriculum required124 GPA credits and 6 NGPA credits (a total of 130 credits). The main reason for restructuring the number of credits was to align the degree programme with the Sri Lanka Qualifications Framework (SLQF) [20]. The changes mainly revolved around the subject structure. Hence, the research and development project, with eight GPA credits, and the internship, with six NGPA credits, were kept in the same semesters as the other 2 curricula.

The 2025 curriculum was developed based on the BPM lifecycle introduced by Dumas et al. [18] and the Association of Business Process Management Professionals (ABPMP) International CBOK [21]. The 2025 curriculum introduced several foundational modules in the first two semesters. The subject structure is given in Table 1, which shows how modules are clustered as supporting and core modules under the above-mentioned pillars.

Figure 1 shows the conceptual framework of the curriculum development of the 2025 revision. As explained in Sect. 2.3, the foundation of the 2025 revision is the BPM lifecycle by Dumas et al. [18] and ABPMP CBOK [21]. The core academic disciplines included in the curriculum to produce a graduate with the BPM skill set to execute the BPM lifecycle are Process Thinking, Management, Information Systems, Data Science and Customer Orientation. While learning the theoretical aspects of BPM, students get experiential learning through real-world problem solving through industry visits, short-term volunteer BPM projects from industry partners, long-term research-oriented projects which span across Semesters 7 and 8, and a minimum 16-week internship at a business organisation. In addition, many modules in the curriculum are designed to engage the students in a real-world business scenario as a full or part of their continuous assessment component. For example, in the *Business Process Simulation* module,

Table 1. Subject structure mapped to core BPM concepts by Dumas et al. [18] and the Association of Business Process Management Professionals [21]

Semester	Focused Area	Core modules	Supporting modules
1	BPM lifecycle stage 1 - process identification, and BPM CBOK Knowledge Area 1 *Introducing Business Process Management*	• Business Process Management (BPM) Foundations	• Foundations of Management • Business Statistics and Software Applications – I • Mathematics for BPM • Principles of Programming
2	BPM lifecycle stage 2 - process discovery, and BPM CBOK Knowledge Area 2 *Introducing Business Process Modelling*	• Business Process Modelling • Business Process Management in Practice	• Managing People and Markets • Business Statistics and Software Applications – II • Database Management
3	BPM lifecycle stage 3 - process analysis, and BPM CBOK Knowledge Area 3 *Introducing Process Analysis*	• Enterprise Lean Six Sigma - I • Business Process Architecture and Maturity Models • Process Analysis and Diagnosis • Work Study and Ergonomics	• Operations Management • Business Analysis
4	BPM lifecycle stage 4 - process redesign, and BPM CBOK Knowledge Area 4 *Introducing Process Design*	• Workflow Management Systems • Enterprise Lean Six Sigma - II • Value Stream Mapping • Process Improvement Project	• Supply Chain Management • Customer Experience Management
5	BPM lifecycle stages 5/6 - process implementation and process monitoring/controlling, and BPM CBOK Knowledge Areas 5 and 9 *Introducing Process Performance Management BPM Technologies*	• BPM Change Management • Project Management • Industry 5.0 and The Future of BPM • ERP Systems and Applications [ERP Minor]	• Quality Management • Operations Research

(*continued*)

Table 1. *(continued)*

Semester	Focused Area	Core modules	Supporting modules
6	BPM CBOK Knowledge Area 6 and 9 *Introducing Process Transformation* *BPM Technologies*	• Business Process Simulation • Intelligent Process Automation • Process Mining • ERP Best Business Practice Models [E] [ERP Minor] • ISO Management Systems and Awards [E]	• Business Research Methods
7	BPM CBOK Knowledge Area 7 and 9 *Introducing Process Organization* *BPM Technologies*	• Business Process Outsourcing and Shared Services [E] • ERP Solutioning [E] [ERP Minor] • Assuring Service Excellence in Organisations [E]	• BPM Research Project • Business Strategy and Managing Performance • Design Thinking [E]
8	BPM CBOK Knowledge Area 8 and 9 *Introducing Enterprise Process Management* *BPM Technologies*	• Enterprise BPM • The ERP Implementation Process [E] [ERP Minor] • Managing Processes in Software Companies [E]	• BPM Research Project • Technology and Innovation Management

students are asked to identify a suitable business process for improvement in a real business and use simulation as a quantitative process analysis technique to identify potential improvement opportunities.

2.4 University of Moratuwa Graduate Attributes and the BPM Education Programme

In all revisions, the Department focused on creating a graduate with the attributes defined by SLQF for a bachelor's degree programme [20, pp. 19] as well as the University of Moratuwa attributes [22]. By benchmarking the global standards via an external reviewer and industry consultant, the degree programme enables the students to become exquisite scholars with the foundation for lifelong learning. We centered each curriculum on core BPM theories and concepts, so students are well-educated about their field of study. Furthermore, by embedding experiential learning in numerous ways (continuous assessments, lab and tutorial sessions, field visits, class discussions, and group activities), the

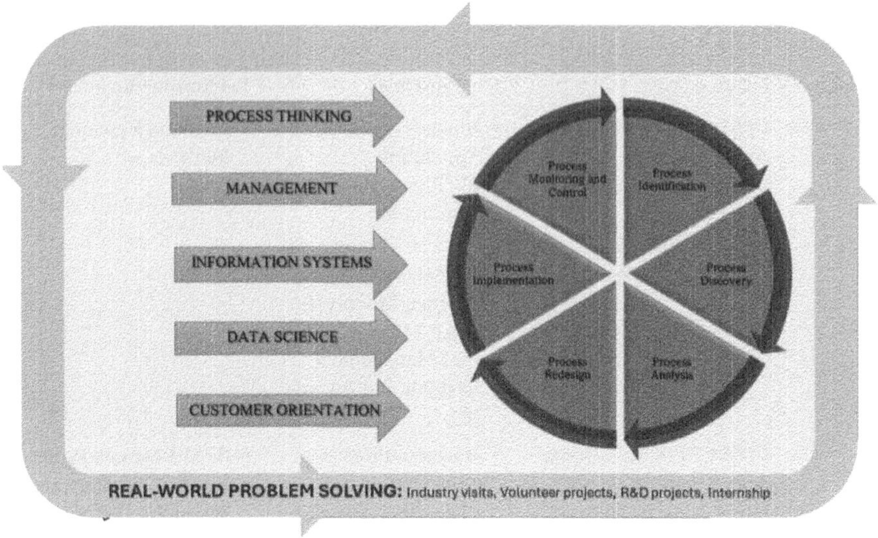

Fig. 1. The conceptual framework of the curriculum development in the 2025 revision. Source: Developed by the authors

students are well-informed about their capabilities as a BPM graduate. Continuous exposure to the industry through various ways, such as guest lectures, seminars, workshops, short-term and long-term projects, and internship, the students to become industry-ready not only with technical skills but also with soft skills. Furthermore, they are expected to learn to challenge themselves, be accountable and responsible for their own decisions, and become up-to-date practitioners, researchers and/or educators. Therefore, ultimately, they are expected to be highly employable in the industry sectors relevant to them. As of 2025, three batches have graduated from our programme, with the employability rates increasing gradually: 90%, 95.5% and 100% in 2022, 2023 and 2024 graduation ceremonies, respectively.

2.5 Key Changes of the Three Curricula

Modules offered within the BPM specialisation have undergone a significant evolution across the 2016, 2021, and 2025 curricula. While strengthening the core focus on BPM, the content reflects changing industry demands and technological advancements. Several key changes can be noted as follows:

- Introducing core BPM Concepts from the Start of the Programme:

This change shows a clear trend of integrating core BPM concepts much earlier in the degree program, moving away from a purely generic business foundation in the initial semesters (see the core modules in Semesters 1, 2 and 3 in Table 1).

- Increased Emphasis on BPE Methodologies:

The strong focus on BPE is continued in the 2025 curriculum with some additional modules, such as Quality Management and ISO Management Systems and Awards. Therefore, there was a significant deepening and broadening of coverage for process improvement methodologies, moving from general operations topics to specific and widely used frameworks like Lean, Six Sigma, Total Productive Maintenance, and Value Stream Mapping (see the core modules in Semesters 3 and 4 in Table 1).

- Expanding the Focus on ERP Systems:

The coverage of ERP systems progresses from a basic introduction and analytics in 2016 to offering a specialised minor option with various related electives in 2021 onwards, and finally integrating several core ERP modules as standard specialisation requirements in 2025. This indicates a growing recognition of ERP systems as integral tools in BPM (see the core modules in Semesters 5, 6, 7 and 8 in Table 1).

- Inclusion of Modern Technologies and Automation:

Though the 2016 curriculum included modules on data and IT, those modules did not deliver the application of knowledge in the BPM context. The 2021 curriculum introduced BPM-related technological modules alongside Data Science for BPM. The 2025 curriculum strengthened this pillar by incorporating advancements in the fields. Modules on cutting-edge technologies highly relevant to modern BPM were introduced to the curriculum, reflecting the digital transformation trend (see core modules in Semesters 7 and 8 in Table 1).

- Consistency and Evolution of Core Business and Research Modules:

Modules covering fundamental business functions like Operations Management, Supply Chain Management, Human Resource Management, Marketing, and Project Management are present across all three versions, although their specific names or placement have changed. Research-related modules, including Business Research Methods and Research Project, remain the same, underlining the importance of analytical and research skills. Therefore, core business areas and research components remain foundational while maintaining their prominence in the curriculum.

- Evolution of Seminar Modules:

The 2016 curriculum included seven seminars across the semesters, covering a wide range of topics from general business exploration to specific IT trends. The 2021 curriculum narrowed the seminar focus to two modules with the intention of bringing industry exposure to the degree. The 2025 curriculum keeps the first seminar module named Business Process Management in Practice and introduces a new seminar module named Process Improvement Project, enabling the students to apply their learnings from the first four semesters. The seminar modules have shifted to a more concentrated focus on practical BPM and BPE applications in 2021, with a specific focus on a small-scale process improvement project in 2025.

Overall, the evolution of the modules in the BPM curriculum from 2016 to 2025 demonstrates a shift towards a more specialised and contemporary focus from the early stages of the degree programme. There is a clear increase in emphasis on BPE methodologies, a significant expansion in the coverage of ERP systems, and the integration of modules related to modern process technologies like Process Mining and Automation. Besides, the presence of sustainability concepts was enhanced along the three curricula. For example, the module Industrial Revolution and Future of BPM was renamed as Industry 5.0 and The Future of BPM, to emphasize human-centricity, sustainability and resilience.

3 U-I Collaborations for Curriculum Design and Delivery

The higher education authorities in the country encourage U-I collaborations [23]. Hence, the Department used various types of U-I collaborations proposed by Meerman et al. [24] as discussed below, to achieve the graduate attributes defined in SLQF [20] and develop an industry-fit BPM curriculum.

- Feedback sessions on content development, graduate performance and skill enhancement (Education-related and Management-related collaborations).

The Department conducts an annual Department-Industry Collaborative Board (DICB) to get feedback from the industry on the content of modules, especially technical and soft skills development. The constructive feedback we receive from industry partners has helped us to fine-tune the curriculum by focusing on the industry requirements. Further, they have commented on the performance of graduates and the skills lacking in them. Accordingly, skill enhancement activities have been incorporated into modules, and industry partners have voluntarily conducted skill development workshops.

- Collaborations to obtain academic software licenses (Management-related collaborations).

These collaborations aim to obtain academic licenses for well-recognised software applications (e.g., IFS, UiPath, Pega, Zoho, Workhub24, Blue Lotus 360) to develop and enhance the technological skills of the students by giving them hands-on experience. Students have secured better career opportunities with such exposure. Furthermore, with the successful completion of the software applications-embedded modules, students can obtain certifications from those software application providers.

- Collaborations for module delivery (Education-related collaborations).

Some of the modules in the curriculum are highly practical, and their content gets updated frequently. Thus, we collaborated with industry partners to deliver highly practical modules (both delivery of the content and assessments). A few such examples are the modules Customer Experience Management, ERP Systems and Applications, and Intelligent Process Automation. In addition, these industry partners provide training to the academic and academic support staff of the Department with updates related to the modules. With this collaboration, the main challenges were lecture scheduling issues and low student engagement with online sessions. As a remedy, the Department is training internal staff to deliver these modules in the long run.

- Experience sharing sessions (Education-related collaborations).

The Department collaborated with industry partners to share their experiences via guest talks, enabling students to learn about applications of the modules they learn, industry expectations, and skills, to ultimately become industry-fit graduates. In addition to these academic-related sessions, industry partners conduct soft-skill development seminars and workshops. Furthermore, leading manufacturing and service organisations in the country have facilitated field visits for students to obtain industry exposure at the early stages of the degree programme.

- Collaborations with professional associations (Education-related collaborations).

The Department constantly works together with professional associations for knowledge sharing and networking purposes. For example, the International Institute of Business Analysis (IIBA) Colombo Chapter works closely with the Department to conduct a series of lectures to share the practical aspects of Business Analysis.

- Project initiation and supervision (Research-related collaborations).

To enhance student engagement with industry-related activities throughout the degree programme, our industry partners allow students to get involved in their operations. Short-term projects, identified as volunteer projects, allow students to engage with the industry for 1–2 months, where they gain basic BPM skills such as process visualisation and waste identification. Long-term projects (i.e., Research Projects) in semesters 7 and 8, providing an opportunity for students to identify an industry problem and propose practical solutions. In these projects, senior management of respective industry partners is involved as industry supervisors in guiding and providing direction. In this collaboration, the main challenges in long-term projects were protraction with finalizing the research problem and objectives with the industry supervisor and a lack of theoretical rigour compared to fundamental research. As a solution, the Department increased the percentage of fundamental research undertaken as research projects.

4 Reflections on Overcoming Challenges and Developing a Highly Demanded BPM Education Programme in Sri Lanka

Designing and delivering a BPM curriculum has been a challenge, not only in Sri Lanka [16] but also in the global context [8]. Like many other countries, Sri Lanka has a mixed economy with the manufacturing sector contributing around 25% to the national GDP and the service sector contributing over 60% to the GDP [25]. Therefore, we aimed to design and deliver a BPM education programme that suits this setting.

4.1 BPM Adoption in Sri Lankan Businesses and Its Impact on Curriculum Development

BPM adoption in Sri Lanka has been increasing, especially in industries aiming to improve efficiency, reduce costs, and ensure compliance with international standards. The present Sri Lankan government has identified significant opportunities to leverage BPM efforts in the public sector to improve citizen services and administrative efficiency under digital transformation programs [26]. With the recent economic and political transformations in Sri Lanka, many businesses recognise the need for business process improvement and transformation, triggering a growing interest in BPM initiatives, mainly among SMEs [14].

In this context, the curriculum revision in 2021 aimed at bridging the gap of lacking process-based knowledge and expertise among the industry community by producing BPM graduates with industry-required skills who can support transforming businesses to excellence. The growing demand for our graduates for employment and for undergraduates to engage in industry projects has proven the success of this aim.

With the latest curriculum revision, we focused on further enhancing the programme by aligning the modules to core BPM concepts, enriching the content with up-to-date BPM topics, and introducing the foundational knowledge on technology and mathematics domains, thereby targeting the development of a highly employable graduate. Table 2 summarises the evolution of the BPM curriculum over revision, comparing the features, strengths and weaknesses.

4.2 Methods Used to Overcome the Challenges in Developing the BPM Curriculum

With a minimal understanding and application of BPM in Sri Lankan industries, we focused on designing a curriculum which creates a BPM graduate with solid knowledge and experience in core BPM skills since 2016, so they can become valuable assets for the Sri Lankan industries to boost their organisational performance. Therefore, our curriculum gives significant attention to core BPM skills such as process modelling and analysis, ERP applications and process improvement. Nevertheless, we offer the foundational theoretical aspects of the trending BPM concepts through stand-alone modules.

The need to deliver industry-fit knowledge and skills led us to obtain the services of an industry expert as a consultant in the first curriculum revision. The consultant brought significant industry insights into the curriculum by suggesting specific modules, as well as assisting us with aligning the content of these modules to be in line with the industry-fit skills and knowledge.

As an initial response to the need for computer applications, the Department chose free-of-charge options despite their restricted functionality. However, with improved U-I collaborations the Department managed to obtain both locally and globally recognised applications. Several Memoranda of Understanding (MoUs) were signed with industry partners who provide licensed software with datasets for teaching.

While growing as a BPM-specialised department, the academics constantly work on their CPD (continuous professional development) to upgrade their BPM skills and knowledge. Involved in international research projects, attending workshops conducted by the global BPM community, hands-on training on software applications, and earning professional certifications help them to develop knowledge and skills in the field.

To spread the word about the availability of BPM graduates in the labour market among the industry community, the Department uses multiple formal and informal channels. Using personal connections and through the BPM alumni, the Department managed to increase the awareness of BPM among the industry community. With this expanded network, the Department is currently in a position where industry professionals reach out and ask for collaborations, recruitment and involvement in industry projects.

Table 2. Strengths and limitations of the three curricula.

	Features	Strengths	Limitations
Initial programme	• 130 GPA credits + 20 NGPA credits • Common modules for all 3 specialisations during the first two years • Specialising students from year 3 (Semester 5) onwards	• First BPM undergraduate education programme in Sri Lanka and South Asia	• No BPM-specific modules in the first two years of the degree programme • More theory-oriented than practical-oriented content in modules • Lack of BPM focus in the commonly offered modules • Less connectivity to the Sri Lankan context
Revision 1	• 120 GPA credits + 10 NGPA credits • Specialising the students from Semester 1 onwards	• More focus on process excellence methodologies relevant to the Sri Lankan context • Addition of modules from the technology domain • Offering an ERP (Enterprise Resource Planning) Minor to incorporate current industry needs • Introduction of seminar modules to give early exposure to industry best practices • Inclusion of practice-based learning outcomes	• Misalignment in the flow of modules • Inadequate foundational knowledge of technology and mathematics
Revision 2	• 124 GPA credits + 6 NGPA credits • Specialising the students from Semester 1 onwards • Alignment with core BPM theoretical concepts from [18] and [21]	• Introduction of foundational modules from the technology and mathematics domains • Alignment with the best-performing BPM education programmes worldwide	

In summary, through the experiences and lessons we learnt in delivering the degree programme for 10 years with two revisions, we have managed to upgrade it to a highly demanded undergraduate degree programme in Sri Lanka.

5 Conclusion

In designing and delivering this BPM education programme, first, in 2016, we focused only on the Sri Lankan context. Second, in 2021, we wanted to contribute towards the business transformation and support Sri Lankan companies to overcome the economic hardships they faced by producing graduates with industry-required skills. Finally, in 2025, we aimed to make it on par with global standards, so it may have the potential to deliver the programme to the South Asian region. The curriculum development journey at the Department shows that a successful curriculum can be designed and delivered over time with a dedicated and innovative multidisciplinary academic team. It begins with an idea of what the course should deliver, then an arc for the course is formed through a series of modules with a gradual deepening of the content.

However, designing a BPM curriculum was a complicated process due to the difficulty of aligning with industry requirements and addressing the low BPM maturity among industry sectors. We overcame those challenges mainly via various U-I collaborations. These collaborations were strengthened with global academic collaborations, securing funds for curriculum development, CPD of the academic team, and BPM education research.

Looking forward, we aim to leverage the technology component of the curriculum, such as the incorporation of AI-driven process tools and introduce micro-credentials (e.g. short courses) into our BPM programme to ensure future relevance. Further, we aim to introduce a Master's programme related to the BPM discipline to spread the BPM education in the Sri Lankan context. The Department has already initiated developing the curricula for these programs with the feedback from stakeholders.

References

1. Reijers, H.A.: Business process management: the evolution of a discipline. Comput. Indust. **126**, 103404 (2021). https://doi.org/10.1016/j.compind.2021.103404
2. vom Brocke, J., Rosemann, M.: Business process management. In: vom Brocke, J. (ed.) Association for Information Systems. Reference Syllabi. Eduglopedia.org (2015)
3. Grisold, T., Wurm, B., vom Brocke, J., Kremser, W., Mendling, J., Recker, J.: Managing process dynamics in a digital world: integrating business process management and routine dynamics in IS curricula. Commun. Assoc. Inform. Syst. **51**, 637–656 (2022). https://doi.org/10.17705/1CAIS.05127
4. Bandara, W., Velmurugan, M., Leemans, S.: Building 'holistic' Business process modelling skills for IS graduates. In: Blending the Local and the Global. ICIS 2020 Proceedings, vol. 9 (2020). https://aisel.aisnet.org/icis2020/digital_learning_env/digital_learning_env/9
5. Pridmore, J., Godin, J.: Business process management and digital transformation in higher education. Issues Inform. Syst. **22**(4), 168–177 (2021). https://doi.org/10.48009/4_iis_2021_180-190
6. Senkus, P., et al.: Bridging the gap: an evaluation of business process management education and industry expectations – the case of Poland. In: Köpke, J., et al. (eds.) Business Process Management: Blockchain, Robotic Process Automation and Educators Forum. BPM 2023. LNBIP, vol. 491. Springer, Cham (2023). https://doi.org/10.1007/978-3-031-43433-4_18
7. Moormann, J., Bandara, W.: Editorial: learning, teaching and disseminating knowledge in business process management. Knowl. Manage. E-Learn. **4**(4), 390–394 (2012). https://doi.org/10.34105/j.kmel.2012.04.030

8. Syed, R., Wynn, M.T., & Bandara, W.: The design and delivery of a holistic BPM education programme. In: Köpke, J., et al. (eds.) Business Process Management: Blockchain, Robotic Process Automation and Educators Forum. BPM 2023. LNBIP, vol. 491. Springer (2023). https://doi.org/10.1007/978-3-031-43433-4_13
9. vom Brocke, J., & Badakhshan, P.: Where to study Business Process Management – an update based on the EDUglopedia.org platform. Class Notes (2018). https://bptrends.info/wp-content/uploads/10-02-2018-COL-vom-Brocke-Badakhsshan-Class-Notes-Where-to-Study-BPM.pdf
10. Koch, J., Koch, J., Sträßner, M., Coners, A.: Theory and practice - what, with what and how is business process management taught at german universities?. In: Di Ciccio, C., Dijkman, R., del Río Ortega, A., Rinderle-Ma, S. (eds.) Business Process Management. BPM 2022. LNCS, vol. 13420. Springer (2022). https://doi.org/10.1007/978-3-031-16103-2_4
11. Hettiarachchi, P. D., Bandara, W., Yapa, T.S.D., Chukwudi, I., Wijekoon, A.: Developing business process improvement skills using a case study of a hospital outpatient department. ICIS 2023 Proceedings, vol. 18 (2023). https://aisel.aisnet.org/icis2023/learnandiscurricula/learnandiscurricula/18
12. vom Brocke, J., Mendling, J. (eds.): Business process management cases. Digital Innovation and Business Transformation in Practice. Springer (2018)
13. Wickramasinghe, V.: Higher education in state universities in Sri Lanka – review of higher education since colonial past through international funding for development. Int. J. Educ. Manag. **32**(3), 463–478 (2018). https://doi.org/10.1108/IJEM-01-2017-0028
14. Chandrasiri, S., Gunatilaka, R.: Impact of multiple crises on Sri Lanka's micro, small and medium-sized enterprises. Colombo: International Labour Office (2023). https://www.ilo.org/sites/default/files/wcmsp5/groups/public/@asia/@ro-bangkok/@ilo-colombo/documents/publication/wcms_901205.pdf
15. Rosemann, M., de Bruin, T., Hueffner, T.: A model for business process management maturity. ACIS 2004 Proceedings, vol. 6 (2004). https://aisel.aisnet.org/acis2004/6
16. Bandara, W., Syed, R., Kapurubandra, M., Rupasinghe, L.: Building essential BPM capabilities to assist successful ict deployment in the developing context: observations and recommendations from Sri Lanka. GlobDev 2012, vol. 14 (2012). http://aisel.aisnet.org/globdev2012/14
17. Bandara, W., et al.: Business process management education in academia: status, challenges, and recommendations. Commun. Assoc. Inform. Syst. **27**(1), 743–776 (2010). http://aisel.aisnet.org/cais/vol27/iss1/4
18. Dumas, M., Rosa, M.L., Mendling, J., Reijers, H.A.: Fundamentals of business process management. Springer (2018)
19. Queensland University of Technology: QUT science process science. https://research.qut.edu.au/bpm/. Accessed 21 May 2025
20. Sri Lanka Qualifications Framework.: SRI LANKA QUALIFICATIONS FRAMEWORK (2015). Ministry of Higher Education. 1156_SLQF_2016_en.pdf
21. Association of Business Process Management Professionals.: ABPMP CBOK. Association of Business Process Management Professionals International (2020). https://www.abpmp.org/page/guide_BPM_CBOK
22. Faculty of Business.: Student Handbook – Intake 2023 (2023). FOB Student Handbook - Intake 2023 FINAL_0.pdf. https://uom.lk/sites/default/files/business/files/FOB%20Student%20Handbook%20-%20Intake%202023%20FINAL_0.pdf
23. Wickramasinghe, V.: Research and innovation: the application of social network analysis in university-industry collaborations. In: De Silva, P., Hewamanage, W., Fernando, A.N., Ananda, L. (eds.), Humanities and Social Sciences Education in Sri Lankan Universities: Past, Present, and Future, vol. III, pp. 226–239. University Grants Commission, Colombo (2022)

24. Meerman, A., Galan Muros, V., Davey, T., Orazbayeva, B., Baaken, T.: The state of university-business cooperation in Europe – final report (2018). https://data.europa.eu/doi/10.2766/676478
25. Central Bank of Sri Lanka.: National Accounts Estimates – 2023 (2023). https://www.cbsl.gov.lk/sites/default/files/cbslweb_documents/national_account_estimates_2023_annual_e.pdf
26. Ministry of Development Strategies and International Trade: national export strategy of Sri Lanka information technology strategy 2018–2022. Colombo (2022). https://www.srilankabusiness.com/pdf/nes/sri-lanka-it-bpm-5-web.pdf

Teaching Case: Process Mining for Predicting Failure to Rescue

Reyes Grangel[1,4(✉)], Noelia Palomares[1], María Sánchez-Galán[2], David Luna-Aleixos[3,4], and Irene Llagostera-Reverter[2,4]

[1] Knowledge Engineering Group, Universitat Jaume I, Castellón de la Plana, Spain
{grangel,al063997}@uji.es

[2] Nursing Research Group (GIENF), Universitat Jaume I, Castellón de la Plana, Spain
{al452557,llagoste}@uji.es

[3] eNURSYS Group, FISABIO, Valencia, Spain
luna_davale@gva.es

[4] NURSIA Joint Research Unit, FISABIO - Universitat Jaume I, Valencia, Spain

Abstract. A Failure To Rescue can be defined as a clinical deterioration and/or functional failure of a patient, which could have been prevented by early detection and appropriate intervention. It represents a major challenge in clinical practice, often arising from complex care processes where multifactorial variables are not fully integrated. To address this issue, this Teaching Case explores how Process Mining and Predictive Process Monitoring techniques can be applied to hospital data to identify critical pathways, predict adverse outcomes, and ultimately reduce the FTR rate.

The article presents a comprehensive Teaching Case that includes both the case study itself, along with the questions and tasks posed to students, and a complementary link to the Teaching Notes, which serve as a support resource for instructors in guiding the implementation of the case in educational contexts.

The Teaching Case is based on real, pseudoanonymized clinical data from a medium-sized hospital and is aimed at postgraduate students in disciplines such as Artificial Intelligence, Business Process Management, or Medical Informatics. It enables learners to develop core BPM competencies in Process Discovery, Conformance Checking, Predictive Process Monitoring, and process redesign within a critical healthcare context.

Keywords: BPMN · Data Quality · Process Mining · Predictive Process Monitoring · Failure To Rescue · Teaching Case

1 Introduction

Failure To Rescue (FTR) refers to the clinical deterioration and/or functional failure of a hospitalized patient that could have been avoided through early

M. ER et al. (Eds.): BPM 2025, LNBIP 565, pp. 417–426, 2026.
https://doi.org/10.1007/978-3-032-02936-2_29

recognition and timely intervention [14]. This phenomenon represents a significant challenge in clinical practice, often leading to severe complications, irreversible damage, or even death. This scenario is described as FTR, and it is generally known that an FTR could be prevented if detected at an earlier stage. The challenge is that FTR is still a very broad concept, and there is no clear way to detect it. FTR is a multifactorial issue, influenced by various interrelated variables and characterized by the absence of a consolidated definition or detection framework.

1.1 Motivation

This Case Study emerges from an interdisciplinary collaboration between nursing and computer science researchers aiming to explore how Process Mining can support the detection and prevention of FTR events. The goal is to identify when and where FTR may occur by analyzing real-world hospital data, modeling the care process, and applying predictive techniques to anticipate clinical deterioration.

FTR is considered a complex and partially unexplained phenomenon due to several reasons:

- The definition of FTR is not consolidated in the literature yet.
- Depending on the definition, the nature of the variables to determine whether there is an FTR or not may vary.
- There are many logical pathways since a patient is admitted to the hospital, and multiple data inputs exist along this process.
- The data availability and consistency may cause a negative impact in the FTR diagnosis.

To address these challenges, the Case Study focuses on the following lines of analysis:

1. **BPMN for Clinical Process Modeling:** Modeling patient care pathways from admission to outcome to identify where FTR may occur.
2. **Data Quality:** Cleaning and preparing clinical datasets for analysis through Process Mining techniques.
3. **Process Mining in Healthcare:** Analyzing historical event data to uncover patterns and deviations associated with FTR in a retrospective approach.
4. **Predictive Process Monitoring:** Training models to help identify an FTR diagnosis in a proactive approach

1.2 Background

In the era of digital health, the data is a key asset to any industry if you know how to get value from it. "Event data is the most important source of information" [1]. Events may refer to life events that happen everywhere (e.g. inside a hospital to monitor the patient's vital signs for the healthcare industry). By using event logs, Data Science aims to answer questions grouped in four categories [2]:

- Reporting: *What happened?*
- Diagnosis: *Why did it happen?*
- Prediction: *What will happen?*
- Recommendation: *What is the best that can happen?*

Multiple techniques are available to provide a response to all these questions. However, our responses will depend on the quality of the underlying data. In this regard, Data Quality will be an important asset. That is why, already in 2006 Clive Humby (co-founder of Dunnhumby) declared: "Data is the new oil"; therefore, "Fancy analytics without suitable data are like sports-cars without petrol" [1].

To turn data into value, "Process Mining aims to discover, monitor and improve real processes by extracting knowledge from event logs readily available in today's information systems" [3]. There are three types of Process Mining that can be derived from event logs: (1) Discovery, which takes an event log and produces a model without any additional information; (2) Conformance, which compares an existing process model with an event log of the same process to double-check the deviation between expectation and reality; finally, (3) Enhancement, which intends to extend or improve an existing process model based on the information recorded from the same process event log [2,17].

This Case Study aims to focus on Conformance Checking and Process Enhancement. For that purpose, we will use Business Process Management (BPM) in combination with Process Mining. BPM is a systematic approach to analyzing, designing, implementing, monitoring, and optimizing business processes to improve organizational performance [16]. The connection point between both is that Process Mining helps uncover how processes actually work (AS-IS), while BPM helps describe how processes should work (TO-BE).

Regarding Conformance Checking, given the event log of the case and the corresponding model performed in Business Process Model and Notation (BPMN) [15], the objective is to diagnose the deviation between both the observed and modeled behavior. For the healthcare scenario, and focusing on detecting when and where an FTR may occur, the BPMN model will be created by the data scientists with the input from the nursing team as Subject-Matter Experts (SMEs). The BPMN model will reflect all the different logical pathways that can be adopted since a patient is admitted to the hospital. In addition, and using the event log from the same process as an input parameter for the Process Mining techniques, we will identify and understand bottlenecks, inefficiencies, deviations, and risks between the model and the reality.

As mentioned above, our results will depend on the quality of the underlying data, that is why the dataset will be cleansed and prepared in advance to maintain a high data quality standard [13]. As a result, we will provide a report (*What happened?*) and a diagnosis (*Why did it happen?*) based on our key findings to help the nursing team understand better when the FTR diagnosis may happen, and as a consequence, improve the care process to assist the patients in the hospital.

For process enhancement, given the same event log, and with the intention to provide a prediction (*What will happen?*) and/or a recommendation (*What is the best that can happen?*), we will investigate how the care process could be improved to avoid an FTR diagnosis when possible. In this sense, we will use Predictive Process Monitoring, as a branch of Process Mining that aims at predicting the future of an ongoing (uncompleted) process execution [4,9]. The predictions will be provided in three different formats:

- Outcome-based predictions: predictions related to predefined categorical or boolean (*Will the patient incur into exitus?*).
- Numeric value predictions: predictions related to measures of interest taking numeric values (*What will be the FTR percentage for this patient?*).
- Next event predictions: predictions related to sequences of future activities and related data payloads (*What could happen to the patient from now on?*).

2 Case Study: FTR

In the following sections, the complete information on the Case Study on Failure To Rescue (FTR) is presented according to the instructions given in [8]. First, a brief description of the Case Study is provided, explaining the context and also its limitations. Secondly, a full description of the Case Study is given by means of some examples. Finally, a summary is pointed out as a conclusion.

2.1 Context and Scope

Patient safety is a fundamental pillar of healthcare. A report by the European Commission titled Costs of Unsafe Care and Cost-Effectiveness of Patient Safety Programs reveals that Spain has the fifth-highest costs attributed to clinical safety failures in the healthcare system of any European Union country, at €1.386 billion per year [19]. Previous studies have shown that hospital complications are primarily influenced by patient- and procedure-related factors [6]. Designing strategies to detect potential complications during hospital care processes can improve healthcare outcomes and optimize the use of clinical resources [11].

To provide more context, the case focuses on a hospital in the province of Castellón (Spain), which is organized into three Health Departments. These departments are equivalent to an organizational structure belonging to the Valencian Community's Ministry of Health, responsible for providing healthcare services to a designated population within a specific geographic area. Together, the three departments serve a total population of 603,952.

Specifically, the case takes place at the La Plana University Hospital, a public center located in Vila-real (Castellón), belonging to the Department of Health of La Plana, which provides specialized coverage to more than 190,000 people. It has 258 beds distributed across services including emergency care, intensive care, rehabilitation, home hospitalization, outpatient consultations, and eight inpatient units organized by medical specialty.

Each unit has 30 beds and is staffed by 12 registered nurses and nine nursing assistants, who work rotating shifts to ensure continuity of care. These units operate without interruption, necessitating constant coordination, information management, and clinical decision-making.

Every hospital process involves a level of organizational complexity that is not always apparent to outsiders. Every day, hundreds of professionals carry out their duties in a dynamic environment that requires quick, coordinated decisions based on principles of safety and efficiency [10]. All relevant patient data is recorded in the hospital's Electronic Health Record (EHR) system to support this.

The implementation of EHRs has significantly improved care delivery by enhancing accessibility to information, communication among professionals and patient participation and safety [18]. However, in certain cases, healthcare teams fail to recognize the early signs of a patient's clinical deterioration, even when clinically relevant information is recorded in the EHR. This phenomenon, known as Failure To Rescue (FTR), refers to a healthcare system's inability to intervene promptly when the first signs of deterioration appear in hospitalized patients. Such events are considered preventable if early identification of clinical or functional decline is followed by an appropriate and timely response [14].

If intervention fails to occur, the patient's condition may deteriorate, resulting in serious complications, irreversible damage, or even death despite receiving medical care [5]. It is important to note that these situations are not the result of individual errors, but rather reflect structural limitations in the care delivery process, particularly with regard to the system's ability to convert clinical data into effective early-warning triggers.

To address these concerns, the VALENF-FTR Project was launched as a preparatory action in collaboration with the Universitat Jaume I[1] and FISABIO[2]. The project will examine how clinical information is integrated into the care process and evaluate whether specific physiological and functional parameters could serve as early warning signs of risk before situations become irreversible. Process Mining techniques will analyze and reconstruct the delivery of hospital care in practice with a view to identifying bottlenecks and discontinuities.

This Case Study encourages us to consider how hospital care is organized in busy units. Analyzing care delivery from a process management perspective can reveal deviations from expected clinical pathways and pinpoint critical points where continuity of care may be compromised. Precisely tracing the actual care process based on records, transitions, and timing provides a powerful tool for understanding what is happening and anticipating where critical failures might emerge.

In the context of a FTR, this analysis highlights the importance of reviewing how care processes are designed, ensuring that relevant information is neither overlooked nor interpreted too late. Two key questions arise from this reflection.

---

1 https://www.uji.es/.

2 https://fisabio.san.gva.es/en/.

First, *how can care processes be optimized to ensure that recorded information leads to timely clinical action?* Secondly, *where does the traceability fail that should enable clinical teams to take action in advance upon detecting potentially critical data?*

2.2 Problem Description

To better understand the practical implications of these questions, let us consider a real case that illustrates how information flow and clinical decision-making unfold in an actual healthcare setting. Imagine a patient: an 81-year-old woman. She was admitted to a medical-surgical inpatient unit for scheduled abdominal surgery within the general surgery department. At the time of admission, she was an independent woman with a history of hypertension and type 2 diabetes. She lived alone and showed no signs of cognitive decline. During the first 48 h following surgery, her condition appeared stable. As is standard practice in such units, her vital signs were monitored and recorded in the Hospital Information System (HIS) at each shift change. While some values showed minor fluctuations, such as slightly low blood pressure, decreased urine output, and increased drowsiness, no specific protocol was activated. Everything was considered within normal limits.

During the morning handover, a nurse told a colleague, "I do not like the appearance of the patient. Her vitals are not right, but there is nothing concrete.". Concerned, the colleague checked her vital signs more frequently than usual, but only subtle changes were noted. By the afternoon, the patient began to show signs of confusion, fever, and tachypnea (a respiratory rate above 20 breaths per minute). The care team responded immediately by monitoring her vital signs and administering treatment to stabilize her condition.

Although she was clinically stable that night, her prognosis did not improve, and she was transferred to the Intensive Care Unit (ICU) with a diagnosis of abdominal sepsis, a severe immune response to infection. Despite the medical team's efforts, her condition deteriorated, and she passed away three days later.

A review of the case revealed that signs of clinical decline had been evident in the patient's records for at least 24 h before the transfer to the ICU. While the necessary information, such as frequent vital sign checks and additional tests, was present, the pattern went unrecognized. This was not due to an individual error. Rather, it reflected a systemic failure to interpret the care process in time.

Cases like this are not isolated incidents. Instead, they are part of a recurring pattern involving different patients and units. At La Plana University Hospital, the clinical pathway is structured and follows an established process. In some cases, it begins in the Emergency Department, where triage determines the level of care required. Depending on the severity of their condition, patients may return to the waiting area, be admitted to non-monitored beds or observation rooms, or proceed directly to the vital signs stabilization bay. From there, they may be transferred to inpatient units or the ICU, or referred to a higher-level hospital. If emergency surgery is required, patients are taken to the operating theater and may continue their recovery in an inpatient unit or ICU. Alternatively, if their condition deteriorates, they may die during or after the procedure. If the emergency care resolves the issue, however, patients may be discharged home directly.

Patients may also be admitted to hospital as scheduled. In surgical cases, patients first undergo a preoperative assessment on an inpatient unit before being transferred to the operating theater. After surgery, they are monitored either on the inpatient ward or in the ICU, depending on their condition. Possible subsequent outcomes include discharge home, home hospitalization for continued care, transfer to another hospital or long-term care facility, or death.

Another entry point is outpatient consultations for medical cases. If hospital care is deemed necessary during the consultation, the patient may be admitted. As this example shows, admission to hospital begins a continuous cycle of care involving complex pathways that require coordinated action by multiple professionals.

During inpatient stays, vital signs are systematically recorded in the HIS at the start of each shift. Nursing care, clinical examinations, laboratory tests and structured clinical assessments, such as those provided by the VALENF instrument [12], are carried out. However, the system is fragmented: clinical notes are recorded separately as free text, assessments occur at different times, and key information is often transmitted verbally during handovers. This complexity is further increased by rotating staff, time constraints and high patient volumes.

The result is a process in which multiple data inputs exist alongside one another without being integrated or interpreted. Although every professional performs their tasks correctly, the current system design does not facilitate the early detection of progressive clinical deterioration. By the time a response is triggered, it may be too late for intervention.

This raises an important question: if all procedures are followed and data are recorded, why is clinical deterioration not anticipated? The issue lies in the need to revisit the design of the overall care process to understand how care is actually delivered. The hospital has the right documentation tools, a clearly defined structure and trained professionals. However, there is a disconnect between data collection and clinical decision-making. Studying the process as a continuous sequence rather than a series of isolated actions enables us to trace how care actually flows from admission to discharge or death and identify potential deviations from the expected clinical pathway.

This perspective does not seek to assign blame to individuals, but rather to understand how the healthcare system functions as a whole, including all its professionals, shifts, and transitions. The objective is to analyze the clinical process by mapping what actually happens, identifying inefficiencies, and suggesting improvements to enhance patient safety.

2.3 Conclusion on Case Study

At this point, we hope you can understand the relevance of this healthcare scenario. We give you the opportunity to dive deep in the Data Science sea and be part of this significant mission. To help you decide, please let us summarize the criticality of your participation in the FTR research:

- The importance of reviewing the care processes design and execution to ensure that data moves seamlessly from collection to action.
- The significance of accurate and reliable patient data for clinical decision-making, considering:
  - Data Capture Issues, such as incomplete or delayed documentation, free-text entries or use of non-standard terminology.
  - System Integration Gaps, such as disconnected systems, lack of interoperability or manual data transfers.
  - Workflow Misalignments, such as unclear protocols for follow-up or time lags in interdisciplinary communication.
- The critical role of traceability in clinical procedures.
- The opportunity to use the innovative techniques of Process Mining to provide predictions and/or recommendations to help the nursing team assist the patients in the hospital and avoid the FTR diagnosis.

3 Learning Opportunities

This Case Study can be highly attractive for students, since it is a vast and complex process that depends on many parameters. In addition to the challenge of investigating how such parameters cause FTR, the practical utility of its results can become really motivating for students' learning. Therefore, it provides interesting learning opportunities that could be structured according to three Case Study perspectives proposed by [7]. In this sense, we raise the following questions for each perspective:

- Problem Diagnosis.
  - What alternative pathways exist in the hospitalization process?
  - Where in the clinical pathway are bottlenecks or delays most likely to occur?
  - Which data signals (e.g., vital signs) might indicate a deteriorating patient condition?
  - What are the possible causes of a FTR in this process?
  - What data should be collected to train predictive models?
- Evaluation.
  - Evaluate the completeness and accuracy of the BPMN model. Are all relevant pathways and decision points captured?
  - What would you consider as an appropriate data quality strategy for the present healthcare scenario?
  - Which data quality checks would you include for the present healthcare scenario?
  - Which tools would you use to support your data quality strategy?

  - Compare the AS-IS model obtained with Process Mining techniques with the TO-BE model of the process in BPMN.
  - Assess the data available for predictive modeling. Is it sufficient to train reliable models? What are the limitations?
  - Which Process Mining approach works better to discover the patient care process in the hospital?
  - Compare different predictive techniques (e.g., decision trees vs. neural networks) in terms of interpretability and accuracy for this context.
  - How effective would process automation or real-time alerts be in preventing FTR in this scenario?
  - How does the proposed process redesign align with existing clinical guidelines and patient safety protocols?
- Decision-Making.
  - How would you propose to monitor and maintain a high data quality standard?
  - Based on the predictive analysis, where should decision support alerts be integrated into the process to prevent FTR?
  - If you had to redesign the process, which steps would you modify, remove, or automate?
  - How can variability and clinical uncertainty be modeled?
  - What would you present to the hospital board to justify investment in this BPM and AI initiative?

Finally, for advanced students, there are more opportunities if you would like to become a senior data scientist trying to work on the same Case Study using complementary techniques in Machine Learning, for example.

The accompanying Teaching Notes, which support instructors in guiding these learning activities, are available online[3].

References

1. Aalst, W.M.P.: Data scientist: the engineer of the future. In: Mertins, K., Bénaben, F., Poler, R., Bourrières, J.-P. (eds.) Enterprise Interoperability VI. PIC, vol. 7, pp. 13–26. Springer, Cham (2014). https://doi.org/10.1007/978-3-319-04948-9_2
2. Aalst, W.M.P.: Process mining in the large: a tutorial. In: Zimányi, E. (ed.) eBISS 2013. LNBIP, vol. 172, pp. 33–76. Springer, Cham (2014). https://doi.org/10.1007/978-3-319-05461-2_2
3. van der Aalst, W.: Process Mining: Data Science in Action, 2nd edn. Springer, Heidelberg (2016)
4. van der Aalst, W., Carmona, J. (eds.): Process Mining Handbook. Springer, Heidelberg (2022)
5. Burke, J., Downey, C., Almoudaris, A.: Failure to rescue deteriorating patients: a systematic review of root causes and improvement strategies. J. Patient Saf. **18**(1), E140–E155 (2022). https://doi.org/10.1097/PTS.0000000000000720

[3] https://github.com/grangelq/teachingCaseFTR.

6. Duke, G.J.: Hospital-acquired complications: the relative importance of hospital- and patient-related factors. Med. J. Aust. **216**(5), 242–247 (2022). https://doi.org/10.5694/mja2.51375
7. Ellet, W.: The Case Study Handbook, Revised Edition: A Student's Guide. Harvard Business Review Press, Boston (2018)
8. Farhoomand, A.: Writing teaching cases: a quick reference guide. Commun. Assoc. Inf. Syst. **13**, 103–107 (2004)
9. Francescomarino, C.D., Ghidini, C.: Predictive process monitoring. In: van der Aalst, W.M.P., Carmona, J. (eds.) Process Mining Handbook, Lecture Notes in Business Information Processing, vol. 448. Springer, Cham (2022). https://doi.org/10.1007/978-3-031-08848-3_10
10. Júnior, M.A.C., Palmeira, S.P., Martins, S.R., Ribeiro, M.T.P., Abuhid, M.D.: Fundamental principles of health management: integrated strategies for efficiency, quality, and sustainability in healthcare. IOSR J. Humanit. Soc. Sci. **29**(9), 17–21 (2024). https://doi.org/10.9790/0837-2909091721
11. Lagoe, R.J., Westert, G.P.: Evaluation of hospital inpatient complications: a planning approach. BMC Health Serv. Res. **10**(1), 200 (2010). https://doi.org/10.1186/1472-6963-10-200
12. Luna-Aleixos, D., et al.: Development and validation of a meta-instrument for the assessment of functional capacity, the risk of falls and pressure injuries in adult hospitalization units (VALENF instrument) (Part II). Int. J. Environ. Res. Public Health **20**(6), 5003 (2023)
13. Marin-Castro, H.M., Tello-Leal, E.: Event log preprocessing for process mining: a review. Appl. Sci. **11**(22), 10556 (2021). https://doi.org/10.3390/app112210556
14. Mushta, J., Rush, K.L., Andersen, E.: Failure to rescue as a nurse-sensitive indicator. Nurs. Forum **53**(1), 84–92 (2018). https://doi.org/10.1111/nuf.12215
15. Object Management Group: Business Process Model and Notation (BPMN) Version 2.0.2. https://www.omg.org/spec/BPMN/2.0.2 (2014)
16. Pufahl, L., Zerbato, F., Weber, B., Weber, I.: BPMN in healthcare: challenges and best practices. Inf. Syst. **107**, 102013 (2022). https://doi.org/10.1016/j.is.2022.102013
17. Turner, C., Tiwari, A., Olaiya, R., Xu, Y.: Process mining: from theory to practice. Bus. Process. Manag. J. **18**(3), 493–512 (2012). https://doi.org/10.1108/14637151211232669
18. Vos, J.F.J., Boonstra, A., Kooistra, A., Seelen, M., Offenbeek, M.V.: The influence of electronic health record use on collaboration among medical specialties. BMC Health Serv. Res. **20**(1), 676 (2020). https://doi.org/10.1186/s12913-020-05542-6
19. Zsifkovits, J., Zuba, M., Geissler, W., Lepuschütz, L., Pertl, D., Kernstock, E., Ostermann, H.: Costs of unsafe care and cost effectiveness of patient safety programmes. https://www.semanticscholar.org/paper/Costs-of-unsafe-care-and-cost-effectiveness-of-Zsifkovits-Zuba/3676efdc7bbeef904beb327697ce5260adfbcd2c (2016)

Education as an Affordance Actualization Driver - Investigating Learning Data from a Process Mining Ecosystem

Adrian Joas(✉)

Celonis ApS, Lautrupsgade 13, 2100 Copenhagen, Denmark
a.joas@celonis.com

Abstract. Process Mining (PM), a technique to discover and analyze business processes, has sparked interest in both practice and research, though working with the technology is perceived as challenging. As early studies mostly explore the technical integration of PM, we examine the PM ecosystem, focusing on the diverse expectations of stakeholders. The study uses an ecosystem perspective and the theoretical lens of affordances with an emphasis on education as an affordance actualization driver. Building on conceptual considerations and findings from analyzing large scale education data from a PM vendor, we propose a refined classification of user group specific affordances for PM and highlight the role of PM education as a key affordance actualization driver for different PM stakeholders. Our study makes three main contributions: first, it creates transparency on differing affordances of the user groups, second, it highlights education's dual role in shaping affordances and driving their actualization, and third, it advances the discourse on the use of PM in ecosystems.

Keywords: Process mining · ecosystem · education · affordance · user group

1 Introduction

Process Mining (PM), a big data analysis technique for business processes, is a rising trend in practice and research. It allows organizations to derive insights from event logs [30], and thereby, enhance process efficiency, leading to monetary and non-monetary improvements [3].

Since the emergence of PM, researchers have mostly investigated the technical artifact with its underlying algorithms and logic [21]. In contrast, there is still a limited, but continuously growing understanding of the business aspects and organizational impacts the use of the technology has [2, 18, 34]. Key research interests address the individual level with the users' perception of the technology, the group level with the effects of PM on people's interaction, the organizational level with questions around the value creation via PM, and the ecosystem level with the effects of PM on networks [34]. Studies highlight the potential of process mining, for example, increased process transparency leading to cost savings and optimized processes [3, 20]. However, potential challenges

M. ER et al. (Eds.): BPM 2025, LNBIP 565, pp. 427–441, 2026.
https://doi.org/10.1007/978-3-032-02936-2_30

like employee privacy concerns, technical barriers, and lacking technical capabilities must be taken seriously in PM implementations [6, 10, 20]. Joas and Matzner [10] have shown that education might be a relevant lever to reduce some of the challenges. To unlock the potential of PM, these benefits and challenges have to be balanced carefully in PM implementations to increase adoption.

Looking at the adoption of PM, one must first understand the affordances the different involved user groups have, which describe the underlying, perceived motivations each user group might associate with the technology [8]. Affordances are context-specific and user-specific and form an action potential on how the user might use the technology [19]. In the field of PM, the affordances for PM users in implementing organizations have recently been investigated [3]. However, considering the overall ecosystem of PM, defined as a view of related individuals or organizations engaged in the technology and its implementation [29], there are additional stakeholders on an inter-organizational level with affordances potentially differing from the ones of the implementing organization. We define the expression "ecosystem" as a view of related individuals or organizations engaged in the technology PM and its implementations in practice. This includes a) the organizations as stakeholders and b) the inter-organizational relationships within PM implementations [26]. Thus, the first objective of this study is to uncover the affordances for different ecosystem players, looking beyond the implementing organizations.

Second, to actualize these affordances and to let the user groups benefit from the technology, affordance actualization drivers are critical [23]. These actualization drivers ensure acknowledgment of the presence of affordances. As prior studies suggest, education can facilitate IS adoption and can serve as an actualization driver. In this context, the first studies shed light on the role of education in the successful implementation of PM [10]. We want to build upon this literature and focus on the role of PM education to actualize user group-specific affordances. Therefore, we pose the following research questions.

RQ1: *What are the affordances of different process mining stakeholder archetypes and user groups in the PM ecosystem, based on their educational journey about PM?*

RQ2: *How can process mining education drive their actualization?*

Answering these research questions, we contribute to the literature in the following manner: First, we derive affordances for the different PM ecosystem stakeholders, based on their educational journey, helping to understand how diverse user groups interact with PM. Second, we provide insights into the role of education as an actualization driver for the different affordances. We introduce PM education archetypes as a conceptualization of our findings. Finally, from a broader perspective, this study adds to the discourse on the organization and ecosystem perspective of PM. Furthermore, the study holds relevant implications for practice by guiding vendors and users of PM concerning PM education.

The paper is structured as follows: First, we outline the technology of PM, introduce our theoretical lens of affordances, and PM education as an affordance actualization driver. Next, we describe the method that uses digital trace data with inductive reasoning based on an extensive learning data set from a PM vendor, which helps in identifying the specific affordances of the diverse PM stakeholders in the PM ecosystem. We shed light on the affordances of the varying user groups for PM, based on their educational journey, and define archetypes in PM education. Finally, we discuss our findings and point out the next steps, highlight the contribution, and point out avenues for future research.

2 Background

2.1 The Ecosystem of Process Mining

Process Mining (PM) has its origins in Business Process Management (BPM), where PM enables the discovery and analysis of business processes, setting the path to a followed optimization of those. A process, in this setting, is defined as a sequence of activities [30]. These activities can be extracted from various source systems, in which activities are recorded, carried out by people, machines, or software, and can be transformed into a so-called event log. The minimum amount of information needed to perform PM is a timestamp, a unique identifier, and an activity name [30]. A higher granularity of information and analysis options is reached by enriching an event log with additional metadata.

PM has major implications in companies, from the technical, individual, group, organizational, and ecosystem perspectives [34], leading to a high interest in further exploring challenges and opportunities in each perspective. Based on the technology concept, PM creates transparency in actual process flows, which, in turn, creates the opportunity to derive implications for business [30]. PM is considered a key technology for digital transformation [13]. With the majority of studies focusing on the technical development and advancement of algorithms and features rather than business-related challenges, the ones existing act as valuable anchor elements in the pursuit of a better understanding of management & organizational understanding [2–4, 18, 20, 21, 34].

Research on personal technology acceptance, usage, and resentment around the technology provides ample room to apply well-established research frameworks like the Technology Acceptance Model [31] or Unified Theory of Acceptance and Use of Technology [32] which are just starting to be applied in the context of PM use [2]. The group perspective on PM research focuses on stakeholders in the PM ecosystem and the major roles they take within an organization applying PM or in external organizations involved in the roll-out and operation of PM in organizations. The most common groups within an organization include process participants, process owners, process managers, and process experts [34]. Additional stakeholder groups outside of the organization can be found in e.g. software vendors, consultancies, and academia [7, 10, 24]. Each stakeholder group has its underlying motivation and reason to be involved in PM, and needs to be taken into account when discussing PM.

The organizational perspective builds on a large body of IS research and may provide insights into adoption, actual use, or acceptance. Several studies derive challenges for PM in organizations [10, 20], as well as success factors for PM implementation [18].

The ecosystem perspective, extending the scope of view onto all stakeholders involved with a PM project, is relevant, as PM, with its underlying element of process transparency, lends to inter-organizational collaboration [26]. From a commercial lens, collaboration in ecosystems can be observed [1, 14, 25]. Besides this practical motivation, studies like Vom Brocke et al. [34] call for research on this aspect, as most previous research has been limited to the central, implementing organization. In this study, we focus on the ecosystem perspective as the underlying framework and leverage the group perspective to define affordances and further refine the different stakeholders found in the ecosystem.

2.2 Affordance Lens to Investigate User Groups in the PM Ecosystem

To better understand the different stakeholder perspectives and user groups considering PM, we make use of the affordance lens. Affordances hereby are perceived action potentials that are formed within the relationship between a user of technology, who pursues a certain goal, and the technology in itself (e.g. [23, 28]. The concept is based on Gibson's ecological psychology [8], where affordances are a non-dualist way and every object has specific properties, which offer various possibilities for action, based on the relative viewpoint of the observer [27]. Based on the relationship between the actor and a part of their environment (e.g. a tool), an action potential (affordance) is created, which can be actualized by the actor.

Put into an IS context, affordances are emerging from the relationship between organizational systems and IT systems [19] and are defined as the potential for action associated with achieving an immediate concrete goal, arising from the stated relation between an object (IT artifact) and a goal-oriented actor [33]. The actor has their interpretation of what to achieve when they interact with technology, yet their options are constrained by the nature (e.g. features and capabilities) of the IT artifact. These features are considered relatively stable across context and time [15], while affordance theory still instills a dynamic nature of the relationship between an IT artifact and its users [5]. The younger the technology is for the respective user, the more likely it is for that person to experiment with the technology and explore different action potentials [12]. With more mature users, the relationship with the technology becomes more stable [11], which essentially describes a learning process of users understanding the link between a certain goal and corresponding actions that lead to achieving said goal [33]. Affordances provide an appropriate lens for organizational change through IT applications [17, 19, 33]. There has been ample research on various aspects of the affordance theory in IS, like the differentiation of individual elements, perception and actualization of affordances [16], the emergence of affordances based on specific user groups [17], and influences on their actualization potential [28].

Affordances do not only have to be perceived; they also have to become actualized. The actualization of an affordance can be defined as "the actions taken by actors as they take advantage of one or more affordances through their use of technology to achieve immediate concrete outcomes in support of organizational goals" [28, p.15]. As Ostern and Rosemann [23] propose in their affordance framework (see Fig. 1), an affordance actualization driver is a "measurable resource that triggers the actualization of or design for a specific affordance(s)" ([23] p. 5). Prior studies suggest that three components drive the actualization of an affordance: the abilities and preferences of the individual, the features of the system, and the characteristics of the work environment [28]. We focus on the first aspect, which is the ability and willingness of a user group to use PM for their user group-specific purposes.

Within PM literature, PM education is considered an important driver to increase PM adoption within organizations and to solve challenges associated with the successful implementation of PM as a technology [10, 20]. To create value with PM, it is not only necessary to have clear responsibilities for tool promotion and usage, but also imperative to create a functioning business-IT partnership within the PM ecosystem. Those skill sets include not only the technical skills but also process and business expertise [3]. Hence, we are interested in investigating the user group-specific affordances of PM and the role of education as an affordance actualization driver.

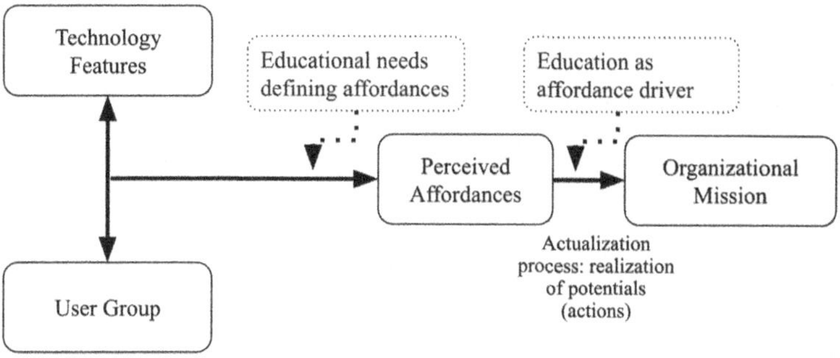

Fig. 1. Education defines affordances and acts as an affordance actualization driver (adapted from [28])

3 Methodology

3.1 Using Learning Data as Digital Traces

To investigate both RQs, we opted to use learning data as digital trace data. Digital traces, such as online records of learning activities, allow us to triangulate the complex reality of the players in the PM ecosystem in practice beyond subjective hypothetical biases that could arise in interview situations [9]. Following a positivistic stance, these trace data reliably represent what learners do in the system and what skills they want to develop [22], uncovering both hidden and perceived affordances in a neutral manner. Viewing PM education as a driver for affordance actualization [23] (see Sect. 2.2), we argue that if individuals take training, they aim to actualize an affordance, therefore making the collected learning data an appropriate proxy to further shed light on the user group-specific affordances and the role of education in actualizing them.

The used digital trace data stems from a leading PM vendor's database, which contains rich data from approximately 210.000 PM learners between the years 2017 – 2022. These learners have various different backgrounds, and they have been active on the PM vendor's e-learning platform, which is available to access with free registration. For each of the 210.000 learners, data has been extracted directly from the e-learning platform, and information on their learning progress (started and completed courses) and content level has been used in the analysis (see Table 1). Given the global reach of the vendor, the learners stem from all parts of the world and levels of involvement in PM. These data allow us to follow individuals' learning journey and create stakeholder archetypes and more nuanced user groups.

The data indicate that, on average, individuals finish 16.5 courses. Triangulating the field data of self-identified groups (required during sign-up) with prior studies on PM stakeholders (see Sect. 2) led to the four conceptual stakeholder archetypes: employees of the PM vendor (~4.000 learners), implementing organizations (~43.000 learners), third-party support (individuals not affiliated with the technology provider or implementing organization but directly involved in PM implementation, e.g. consultancies (~49.000 learners), and academics (~115.000 learners). This rich, balanced dataset allowed us to investigate the entire PM ecosystem.

Table 1. Example of data points extracted from e-learning platform.

Learner Information	
Learner E-Mail & Name	Anonymized
Country	e.g. India
Organizational Name	e.g. ABC Inc
Self-identified Stakeholder Group	PM Vendor/Customer/Third-Party/Academic/General Interest
Creation Date of Account	01–01-2022
Courses Taken/Competed	Individual Courses (started/completed)
Learning Content Information	
Course Name	e.g. Configure Tables and Charts in Analysis
Course Description (incl. Learning Goals)	e.g. Configure Analysis tables and charts with dimensions and KPIs You'll learn: - Configure dimensions and KPIs for charts and tables - …
Course Length	e.g. 45 min

3.2 Deriving Affordances from Learning Data

We followed a qualitative approach when analyzing the trace data, including deduction and induction (see Fig. 2). In the deduction step, we first looked at the learning content information and deductively derived learning goals for each course, abstracting the learning goals of the individual courses given by the PM vendor (e.g. "Conceptualize your process in terms of activities and cases" and "Interact with charts and tables to verify your assumptions about how the process is performing…") and going through the actual learning content as learners ourselves. This enabled us to understand what information is covered by the learning elements and what it enables the learner to do with PM. We then looked at the learners' information and their courses taken. Triangulating learning goals and courses taken, we inductively derived affordances. The affordances could be mapped to user groups, which, in return, could be aggregated into four different stakeholder groups.

The overall data analysis was characterized by iterative consultations between the researchers and stakeholders in the PM ecosystem, as well as engagement with current literature on PM affordances and PM user groups. Furthermore, we intended to follow principles for trace data research, such as triangulating different datasets, establishing a fit between the trace data and the real world, and developing situated explanations that go beyond the trace data themselves [9].

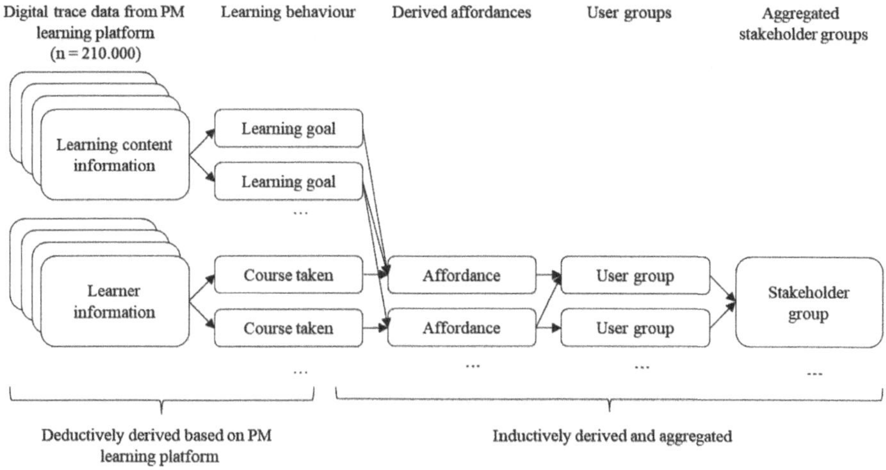

Fig. 2. Methodological Process.

4 Findings

4.1 Conceptualizing the PM Ecosystem

As established in Sect. 2, the PM requires a multi-dimensional view of the affordances users have within their relationship with the technology. Due to the dynamic developments on the one side and the technical complexity of PM on the other side, PM in practice is heavily reliant on functioning ecosystem structures [1, 14, 25] as well as on preceding education [10, 20]. In addition, the learners' self-identified stakeholder groups add another data point for our proposal of archetypes.

Within the implementing organizations, but also the broader business ecosystem, individuals can be clustered into groups, following similar task profiles and interactions with the technology, depicted by clustering the types of training courses taken, and the affordances that correspond to the training courses. We leverage those two aspects as the base for our proposal of archetypes within PM. Given that affordances are tied to an IT artefact (see Sect. 2), we see the technology providers (software vendors) at the core of the ecosystem, as these groups create and shape the IT artefact the affordances are based (visible by the group Product Development). Their main stakeholders are the organizations that implement the technology as part of their technology stack, given that this archetype contains the most end-users of the implemented PM solution (groups Process Participant, Process Manager). These organizations are typically supported by third parties that are external to the implementing enterprise, typically in the form of consulting services firms or system integrators, while not present in all PM implementations, their knowledge is leveraged by the implementing organizations [7, 25], fullfilling a similar role than direct implementation services of the PM vendors. Around those three core stakeholders, we see the academic space. Groups in this space are independent of specific PM implementations. The training taken by the groups is more theoretical and technology-focused, for some groups, more fundamental for the student group, focusing on technology advances as well as on future employability (see Fig. 3).

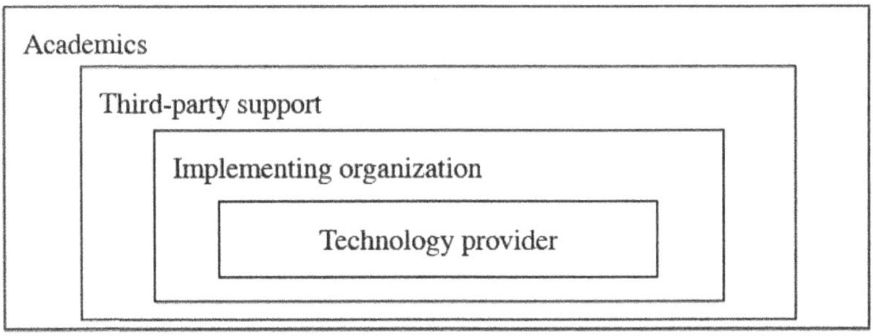

Fig. 3. PM Ecosystem.

4.2 Affordances of PM Vendors' Employees

To start at the center of the ecosystem, the first stakeholder archetype includes individuals working directly at PM technology providers. The focus is on roles involved in developing PM software or supporting the broader ecosystem, particularly implementing organizations. Looking at their taken courses, PM affords them to support implementing organizations in finding and realizing business value through the provider's software solution. The four sub-groups within this archetype are product development, technical sales, data analysts, data engineers, and value architects.

Product development professionals are engaged in the research and development of PM software. With a strong technical background, their motivation is to create technically robust and user-friendly solutions. In their learning choices, we see them learning about technical skills related to software development and data ingestion, to be able to actualize the affordance of advancing the technology. Technical sales teams work with potential customers to position PM and showcase the software's value. They require a blend of methodological and technical learning, supplemented by internal best practices, to effectively demonstrate the product's benefits. This helps to actualize their affordance to convince potential customers of PM's advantages. Data analysts and data engineers at technology providers perform roles similar to process experts and system engineers in implementing organizations, akin to third-party support. Their learning data and affordance mirror that of their counterparts–build process visualizations and KPIs and ensure high-quality data availability and system uptime–with an additional focus on methodology and change management. Value architects assist customers with business-related queries and serve as long-term advisors to ensure that organizations can effectively leverage PM software. Their focus is on business and change management, though they also receive basic technical learning to fully support clients, supporting their affordance of making customers successful in using PM solutions.

4.3 Affordances of Implementing Organization

We define the stakeholder archetype users at the implementing organization as all individuals who are involved in initiatives around the usage or implementation of PM. The goals of these user groups generally align with the organization's overarching objectives, such as revenue increase, cost reduction, and customer satisfaction [3]. As our learning data shows, the implementing organizations hold the core affordance of understanding and optimizing their processes when using PM. This core affordance is complemented by specific affordances for more specific user groups: process participants, process managers, process owners, process experts, and process engineers (cf. [34]).

Within the implementing organization, process participants work within the processes where PM is implemented, typically using enterprise systems like ERP systems on an operational level. They rely on PM insights to improve their work, possess domain expertise rather than technical skills, and often pursue usage-oriented learning assets. Their central affordance is using PM to make data-driven decisions. Process managers focus on the overall performance of business processes, aiming to improve key performance indicators from a more managerial perspective. They drive analysis strategies and decisions, often engaging in analysis-focused learning, to actualize their affordance of understanding their process and making relevant decisions. Deriving from the learning data, those two user groups only use PM or the results of PM, rather than coordinating, setting up, and managing the PM implementation. Their knowledge of the technical details and challenges is limited. The process owner can be considered the project manager for PM implementations. Their task is to bring the managerial domain and technical requirements together. Change management and cultural shifts due to PM are also considered among the key tasks of a process owner. They must understand the business and organizational needs and how to best translate those into clear technical requirements for the technical user groups. Due to their skillset, they can be considered the jack-of-all-trades, which is also reflected in taking learning assets in all areas of PM. We derive that their specific affordance is to deliver successful PM projects in the organization. Process experts bridge PM and application domains. Guided by process owners, they translate business needs into analytical queries and visualizations, based on the process event data. Key learning assets taken are around querying languages and front-end visualization capabilities, affording them to build process visualizations and KPIs. The system engineers ensure the connection between enterprise source systems and PM tools. They manage data quality, prepare data, and maintain the event log and data model, often administering the PM software and addressing data quality issues.

Their learning revolves around data preparation and administrative usage. These learning assets help them realize their affordance of ensuring high-quality data availability and system uptime.

4.4 Affordances of Third-Party Support

We define the stakeholder archetype of third-party support to include individuals not affiliated with the technology provider or implementing organization but directly involved in PM implementation, e.g. consultancies. These experts provide services either independently or in collaboration with technology providers. The main goal of any individual in this archetype is to add value to the implementing organization while benefiting their own organization. Based on the learning data, we see that user groups in this archetype share the core affordance of ensuring a successful collaboration around PM usage with an implementing organization. These user groups are involved in various stages, from pre-adoption through full implementation, technology diffusion, and ongoing maintenance of PM technology [10]. The key roles identified include value architect, data analyst, data engineer, and project manager.

Value architects handle business-focused tasks, aiming to identify and define the benefits of PM for client organizations. Their expertise lies in management, translating PM's technical aspects for non-technical audiences. We find that they focus on sales and usage-oriented learning. Their central affordance is leveraging the technology to find and define the benefits for the client organization. Data analysts and data engineers fulfil roles similar to process experts and system engineers in implementing organizations, but as external contractors. Their conducted learning aligns closely with that of their counterparts within the implementing organization. Hence, their core affordance is to build process visualizations and KPIs and ensure high-quality data availability. Project managers oversee business-related aspects of PM implementation, focusing on realizing and extending business value, managing change, and supporting organizational adoption, as reflected in their educational background, by taking methodological learning assets. While they do not use PM personally and therefore do not take usage-oriented learning elements, PM affords them to deliver a successful project.

4.5 Affordances of Academics

This stakeholder archetype encompasses user groups from academia. Their primary motivation is to understand the theory behind PM, explore its application, compare different technologies, and advance algorithms. The core affordance is understanding the technical complexity and application possibilities of PM. We identify three sub-groups within this archetype: teachers, researchers, and students.

Teachers aim to integrate PM into their curricula, teaching the intersection of information systems, data science, and business. PM affords them to equip students with a foundational understanding of PM and related topics through theory-based courses. The data suggest that teachers, therefore, take core courses, e.g., theory-based courses and fundamentals on PM and automation. Researchers use PM for data analysis or seek to advance the PM field itself. Both approaches require a deep technical understanding of PM, leading them to pursue learning across the entire PM process, from data ingestion to result interpretation, as seen in their learning choices. Their core affordance is to advance the understanding of the technology and its complexity. Students see PM as offering valuable educational and career opportunities. With PM skills in high demand, students focus on learning PM fundamentals, often engaging in theory-heavy courses similar to those of teachers. Additionally, they may participate in student consulting programs that enhance employability. PM offers the action potential to increase their chances in the job market based on PM understanding.

5 Discussion

5.1 Reflection on Stakeholder Archetypes' Affordances

We set out to answer what the process mining affordances of different stakeholder archetypes are and how process mining education can drive their actualization. In the discussion, we first reflect on the similarities and differences between the PM archetypes' affordances, which are mirrored in their learning behavior.

As outlined in Sect. 4, we define four archetypes with 17 user groups that are stakeholders of PM and part of the overall PM ecosystem (see Table 2). With this nuanced conceptualization of the PM ecosystem, our findings extend prior works focusing on the affordances of PM users [2, 3]. We find that, while each of those archetypes has a distinct core affordance, we can see similar affordances in some user groups. Looking into the description of those types, they set out to perform the same tasks with PM, but with different backgrounds, which are linked to their distinct core affordances. Specifically, the affordances of the third-party support and employee of technology provider archetypes highlight an interrelated nature of affordances and value co-creation within the PM ecosystem. This finding suggests that, even though more peripheral user groups may have more indirect goals when engaging in PM, they still share similar affordances. This similarity of affordances in the ecosystem forms a solid foundation for value co-creation with the PM ecosystem.

However, some affordances are highly user group-specific [17]. For example, the affordance of the academics archetype stands independently in its relation to the other archetypes, which can be explained by the absence of a profit-oriented usage of PM and a rather theoretical perspective on the topic as part of their curricula and research endeavours.

5.2 Reflection on Education as an Affordance Actualization Driver

We furthermore collected initial insights on RQ2 – the role of PM education in actualizing user group-specific affordances. Our definitions of the archetypes and the user groups are the result of triangulating existing literature with insights from digital trace data from the PM vendor's learning platform. The relevant analyzed courses are available for all learners regardless of their organizational background, and they are free to access and choose what learning they deem relevant for their function in relation to their usage of PM. The derived affordances are aligned with the learning goals, content of the courses, and the stakeholder archetypes identified. Hence, we deem it a suitable approximation

Table 2. Affordances per Role

Archetype	Role	Affordance
PM Vendor	Product Development	Advancing the technology
	Technical Sales	Convince potential customers of PM's advantages
	Data Analyst	Build visualizations and KPIs
	Data Engineer	Ensuring high-quality data availability
	Value Architect	Making customers successful in using PM solutions
Implementing Organization	Process Participant	Using PM to make data-driven decisions
	Process Manager	Understanding their process and making relevant decisions
	Process Expert	Build visualizations and KPIs
	System Engineer	Ensuring high-quality data availability and system uptime
	Process Owner	Deliver successful PM projects in the organization
Third-party support	Value Architect	Leveraging the technology to find and define the benefits for the client organization
	Data Analyst	Build process visualizations and KPIs
	Data Engineer	Ensuring high-quality data availability
	Project Manager	Deliver a successful project
Academics	Teacher	Equip students with a foundational understanding of PM
	Researcher	Advancing the understanding of the technology and its complexity
	Student	Increasing their chances in the job market based on PM understanding

of their affordances in the relationship between PM as a technology with its features and the individuals who choose to work with PM for their needs.

The data collected across the learners indicates a proactive interest from individuals to understand how they can actualize their affordances (compare Fig. 1). This is in line with literature on the relevance of education for successful PM projects [10, 20], both on technical as well as on business/methodological skills [3, 10]. A split of relevant education based on the user group also became visible [7, 10, 24], supporting the relevance of user group-specific education to increase the ability of individuals to actualize their affordances, based on their unique preferences [28]. We, therefore, argue that education can be seen as an affordance actualization driver for IS, specifically PM [23]. Taking this retrospective view on affordances through the affordance actualization driver of education highlights their inherent drive to find ways of reducing their effort on affordance actualization. In turn, this leads to a more stable and mature relationship over time.

6 Conclusion

We set out to explore the affordances of the different stakeholders and user groups in the PM ecosystem and investigate how PM education can drive the actualization of these affordances (see Fig. 1). With the collected learning data, we define four stakeholder archetypes with an overall number of 17 user groups in the overall PM ecosystem. While each stakeholder archetype has a distinct core affordance, we can see similar affordances in some user groups. Extending the insights by Badakhshan et al. [3], one can assume that those roles and the affordances are interrelated. Furthermore, we illustrate how user groups intend to actualize their affordances by utilizing specific educational material.

With that, the study's contribution is threefold. First, we extend the literature on the organizational embedding of PM, by examining user groups beyond the ones within the implementing organization, including major stakeholders in the PM ecosystem. We map their affordances, informed by courses taken, and explore first insights on the interaction and co-dependency of user groups. Second, we explore how education drives the actualization of affordances. Analyzing real user behavior via a vast learning data set, we confirm the practical applicability of PM education as an affordance actualization driver. These findings underline the importance of user group-specific PM education to facilitate PM success. Third, these findings extend prior technology-centric research on PM, opening the field to research on value co-creation within the PM ecosystem.

For practitioners, this study offers valuable insights as well. We increase awareness of roles within a PM ecosystem to support collaboration and joint value creation through PM. Following our archetypes, vendors of PM solutions can target their education more precisely to the groups, and users of PM can benefit from personalized programs that help in achieving their goals with PM.

Still, the study contains certain limitations. Concerning the method, digital trace data research is a promising yet developing research method that builds on the positivist ontology, assuming that reality can (partly) be captured in log data. While we have a large-scale data set to base our insights on, this data set is based on one technology provider and might not be applicable to all PM software providers and their offerings. We highly encourage future studies to go beyond our findings by triangulating them with interview material or applying them in the context of other PM vendors, following a more interpretivist paradigm.

Looking at the conceptualization of education as an affordance actualization driver, we only consider formal learning, as there is no data source available on informal learning on PM, which might happen through interaction with other individuals or based on work experience of individuals. While we see education as an important driver for affordance actualization, future research is advised to investigate other drivers influencing the affordance actualization of individuals in their relationship with PM.

Finally, with regard to the nature of affordances, we see a first indication of the relations between user groups based on their affordances. However, we have not been able to generate insights into how those similarities and synergies between affordances affect the collaboration between different stakeholders in the PM ecosystem. Thus, we call for future research on the interaction between those stakeholders and their various role profiles to better understand the value co-creation. We propose a longitudinal study on changes in affordances over time to better understand how the relationship between technology and users matures and if education has an ongoing effect as an affordance actualization driver.

References

1. Accenture. Accenture and celonis form strategic alliance to help clients unlock new value in business processes (2022). https://newsroom.accenture.com/news/accenture-and-celonis-form-strategic-alliance-to-help-clients-unlock-new-value-in-business-processes.htm. Accessed 4 March 2025
2. Ammann, J., Lohoff, L., Wurm, B., Hess, T.: How do process mining users act, think, and feel? Bus. Inform. Syst. Eng. 1–23 (2025)
3. Badakhshan, P., Wurm, B., Grisold, T., Geyer-Klingeberg, J., Mendling, J., Vom Brocke, J.: Creating business value with process mining. J. Strat. Inf. Syst. **31**(4), 101745 (2022)
4. Beverungen, D.: Seven paradoxes of business process management in a hyperconnected world. Bus. Inform. Syst. Eng. **63**(2), 145–156 (2021)
5. Cook, S.D.N., Brown, J.S.: Bridging epistomologies: the generative dance between organisatoral knowledge and knowing. Organ. Sci. **10**(4), 381–400 (1999)
6. Eggers, J., Hein, A., Böhm, M., Krcmar, H.: No longer out of sight, no longer out of mind? how organizations engage with process mining-induced transparency to achieve increased process awareness. Bus. Inf. Syst. Eng. **63**(5), 491–510 (2021)
7. Engert, M., Chu, Y., Hein, A., Krcmar, H.: Managing the interpretive flexibility of technology: a case study of celonis and its partner ecosystem. In: International Conference on Information Systems (2021)
8. Gibson, J.J.: The theory of affordances. Hilldale, USA (1977)
9. Grisold, T., Kremser, W., Mendling, J., Recker, J., vom Brocke, J., Wurm, B.: Generating impactful situated explanations through digital trace data. J. Inf. Technol. **39**(1), 2–26 (2024)
10. Joas, A., Matzner, M.: Unleashing digital process innovation with process mining: designing a training concept with action design research. In: European Conference on Information Systems (2022)
11. Karahanna, E., Xu, S.X., Xu, Y., Zhang, N.A.: The needs–affordances–features perspective for the use of social media. MIS Q. **42**(3), 737–756 (2018)
12. Keller, R., Stohr, A., Fridgen, G., Lockl, J., Rieger, A.: Affordance-experimentation-actualization theory in artificial intelligence research – a predictive maintenance story. In: International Conference on Information Systems (2019)
13. Kerremans, M., Sugden, D.: Magic quadrant for process mining platforms (2024). https://www.gartner.com/doc/reprints?id=1-2HGMM7VN&ct=240502&st=sb. Accessed 4 March 2025
14. Kpmg. Process Mining Services (2025). https://home.kpmg/nl/en/home/services/advisory/technology/data-and-analytics/analytics/data-process-mining.html. Accessed 4 March 2025
15. Lehrer, C., Wieneke, A., Vom Brocke, J.N., Jung, R., Seidel, S.: How big data analytics enables service innovation: materiality, affordance, and the individualization of service. J. Manage. Inform. Syst. **35**(2), 424–460 (2018)
16. Leidner, D.E., Gonzalez, E., Koch, H.: An affordance perspective of enterprise social media and organizational socialization. J. Strat. Inf. Syst. **27**, 117–138 (2018)

17. Leonardi, P.M.: When flexible routines meet flexible technologies: affordance, constraint, and the imbrication of human and material agencies. MIS Q. **35**(1), 147–167 (2011)
18. Mamudu A., Bandara W., Wynn M.T., Leemans S.J.J.: Process mining success factors and their interrelationships. Bus. Inform. Syst. Eng. 1–20 (2024)
19. Markus, M.L., Silver, M.S.: A foundation for the study of IT effects: a new look at desanctis and poole's concepts of structural features and spirit. J. Assoc. Inform. Syst. **9**(10/11), 609–632 (2008)
20. Martin, N., et al.: Opportunities and challenges for process mining in organisations results of a delphi study. Bus. Inf. Syst. Eng. **63**(5), 511–527 (2021)
21. Mendling, J., Berente, N., Seidel, S., Grisold, T.: The philosopher's corner: pluralism and pragmatism in the information systems field: the case of research on business processes and organization routines. ACM SIGMIS Database: Database Adv. Inform. Syst. **52**(2), 127–140 (2021)
22. Østerlund, C., Crowston, K., Jackson, C.: Building an apparatus: refractive, reflective, and diffractive readings of trace data. J. Assoc. Inf. Syst. **21**(1), 1–22 (2020)
23. Ostern, N.K., Rosemann, M.: A framework for digital affordances. In: European Conference on Information Systems (2021)
24. Ritz, E., Joas, A., Wambsganss, T., Rietsche, R., Leimeister, J.M.: How a Swiss luxury retailer implements process mining to improve data-driven customer excellence. J. Inform. Technol. Teach. Cases 1–9 (2024)
25. Roboyo Group Limited. Roboyo announces global Partnership with Process Mining Leader Celonis (2022). https://roboyo.global/all-insights/press-release/roboyo-announces-global-partnership-with-process-mining-leader-celonis/. Accessed 4 March 2025
26. Rott, J., Böhm, M., Krcmar, H.: Laying the ground for future cross-organizational process mining research and application: a literature review. Bus. Process. Manag. J. **30**(8), 144–206 (2024)
27. Stendal, K., Thapa, D., Lanamäki, A.: Analyzing the concept of affordances in information systems. In: Hawaii International Conference on System Sciences, pp. 1–9 (2016)
28. Strong, D.M., et al.: A theory of organization-HER affordance actualization. J. Assoc. Inf. Syst. **15**, 53–85 (2014)
29. Teece, D.J.: Explicating dynamic capabilities: the nature and microfoundations of (sustainable) enterprise performance. Strateg. Manag. J. **28**(13), 1319–1350 (2007)
30. Van der Aalst, W.: Data science in action. In: Process Mining, pp. 3–23. Springer Berlin Heidelberg, Heidelberg, Berlin (2016)
31. Venkatesh, V., Bala, H.: Technology acceptance model 3 and a research agenda on interventions. Decis. Sci. **39**(2), 273–315 (2008)
32. Venkatesh, V., Thong, J.Y.L., Xu, X.: Consumer acceptance and use of information technology: extending the unified theory of acceptance and use of technology. Manag. Inf. Syst. Q. **36**(1), 157–178 (2012)
33. Volkoff, O., Strong, D.M.: Critical realism and affordances: theorizing IT-associated organizational change processes. Manag. Inf. Syst. Q. **37**, 819–834 (2013)
34. Vom Brocke, J., Jans, M., Mendling, J., Reijers, H.A.: A fivelevel framework for research on process mining. Bus. Inf. Syst. Eng. **63**(5), 483–490 (2021)

Bridging the BPM Education Gap: Reflections on Affordances of BPM Techniques and a Call for Collaboration

Mahendrawathi ER(✉) and Avisha Nathania

Institut Teknologi Sepuluh Nopember, Surabaya, Indonesia
mahendra_w@is.its.ac.id

Abstract. This paper reflects on our experience in Indonesia, where we face a challenge to align the Business Process Management Education with practice. Despite the effort to include the BPM course in the curriculum, in practice, BPM is applied in a fragmented way and not as a holistic management approach. The paper uses affordance theory as a theoretical lens to examine what BPM techniques are applied by alumni in their workplace and what justifies their use. A preliminary study involving alumni interviews (16 alumni) and a survey (52 alumni) showed different uses of BPM techniques depending on each technique's affordance existence, perceived utility, and ease of application. Automation's strong affordance and contextual fit led to consistent actualization. Root Cause Analysis and qualitative tools were flexibly applied. BPMN as a modeling standard, despite being taught, was rarely realized. The findings highlight that alumni's ability to apply BPM remains limited by many factors such as job roles and positions, organizational awareness and culture. We invite further discussions and collaboration to address these challenges to design BPM education that is flexible and context-sensitive, so our discipline can be inclusive and bring meaningful societal impact globally.

Keywords: BPM Education · BPM Practice · BPM Technique · Affordance

1 Introduction

Business Process Management (BPM) is a holistic management discipline that integrates a range of methods, techniques, and tools to support the effective governance of business processes across their entire lifecycle from initial discovery through analysis, execution, and monitoring [18]. Previous studies have shown that organizations with strong process orientation tend to perform better compared to those that do not adopt such an approach [2, 11, 20]. Furthermore, the implementation of BPM practices has been positively associated with improvements in organizational performance [16, 20].

The Information Systems Department at Institut Teknologi Sepuluh Nopember (IS-ITS) was initially established with a curriculum deeply rooted in computer science. A tracer study conducted as part of the curriculum updates every five years revealed that many IS-ITS alumni are employed in roles in ERP implementation. Based on this,

M. ER et al. (Eds.): BPM 2025, LNBIP 565, pp. 442–455, 2026.
https://doi.org/10.1007/978-3-032-02936-2_31

IS-ITS introduced Business Process Design and Management in 2014 curriculum, then Business Process Management course in 2018 and 2023 curricula.

Despite the effort from the academic side, BPM adoption in practice remains limited in the Indonesian context. A survey conducted by BPTrends in 2020 highlights that most BPM initiatives are concentrated in Europe, North America, and Australia, with Asia, particularly Indonesia, being underrepresented [6]. While ERP adoption in Indonesia continues to grow, the overall market value of ERP systems is projected to increase at a compound annual growth rate (CAGR) of 7.01% between 2022 and 2028 [9], this is not followed by widespread uptake of BPM as a comprehensive management approach.

These conditions have both micro and macro implications for BPM education. On the micro level, graduates from institutions like ITS often find it difficult to apply BPM concepts in a structured and integrated way at work. This also points to how BPM education has not been inclusive or well-adapted globally. It calls for BPM education that goes beyond just theory and technical skills. It should also help students understand and deal with the practical and organizational challenges of applying BPM in their local settings.

This paper uses affordance theory as a theoretical lens to examine what BPM techniques are applied by alumni in their workplace and what justifies their use. The findings contribute to BPM literature by showing how the use of different BPM techniques is influenced by their perceived usefulness, ease of application and contextual factors in the workplace.

This paper addresses the Forum themes "Teaching BPM for underrepresented groups" by focusing on a group of learners whose contexts differ from those in mainstream BPM curricula. The goal of this paper is to raise awareness of this gap, share preliminary findings from Indonesia, and initiate a collaboration on curriculum adaptation and pedagogical strategies that cater to diverse socio-economic and organizational settings.

The remainder of this paper is organized as follows. In the following sections, the background work on BPM education is provided. This is followed by a detailed description of the BPM practice and education in Indonesia. In Sect. 4, we discussed preliminary research on the affordance of BPM techniques, followed by a discussion on the findings' implications in Sect. 5. We describe our call for collaboration in Sect. 6 before closing with a brief concluding remark in the final section.

2 Previous Studies on BPM Curriculum and Industry

The issue of aligning academic instruction and industry needs has been the focus of several studies. According to a study in Germany, only a small percentage of academic programs incorporate BPM content. BPM courses are often introduced late in the curriculum and primarily focus on modelling methods such as Event-driven Process Chains (EPC) and Business Process Model and Notation (BPMN) [10]. Another study in Poland found that BPM graduates often lack practical competencies, such as process modelling and proficiency with digital tools needed in BPM implementation [19]. A study in Australia emphasizes the need for a comprehensive BPM curriculum that incorporates managerial and technical skills, particularly due to ongoing digital transformation challenges [21].

These studies emphasize the need for developing BPM education that is comprehensive and considers different contextual issues. A collaboration between academia and industry is needed to ensure graduates are equipped with relevant skills and conceptual frameworks that can be applied in practice. However, these studies are largely based in countries where BPM is already well established. More attention is needed to adapt and align BPM curricula in regions where the discipline is still emerging.

3 Context: BPM Practice and Education in Indonesia

3.1 BPM Education in Information Systems ITS (IS-ITS)

The Department of Information Systems at Institut Teknologi Sepuluh Nopember (IS-ITS) was established in 2001 with its roots in computer science. The curriculum is reviewed every five years to ensure its relevance to the market demands. Table 1 shows the evolution of IS-ITS, the graduate profiles and the position of BPM within the curricula. Since 2004, Enterprise Resource Planning (ERP) has been included in the IS-ITS curriculum, recognizing ERP as a specific type of enterprise information systems.

Table 1. BPM Position in IS-ITS Curricula

Curriculum	Graduate Profile	BPM Course
2009	Academics, Consultants, Managers, IT Professionals, Technopreneurs	Not part of the curriculum
2014	IS Developer, Consultant & System Integrator, Database Specialist, IS Academics	Business Process Design and Management (BPDM)
2018	Data & Information Engineer, System Analyst, Project Manager, Business Information Manager, Business Analyst, IT Business Initiator	Business Process Management (BPM)

A tracer study as part of the curriculum review in 2014 indicated that many of the IS-ITS alumni work as consultants, particularly in ERP implementation. At the same time, benchmarking with global trends shows an increasing relevance of BPM and ERP implementation is considered a form of process automation within the broader Business Process Management (BPM) lifecycle. To accommodate these market signals, IS-ITS introduced the BPM course in the curriculum, i.e., Business Process Design and Management (BPDM), in the curriculum in 2014, and Business Process Management (BPM) in the curricula in 2018 and 2023. These courses are based on the widely accepted textbook, Fundamentals of Business Process Management [3]. The BPDM course in curriculum 2014 focuses on the modeling, analysis and redesign of the BPM lifecycle. The BPM course in the later curricula focuses on the BPM lifecycle stages, including process identification, discovery, analysis, redesign, and implementation. It aims to develop student competencies in identifying, discovering and modeling business processes, analyzing them both qualitatively and quantitatively, and utilizing BPM tools to simulate or automate process models.

3.2 BPM Practice

Despite the academic effort to include BPM in the curriculum, the BPM practice remains limited. Many organizations in Indonesia started to implement ERP systems at the turn of the Millennium. This allowed us to conduct investigations using process mining in several organizations between 2010 and 2018. For example, our study in a shoe manufacturing company showed the disconnect between procurement, production and delivery processes even after they implement ERP systems [4]. Another study in a telecommunications company showed a very low completion rate (only 8%) because field officers faced difficulty in processing orders, and their resolutions are recorded manually or in different systems [5]. A fertilizer company faced procurement inefficiencies due to manual workflows and data migration issues [7]. These cases emphasize IT and even ERP systems implementation focus more on the automation of tasks and have not fully embraced the other aspects of BPM.

The move toward Electronic-Based Government Systems under Presidential Regulation No. 95/2018 pushed the business process agenda in the government sector [15]. Under Regulation of the Minister of Administrative and Bureaucratic Reform No. 59/2020, every government agency must prepare its Business Process Maps [12]. In practice, IS-ITS supported Surabaya's COPACR in developing business process architecture and identifying improvement areas [1, 22]. Despite this positive trend, a 2020 industry survey of 190 firms conducted by the first author found that while many Indonesian organizations recognize and use process-related techniques, such as continuous improvement, benchmarking, business process re-engineering and value-added analysis, most are still unfamiliar with the term BPM indicating that most Indonesian organizations have not yet adopted BPM.

4 Preliminary Research on Affordance of BPM Techniques: Insights from IS-ITS Alumni

Section 3 shows a challenge for BPM educators in Indonesia. On the one hand, there is demand and opportunity for graduates with BPM skills. But, on the other hand, the adoption of BPM remains slow and fragmented. To understand what causes this disconnect we use the general adoption view and affordance as a theoretical lens.

4.1 Theoretical Lens: BPM Adoption and Affordance

Figure 1 presents the conceptual framework, which combines the adoption perspective with affordance as the theoretical lens guiding this preliminary study.

First, align with [17], we conceptualized that organization-wide BPM adoption tends to go through multiple stages, such as (1) awareness and understanding of BPM, (2) desire to adopt BPM, (3) setting up, executing and monitoring BPM projects, (4) converting

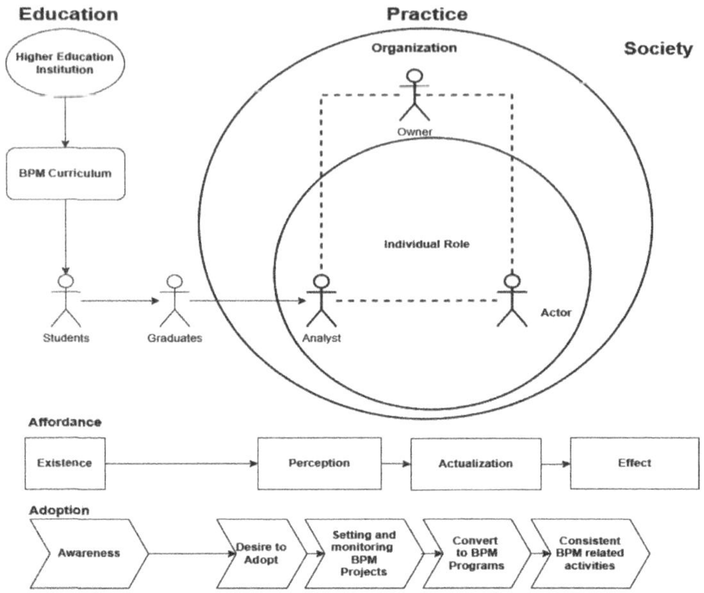

Fig. 1. Conceptualization of BPM Education from Adoption and Affordance Theory

BPM projects into a BPM program, and (5) ensuring that all BPM-related activities are consistently delivered in a cost-effective way. To move from the initial "awareness" to the "desire to adopt" stage means organizations must recognize the affordance of BPM and perceive its potential benefits. This perception is a critical point to the implementation of BPM initiatives.

This adoption process is further conceptualized using the affordance theory. Affordance is the potential actions or interactions that the environment enables between individuals and objects. These affordances represent the opportunities or resources that the environment offers to support behaviors [8]. Information systems researchers adopt the theory to develop an IT affordance framework, which includes four dimensions: existence, perception, actualization, and effect [13, 14]. Affordance existence is a cognitive process because of the interaction between the IT artefact and the organization. The process of recognizing the existence of affordance is known as affordance perception, and it is impacted by the following factors: (1) the features of the object; (2) the actor's capabilities; (3) the actor's goal; and (4) outside information [14]. Affordance actualization refers to the ways organizations use IT to achieve their goals and act in accordance with these perceptions. The tangible results of these actions, tangible behaviors that produce immediate and quantifiable results, are known as affordability effects.

In BPM, affordance begins with a person understanding the existence of a capability, like a tool or method, and recognizing its potential. However, simply being aware of the existence of BPM concepts or tools' affordance is not enough. Organizations must recognize how these BPM capabilities could be used to improve processes. In other words, transferring existence to perception. Finally, affordance actualization is when people in the organization apply their recognition of BPM capability and use it within their specific situation. This actualization is what leads to **affordance effects**, real, visible outcomes like more transparent, accurate or efficient processes.

Higher education institutions (HEIs) like IS-ITS play a critical role in enabling BPM adoption by contributing to the affordance dimensions of existence and perception. Through structured curricula, we introduced students to BPM concepts, tools, and techniques, thereby establishing the availability of BPM knowledge (existence) and fostering recognition of its practical use (perception). This educational foundation supports the development of BPM capabilities that graduates carry into professional environments. Upon entering the workforce, graduates may assume roles such as process analysts, actors, or owners, each contributing to BPM implementation. These roles operate within organizational settings but should ultimately extend to broader societal contexts, highlighting the wider impact of BPM education.

Based on this conceptualization, we derived two research questions: To what extent does our BPM education lead to actualization? What BPM techniques are used by our alumni in the workplace, and what justifies their selection? By understanding the drivers and barriers to the actualization of BPM techniques, we can better align our teaching with the practical needs of Indonesian organizations and more effectively inspire BPM adoption.

To answer the research questions, we conducted preliminary research using two data collection methods, i.e., interviews and a survey with IS-ITS alumni from different curriculum years (2009, 2014, and 2018). The findings will be discussed in the following sections.

4.2 Alumni Interviews

The alumni interviewed are purposefully sampled to represent several key areas: the curriculum under which the alumni studied, their exposure to BPM during their academic program, and their current professional roles. We could not find any alumni who work specifically in BPM-specific roles. So, we chose related roles such as ERP specialist, IT specialist, etc.

The interview involves 16 alumni from an Information Systems program across three curriculum cohorts. These alumni currently hold a range of roles, including Enterprise Systems Engineer (6 alumni), Agile IT Specialist (4), Data Analyst (3), Business Analyst (1), and two in project-based or mixed roles related to business processes. The alumni involved in the interviews have different exposure to BPM, as 13 of the 16 alumni completed final projects related to BPM, ERP, or related process topics. Additionally, 7 alumni had internships involving BPM work, and 4 served as teaching assistants for BPM or Enterprise Systems courses. Information about the alumni involved in the interviews can be found in the supplementary material (link in the appendix).

During the interview, the alumni were asked to identify which BPM techniques, including BPMN, qualitative process analysis (value-added and waste analyses), root cause analysis, quantitative process analysis, and automation, they had applied in practice. These techniques were selected because they are the ones taught in depth during the course. They were asked whether each technique had been used in their professional context and the reasons for the use or non-use. Each interview lasted approximately 30 min on average.

4.3 Alumni Survey

A follow-up survey was conducted to obtain more data regarding the use of BPM techniques in the alumni workplace. The same questions as in the interviews are given to the participants. A total of 52 alumni participated in the survey, grouped based on the curriculum they followed during their studies. Eight alumni studied under the 2009 curriculum, which did not include a BPM course. These alumni were exposed to BPM through work experience, internships, or independent learning, such as a thesis or certifications. Twenty-two alumni studied Business Process Design and Management in the 2014 curriculum. These alumni studied BPM in a more structured way, with additional experience from serving as teaching assistants, working on BPM-related final projects, or applying BPM during internships. The rest twenty-two alumni studied under the 2018 curriculum, which included a BPM course. These alumni have additional experience as teaching assistants, worked on BPM-focused final projects, and applied BPM knowledge during internships or in their current jobs. The alumni worked in different roles such as ERP Consultant, Business Analyst, Data Analyst, Software Engineer, QA Engineer, Project Manager, and Auditor (More details can be found in the supplementary materials).

4.4 Findings from Alumni Interviews and Survey

Interviews and survey responses were analyzed through the affordance lens. First, we transcribe the interview results. Then, for each informant, we read the answers line-by-line and map them with the three phases of affordance: existence, perception and actualization. An example of our analysis of Alumni 1, who studied under the 2009 curriculum, is shown in Fig. 2. When asked whether she used BPMN in the workplace, she immediately recalled that she was not taught about it during her undergraduate studies. But she is aware of BPMN's existence as a technique to visualize a process model.

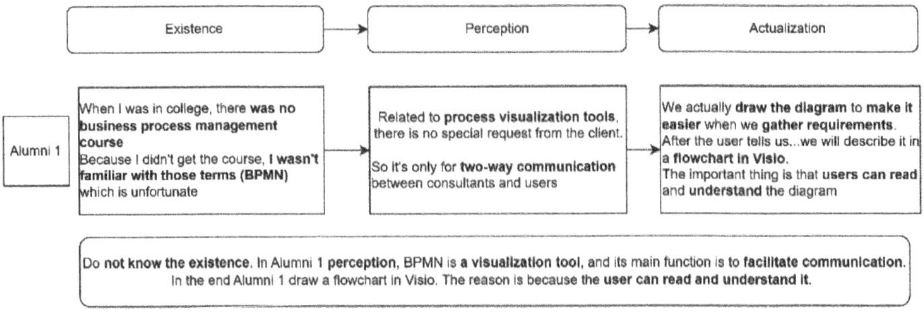

Fig. 2. Example of BPMN Affordance Analysis

Her perception is that BPMN is mainly a process visualization tool, so it serves as a two-way communication between consultants and users. She does not use the technique because she only draws diagrams to make it easier to gather requirements from the user. To do that, she feels that a flowchart is easier to understand by the users. Similar analyses were conducted for the rest of the alumni and all the BPM techniques. More detailed analysis of BPMN affordance and other techniques from the alumni interviews and survey can be found in the supplementary material.

We will now discuss the key findings regarding the affordance of BPM techniques. Overall, the use of BPM techniques can be seen in Fig. 3. The most widely used technique is automation, with 76.47% of alumni claim to use it in their workplace. BPMN is used by 55.48% of alumni. Interesting finding is for root-cause analysis. While most alumni claim to use it, only 50% use it formally, while 38.24% claim to use it conceptually, i.e., not drawn in a formal depiction.

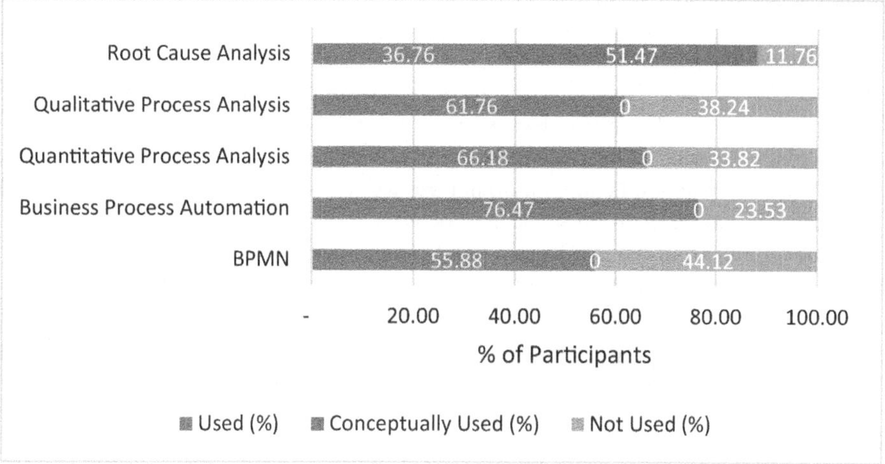

Fig. 3. Use of BPM Techniques among IS-ITS Alumni

Further analysis with the affordance lens allows us to gain more insights into the existence (cognitive knowledge), perception (recognition) and actualization of each BPM technique.

BPMN *existence* is found among alumni from the 2014 and 2018 cohorts because they are formally trained with it during BPM courses. Alumni from the 2009 curriculum had limited exposure, and those who know its *existence* often learn it informally through workplace demands. Despite the difference in *existence*, *perceptions* of BPMN are consistent across all groups. BPMN was recognized mainly as a visualization tool. Very few alumni perceive its value to enable process simulation and automation. In general, alumni view BPMN as a useful tool for mapping business processes, facilitating team communication, and providing a clear and structured view of processes. However, some alumni do not apply BPMN mainly because their work focuses more on technical tasks, the company already has its own modeling standards (such as flowcharts), or there is a lack of understanding among stakeholders.

The *existence* of waste and value-added analysis as a qualitative technique is high across all alumni in different cohorts, including those without formal instruction. However, alumni have different *perceptions*. Some alumni consider qualitative process analysis important for understanding the added value of ongoing processes, assessing efficiency, improving processes, and supporting decision-making. However, some alumni find this technique less relevant because it is outside their job scope. In *actualization*, qualitative process analysis is mostly used to assess efficiency, identify bottlenecks, recognize value-adding activities, and provide improvement recommendations. However, some alumni do not use it due to roles or job description limitations, a stronger focus on quantitative analysis, and no formal need to do that analysis.

Root cause analysis (RCA), as another qualitative technique, is widely known among alumni, indicating a *strong existence*, regardless of the curriculum. Alumni view RCA as an important technique to identify the root causes of problems so that the solutions provided are effective and prevent recurrence, whether in technical contexts, incidents, or business processes. However, this *strong perception* did not always translate into *actualization*. As can be seen in Fig. 3 some alumni (mostly from later cohorts) use it with formal tools such as fishbone diagrams and 5 Whys, while earlier alumni used the method conceptually. Some alumni do not use RCA due to the irrelevance of their job roles, and it's outside their job scope and focus.

Most alumni from the 2014 and 2018 curricula are aware of quantitative process analysis techniques, indicating their strong *existence*. In terms of *perceptions*, alumni consider quantitative process analysis as a useful technique to measure performance, cost, and efficiency, and to support data-driven decision-making. However, some alumni perceive the technique as time-consuming, requiring deep understanding, and less relevant to their daily work. The varying perception leads to different *actualization*. Some alumni use it for performance evaluation, bottleneck identification, SLA measurement, and supporting automation development. Those who do not use this technique cite the reasons as technical job focus, lack of necessity, highly dynamic processes, or because it is outside their job scope.

Automation is the most applied technique across all alumni groups. Most alumni from the 2014 and 2018 curricula are familiar with and use automation in their work. Although alumni from the 2009 curriculum did not formally take the BPM course, some of them have also implemented automation in practice, especially related to reducing manual tasks and improving operational efficiency. The perception of automation across alumni depends on their job roles. Some perceive automation as effective in speeding up tasks, reducing manual errors, improving efficiency, and supporting interdepartmental integration. However, some feel that automation is less relevant to their job focus and not particularly necessary in their work. Automation is widely used by alumni to speed up repetitive tasks, reduce manual errors, optimize resources, and improve operational efficiency by utilizing tools such as ERP and RPA. However, some do not use automation because it is outside their job scope, and their business processes currently don't require automation.

By interpreting the interview data through the lens of affordance theory, two thematic findings emerged: technique-specific affordances and contextual enablers and constraints:

1. **Technique-Specific Affordances.** Alumni showed different uses of BPM techniques depending on each technique's existence, perceived utility, and ease of application. Automation exhibited the strongest affordance. It was widely recognized (existence), consistently valued for its operational impact (perception), and frequently implemented (actualization), especially in ERP-supported tasks like approvals and integrations, regardless of formal BPM education. In contrast, BPMN had a weak affordance. Alumni in the 2014 and 2018 curricula were taught this standard, so the *affordance existence* of BPMN among these alumni should be high. But the perception of BPMN is primarily as a visualization tool and was perceived as complex and misaligned with typical workplace communication styles (perception). As a result, actualization of it is not as high, often replaced by simpler visual tools such as flowcharts. RCA, Value-added and waste, and quantitative analyses showed moderate affordances. Alumni know their existence and perceive them as useful in certain contexts. RCA is often used conceptually, while quantitative techniques are more common in data-driven roles.
2. **Contextual Enablers and Constraints.** The findings suggest that affordance actualization depended heavily on context. Alumni often began careers in roles where BPM as a holistic concept was not seen as relevant. Organizational culture also influenced tool uptake. For example, environments that were not familiar with BPMN discouraged its use.

The findings suggest that consistent actualization is closely tied to a technique's affordance and its alignment with organizational context. Automation, with strong affordance and contextual fit, was consistently actualized. RCA and qualitative tools showed adaptable affordance, allowing use even without formal structures. In contrast, BPMN, though included in education, was rarely actualized, indicating a gap between what is taught and what organizational contexts afford.

5 Discussion and Proposed Actions

This study looks at the problems and opportunities of teaching BPM in an Information Systems (IS) study program in Indonesia. While the curriculum equips students with a set of conceptual and technical skills, not all of them are applied in the workplace, especially in environments where BPM is not yet a recognized or institutionalized practice. These findings are relevant beyond Indonesia, especially in contexts with similar cultural or institutional barriers. In this section we discussed the key findings and presented several proposed actions for other educators.

5.1 BPM in the IS Context

The varied use of BPM technique reflects the structure of the IS -ITS study program and the realities of the job market. In Indonesia, roles specifically dedicated to BPM remain limited. A quick LinkedIn search on "business process" only returns 24 jobs, which, after careful investigation not all related to BPM. BPM is frequently one of several specializations offered in IS programs, and the career paths of graduates differ greatly. So, IS-ITS alumni often work on various roles that do not explicitly focus on BPM, such as software engineer, auditor, data analyst, or IT support. Consequently, techniques like process modelling, RCA, and quantitative process analysis are underused, even though graduates are equipped to apply them.

This underutilization is not due to a lack of skill but to a misalignment between what is taught and what is feasible in the early stages of alumni careers. Many alumni start their careers in operational or technical roles, where they do not have the authority to make changes at the process level. However, as they progress into more strategic positions, latent BPM affordances can be reactivated. BPM education, therefore, serves as a long-term investment, preparing graduates for future responsibilities in process leadership and transformation. Its impact becomes more visible as alumni advance in their careers and are better positioned to introduce change.

5.2 Technique Specific Affordance

Our findings suggest that not all BPM techniques are actualized equally. Automation has high affordance and contextual fit, making it more widely applied. Other techniques, particularly BPMN, are taught but rarely used. Despite training, graduates report limited use of BPMN due to misalignment with workplace needs, perceived complexity, or lack of organizational support. This highlights the need for instruction that goes beyond tools and introduces frameworks for selecting and adapting techniques based on context.

Moreover, many organizations lack awareness of the full range of BPM capabilities graduates bring. The use of structured BPM techniques in practice is restricted by informal and legacy methods. BPM education must therefore equip students to be internal change agents who can advocate for greater BPM maturity when organizational circumstances permit, in addition to carrying out BPM tasks.

5.3 Proposed Actions.

To address the identified gaps, BPM education must be both strategic and responsive. The following actions are proposed:

1. **Embed Strategic Framing into Course Materials**. Revise course content to present BPM as a capability for organizational change, and not just a set of tools. This includes integrating case studies where BPM played a strategic role in process improvement or digital transformation.
2. **Introduce a Technique-Context Mapping Module**. Develop a dedicated module that trains students to assess organizational context and select appropriate BPM techniques accordingly. Students should learn to identify when and how techniques like BPMN or quantitative analysis are most useful depending on process maturity and stakeholder readiness.
3. **Incorporate Role-Specific BPM Scenarios.** Design assignments and simulations based on job roles that graduates are likely to encounter (e.g., software developer, data analyst, IT support). This helps students see how BPM competencies apply in less formal or indirect ways, even when BPM is not their main responsibility.
4. **Develop BPM Advocacy Skills Through Communication Practice**. Include classroom activities that build students' ability to explain, justify, and promote BPM techniques to non-expert stakeholders. This could include building a shared vocabulary

to engage stakeholders with differing levels of BPM familiarity, writing business memos, delivering elevator pitches, or conducting stakeholder role-plays.

5. **Curriculum Integration with Industry Realities**. Partner with local companies to co-develop projects, guest lectures, or mentorship programs. This ensures students gain insights into actual BPM practices and limitations in Indonesian organizations, while helping industry partners better understand the potential of BPM-trained graduates.
6. **Long-Term Career Alignment.** Recognize that the actualization of BPM knowledge may be delayed. Embed reflective activities that help students connect BPM concepts with long-term professional development.

6 Call for Collaboration

Although the Indonesian experience serves as the foundation for our work, many educators worldwide may encounter similar challenges in bridging the gap between what we teach and what our graduates face in the workplace.

To validate our findings and translate our proposed actions into concrete steps, we would like to have a more in-depth discussion with other researchers and educators. The following are some of the questions we would like to investigate further.

1. Do you encounter similar challenges at your institution or in your region?
2. How do you modify your BPM curriculum considering informal or fragmented industry practices?
3. Which instructional techniques are most effective when BPM is not yet completely appreciated or understood?
4. How can we make our BPM education more flexible, without losing the fundamentals?

One area that we would like to investigate further is the affordance of BPM techniques and discipline. According to our preliminary research, some techniques, like automation tools, are easily adopted in practice. But others, like BPMN or qualitative analysis, are more difficult to implement unless the organizational culture is supportive. We invite further investigation and discussion of the ways in which BPM affordances manifest in various contexts. For instance:

1. What are the factors that facilitate or impede the implementation of BPM affordance in entry-level positions?
2. How do graduates start to actualize what they have learned into practice as they advance and move up in the corporate ladder?
3. How do organizational or regional variations affect the range of possible practices?

We also welcome collaboration on developing **more contextual teaching materials,** such as case studies, assignments, and simulations that reflect the diversity of BPM maturity and practice across sectors and geographies.

7 Concluding Remarks

This paper contributes to the BPM Educator Forum by highlighting the underrepresented group of learners whose contexts differ from those in mainstream BPM curricula. While we, as BPM educators, designed a curriculum with what we think is the necessary knowledge and skills for the students, their ability to apply BPM remains limited by many factors such as job roles and positions, organizational awareness and culture. We invite further discussions to address these challenges collaboratively to design BPM education that is flexible and context-sensitive, so our discipline can be inclusive and bring meaningful societal impact globally.

Acknowledgement. The authors gratefully acknowledge financial support from the Institut Teknologi Sepuluh Nopember for this work, under project scheme of the Publication Writing and IPR Incentive Program (PPHKI) 2025.

Disclosure of Interests. The authors have no competing interests to declare that are relevant to the content of this article.

Appendix: Supplementary Materials

Supplementary material for this article is available online at https://bit.ly/Eduforum2025.

References

1. Ambarawati, H.M., Er, M., Wibowo, R.P., Sonhaji, A.I.: Business process architecture, data and information architecture: a public sector case. In: Proceedings - ICACSIS 2022: 14th International Conference on Advanced Computer Science and Information Systems, pp. 241–246 Institute of Electrical and Electronics Engineers Inc. (2022)
2. Davenport, T.H., Short, J.E.: The new industrial engineering: information technology and business process redesign. Sloan Manage Rev. **31**(4), (1990)
3. Dumas, M., La Rosa, M., Mendling, J., Reijers, H.A.: Fundamentals of business process management. Springer, Berlin Heidelberg, Berlin, Heidelberg (2018)
4. ER, M., Astuti, H.M.: A case study on process mining implementation in modelling supply chain business process : a lesson learnt. In: 6th International Conference on Operations and Supply Chain Management, vol. **2014**, pp. 808–819. Bali (2014)
5. ER, M., Astuti, H.M., Nastiti, A.: Analysis of customer fulfilment with process mining: a case study in a telecommunication company. Procedia Comput Sci. **72**, 588–596 (2015)
6. E.R, M., Hanggara, B.T., Astuti, H.M.: Model for BPM implementation assessment: evidence from companies in Indonesia. Bus. Process Manage. J. **25**(5), 825–859 (2019)
7. ER, M., Zayin, S.O., Pamungkas, F.J.: ERP post implementation review with process mining: a case of procurement process. Procedia Comput Sci. **124**, 216–223 (2017)
8. Gibson, J.J.: The ecological approach to visual perception. Lawrence Erlbaum Associates (1986)
9. Inkwood Research: Indonesia Enterprise Resource Planning (ERP) Market Forecast 2022–2028. https://inkwoodresearch.com/reports/indonesia-enterprise-resource-planning-erp-market/. Accessed 7 June 2024
10. Koch, J., Koch, J., Sträßner, M., Coners, A.: Theory and practice - what, with what and how is business process management taught at german universities? In: Lecture Notes in Computer Science (including subseries Lecture Notes in Artificial Intelligence and Lecture Notes in Bioinformatics), pp. 34–39. Springer Science and Business Media Deutschland GmbH (2022)
11. McCormack, K.P., Johnson, W.C.: Business process orientation. CRC Press (2001)

12. Minister of Administrative and Bureaucratic Reform: Regulation of the Minister of Administrative and Bureaucratic Reform Number 59 of 2020 (2018)
13. Pedrola, F., Vitari, C., Raguseo, E., Pigni, F.: An organization-and process-based framework of IT an organization-and process-based framework of IT affordances for the IS discipline. In: Mediterranean Conference on Information Systems (MCIS), p. 2023 (2023)
14. Pozzi, G., Pigni, F., Vitari, C.: Affordance theory in the IS discipline: a review and synthesis of the literature. In: AMCIS 2014 Proceedings (2014)
15. President of the Republic of Indonesia: presidential regulation number 95 of 2018 (2018)
16. Reijers, H.A.: Implementing BPM systems: the role of process orientation. Bus. Process. Manag. J. **12**(4), 389–409 (2006)
17. Rosemann, M.: The service portfolio of a BPM center of excellence (2008)
18. Rosemann, M., Vom Brocke, J., Looy, A. Van, Santoro, F.: Business process management in the age of AI-three essential drifts. Inform. Syst. e-Bus. Manage. 415–429 (2024)
19. Senkus, P., et al.: Bridging the gap: an evaluation of business process management education and industry expectations – the case of Poland. In: Lecture Notes in Business Information Processing, pp. 250–262. Springer Science and Business Media Deutschland GmbH (2023)
20. Škrinjar, R., Bosilj-Vukšić, V., Indihar-Štemberger, M.: The impact of business process orientation on financial and non-financial performance. Bus. Process. Manag. J. **14**(5), 738–754 (2008)
21. Syed, R., Wynn, M.T., Bandara, W.: The design and delivery of a holistic BPM education programme. In: Lecture Notes in Business Information Processing, pp. 201–210. Springer Science and Business Media Deutschland GmbH (2023)
22. Viontita, S.C., Er, M., Nurkasanah, I., Sonhaji, A.I.: Determining business process improvement priorities at surabaya city office for population administration & civil registration. In: Proceedings - ICACSIS 2022: 14th International Conference on Advanced Computer Science and Information Systems, pp. 211–218. Institute of Electrical and Electronics Engineers Inc. (2022)

The Envisioned Global Process Institute: Transforming BPM Education Through EaaS and Process Learners in a B2P Model

Andréa Magalhães Magdaleno[1] (✉) and Michael Rosemann[2]

[1] dheka – BPM Consulting and Academy, Rio de Janeiro, RJ, Brazil
andrea@dheka.com.br

[2] Centre for Future Enterprise, Queensland University of Technology, Brisbane, Australia
m.rosemann@qut.edu.au

http://www.dheka.com.br

Abstract. Using prospective theorizing, this paper presents a visionary approach to Business Process Education (BPE) through the proposal of a Global Process Institute (GPI), a two-sided market model for BPE services distinguished by two key innovations: (i) a continuous BPM Education-as-a-Service (EaaS) model leading to upgradeable certificates and (ii) the treatment of business processes as autonomous learners. The latter introduces a novel B2P (Business-to-Process) model that integrates business processes, as active participants in the learning economy. We discuss the service portfolio opportunities tailored to different stakeholders, explore the implications and challenges of this speculative design, and provide a business model to integrate and summarize the distinct features of the proposed Global Process Institute.

Keywords: Global Process Institute (GPI) · Business Process Education (BPE) · Business-to-Process (B2P) · Process Learners · Education-as-a-Service (EaaS)

1 The Rise of Business Process Education (BPE)

Education and learning hold immense significance for contemporary societies and have been foundational for individual, organizational, and national progress, and the advancement of disciplines such as Business Process Management (BPM). In an environment of accelerated change, the ability to unlearn, relearn, and learn new knowledge and skills is paramount [1]. In fact, Arie de Geus argued that *"The ability to learn faster than your competitors may be the only sustainable competitive advantage"* [2].

Nowadays, it's rare for individuals to stick to a single, unchanging career; instead, they will likely pursue a diverse professional trajectory. As we transition between careers throughout our lives, much of the required knowledge will not be what we learned in secondary and tertiary education systems [3]. Whether with one or multiple careers, we must continuously acquire new knowledge and experiences. This makes the future of learning as important as the much more frequently discussed future of work.

M. ER et al. (Eds.): BPM 2025, LNBIP 565, pp. 456–470, 2026.
https://doi.org/10.1007/978-3-032-02936-2_32

Typically, BPM education equips individuals with the necessary knowledge and experiences to analyze, design, implement, and improve business processes [4]. However, due to the field's rapid development (e.g., rising capabilities of AI, impact of new regulations), the knowledge embedded in BPM curricula risks becoming obsolete [5].

Knowledge is an intellectual asset built with significant commitment and investment (time, energy, and finance). Like all assets, the value of knowledge depreciates over time, even if we are unconscious of that. Experience is reflective knowledge gained through practical application. The greater our experience, the more opportunities we have had for reflection and continuous improvement. However, in a changing environment, the need for new experiences increases, and the value of old experiences decreases [1]. Therefore, Business Process Education faces a critical turning point, and we need to rethink how knowledge and experiences are delivered, updated, and applied.

The urgency of this situation is further amplified by a learning economy [1] which emphasizes local service provision over a defined period (number of semesters) only. This means students enroll in a course with BPM content and study until a degree is awarded. Such a model prevents a global best-of-breed approach available on demand, common to other forms of content (e.g., entertainment), and assumes a formal endpoint for structured learning.

Even further, this human-centered approach to learning has now been complemented by an additional type of learner: Machine Learning (ML). Advances in Artificial Intelligence (AI) have led to the maturation of the approaches used for supervised, unsupervised, and reinforcement learning, and with this, catapulted ML to an equal counterpart in the learning economy. Machines as learners, however, have not yet made it into the classroom. The established and well-institutionalized form of learning provides for human learners only; a comparable educational system does not exist for non-humans such as robots, chatbots, and other AI-empowered agents. The uptake of AI and ML in BPM has now elevated business processes to become agents on their own with an increased level of autonomization [6, 7]. This leads to the question of how business processes can go beyond ML from local data sets and learn from global educational assets.

These trends on the future of learning and the best way to cater to machines are significant to Business Process Education. In BPE, traditionally, humans in roles such as process analysts or process owners have undergone regular training to gain their contemporary knowledge and expertise. This demand for BPE is increasing, which is also evidenced by the range of BPM educational service providers, such as universities [8, 9] and corporate training offerings (ABPMP, etc.). However, despite the sense of urgency, BPE, unlike BPM, remains scattered and lacks attention in real-world education and training [5].

Motivated by the need to design an expanded model for BPE, this paper proposes the notion of a Global Process Institute (GPI) as a template for a matured, contemporary approach to orchestrate ongoing BPM education, addressing process professionals as well as business processes themselves. Thus, we ask the following research question:

How can a Global Process Institute ensure continuous BPM education targeting BPM professionals and business processes?

Adopting prospective theorizing [10], we develop a possible future model of a new BPM learning environment in which process professionals and business processes can coexist in a global learning system. Anchored in values drawn from the new learning economy [1]—personalization, continuity, and innovation—we propose a desired future with a novel B2P (Business-to-Process) educational business model. In line with the criteria of speculative rigor [10], we imagine a coherent and plausible vision that not only expands the conceptual boundaries of BPE but also inspires experimentation and future research into its practical implementation.

Our envisioned but non-existent-yet future is a global BPE marketplace, where educational service providers can offer award and non-award courses, hands-on training, related services, and content in a two-sided market model. It is based on a continuous Education-as-a-Service (EaaS) model, a subscription service allowing the ongoing upgrade of degrees and certificates awarded, instead of one-time, static courses. Finally, the GPI is tailored to the shift from human-only learning to "process learners", i.e., business processes as autonomous learners, applying ML and AI techniques. Empowered by advanced technologies, processes are increasingly capable of self-analyzing, self-monitoring, self-improving, and even self-innovating.

2 The Global Process Institute

The Global Process Institute (GPI) is grounded in the methodological lens of prospective theorizing [10]. Rather than extrapolating from existing realities, this approach encourages the imaginative construction of desirable futures. The design of the GPI was guided by the three core criteria of speculative rigor from prospective theorizing [10]: plausibility, desirability, and generativity. The proposal is plausible since it builds upon existing technological capabilities, such as AI, ML, process mining, and known two-sided market models. It is desirable because it addresses recognized limitations in Business Process Education, such as fragmentation and obsolescence, offering a more adaptive and continuous alternative. Finally, it capitalizes on the generativity that comes with platform business models. Together, these criteria ensure that the GPI is not merely imaginative but intellectually robust and strategically actionable.

We introduce the GPI as a desired but non-existent future global ecosystem for BPE. GPI aims to bring together universities, providers of BPM educational services, products and services, professionals, students, and also business processes themselves within a global hub for curated BPM education services tailored for a worldwide audience of learners motivated by awards as much as the desire to gain relevant BPM skills. The GPI aims to become a trusted partner for the educational well-being of BPM professionals and a provider of continuous learning in the form of updates, new BPM training data, or access to proactive process improvement proposals sourced globally and contextualized to the individual process characteristics.

GPI focuses on continuous BPM education, defined as the ongoing, lifelong [11] process of acquiring, updating, and applying BPM knowledge and skills throughout professional careers. It contrasts with one-off or degree-based learning by emphasizing sustained development. Continuous BPE involves reskilling, upskilling, unlearning, and maintaining knowledge relevance.

The GPI aims to adopt a collaborative and merit-based content acquisition process. Periodically, the Institute will launch public calls, outlining specific BPM topics and learning needs identified by the community or emerging from market trends. BPM educators are invited to submit proposals for developing learning materials aligned with these themes. Submissions are evaluated by a multidisciplinary review panel based on criteria such as pedagogical quality, practical applicability, and originality. Selected proposals for each theme will be produced and integrated into the BPE's knowledge base, ensuring that learners receive high-quality, up-to-date content developed by recognized experts in the field.

BPM educators retain authorship and creative rights over their materials. In recognition of their contribution, BPM educators are commissioned based on the usage of their content, following a streaming-based model inspired by platforms such as Spotify. This ensures a scalable remuneration: the more the content is accessed, viewed, and valued by learners, the higher the educator's earnings.

2.1 Stakeholders and Services of the GPI

The GPI connects multiple audiences and creates value by mediating interactions between them. Each stakeholder group engages with the Global Process Institute to pursue distinct goals and interests, as summarized in Table 1. This diversity of interests is fully addressed in the GPI's design. The Institute would be inclusive and not limited to serving a single group, a specific geographic region, or a single BPM course.

GPI offers tailored educational services to its multi-faceted community of stakeholders. Professionals and students interested in BPE can get an overview of what is available, helping them decide which options might best suit their purposes. The envisioned initial set of initiatives addresses the unique needs of each stakeholder group. It fosters continuous education, career advancement, and process excellence within the global BPM community, positioning the GPI as a catalyst for the future of BPM learning.

At the GPI, universities play a crucial role as the main strategic partners and providers of validated BPE content. Their primary objectives include joining an international network, expanding their academic reputation, and exploring new revenue sources and distribution channels. Through the GPI, universities extend their reach to a global student body and have the potential to convert learners and the organizations they are working for into consumers of university services. Therefore, GPI's Global University Partnership (GUP) invites universities worldwide to join a network of BPM education providers. Partner universities can receive the GUP seal and gain early access to state-of-the-art materials and benchmarks, collaborate in academic committees, positioning themselves as global leaders in BPE.

Companies engage with the Global Process Institute to upskill their workforce, access fresh and trained BPM talent, and stay aware of innovations applicable to their processes. Their goals revolve around improving organizational process maturity, fostering process awareness, and connecting with a highly skilled talent pool. Companies benefit from access to carefully curated BPM training and the assurance that their BPM professionals possess contemporary BPM knowledge and skills. Their interests lie in maximizing ROI through process improvements, maintaining competitiveness, and achieving sustainability and compliance objectives through BPM.

The BPM Corporate Excellence Program is a BPM training and transformation program designed for companies. Many educational offerings and learning experiences can be delivered, including micro training courses, gamified team training, and tailored case studies. The program culminates in a BPM Corporate Certification that recognizes the organization's commitment to BPM and investment in process capability building.

Table 1. Stakeholders' objectives, benefits, and interests.

Stakeholders	Objectives	Benefits	Interests
Universities and providers of BPE services	Additional distribution channel for BPE content, new revenue source, reputational gain, and to acquire students	Co-creation, funding, visibility	Global impact, partnerships
Companies	Continuous upskilling of the workforce, and regularly updated business processes	Training, real-world projects	ROI, competitiveness
Individual BPM Educators	Personal profiling, additional revenue source, flow-on service income (consulting, mentoring)	Content, certifications, teaching methods	Engagement, professional development
BPM Professionals	Career evolution, remaining competitive, the immediate application of new skills	Certifications, networking	Performance, growth
BPM Students and Learners	BPM foundation, job preparedness	Courses, scholarships	Job market entry, modern education
Machine Process Learners	Self-optimization	Input data, benchmarks	Performance, compliance

BPM educators (professors, trainers, mentors, and coaches) find an environment in the GPI to enhance global recognition. They can produce and deliver BPE content that is aligned with emerging BPM trends and stay relevant to new generations of learners. Educators benefit from participation in the BPM Educator Academy, a specialized train-the-trainer program that empowers them to modernize their teaching practices worldwide by offering instructional design training, ready-to-use educational resources (slides, books, articles, teaching cases, syllabi, templates, exercises, simulators, webinars, and software, to name a few), and hands-on training on the platform usage. Educators who complete the program receive a credential and join an exclusive community. The continuity is promoted through education development associated with badges and educator levels. The educators' interests center on continuous professional development, increasing student engagement, and driving innovation in BPE.

BPM professionals, such as process analysts, managers, specialists, and consultants, look to the Institute for continuous professional growth and practical application of BPM methodologies. They aim to enhance their skills, achieve globally recognized certifications, and bring new insights and tools into their organizational contexts. Through a modular program covering different theoretical and practical subjects, professionals complete a real-world project. GPI equips BPM professionals with advanced knowledge and practical experiences to drive organizational transformation. They benefit from personalized learning pathways, access to a global professional network, and tangible credentials such as diplomas and digital badges. Their primary interests include career advancement, enhanced performance in their roles, and broader recognition within the BPM community.

Students and BPM learners—whether undergraduates, graduates, or lifelong learners—are at the heart of the Institute's mission to democratize BPM education. Their objectives include building a strong foundational knowledge of BPM, preparing for successful careers, and gaining practical experience through real-world projects and internships. Students benefit from free or affordable access to BPE content produced by specialists worldwide and concentrated in a single platform. Their primary interests are securing employability, experiencing modern, flexible learning environments, and earning globally respected BPM credit points.

The Process Futures Program is tailored to students and emerging BPM professionals aiming to fast-track their careers. By combining intensive learning sprints, hands-on mini-projects with companies, career mentoring, and scholarship opportunities for standout students, the program prepares participants for immediate impact in the BPM job market and future leadership roles.

The program starts with a learning diagnostic, which allows individuals to assess their current BPM knowledge, skills, and experience through a structured self-diagnosis process. Learner profiles encompass many characteristics, including learning preferences, interests, and strengths. Based on the results, BPM learners receive personalized pathways tailored to accommodate their needs, preferences, learning styles, career goals, and maturity levels. This approach transforms BPE from a one-size-fits-all model to a dynamic, learner-centric journey, ensuring that every participant advances through the most relevant and impactful learning experiences. One could even imagine that the use of generative AI will soon lead to precision education [6].

As part of this personalized journey, each learner gains access to a customized educational playlist—a Learn List—that curates the most relevant BPM content based on their individual profile. This Learn List could be curated by a BPM expert, a co-leader, or dynamically generated, guiding the learning experience according to diagnostic results. Upon completing each module, learners can earn academic credit points recognized by leading universities worldwide, enabling them to build a globally respected learning record. In this way, students are not just studying BPM—they are learning directly from the world's experts in each BPM topic, within a flexible, scalable, and internationally validated educational ecosystem.

In addition to being a global marketplace for BPE services as outlined above, the GPI is envisioned as a continuous EaaS [12] model for BPM content delivery. It replicates the well-known Software-as-a-Service (SaaS) model from typical information systems (like

ERP and CRM) to the BPE context. It enables BPM educators, universities, and training providers to update and launch BPM learning material in a modular, version-managed, and collaborative way. The upgradeable nature of the award means that GPI graduates receive version-managed certifications (e.g., a degree in RPA) and are proactively notified when upgrades to that certification become available. Process learners (humans or machines) can access content on demand, subscribe to updates, and ensure their skills and knowledge remain current as BPM practices and technologies evolve. A continuous EaaS [12] model is relevant for BPM education because of rapid methodological and technological changes that characterize the field. Training approaches completed once and never updated become outdated, leaving professionals unprepared for new trends like AI.

3 The Business-To-Process (B2P) Model

The last stakeholder group of GPI is process learners, which represents autonomous business processes capable of learning, adapting, and evolving. Their objective is to continuously self-assess their own performance and identify demands for an educational uplift to improve operational outcomes. The processes' core interests align with maintaining high performance and conformance. This capability is central to the B2P model, where processes become active participants in the learning ecosystem.

3.1 The Building Blocks of B2P Education

Traditional education and service delivery models are often framed within two dominant paradigms: Business-to-Consumer (B2C) [13, 14] and Business-to-Business (B2B) [15, 16]. The B2C model refers to transactions where products or services are delivered directly from an organization to individual consumers. In the education sector, this includes platforms like Coursera or Udemy, which offer standardized courses, certifications, or learning content to a broad audience. In contrast, the B2B model centers around interactions between organizations. Within education, this includes corporate training providers, enterprise Learning Management Systems (LMS) or Learning Experience Platforms (LXP) (e.g. Go1) platforms, or consulting firms that design and deliver customized learning solutions to client organizations.

The Business-to-Process (B2P) model proposed within the GPI represents a novel approach. Here, not only humans but also business processes are students. Business processes themselves become the learners—autonomous entities capable of subscribing to services, selecting content, and allocating resources to improve performance. Rather than targeting only individuals or institutions, B2P positions the process as the recipient and consumer of educational resources, enabled by AI, ML, and process mining [17]. Process learners can autonomously (or semi-autonomously) acquire services, software, educational content, automations, analytics, predictive capabilities, or proposals for process improvements.

In this way, human learners and process learners co-exist in GPI. Human learners acquire the skills to design, configure, and oversee process learning capabilities. At the same time, process learners autonomously consume educational services to optimize performance. This dynamic creates feedback loops: humans analyze process performance data and update learning content. Meanwhile, machine process learners generate operational insights. By supporting both types of learners, the GPI aims to maintain alignment between human expertise and process performance in the BPM landscape.

A requirement of the B2P model is the development of educational content specifically tailored for processes rather than humans. While traditional instructional material for human learners focuses on cognitive understanding and pedagogical design, educational assets intended for process learners in the B2P model must instead be machine-readable, modular, and actionable to enable autonomous consumption and execution. This includes structured decision rules, simulation models, performance benchmarks, and optimization routines that a process can autonomously interpret and apply. Content must align with process-level goals, maturity stages, and execution contexts, allowing, for example, a procurement process to subscribe to a cost-reduction or a new legislation training module. Designing such content demands a new instructional paradigm that bridges process logic, data analytics, and operational semantics.

One of the foundational mechanisms of the B2P model is the concept of Process Subscription Services (PSS), a recurring revenue model based on individual subscriptions paid by each process learner. An onboarding workflow, order-to-cash cycle, or customer support process could autonomously subscribe to services tailored to its operational profile, maturity level, and performance gaps. These services might include diagnostics, simulation tools, training modules, predictive analytics, or micro-automations. The subscription would be guided by the process's contextual needs and continuously updated as conditions and goals evolve. With PSS, each process will be able to self-analyze (identify bottlenecks, inefficiencies, and variances), self-monitor (evaluate its performance in real time), self-improve (implement improvements incrementally and adaptively), and self-innovate (explore new ways to achieve results).

These learning capabilities would be offered within a marketplace of BPE services, where companies, academic institutions, and solution providers offer services and content directly to subscribing process learners. Unlike traditional marketplaces that target human decision-makers, this digital environment would be machine-readable, interoperable, and structured for automated consumption. Service descriptions must be semantically standardized, and APIs would facilitate secure, trackable interactions between providers and processes.

The autonomy and intelligence of the process are central to this model. Guided by predefined objectives—such as reducing cycle time, increasing quality, or achieving compliance—the process can identify its own performance deviations or improvement opportunities. Based on real-time monitoring and historical data, it may autonomously initiate subscriptions to services that address emerging needs, correct inefficiencies, or explore alternatives. This capability is enabled by integrating process mining [17] and AI technologies to support reactive and proactive behaviors.

To enable this self-service, each process would be equipped with a Process Wallet—a virtual budget assigned to the process for educational purposes, with limits established by the process owner. Within these predefined limits, the process could independently select educational services from a catalog of offerings, using rule-based decision logic or AI-supported algorithms. This wallet system introduces financial accountability and boundary control, while allowing for a level of autonomy that supports continuous adaptation and incremental improvement. Human oversight remains present to define constraints, monitor performance, and intervene when needed, ensuring governance and strategic alignment.

The B2P model enables a subscription-based revenue tailored to process-level consumption. Processes could be charged through recurring subscriptions, pay-per-use arrangements, or bundled service offerings with upsell and cross-sell opportunities. This allows service providers to scale their offerings to thousands of processes within or across organizations, opening a new frontier in BPE service monetization. Furthermore, micro-transactions enable processes to access low-cost, one-off services, such as a compliance check or optimization suggestion, without requiring extensive approval chains, thereby promoting agility.

An additional layer of sophistication can be introduced through value-based pricing mechanisms. Service fees could be partially or entirely based on their performance impact, measured in improved KPIs or reduced costs, thereby aligning provider incentives with client outcomes. This value-based pricing model drives service quality, forming the basis for a more transparent and accountable B2P economy.

3.2 Advantages and Challenges of B2P Education

The B2P model introduces a set of competitive advantages that transcend traditional B2C and B2B frameworks by leveraging process-level autonomy. One of its primary strengths is scalability and automation. Instead of relying on human-driven purchasing, processes themselves can subscribe to education services, enabling scalable service consumption without increases in sales or administrative overhead.

B2P promotes continuous innovation by establishing a feedback loop between process behavior and service design. Since processes consume services in response to performance metrics, providers are incentivized to develop increasingly adaptive and targeted offerings. This drives rapid cycles of experimentation and refinement, creating a marketplace that evolves organically in response to real operational needs.

Another significant advantage lies in accuracy and objectivity in decision-making. Unlike human-led decisions, which may be influenced by biases or incomplete information, autonomous processes rely on real-time data, historical patterns, and algorithmic optimization to select services that best fit their context, without human intermediaries. This increases the likelihood of impactful interventions and minimizes inefficiencies caused by subjective decision-making.

Additionally, B2P enables a new level of precision and personalization in service consumption. Each process can identify its own performance gaps and select services that address its specific context, constraints, and goals. This level of granularity—unachievable in traditional organizational learning and improvement models—allows for highly tailored interventions without requiring one-size-fits-all solutions.

The model unlocks significant economic potential and scalability. By transforming business processes into autonomous service consumers, B2P creates a new market of digital demand, composed not of individuals or corporations, but of operational workflows. This unlocks new value streams through subscription-based revenue, micro-transactions, and value-based pricing tied directly to process-level outcomes. Data-driven micro-decisions made directly by processes replace traditional improvement cycles (often lengthy and reactive). It also enables service providers to design scalable, modular offerings deployed across diverse industries without customization overhead.

However, implementing B2P also brings challenges and risks that must be addressed. Introducing autonomous or semi-autonomous processes within the B2P model raises questions about governance and accountability. As processes begin to make decisions, such as selecting and purchasing educational services, there must be mechanisms to define the responsibility and accountability for those actions. Human oversight becomes essential to monitor outcomes and intervene when necessary to avoid the risk of unintended consequences or misaligned decisions.

A further challenge of B2P concerns the organizational context in which processes operate. While business rules and regulatory requirements can be formalized, modeled, and supplied as inputs to guide autonomous processes, the more nuanced variations of organizational culture, informal practices, and local management styles are more difficult to capture and standardize. Therefore, human oversight and organizational contextualization are still needed to ensure that process learning aligns with local requirements and stakeholder expectations.

The ethical challenges [18] introduced by the B2P model mirror many of the dilemmas already present in human-executed processes. Issues such as prioritizing efficiency over fairness, bias in decision-making, or lack of transparency are common in current organizational routines. What changes with B2P is the potential opacity introduced by algorithmic logic. As processes gain autonomy, organizations must ensure that ethical standards are translated into machine-executable principles. It must safeguard against bias, misuse of resources, and unintended discrimination, ensuring that the behavior of autonomous processes aligns with organizational values and societal expectations.

Implementing B2P requires advanced technical infrastructure supporting secure, reliable, transparent automated transactions. Processes must be able to authenticate, evaluate service offerings, make informed choices, and execute decisions. This level of automation demands integration of different and secure technologies. Building such a platform presents a non-trivial development challenge.

Finally, one of the most profound challenges of the B2P model lies in its cultural and organizational acceptance. The idea of processes acting with a degree of autonomy—subscribing to educational services or making micro-decisions—may conflict with traditional hierarchical structures and command-and-control mindsets. For B2P to succeed, organizations must undergo a mindset shift, embracing new paradigms of distributed intelligence and trust in process autonomization [6, 7].

As a GPI and B2P boundary, we explicitly do not include manual or human-driven processes within autonomous learning. Such processes will remain under traditional governance and management practices, including human supervision, audits, and organizational quality controls. In this way, our envisioned platform acts as a complementary solution that extends learning capabilities where automation and monitoring are viable, without replacing existing approaches for manual process management.

4 Business Model

The Global Process Institute combines business model [19] patterns to maximize sustainability, impact, and scalability. At its core, it operates as a two-sided platform connecting the global BPE ecosystem. By facilitating value exchange between BPM learners and educators, the GPI generates direct and indirect network effects. The more learners are involved, the more valuable the platform becomes for everyone, as learner-to-learner interactions facilitate additional insights (direct effects). In addition, there are indirect network effects, i.e., a growing community of BPM learners attracts providers of educational services, and vice versa.

The GPI could follow a freemium model to balance inclusivity and financial sustainability. Foundational content, such as introductory BPM modules, is freely available to all learners. This ensures global reach and democratized access to high-quality BPE. In parallel, premium offerings—including advanced courses, personalized mentoring, customized corporate programs, and exclusive access to expert-led sessions—are monetized through subscription or pay-per-use fees. This approach provides wide accessibility while driving sustainable revenue.

Lastly, the GPI's business model as a marketplace enables implementing a long tail model. Intended to host a high number of educational service providers, allows catering for a broad variety of niche BPM topics, such as Green BPM or BPM for Government, that may not be viable in the curricula of a single university.

The GPI's speculative business model (Fig. 1) was developed based on our experience and previous research in business model design [1]. Using the prospective theorizing [10] method, the components were iteratively identified and refined to address the continuity of learning for different stakeholders' needs in BPE.

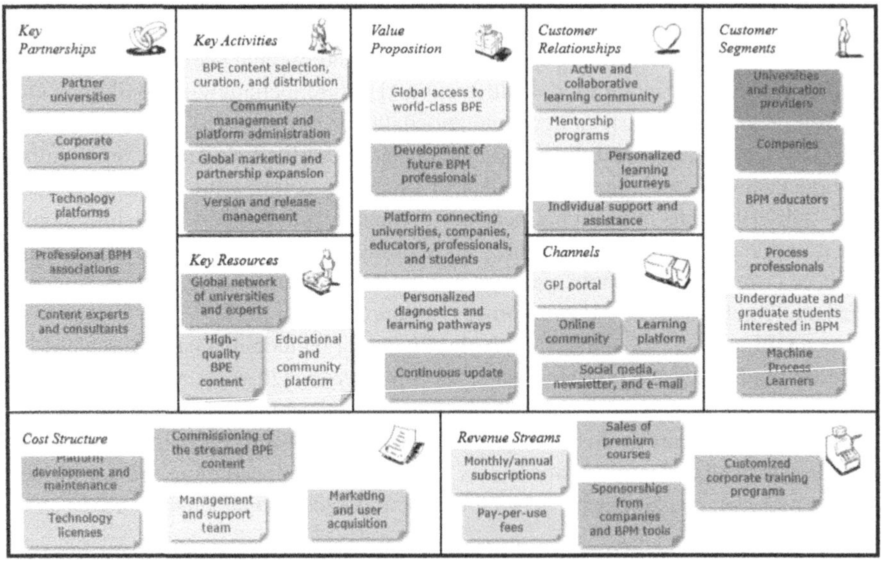

Fig. 1. Global Process Institute's Business Model.

It offers a compelling value proposition: global access to selected world-class BPM education and a platform connecting key educators and learners from academia and industry. Through a combination of personalized diagnostics and learning paths, the Institute prepares professionals and students to thrive in the evolving world of BPM.

The key activities center around selection, curation, and distribution of BPE offerings produced by universities or individual educators, management of the global BPM community of providers and learners, administration of the educational platform, marketing, and version and release management as part of upgradeable certificates.

The GPI serves a wide range of customer segments, including universities seeking greater academic relevance, companies aiming to enhance their BPM capabilities, educators eager to apply and expand BPM knowledge, professionals advancing their BPM careers, and students preparing for the job market. Each segment finds tailored offerings that match their unique goals and needs.

The Institute fosters strong customer relationships through active community engagement, personalized learning experiences, and a blend of automated and human support. Learners are encouraged to participate in mentorship programs, ensuring that their journeys are individualized but also dynamic and interactive.

A variety of channels are used to engage and deliver value to customers. The portal of the GPI acts as the central hub, supported by an integrated learning platform (such as Open edX or Moodle), an interactive online community, and ongoing activities through social media and targeted newsletters.

The GPI relies on key partnerships with universities, corporate sponsors, technology providers, BPM professional associations, and content experts to support its operations and amplify its reach. The GPI's key resources are a worldwide network of universities and BPM experts, a robust educational and community platform, and a continuously updated high-quality library of BPE content.

The revenue streams for the Institute are diversified. They include monthly or annual subscriptions, pay-per-use fees for exams and certifications, customized corporate training programs, and sponsorships from BPM-related companies. This hybrid revenue model supports the accessibility of basic education while sustaining high-value offerings. As a counterpart, the cost structure includes platform development and maintenance, commission of the streamed educational content, costs of management, marketing, and administrative teams, and technology licenses.

Overall, the business model (Fig. 1) serves as a framework that connects the main building blocks of GPI. It lays the foundation for operationalizing the GPI's vision by structuring how value is co-created, delivered, and sustained across the BPE ecosystem.

5 Related Work

Education is undergoing a transformational journey. It is no longer confined to traditional school-based systems but has expanded to lifelong learning [11]. People can no longer navigate their life course using only the skills and knowledge acquired at school, college, or university—they need to learn throughout life. However, the BPM field has not yet significantly embraced the principles of lifelong learning [11], resulting in fragmented efforts and a lack of continuous educational models to meet evolving professional demands.

Although the number of institutions offering BPM education is growing [5, 8], the BPM Community recognizes that contributions and attention to real-world BPM education and training remain limited [4, 5], with only a handful of studies [4, 5, 8, 9, 20–22] explaining the methods and best practices adopted by BPM education providers. Even so, some initiatives must be considered in the context of the GPI.

Previous initiatives have sought to create centralized educational resources for the BPM community. Notably, the EDUglopedia.org platform, discussed by [5], was a social network providing a directory of BPM education programs worldwide, facilitating the sharing and reusing of educational resources. While EDUglopedia represented an important step toward connecting BPM educators and learners globally, the platform is offline and no longer accessible. This highlights the need for a new, sustainable, and continuously evolving ecosystem to support BPM education on a global scale.

The idea of a platform to consolidate a broad and deep set of pedagogical resources, such as teaching cases and exercises, was also advocated by [9]. However, this knowledge base is unavailable, and the cited Process Knowledge Initiative can't be accessed anymore. The existence of previous introductory and comprehensive BPM MOOCs (Massive Open Online Courses) [8], which were discontinued in 2017, also reinforces the BPM community's past interest in sharing educational knowledge.

While existing industry initiatives such as BPMInstitute.org or ABPMP have contributed significantly to disseminating BPM knowledge through articles, webinars, certifications, and professional development, they primarily operate as content providers and practitioner-focused platforms. In contrast, our proposal for the GPI aims to establish a dynamic multi-stakeholder educational ecosystem. Rather than focusing solely on practitioner resources, the GPI distinguishes itself by emphasizing educational content, applied collaboration, personalized learning pathways, and aiming to advance BPM education as a continuous, interdisciplinary, and innovation-driven endeavor.

Syed *et al.* [4] highlight the need to consider technical and managerial competencies in a BPM education program. They claim it is essential to ensure that content aligns with the industry's multidisciplinary skill demands and provides learners with opportunities to apply their knowledge in practical settings. Maintaining industry relationships, through real-world BPM experiences, is also emphasized in other works [8, 9, 21]. Therefore, these two requirements need to be covered in the proposed GPI.

In summary, previous initiatives demonstrate a growing recognition of the importance of BPM education, yet they reveal a fragmented landscape marked by short-lived platforms. While these early efforts reflect community interest, their discontinuation signals the need for a more sustainable approach. Industry-focused platforms (such as BPMInstitute.org or ABPMP) provide valuable practitioner content but do not address the broader educational ecosystem. In contrast, the GPI is designed to bridge these gaps by offering a multi-stakeholder and technology-enabled environment for continuous BPM education, considering both human and machine process learners. Its integrated and innovation-driven model responds directly to the limitations of past efforts while envisioning a sustainable path for global BPE.

Advancements in digital technologies strongly support the viability of the GPI. LMS, LXP, and Adaptive Learning technologies [23] facilitate personalized learning journeys. AI, ML, and Process Mining [17] empower continuous skills diagnostics, predictive analytics, and real-world process insights. Furthermore, collaborative technologies and social learning tools enhance global interaction among diverse stakeholders.

Finally, the concept of B2P represents a novel perspective in the BPM and educational domains. While B2B [15, 16] and B2C [14] models are well-established, processes acting as autonomous subscribers to services, such as educational content, analytics, or automation, are still unexplored in academic literature. To the best of our knowledge, no existing studies envision, conceptualize, or operationalize the notion of processes as learning agents with purchasing power and decision-making autonomy. This positions B2P as a groundbreaking contribution that invites further investigation and experimentation at the intersection of BPM, AI, and digital education.

6 Conclusion

This paper uses prospective theorizing to propose a possible, in our view, desirable future of BPE, labelled the Global Process Institute (GPI). GPI integrates the principles of lifelong learning with digital technologies and aims to foster a global ecosystem for BPE. Among its key contributions are: (i) a two-sided marketplace of BPE services connecting providers and learners on a subscription-based model that enables learners to continuously evolve their BPE credentials; and (ii) the introduction of the B2P model, which reimagines business processes as active and autonomous learners.

The GPI triggers a rich agenda for future research and experimentation. Open questions remain around the governance and ethical oversight of autonomous process learners, mechanisms for equitable budget allocation, and the design of transparent and auditable learning. This also includes how to synchronize human and process learning. Empirical studies are needed to assess the organizational readiness for B2P adoption, the effectiveness of process-driven learning pathways, and the economic models that can sustain such a platform at scale. Exploring the intersections between BPM, AI, process mining, and digital education will be essential for advancing the concept.

The potential impact of the GPI extends beyond the academic and educational spheres—it can transform how organizations perceive, develop, and invest in BPM capabilities. By enabling continuous, contextualized, and autonomous learning at both the human and the process level, the GPI promotes BPE as a new capability of BPM.

The BPM community is well-positioned to adapt to emerging trends and advancements while continually enhancing its intellectual core. Therefore, we invite the global BPM community to actively engage in the GPI development. This initiative is not a finished product but a shared vision that requires collective intelligence, co-creation, and long-term collaboration. By aggregating knowledge, sharing experiences, piloting ideas, and shaping governance models, stakeholders can ensure that the GPI becomes a living ecosystem for BPE. Together, we can redefine how BPM is taught and learned in a dynamic, interconnected, and increasingly autonomous world.

References

1. Betts, M., Rosemann, M.: The new learning economy: thriving beyond higher education. Routledge (2022)

2. de Geus, A.: Planning as learning. Harvard Bus. Rev. 70–74 (1998)
3. Brown, J.S., Adler, R.P.: Minds on Fire: Open Education, the Long Tail, and Learning 2.0. Educ. Rev. 43, 16 (2008)
4. Syed, R., Wynn, M.T., Bandara, W.: The design and delivery of a holistic BPM education programme. In: Business Process Management: Blockchain, Robotic Process Automation and Educators Forum, pp. 201–210. Springer (2023). https://doi.org/10.1007/978-3-031-43433-4_13
5. vom Brocke, J.: Where to study business process management? A Global Perspective Based on EDUglopedia.org. BPTrends (2017)
6. Rosemann, M., et al.: Business process management in the age of AI – three essential drifts. Inf. Syst. E-Bus. Manag. **22**, 415–429 (2024). https://doi.org/10.1007/s10257-024-00689-9
7. Jaenisch, C., Kowalkiewicz, M., Rosemann, M.: Process Autonomization. Rethinking Business Process Management. In: Senderovich, A., Cabanillas, C., Vanderfeesten, I. (eds.) Proceedings of the BPM Forum. Seville, Spain, 31 August-5 September 2025
8. vom Brocke, J., Seidel, S., Tumbas, S.: Class notes: the BPM curriculum revisited. BPTrends (2015)
9. Recker, J.: How was school today? BPTrends. **9**, 1–6 (2012)
10. Gümüsay, A.A., Reinecke, J.: Imagining desirable futures: a call for prospective theorizing with speculative rigour. Organ. Theory. **5**, 26317877241235940 (2024). https://doi.org/10.1177/26317877241235939
11. UNESCO: Lifelong learning for all. https://www.unesco.org/en/lifelong-learning/need-know. Accessed 28 April 2025
12. Prifti, L.: Emerging business models in education provisioning: a case study on providing learning support as education-as-a-service. Int. J. Eng. Pedagogy **7**, 92–108 (2017). https://doi.org/10.3991/ijep.v7i3.7337
13. Kotler, P., Kartajaya, H., Setiawan, I.: Marketing 4.0: moving from traditional to digital. Wiley (2016)
14. Solomon, M.: Consumer behavior: buying, having, and being, Global Edition. Pearson Education Limited, Harlow, England (2019)
15. Coe, J.: The fundamentals of business-to-business sales & marketing. McGraw Hill (2003)
16. Hutt, M., Speh, T., Hoffman, D.: Business marketing management: B2B. Cengage Learning, Australia (2023)
17. van der Aalst, W.M.P.: Process mining: data science in action. Springer International Publishing (2018)
18. Santoro, F.M.: Ethics at the core: the future of BPM. In: Marella, A., et al. (eds.) Proceedings of the 22nd International Conference on Business Process Management (BPM 2024). Krakow, Poland, 1–6 September 2024, pp. xxiii–xvi (2024)
19. Osterwalder, A., Pigneur, Y.: Business model generation: innovation in business models. Alta Books, Rio de Janeiro (2011)
20. Grzesiak, M., et al.: Beyond traditional learning: the LLM revolution in BPM education at university. In: Business Process Management: Blockchain, Robotic Process Automation, Central and Eastern European, Educators and Industry Forum, pp. 406–415. Springer Nature (2024). https://doi.org/10.1007/978-3-031-70445-1_29
21. Nobre, M.R.B., Vilela, J.F.F.: Content and skills for teaching BPM in computer science courses: a systematic mapping study. In: Proceedings of the 16th International Conference on Computer Supported Education (CSEDU) (2024). https://doi.org/10.5220/0012629500003693
22. Pridmore, J., Godin, J.: Business process management and digital transformation in higher education. Inf. Syst. **22**, 168–177 (2021). https://doi.org/10.48009/4_iis_2021_180-190
23. Capuano, N., Caballé, S.: Adaptive learning technologies. AI Mag. **41**, 96–98 (2020). https://doi.org/10.1609/aimag.v41i2.5317

Author Index

M. ER et al. (Eds.): BPM 2025, LNBIP 565, pp. 471–472, 2026.
https://doi.org/10.1007/978-3-032-02936-2

The manufacturer's authorised representative in the EU is Springer Nature Customer Service Centre GmbH, Europaplatz 3, 69115 Heidelberg, Germany. If you have any concerns regarding our products, please contact ProductSafety@springernature.com

Printed and bound by CPI Group (UK) Ltd, Croydon, CR0 4YY
30/09/2026
02225234-0016